ANNUAL REVIEW OF NUCLEAR AND PARTICLE SCIENCE

ANNUAL REVIEW OF NUCLEAR AND PARTICLE SCIENCE

J. D. JACKSON, *Editor*
University of California, Berkeley

HARRY E. GOVE, *Associate Editor*
University of Rochester

ROY F. SCHWITTERS, *Associate Editor*
Stanford Linear Accelerator Center

VOLUME 28

1978

ANNUAL REVIEWS INC.　　　4139 EL CAMINO WAY　　　PALO ALTO, CALIFORNIA 94306

ANNUAL REVIEWS INC.
Palo Alto, California, USA

REPRINTS   The conspicuous number aligned in the margin with the title of
each article in this volume is a key for use in ordering reprints. Available
reprints are priced at the uniform rate of $1.00 each postpaid. The minimum
acceptable reprint order is 5 reprints and/or $5.00 prepaid. A quantity discount
is available.

International Standard Serial Number: 0066-4243
International Standard Book Number: 0-8243-1528-6
Library of Congress Card Number: 53-995

FILMSET BY TYPESETTING SERVICES, LTD, GLASGOW, SCOTLAND
PRINTED AND BOUND IN THE UNITED STATES OF AMERICA

PREFACE

This volume marks several departures. A new or at least augmented name, *Annual Review of Nuclear and Particle Science*, has been chosen the better to convey the variety of reviews offered and the audience for whom they are intended. The new complement of editors, anticipated two years ago by Emilio Segrè's valedictory in Volume 26, is now a reality, although in the interval Segrè and his colleagues, Grover and Noyes, edited Volume 27. The present volume, too, bears the imprint of the outgoing editors, as well as the new. The listing opposite the title page reminds readers that Volume 28 was planned by the Editorial Committee of 1976. The new editors have merely carried those plans to completion.

A survey of the contents shows that the present volume continues the tradition of the *Annual Review of Nuclear Science* of Beckerley and Segrè. Theoretical particle physics and nuclear physics get their due in the reviews by Appelquist, Barnett & Lane, by Greenberg, and by Quentin & Flocard. The relatively new field of very-high-energy collisions between nuclei is surveyed by Goldhaber & Heckman. Interdisciplinary topics, mixtures of nuclear and/or particle physics with other fields, include hypernuclei by Povh, muonic chemistry and condensed matter physics by Brewer & Crowe, isotopic anomalies by Clayton, and nuclear astrophysics by Rolfs & Trautvetter. The technical side of our subject is represented by the review of intense neutron sources and their applications by Barschall. Somewhat special is the chapter on synchrotron radiation research by Winick & Bienenstock. Although the uses of synchrotron radiation lie almost entirely outside the scope of these volumes, the close historic connection of the sources with particle physics amply justifies inclusion of such a review.

The above catalogue of chapters indicates the diversity, but the proof of the pudding is in the eating. In an effort to make the articles more digestible, the editors have encouraged authors in a modest innovation, a beginning section entitled Introduction and Summary. Ideally this contains an introduction to the subject in simple terms appropriate for the nonspecialist and also a survey of the main results—a bird's eye view of the topic as well as an entrée to the main body of the review. Not all topics lend themselves to such summary treatment, and not all authors wish to avail themselves of the opportunity. Nevertheless, the editors keep as a primary goal the accessibility of at least the first parts of all reviews to the average reader.

With more than a quarter of its pages devoted to them, this volume will undoubtedly be known among particle physicists as the "Quark Volume" of the *Annual Review of Nuclear and Particle Science*. It is hoped that, in spite of the emphasis on particle physics or perhaps because of it, other readers will find the book useful and rewarding. Future volumes will surely redress the balance toward nuclear physics and particle physics as equals, more or less, with related areas reviewed regularly as well. If there is any departure from past practice anticipated, it is a slight narrowing of the subject matter away from the technical and applied aspects, never a large component anyway. Since these volumes exist to serve the scientific community, the editors solicit comments from readers on what they expect from the review literature and whether the present and proposed format meets those expectations.

My Associate Editors, H. E. Gove and R. F. Schwitters, join me in dedicating this volume to Emilio Segrè, Editor of *Annual Review of Nuclear Science* for 21 years, and to J. Robb Grover and H. Pierre Noyes, long-time Associate Editors. The international prestige and authority of the *Annual Review of Nuclear Science* stem from their enlightened efforts, and those of their predecessors, through the years. May the continuation, *Annual Review of Nuclear and Particle Science*, have as successful a career!

J. D. JACKSON

Annual Review of Nuclear and Particle Science
Volume 28, 1978

CONTENTS

ANNUAL REVIEWS INC. is a nonprofit corporation established to promote the advancement of the sciences. Beginning in 1932 with the *Annual Review of Biochemistry*, the Company has pursued as its principal function the publication of high quality, reasonably priced Annual Review volumes. The volumes are organized by Editors and Editorial Committees who invite qualified authors to contribute critical articles reviewing significant developments within each major discipline.

Annual Reviews are published in the following sciences: Anthropology, Astronomy and Astrophysics, Biochemistry, Biophysics and Bioengineering, Earth and Planetary Sciences, Ecology and Systematics, Energy, Entomology, Fluid Mechanics, Genetics, Materials Science, Medicine, Microbiology, Neuroscience, Nuclear and Particle Science, Pharmacology and Toxicology, Physical Chemistry, Physiology, Phytopathology, Plant Physiology, Psychology, and Sociology. In addition, three special volumes have been published by Annual Reviews Inc.: *History of Entomology* (1973), *The Excitement and Fascination of Science* (1965), and *Annual Reviews Reprints: Cell Membranes, 1975–1977* (published 1978).

Ann. Rev. Nucl. Part. Sci. 1978. 28 : 1–32

HYPERNUCLEI   ✶5591

B. Povh

Max-Planck-Institut für Kernphysik, Heidelberg, Germany, and CERN, Geneva, Switzerland

CONTENTS

1 INTRODUCTION AND SUMMARY

The main objective of hypernuclear research is to study the Λ-nucleus interaction. The Λ particle bound to a nucleus is an excellent probe of nuclear properties; it has mass $M_\Lambda = 1115$ MeV, exceeding the mass of a nucleon by less than 20%. The Λ-nucleus interaction is only slightly weaker than the N-nucleus one. Therefore we can expect the Λ particle in the nucleus to behave very much like a neutron; a neutron, however, with the strangeness quantum $S = -1$ that can be distinguished from other nucleons. The great success in handling strong interacting systems such as the nucleus, solid state, plasma, and very likely also the elementary particles, has resulted from the introduction of the quasiparticle picture in theoretical models. The low excited states of strong interacting systems can be easily reproduced by simple excitation of properly chosen

1

0066-4243/78/1201-0001$01.00

quasiparticles and by collective excitation of the system. But only in the case of the nucleus do we have a suitable probe, the Λ particle, with properties so similar to those of the nucleon as to simulate a distinct nucleon in the nucleus. With such a probe the physical content of the quasiparticle picture can be investigated experimentally.

The present article is devoted almost entirely to the simplest case of hypernuclei, namely the hypernuclei with just one Λ particle bound to a nucleus. So far there have been only two decays of hypernuclei observed in nuclear emulsions that have been interpreted as decays of double Λ hypernuclei. When discussing hypernuclear lifetimes we mention only briefly the interesting notion of producing multiple Λ hypernuclei in relativistic heavy-ion reactions. There is little chance of studying multiple Λ hypernuclei experimentally in the near future; thus we do not discuss them here despite the fact that they are very interesting, as they are the only source of information on the $\Lambda\Lambda$ interaction.

The Λ particle is the lightest strange baryon with a mass $M_\Lambda = 1115$ MeV; it is neutral and has spin-parity $J^\pi = \frac{1}{2}^+$ and isospin $I = 0$. The strangeness of the Λ particle is $S = -1$. Because the strangeness is conserved in strong interactions and the Λ particle is the lightest hyperon, the latter is stable against strong decay in nuclear matter. The free Λ particle decays, via strangeness-nonconserving weak interaction, into a nucleon and a pion, $\Lambda \to N + \pi$, with a lifetime of 2.6×10^{-10} sec. In nuclear matter an additional decay mode (also a weak decay) $\Lambda + N \to 2N$ is possible. In fact this latter mode determines the lifetime of the Λ particle in hypernuclei with $Z > 2$. The lifetime of heavy hypernuclei is estimated to be about 10^{-10} sec. This is long enough to permit the study of strong and electromagnetic properties of hypernuclei to quite the same extent as for short-lived β-unstable nuclei.

The heavier hyperons Σ, Ξ, and Ω, which decay as the Λ particle via the weak interaction, are no longer stable against the strong decay in nuclear matter. They convert to Λ particles by strangeness-conserving strong decay, by the following reactions:

$$\Sigma + N \to \Lambda + N \qquad\qquad\qquad\qquad\qquad 1.$$

$$\Xi + N \to 2\Lambda \qquad\qquad\qquad\qquad\qquad\qquad 2.$$

$$\Omega + 2N \to 3\Lambda. \qquad\qquad\qquad\qquad\qquad\quad 3.$$

The strong decay of hyperons in nuclear matter does not permit a standard spectroscopy of hypernuclei with the exception of the Λ hypernuclei. Therefore it is reasonable to use the term "hypernucleus" when referring to systems with nucleons and Λ particles. In those special cases in which we can expect heavier hyperons to be bound to a nucleus long enough to identify such a system, we shall designate the latter as Σ, Ξ, or Ω hypernuclei.

The Λ particle is not the only heavy baryon that is stable against strong decay in nuclear matter. The lightest charmed baryon Λ_c should also form a bound system with a nucleus, the lifetime of which should be determined by the weak decay of Λ_c. But before speculating further about the properties of charmed nuclei we should wait for Λ_c to be firmly established and at least its main decay modes determined.

The most efficient way of producing hypernuclei is to expose the nuclear targets to low-momentum negative kaons with strangeness $S = -1$. In the reactions

$$K^- + N \rightarrow \Lambda + \pi \qquad\qquad 4.$$

and

$$K^- + N \rightarrow \Sigma + \pi, \qquad\qquad 5.$$

which transfer the strangeness from kaon to nucleon, a fraction of the K^- interactions with the nucleus lead to the formation of a hypernucleus. In nuclear emulsions the hypernuclei can be clearly identified through the characteristic decay fragmentation of its nucleus, in some cases with π emission. The binding energies of Λ particles in light hypernuclei are derived from the kinematical analysis of the decay products of hypernuclei. With the advent of intense K^- beams in the late 1960s, counter experiments on hypernuclei were made possible. Not only bound states but also hypernuclear continuum states have thus become accessible to experiment.

Experimentally investigating the physical content of the quasiparticle picture is quite an ambitious undertaking; the present world production of low-momentum kaons used for hypernuclear physics is only about 10^{11} particles per year. In comparison, this number of particles is supplied within seconds by the standard accelerators used in nuclear physics. In hypernuclear spectroscopy it is consequently impossible to collect a large number of redundant data (as is the case in nuclear spectroscopy) and to systematize them in terms of simple models. The success of hypernuclear spectroscopy depends entirely on the chance of selecting hypernuclear states with configurations simple enough to allow a straightforward and unique theoretical analysis. Such simple states are the hypernuclear ground states. They are states with just an additional Λ particle in the $1s$ state, which is coupled to the nuclear core without disturbing its configuration. The large amount of binding energy of the Λ particle in nuclei collected by the emulsion technique is still the main source of information on the Λ-nucleus interaction.

The second type of states with simple configurations is the one obtained in direct hypernuclear production in the strangeness exchange reaction $K^- + n \rightarrow \Lambda + \pi^-$ by "recoilless Λ production." When the kaon momentum is close to 500 MeV/c and the pion is detected in the forward direction, the Λ momentum can approach zero. Under such kinematical conditions the strangeness exchange reaction just turns a neutron in the nucleus into a Λ particle without changing the nuclear wave function. The Λ particle remains in the same orbit in which the strangeness exchange reaction took place. Such states have been observed experimentally as pronounced resonances in the hypernuclear continuum, and they can be populated selectively by the (K^-, π^-) reaction. The configuration of the "strangeness exchange resonances" populated in recoilless Λ production is the same as that of the target nucleus. In these resonances we thus have an ideal type of hypernuclear excitation, which allows a direct comparison of the Λ-nucleus to the N-nucleus interaction without the complications inherent in the theoretical treatment of the many-body system.

The recoil momentum of the Λ particle can be simply adjusted in the (K^-, π^-) reaction by varying the kaon momentum and the pion emission angle. For recoils of the order of the Fermi momentum in the nucleus, less prominent states than the strangeness exchange resonance will be populated with comparable strength. The strangeness exchange reaction is accompanied by a jump of the Λ particle to one of the neighboring shells without otherwise disturbing the nuclear core. These states have not yet been individually identified by experiment, but there is strong evidence that they can also be detected free of background.

The hypernuclear ground states and the states produced in the (K^-, π^-) reaction with little or no Λ recoil all satisfy our requirement of having simple configurations; "simple" meaning that the Λ-nucleus system can be compared directly to the N-nucleus system without a detailed analysis of the hypernuclear configuration. If all these types of states are identified in single nuclei, there will be sufficient constraint to deduce the properties of the Λ particle in nuclear matter. So far, it is only in $^9_\Lambda\text{Be}$ and $^{12}_\Lambda\text{C}$ that the ground states and the strangeness exchange resonances have been identified in the same hypernucleus thus giving the position of the $1s$ and $1p$ levels of the Λ particle. Heavier hypernuclei are a richer source of information than p-shell hypernuclei, since the Λ particle can occupy many different shells between the ground state and the strangeness exchange resonance. But in these hypernuclei only recoilless Λ production has been observed. Nevertheless, detection of the remaining "interesting" states is also within the reach of present experimental techniques in spite of the rather modest kaon beams available. The search for these states is the main goal of the present generation of hypernuclear experiments.

In Section 2 we briefly discuss the general properties of the ΛN interaction and compare them to those of the NN interaction. The link between the elementary and the effective ΛN interaction in the nucleus can be studied best, if at all, in s-shell hypernuclei. In Section 3 the properties of hypernuclear ground states are reviewed. The basic information on the Λ-nucleus interaction is deduced from the data on the binding energies of the Λ particle in the ground states. In Section 4 we present the recent development of experimental methods relevant to the high resolution spectroscopy of the hypernuclear continuum states leading to the experimental determination of recoilless Λ production in nuclei. It will also be demonstrated that in the (K^-, π^-) reaction we have a powerful tool for selectively populating hypernuclear states with a configuration closely related to that of the target nucleus. Finally, in Section 5 we mention Σ hypernuclei. There is some evidence that the $\Sigma^+ n$ system has a weakly bound state. This finding stimulated the search for further possible "dibaryon" states with strangeness $S = -1$ and even $S = -2$.

2 s-SHELL HYPERNUCLEI

2.1 Λ-Proton Scattering

The only direct information on the low-momentum ΛN interaction stems from two bubble-chamber experiments (1, 2) that measured the differential and total elastic scattering cross section for Λp below 300 MeV/c. In the hydrogen bubble chamber a stopped K^- reacts with a proton giving the source of Λ particles either directly in the $K^- + p \rightarrow \Lambda + \pi^0$ reaction or via the production of Σ particles. Kinematical reconstruction of the Λp scattering event gives the value of the incoming Λ momentum. The measurement has been done in momentum intervals between 120 and 320 MeV/c. The measured cross section is dominated by s-wave scattering. Only at the highest measured momenta is a small forward-backward asymmetry observed, indicating a p-wave contribution to the scattering cross section. It is obvious that measuring the energy dependence of the total scattering cross section of two fermions is not sufficient to determine the scattering parameters even for a pure s-wave interaction. The effective range expansion for s-wave scattering yields for the total cross section

$$\sigma = \frac{3\pi}{\left[\left(-\frac{1}{a_t}\right) + \frac{1}{2r_{0t}k^2}\right]^2 + k^2} + \frac{\pi}{\left[\left(-\frac{1}{a_s}\right) + \frac{1}{2r_{0s}k^2}\right]^2 + k^2}, \qquad 6.$$

where a_t and a_s are triplet and singlet scattering lengths, respectively, and r_{0t} and r_{0s} the corresponding effective ranges. For a weak spin dependence of the Λp interaction, the triplet term dominates the cross section because of its triple statistical weight. Therefore it is useful to try to determine the scattering parameters assuming $a_t = a_s = a$ and $r_{0t} = r_{0s} = r_0$, as suggested by Londergan & Dalitz (3), and to write

$$\sigma = \frac{4\pi}{\left[\left(-\frac{1}{a}\right) + \frac{1}{2r_0 k^2}\right]^2 + k^2} \qquad 7.$$

The fit of experimental data (1, 2) using Equation 7 is shown in Figure 1 yielding the values (3)

$$a = -1.80 \text{ fm}, \qquad r_0 = 3.16. \qquad 8.$$

These values should be a good approximation of the parameters for triplet scattering. We can see that the Λp interaction is somewhat weaker than the NN interaction and that there is no Λp bound state. But the ΛN interaction is not considerably weaker, since the Λpn system is already bound. The binding energy of the Λ particle in $^3_\Lambda$H is $B_\Lambda = 0.13 \pm 0.05$ MeV.

2.2   Ground States of s-Shell Hypernuclei

The most accurate information on the spin-spin part of the ΛN interaction can be obtained from the $A = 4$ hypernuclear system. The two mirror hypernuclei, $^4_\Lambda$H and $^4_\Lambda$He, are known (4) to have $J^\pi = 0^+$ ground states. Recently (5–7) the first excited state at 1.09-MeV excitation was

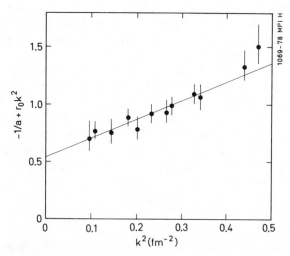

Figure 1 The quantity $-1/a + r_0 k^2$ is plotted against k^2 for the total Λp scattering cross section. The line shows the best fit linear in k^2.

found in $^4_\Lambda$H and $^4_\Lambda$He. This state should have $J^\pi = 1^+$. We can thus readily conclude that the ΛN interaction is stronger in the singlet than in the triplet state.

The configuration of the mass-four hypernuclei is extremely simple in the lowest order. In the ground state the four particles, all in the s-shell, are coupled to $J^\pi = 0^+$ in a configuration that resembles that of the α particle. The Λ particle is distinguishable from the nucleons, and in the s-state the mass-four hypernuclei can also couple to spin $J^\pi = 1^+$. The first excited state in $^4_\Lambda$H and $^4_\Lambda$He is obtained from the ground state by the spin flip of the Λ particle. If this simple model were correct, it would be rather straightforward to deduce the spin-spin part of the ΛN interaction and to separate the scattering lengths a_s and a_t in a combined analysis of the Λp scattering data and mass-four hypernuclei. The problem is more complicated if $^4_\Lambda$H and $^4_\Lambda$He do not have such a simple configuration. Owing to the strong tensor force in the NN interaction, a considerable amount of D-state can be admixed into the ^{3}H and ^{3}He configurations. In this case splitting of the $J^\pi = 0^+$ and 1^+ state in the mass-four hypernuclei does not depend solely on the spin-spin interaction. This and further sources of error in determining the spin-spin interaction from the hypernuclear data are discussed extensively by Gal (8).

The most detailed theoretical analysis of s-shell hypernuclei, including also Λp scattering data, has been made by Herndon & Tang (9, 10) and by Dalitz, Herndon & Tang (11). In this analysis the ΛN interaction has been assumed to have the form

$$U_{\Lambda N} = \tfrac{1}{4}(3 + \boldsymbol{\sigma}_\Lambda \boldsymbol{\sigma}_N) U_t(r_{\Lambda N}) + \tfrac{1}{4}(1 - \boldsymbol{\sigma}_\Lambda \boldsymbol{\sigma}_N) U_s(r_{\Lambda N}) + U_{CSB}(r_{\Lambda N}). \qquad 9.$$

The triplet and singlet potentials U_t and U_s are central outside the hard core of radius 0.45 fm:

$$U_t(r) = \infty, \qquad U_s(r) = \infty, \qquad r < d \qquad\qquad 10.$$

$$U_t(r) = -U_{0t}\, e^{-\lambda(r-d)}, \qquad U_s(r) = -U_{0s}\, e^{-\lambda(r-d)}, \qquad r > d \qquad 11.$$

with the values $d = 0.45$ fm and $\lambda = 3.219$ fm^{-1}. The U_{CSB} potential takes care of a possible charge-symmetry-breaking effect in the ΛN interaction. The NN interaction has been assumed to have the same form as well as the same hard core radius. The parameters have been adjusted in order to give the correct binding energies of s-shell nuclei. This can be achieved without invoking the tensor force in the NN interaction, which, however, is known to be essential if correct nuclear configurations (in particular the D-state contribution) of the s-shell nuclei are to be reproduced. Within the framework of this calculation it was possible to treat the Λp scattering and the s-shell hypernuclei consistently with only the exception of $^5_\Lambda$He.

This was done prior to the recent unique determination of the first excited state in $^4_\Lambda$H and $^4_\Lambda$He (7). It is easy to see, however, that the potential parameters can be modified to reproduce the new experimental data; $^5_\Lambda$H remains the unsolved problem of the s-shell hypernuclei. Most calculations give a Λ binding energy larger than 5 MeV in $^5_\Lambda$He, contrary to the measured value $B_\Lambda = 3.12 \pm 0.02$ MeV, which cannot be accounted for by the present models (8).

2.3   Gamma Spectroscopy

The first counter experiments in hypernuclear physics were performed in order to detect gamma transitions (5). Obviously, gamma spectroscopy offers the simplest way of achieving a good resolution in hypernuclear spectroscopy without using large magnetic spectrometers for charged particles involved in the production of hypernuclei. Hypernuclei produced in the K^- interaction with a nucleus may be produced in excited states; by observing the gamma transitions between the bound states, the level scheme of the hypernuclei can be reconstructed provided that a method is found for identifying the hypernucleus to which the gamma belongs. So far no such ambitious program has been planned. The first experiments were simply based on the fact that K^- interacting in ^{6}Li and ^{7}Li could not produce bound excited states of nuclei, where the gamma decay would be unknown.

The K^- interaction in the nucleus leads to excitation in the continuum followed by the emission of particles. The hypernuclei will probably have at least one or two masses less than the target nucleus when an excitation below the particle emission threshold is reached. At present, little is known about the decay of hypernuclear continuum states. If we find that only small numbers of hypernuclei are populated via the decay of continuum states, gamma spectroscopy may play a very important role in hypernuclear spectroscopy. In this case only a small number of hypernuclei will be populated simultaneously, and disentangling the gamma spectra will become feasible with the present gamma-gamma coincidence method.

Until now, gamma experiments were restricted to the use of ^{6}Li and ^{7}Li targets. Negative kaons stopped in these targets are unlikely to give heavier hypernuclei than $^5_\Lambda$He; the latter does not have any particle-bound excited state. In fact, we would only expect gamma rays from $^4_\Lambda$H and $^4_\Lambda$He. In the first experiment of the CERN-Heidelberg-Warsaw Group (5), two gamma transitions at 1.09 and 1.42 MeV were observed. The 1.09-MeV gamma ray has been assigned uniquely to a transition in the hypernuclear mass-four system, whereas the origin of the 1.42-MeV gamma ray was uncertain. In a recent experiment of the Lyon-Warsaw Group (7) it was possible to show that the 1.09-MeV gamma ray belongs to $^4_\Lambda$H as well as to $^4_\Lambda$He, i.e. within the experimental resolution of about 100 keV the excited states in the two mirror hypernuclei are the same.

To obtain this simple but very important information, i.e. the unique assignment of the two gamma rays to $^4_\Lambda$H and $^4_\Lambda$He, a complicated experiment that made use of the specific properties of the $^4_\Lambda$H and $^4_\Lambda$He decay was necessary. The $^4_\Lambda$H decays predominantly into ^{4}He$+\pi^-$ with monoenergetic pions of 53 MeV. The Lyon-Warsaw Group showed that the 1.09-MeV gamma transition is in fact in coincidence with the 40-MeV π^-, and thus it belongs to $^4_\Lambda$H. Searching for the gamma transition in $^4_\Lambda$He is more complicated, requiring the detection of gamma rays from the π^0 decay with good resolution. The main decay channel is again the two-body decay $^4_\Lambda$He $\to$ ^{4}He$+\pi^0$ with monoenergetic π^0 as well. Again it was possible to demonstrate that the 57-MeV π^0 stemming from the $^4_\Lambda$He decay is also in concidence with the 1.09MeV gamma rays. The combined information on $^4_\Lambda$H and $^4_\Lambda$He from emulsion and gamma work is shown in Figure 2. Notice that the scale of both hypernuclei is given in terms of the binding energies of Λ to the ^{3}H and the ^{3}He core, respectively. The Λ binding energies in $^4_\Lambda$H and $^4_\Lambda$He are quite different. This fact has been taken to be evidence for the charge-symmetry-breaking effect in the ΛN interaction. Both mass-four hypernuclei are mirror hypernuclei, and they should have the same B_Λ, except for the difference in the Coulomb energies of the two mirror hypernuclei. However, this only increases the difference in B_Λ (12). Adding a Λ to ^{3}He will contract the nucleus, thus increasing the Coulomb energy. Adding a Λ to ^{3}H having a single charge $Z = 1$ does not change the Coulomb energy even if the nucleus contracts.

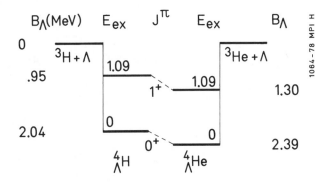

Figure 2 Level scheme of $^4_\Lambda$H and $^4_\Lambda$He.

Therefore we would expect to find B_Λ in $^4_\Lambda$He to be smaller than in $^4_\Lambda$H, just the opposite of the measured values (Figure 2).

The splitting between the 0^+ ground and 1^+ excited states in $^4_\Lambda$H and $^4_\Lambda$He, which at least partially corresponds to the difference in the binding energies between the singlet and triplet states in the Λp and Λn systems, does not depend on the Coulomb force. This splitting does not show any violation of the charge symmetry. Such behavior, i.e. no charge-symmetry violation in the spin-spin part of the ΛN interaction and a strong violation in the spin-independent part, is not easy to reconcile with the present picture of the ΛN interaction.

2.4 ΛN Interaction

Let us briefly sketch the difference in the ΛN and NN interaction at low energies. The long-range NN interaction is dominated by one-pion exchange, which is responsible for a strong spin-spin and tensor force. The spin-orbit force results from exchange of the vector bosons, the iso-singlet ω, and isovector ρ mesons; the strong repulsive core is due to ω exchange. The Λ particle has isospin $I = 0$, and contrary to the NN case the exchange of the isovector bosons such as pions and ρ mesons is not allowed. The lowest order contributions to the long-range ΛN force are believed to stem from a two-pion exchange and a kaon exchange (Figure 3a). It can thus be understood that the ΛN interaction is dominated by the central potential and that there is no strong spin-spin and tensor contribution in the long-range ΛN interaction (13), but it has a hard core comparable to the one in the NN interaction.

It has been suggested by Dalitz & von Hippel (12) that the ΛN inter-action should have quite an appreciable isospin-violating part. The Λ particle does not have a pure $I = 0$ isospin because of the isospin mixing

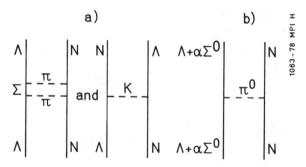

Figure 3 (*a*) The most important contributions to the ΛN potential; (*b*) π^0 exchange responsible for charge-symmetry-breaking effects.

of baryons and mesons within their SU(3) multiplets. The Λ and Σ^0 differ in quantum numbers only by their isospin 0 and 1, respectively. Actually they are not pure eigenstates of isospin, but they have small isospin impurity resulting from mutual mixing. In this case the one-π^0 exchange is allowed to a small extent and contributes to the long-range interaction (Figure 3b). The amplitude of the π^0 exchange has different sign in the Λ-neutron and Λ-proton cases, hence the contribution to the ΛN interaction is charge-symmetry-breaking. As we know from the NN interaction, one-pion exchange leads to the dominance of the spin-spin and tensor forces. Therefore we would expect that the isospin-violating interaction in the ΛN interaction would be best demonstrated in the different splitting of the singlet and triplet states of mirror hypernuclei. The experiments do not seem to support this simple reasoning.

The charge-symmetry-breaking effects are best studied in the lightest pair of mirror hypernuclei, $^4_\Lambda H$ and $^4_\Lambda He$. No difference in the splitting of the singlet and triplet states in $^4_\Lambda H$ and $^4_\Lambda He$ is observed. But there is a large difference in the B_Λ values of the two hypernuclei, $\Delta B_\Lambda(^4_\Lambda He - {}^4_\Lambda H) = 0.34 \pm 0.08$ MeV. The Coulomb difference (10, 12) in the energies of the mirror hypernuclei of $+0.25$ MeV should be added to the measured value, resulting in a total of 0.6 MeV for the energy difference to be accounted for by the charge-symmetry-breaking in the mass-four hypernuclei. It has not yet been estimated how much of this difference stems from direct charge-symmetry-breaking in the elementary ΛN interaction. Such a calculation will not be easy. It must include every possible source of charge-symmetry-breaking effects in nuclei and reconsider the estimated difference in Coulomb energies for realistic configurations of ^3H and ^3He. It is interesting to note that the energy difference of 600 keV in the binding energies of the two mirror nuclei is not completely accounted for theoretically. After taking into account the different Coulomb energies and all other known effects that influence the binding energy of the two nuclei differently, a difference of 100 keV remains unexplained (14).

The theory of the nuclear few-body system aspires to explain the link between the elementary NN interaction and the properties of the lightest nuclei. Powerful mathematical tools have been developed in order to study these systems with sufficient accuracy. The s-shell hypernuclei yield new information on the nuclear few-body system; some of them, for instance the binding energy of $^5_\Lambda He$, are not even qualitatively understood. By including the s-shell hypernuclei in the simultaneous treatment of the few-nucleon problem, further constraints can be put on the theory, which should in turn result in a deeper understanding of the lightest nuclear systems.

3   Λ-NUCLEUS INTERACTION AND HYPERNUCLEAR GROUND STATES

3.1   *Binding Energy of the Λ Particle in Nuclear Matter*

The binding energies of the Λ particle in the nuclear ground states give one of the basic pieces of information on the Λ-nucleus interaction. Most of the observed hypernuclear decays take place from the ground states because the electromagnetic transitions are generally faster than the weak decay of the Λ particle. The only exceptions are a few decays from isomeric states. The kinematical analysis of decay fragments in nuclear emulsions is still the best method for determining the binding energy of the Λ particle in the hypernucleus. Experience shows that only decays with charged mesons and all fragments producing visible tracks can be considered when measuring hypernuclear binding energies. These conditions can only be realized in light hypernuclei; hence the binding energies of hypernuclei with $A < 16$ can be determined by using this method.

The binding energy of Λ in the ground state is defined by

$$B_\Lambda(\text{g.s.}) = M_{\text{core}} + M_\Lambda - M_{\text{HY}}. \qquad 12.$$

The mass M_{core} is merely the mass of the nucleus that is left in the ground state after the Λ particle is removed. The B_Λ values, taken from the updated data of the European K^- Collaboration (15, 16), are summarized in Table 1.

It is very unlikely that $B_\Lambda(\text{g.s.})$ of hypernuclei with $A > 16$ can be determined by analysis of their decays, because the decay of a heavy hypernucleus cannot be identified uniquely. The lack of knowledge of the ground-state binding energy in heavy hypernuclei is very embarrassing. This energy is one of the most important pieces of information of the Λ-nucleus interaction as well as the most natural reference according to

Table 1   Λ binding energies of hypernuclei identified uniquely[a]

B_Λ(MeV)	B_Λ(MeV)	B_Λ(MeV)
$^3_\Lambda$H 0.13±0.05	$^8_\Lambda$Li 6.80±0.03	$^{10}_\Lambda$B 8.89±0.12
$^4_\Lambda$H 2.04±0.04	$^9_\Lambda$Li 8.53±0.15	$^{11}_\Lambda$B 10.24±0.05
$^4_\Lambda$He 2.39±0.03	$^7_\Lambda$Be 5.16±0.08	$^{12}_\Lambda$B 11.37±0.06
$^5_\Lambda$He 3.12±0.02	$^8_\Lambda$Be 6.84±0.05	$^{12}_\Lambda$C 10.76±0.19
$^6_\Lambda$He 4.18±0.10	$^9_\Lambda$Be 6.71±0.04	$^{13}_\Lambda$C 11.69±0.12
$^8_\Lambda$He 7.16±0.70	$^{10}_\Lambda$Be 9.11±0.22	$^{14}_\Lambda$C 12.17±0.33
$^7_\Lambda$Li 5.58±0.03	$^9_\Lambda$B 7.88±0.15	$^{15}_\Lambda$N 13.59±0.15

[a] Heavy hypernuclei $60 < A < 100$: upper limit of $B_\Lambda = 22.7 \pm 0.02$ MeV.

which the energy of the excited states may be measured. Excited states have been observed in many heavy hypernuclei, although their excitation with respect to the ground state can only be guessed at.

The upper limit for B_Λ(g.s.) in heavy hypernuclei can, however, be obtained from the observed decays accompanied by pion emission (17). The K$^-$ interaction with Ag and Br nuclei in emulsions produces hypernuclei in the mass region $60 < A < 100$. It is impossible to identify the particular hypernucleus from which the pion has been emitted, but one can obtain a good estimate of the average Λ binding energy in this mass region. Hypernuclei with $A \approx 100$ are not expected to have a strong variation of B_Λ(g.s.) with A. A small fraction of heavy hypernuclei decay by emission of π^-. Most of π^-, however, are absorbed by the nucleus, so less than 1% of the decays are accompanied by a visible π^- track. A large sample of such hypernuclear decays has been analyzed, and the maximum π^- energy has been determined. The Q value for the $\Lambda \rightarrow p + \pi^-$ decay is 35.7 MeV, which must be shared by the proton and the pion. In the nucleus all proton states are occupied up to the Fermi surface. Because of the Pauli principle the proton produced in the Λ decay cannot remain strongly bound in the nucleus and will jump at least to the lowest unoccupied level. In heavy nuclei the energetically lowest free levels are at the binding energy of about 8 MeV. Taking this into account, the upper limit for Λ binding energy in heavy hypernuclei has been estimated to be $B_\Lambda = 22.7 \pm 0.4$ MeV (17). At present this number is the most important piece of information for determining the depth of the potential well of the Λ particle in the nucleus.

B_Λ(g.s.) in hypernuclei can be well reproduced by assuming that the Λ feels a potential well of radius R approximately equal to that of the nuclear core and a depth D_Λ independent of the hypernuclear mass. In the hypernuclear ground state the Λ particle is always in the $1s$ state, and thus its binding energy increases with A. For heavy hypernuclei with $A \gg 1$ the relation between B_Λ(g.s.) and the potential depth D_Λ is particularly simple since one can just use the square-well potential. The kinetic energy E_{kin} of the Λ particle with mass M in the square well is

$$E_{\text{kin}} = D_\Lambda - B_\Lambda = \frac{\pi^2 \hbar^2}{2MR^2}.$$ 13.

In Figure 4, B_Λ(g.s.) is plotted against $A^{-\frac{2}{3}}$ in order to display the linear dependence of B_Λ on the inverse square radius for heavy hypernuclei. We see that B_Λ extrapolated from light hypernuclei (18, 19) agrees well with the values given by Expression 13 for heavy hypernuclei. A typical value of B_Λ(g.s.) in heavy hypernuclei is about 23 MeV, and the potential depth D_Λ obtained from it is 27 MeV.

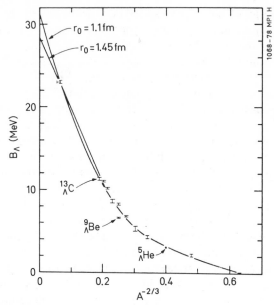

Figure 4 The binding energies B_Λ vs $A^{-\frac{2}{3}}$. The two curves are fitted to the values of B_Λ for the indicated values of r_0.

3.2   *Decay of Hypernuclei and their Lifetimes*

Only very few hypernuclear lifetimes have been determined experimentally. The measured values are plotted in Figure 5 and are compared to the lifetime of 2.6×10^{-10} sec of the free Λ particle. Only s-shell hypernuclei are produced in strangeness exchange reactions with sufficient recoil to distinguish the decay in flight and at rest in the bubble chamber or in nuclear emulsions. The stopping time of light nuclei in the bubble chamber, about 10^{-11} sec, is barely adequate for estimating the lifetime from the ratio of the hypernuclear decay at rest and in flight. This explains the poor accuracy of the lifetime measured with this method.

A promising new possibility of measuring hypernuclear lifetimes emerged when relativistic heavy-ion beams became available. In the interaction of a heavy ion of 2 GeV/nucleon with a target nucleus, a Λ particle can be produced. In such a reaction a hypernucleus can be formed, which moves with a velocity v/c of about 0.95. In a time of 10^{-10} sec the hypernucleus will traverse a distance of a few centimetres before decaying. Interesting events can be well selected by a K^+ signature accompanying the Λ in associated production. The decay length of the hypernucleus can be accurately measured by reconstructing the position of its fragmentation with position-sensitive counters.

A pioneering experiment of this type has recently been done at Berkeley by a Tucson group (20) using a ^{16}O ion beam. The analysis of 22 events yielded a mean lifetime of $0.88 \pm ^{0.33}_{0.26} \times 10^{-10}$ sec for the ^{16}O hypernucleus. This method can be much improved on and extended to heavier hypernuclei.

Intense relativistic heavy-ion beams offer another exciting possibility, namely the production of multiple hypernuclei. In the heavy-ion reaction, two or even more nucleons of the same nucleus can be converted into Λ particles. The probabilities of such processes to be within the reach of present experimental detection are claimed to be sufficiently large (21). Again, a multiple K^+ trigger accompanying such events would help select the proper events.

Because of the small energy release of 35 MeV in the decays

$$\Lambda \rightarrow p + \pi^- \quad \text{and} \quad \Lambda \rightarrow n + \pi^0, \qquad\qquad 14.$$

the momentum of the nucleon is only 100 MeV/c if Λ decays at rest. The probability for these decays in nuclear matter is generally reduced as a result of the Pauli principle, since a large fraction of the final states that can be reached by the decay nucleon are already occupied. Reduction of the π decay rates in $^5_\Lambda$He is already about 0.4, and in $^{13}_\Lambda$C it is 0.14 (22).

In a few suitable cases, however, where the final nucleon can land in the s-shell of the nucleus, the π decay rates can even be enhanced in the nucleus. This is the case for the $^4_\Lambda$H $\rightarrow \pi^- + {}^4$He and $^4_\Lambda$He $\rightarrow \pi^0 + {}^4$He decays from the $J^\pi = 0^+$ ground states. The s-interaction, dominant in the Λ decay, is spin independent, thus the final nucleon has the appropriate configuration to form ^{4}He from the ground states of $^4_\Lambda$H and $^4_\Lambda$He.

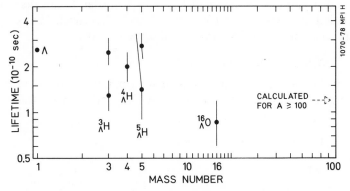

Figure 5 Mean lifetime of Λ particles and hypernuclei vs mass number. A theoretical estimate for $A \geq 100$ is also indicated.

The final-state nucleon is bound more strongly in ^{4}He than is the Λ particle in the mass-four hypernuclei, and the energy release in the decay increases as compared to the free decay. The π decay rates are enhanced by a factor of almost two in spite of some mismatch in the space-wave function of $^4_\Lambda$H, $^4_\Lambda$He, and ^{4}He.

In general, however, the π decay rates will be strongly reduced, so in most hypernuclei the decay is determined by the processes

$$\Lambda + p \rightarrow n + p \quad \text{and} \quad \Lambda + n = n + n \qquad\qquad 15.$$

in which the full energy difference of about 176 MeV between the Λ particle and the nucleon is released. In Table 2 the ratio R of nonmesic decays to π^- decay is given. It can be seen that for hypernuclei with $Z > 2$, nonmesic decay already prevails.

Processes 15 are interesting in themselves, as they present the only strangeness-changing weak interaction readily accessible to observation and involving four strongly interacting fermions. But unfortunately, present experimental techniques are not yet adequate for such a study, as has become apparent from the previous discussion on lifetime measurements.

The nonmesic decay rates have been calculated by Dalitz (22). The rapid increase with mass number A of the observed ratio R between the nonmesic decays and the decays with π^- is primarily due to the rapidly increasing suppression of the π^- decay process by the Pauli principle. The nonmesic decay rates increase rather slowly for 0.45 Γ_Λ for $^5_\Lambda$He to 1.5 Γ_Λ for $^{13}_\Lambda$C to a value of 2.0 Γ_Λ for very heavy hypernuclei with $A \approx 100$. The unit Γ_Λ gives a decay rate of the free Λ particle of 0.375 $\times$ 10^{10} sec^{-1}. The steady increase of the ratio R is caused by the reabsorption effect of pions in the nucleus, which thus reduces the number of observable pions in the decay of heavy hypernuclei. It can therefore be concluded that the decay lifetime of heavy hypernuclei is essentially independent of A and has a value of about 1.2×10^{-10} sec.

3.3   p-Shell Hypernuclei

One of the most important pieces of information on the quasiparticle property of the Λ particle is its effective interaction in nuclear matter.

Table 2 $R = (\text{nonmesic})/(\pi^--\text{mesic})^a$

Hypernucleus	$^4_\Lambda$H	$^4_\Lambda$He	$^5_\Lambda$He	$^7_\Lambda$Li	$^9_\Lambda$Be	$^{13}_\Lambda$C	$^{109}_\Lambda$Ag
R	0.15	0.5	1.8	2.7*	6.2*	11.3*	130*

^{a} The asterisked entries are calculated.

The s-shell hypernuclei are obviously unsuited for such studies. For example, particles confined only to the s-shell cannot give any information on the spin-orbit force, which plays a central role in nuclear physics. The p-shell hypernuclei seem to be more promising, as the amount of data that can be obtained experimentally is sufficiently abundant for such an investigation. Already a large number of Λ binding energies are known for p-shell hypernuclei. Moreover, four pairs of ground states are known to belong to isospin multiplets (Table 1); also two nontrivial spin assignments, both with $J = 1$, for $^8_\Lambda\text{Li}$ (23, 24) and $^{12}_\Lambda\text{B}$ (25, 26) have been uniquely determined. It should also be recalled that the first step in understanding nuclear structure has just been made for p-shell nuclei. In the late 1940s Inglis (27) and later Kurath (28) reproduced most of the properties of the natural parity states [parity $= (-1)^4$] within the inter-mediate-coupling model. Essentially, two parameters were necessary to systematize p-shell nuclei: one parameter measured the central nucleon-nucleon interaction in the p-shell, and the other, the intermediate-coupling parameter, measured the relative strength of the spin-orbit against the spin-spin interaction. It is therefore understandable that much effort has been made to interpret p-shell hypernuclei, using a model similar to the one used for nuclei (29, 30). The nuclear core is treated by the intermediate-coupling model; the Λ particle in the $1s$ state is coupled to it. In the calculation a very general effective two-body and three-body interaction has been considered. The two-body interaction itself has five free parameters; in addition to Interaction 9—which is adequate if the particles are only confined to the s-shell—tensor and symmetric and antisymmetric spin-orbit interactions were taken into account. The measured values of the Λ binding energies in p-shell hypernuclei (Table 1) do not give enough constraint to determine this complicated interaction nor even simplify it by eliminating the superfluous terms. But this is not surprising: Phenomenological models are not suitable for calculating the absolute values of binding energies in nuclei; they are quite successful, however, in reproducing the properties of the excited states in a particular nucleus. Nevertheless, one encouraging result came out of these calculations. It was found that the hypernuclear wave functions for ground states are generally rather pure, i.e. they mostly consist of a Λ particle attached to one particular state of the core nucleus. This is also expected from the lesson we learned long ago about p-shell nuclei. The unnatural parity states, i.e. those in which one nucleon has left the p-shell for the s,d-shell, are well described in the weak-coupling model (31). Their wave functions are in a good approximation, just an s,d-shell nucleon attached to one particular state of the p-shell core. It is thus very likely that with knowledge of the excited hypernuclear state, the effective interaction will easily be isolated, and the shell model for p-shell hypernuclei will be expressed in a simple and elegant form. Only good data on low-lying excited states in the p-shell can help solve the problem. This calls for new experiments in p-shell hypernuclei; and the only suitable ones seem to be the gamma-type experiments discussed above.

4 SPECTROSCOPY WITH THE (K^-,π^-) REACTION

4.1 *Strangeness Exchange Reaction*

For determining the Λ-particle binding energy in the ground state, as discussed above, it was not of great importance to know how the hypernucleus was formed. The hypernucleus was identified through its decay. In counter experiments, however, the production of hypernuclei can be investigated more easily than their decay. Therefore it is essential to understand the reaction mechanism in which hypernuclei are produced. Of the many possible reactions in which hypernuclei can be formed, only those in which the Λ particle gets a small recoil momentum are of interest. The probability of forming a hypernucleus is substantial only if the recoil momentum is comparable to the Fermi momentum of nucleons in the nucleus. Obviously, hypernuclear spectroscopy is possible only if the energy and momentum of the Λ particle can be measured. The best example of such a reaction is

$$K^- + {}^A Z \rightarrow \pi^- + {}^A_\Lambda Z, \qquad\qquad 16.$$

which is excellently suited for the missing-mass type of experiment. Measuring the energy and momentum difference of the incoming kaon and the outgoing pion gives complete information on the hypernuclear system.

Reaction 16 plays a central role in hypernuclear spectroscopy. Let us first consider the reaction $K^- + n \rightarrow \pi^- + \Lambda$ on a free neutron. For pions emitted at 0° the recoil momentum—in this particular case a longitudinal momentum only—of the Λ particle depends on the kaon momentum as shown in Table 3. For kaon momenta between 300 and 1000 MeV/c the recoil of the Λ particle is less than 100 MeV/c under these kinematical conditions. The Fermi momentum of the nucleons in the nucleus is of the order of 250 MeV/c. The Λ recoil is much smaller than the Fermi momen-

Table 3 The recoil momentum in the $K^- + n \rightarrow \pi^- + \Lambda$ reaction of pions as detected at 0°

K^- momentum (MeV/c)	0	100	300	500	700	900
Λ momentum (MeV/c)	250	190	70	0	40	80

tum, and there is an appreciable probability of forming a hypernucleus.

For kaon momenta above 500 MeV/c and for small reaction angles, the transverse recoil momentum of the Λ particle is

$$q_T \approx 2\mathbf{p} \sin \alpha/2, \qquad\qquad 17.$$

where α is the reaction angle and $\mathbf{p} \approx \mathbf{p}_\pi \approx \mathbf{p}_K$. In the forward direction the kinematics of the strangeness exchange reaction strongly resembles the kinematics of elastic scattering (Figure 6). The analogy of the (K^-,π^-) reaction with the scattering is especially apparent if one considers that in the strangeness exchange reaction there is an appreciable probability that the Λ particle just replaces a neutron in the nucleus without otherwise changing its wave function. In such cases we will refer to "recoilless Λ production" in order to emphasize that the whole nucleus rather than Λ takes the momentum mismatch of the (K^-,π^-) reaction. Recoilless Λ production is a coherent process in the same sense as elastic scattering, because we cannot identify the neutron on which the strangeness exchange reaction took place. In elastic scattering the target nucleus remains in its ground state; in recoilless Λ production, the hypernuclear states are populated with the same configurations as that of the target nucleus, but with a neutron replaced by the Λ particle. The possibility of comparing nuclear with hypernuclear states of identical configurations makes the strangeness exchange reaction the central tool of hypernuclear spectroscopy.

The states populated by recoilless Λ production, usually referred to as strangeness exchange resonances, are highly excited and embedded in a continuum. This is easily understood qualitatively. By changing a neutron into a Λ particle (Figure 7), a neutron hole is created. A Λ particle is produced with the spin and orbit quantum numbers of the neutron.

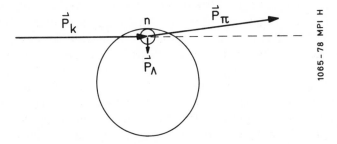

Figure 6 In the forward direction and for kaon momenta between 300 and 1000 MeV/c, the kinematics of the strangeness exchange reaction resembles the kinematics of elastic scattering. Here $\mathbf{p}_K$, $\mathbf{p}_\pi$, and $\mathbf{p}_\Lambda$ are the momenta of the incoming kaon, outgoing pion, and recoiling Λ particle, respectively.

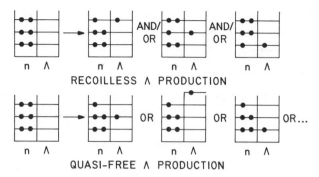

Figure 7 Recoilless and quasifree Λ production is shown schematically.

This is generally not the 1s state corresponding to the hypernuclear ground state. The strangeness exchange resonance is a Λ-particle neutron-hole state, the energy of which is the sum of the neutron-hole and Λ-particle excitations.

The differential cross section for pions from recoilless Λ production should also have the same angular dependence as the elastic scattering on the nucleus, and according to the form factor of the nucleus it should decrease with the momentum transfer.

The major part of the strangeness exchange in the nucleus, however, leads to "inelastic processes." The total cross section for kaons and pions at momenta above 500 MeV/c is about 30 mb, and the absorption is so strong that only a small portion of the nuclear surface contributes to recoilless Λ production. Of all the possible inelastic processes, we will be interested only in those in which the strangeness exchange is accompanied by a jump of the Λ particle into one of the neighboring orbits without any additional interaction of kaons or pions with the nucleus. In this process the recoil is transferred to the Λ particle, and it is obvious that this is no longer a coherent process. In analogy to scattering, Dalitz & Gal (32) introduced the name "quasifree Λ production" for such processes. Both recoilless and quasifree Λ production are one-step processes, i.e. apart from the strangeness exchange there is no additional interaction of either the kaon or the pion with the nucleus, and they are thus simple to treat theoretically. The central question of hypernuclear spectroscopy is whether recoilless and quasifree Λ production can be detected without background. This question can only be answered by experiment.

4.2 Experiments with K^- on Nuclei

Experimental methods used in hypernuclear physics have been developed in the last few years, parallel to a similar development in intermediate energy physics. In both cases experiments are performed on high energy accelerators with protons of a rather large energy spread or, to a much greater extent, with secondary beams of pions and kaons. To obtain a reasonable intensity, secondary beams must have a large momentum bite and emittance.

The direct use of such beams in hypernuclear spectroscopy would give no answer to the relevant problems of hypernuclear physics. The energy resolution of the experiments should suffice to separate at least the most prominent excitations in the hypernucleus. As for the strangeness exchange Reaction 16, where we do not expect many states to be populated, a resolution of about 1 MeV should suffice, and magnetic spectrometers of a rather modest resolution could be used for the momentum analysis of the beam and the reaction products.

In the late 1960s, the first low-momentum (<1 GeV/c) K$^-$ beams were constructed in such a way as to fulfill the minimum requirements for their use in counter experiments. At present the best K$^-$ beams operate at CERN using the 25-GeV Proton Synchrotron and at Brookhaven National Laboratory using the 30-GeV AGS proton accelerator. A further K$^-$ beam of similar quality is in construction at the 10-GeV proton accelerator of KEK (Japan).

The production of K$^-$ by a 25-GeV proton energy is smaller by a factor of about 100 than the production of pions at small angles, which is of interest because of the forward peaking of the particle production. The K/π ratio, however, drops quickly with increasing distance from the production target for low-momentum beams. The lifetime of 1.23×10^{-8} sec of K$^-$ is shorter by a factor of two than the lifetime of π^-. In addition, for momenta below 1 GeV/c the velocities of pions and kaons of the same momentum differ enough to make an appreciable difference in the time dilation for the two mesons. Therefore kaon beams have to be as short as possible, and pions have to be reduced in number by using particle separation.

The main characteristics of hypernuclear experiments can best be deduced by inspecting the experiment presently running at CERN (33), which is schematically shown in Figure 8. The beam transport system refocuses particles of a selected momentum from the production target to the experimental target. This is just long enough to allow the momentum definition of the beam and the use of a mass separator for pion suppression. The particles emerging from the experimental target are analyzed by a rather sophisticated spectrometer (34) that has a large solid angle of 20 msr and a momentum acceptance $\Delta p/p$ of 20%. The resolution of the spectrometer is better than 1 MeV/c for momenta up to 900 MeV/c. The beam itself is designed to have an angular acceptance

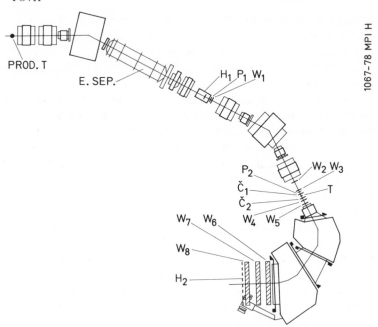

1067-78 MPI H

Figure 8 Kaon and pion momenta are analyzed by use of a magnetic system. The particle trajectories are determined by the hodoscopes H1, H2 and the wire chambers W_1-W_8. Kaons and pions are identified by the Čerenkov counters C1, C2 at the target position and by the time-of-flight measurement.

of 9 msr and to transport a momentum bite of only 3%. Therefore the momentum of the beam particles can be determined with sufficient precision, using just the last part of the beam transport system as a spectrometer. With this beam transport system of only 12 m in length and of 10^{12} protons on the production target, 10^4 K^- of 700 MeV/c reach the experimental target. Obviously, the kaon number on the target is a very steep function of the momentum, because the production rate increases with the kaon energy and because the decay length is velocity dependent. The first-generation experiments (35, 36) have been taken by kaon beams and spectrometers of far more modest but similar design. Thus most of the data on strangeness exchange reactions have been collected at 900 MeV/c in order to obtain sufficient kaon rates on the target, the results of which are shown in Figure 9. The scale is given in the binding energy of the Λ particles in the hypernucleus, with zero indicating the threshold of Λ and the nucleus core in the ground state. The use of this scale in the present stage of development is reasonable, because it depends merely on the known masses of the particles involved in the reaction.

the distortion in the (K^-,π^-) reaction and use the harmonic oscillator potential for the nucleus. Recoilless Λ production corresponds to the recoilless gamma emission of atoms bound in crystal, the quasifree Λ production to the gamma emission accompanied by phonon excitation. The probability of recoilless Λ production is then given by the nuclear Debye-Waller factor (41),

$$P(n_i n_i) = \exp\left\{-(2n_i+1)(\hbar q)^2/2M_\Lambda D\right\}, \qquad\qquad 18.$$

where n_i is the quantum number of the harmonic oscillator state, q the Λ recoil momentum, and D the spacing of the harmonic oscillator levels for a Λ particle in the nucleus. The probability of recoilless and quasifree Λ production adds up to one. Because of the strong absorption of the (K^-,π^-) reaction, this expression can be usefully applied if one takes into account the last filled nuclear shell, just neglecting the inner shells. At 900 MeV/c for kaons, the nuclear Debye-Waller factor gives a value of 40% for recoilless Λ production on an oxygen target and 20% on a calcium target, in agreement with the calculations of Bouyssy. Equation 18 demonstrates the great flexibility of the strangeness exchange reaction for selective population of hypernuclear states. At 700 MeV/c the average Λ recoil momentum is only 40 MeV/c, and the probability for recoilless Λ production is already an order of magnitude larger than that of the quasifree Λ production. By choosing the Λ recoil, strangeness exchange resonances can easily be distinguished from the most prominent quasifree transitions. The most prominent quasifree transitions are those in which the Λ particle jumps from the valence shell to one of the stronger bound shells (Figure 6). If in addition to the strangeness exchange resonance these transitions can be identified, we will have a set of states corresponding to the same nuclear core and a Λ particle in the valence and deeper-lying shells. This set of data would give enough constraint to pin down the properties of the Λ particle in nuclear matter in order to find its effective mass and the potential depth depending on its excitation. This would be the first direct measurement of quasiparticle properties in a strongly interacting medium.

Probably the most impressive feature of the spectra is the narrow width of the dominant level populated in recoilless Λ production (Figure 9). The strangeness exchange resonance lies in a continuum at energies up to 30 MeV for heavy nuclei. Nevertheless, the observed width of the level is less than 2 MeV and is consistent with the resolution of the apparatus. The Λ escape widths of the Λ-particle neutron-hole states have been found to be narrow (42), on the order of several hundred keV. The width of the particle-hole state is in the first approximation the sum of the Λ escape and the neutron-hole width. The narrow width of the strangeness exchange resonance is therefore consistent with the previous statement that its configuration is limited to the particle-hole excitation in the last nuclear shell. The widths of the neutron-hole states in the deeply bound shells are likely to be broad and on the order of several MeV.

Recoilless Λ production can happen to any neutron in the nucleus, thereby producing strangeness exchange resonances. If the resonances are energetically close together, they will mix. There is the possibility of coherent excitation of a particular linear combination of the strangeness exchange resonances, each of which has a different neutron in the nucleus converted into a Λ particle. One such linear combination has been called the strangeness analogue state or resonance (SAR) (43, 44) in analogy with the isobaric analogue states in nuclei:

$$| \text{SAR} \, {}^A_\Lambda Z \rangle = N^{-\frac{1}{2}} | \sum_\alpha c^+_\alpha b_\alpha \text{ ground state } {}^A Z \rangle, \qquad 19.$$

where c^+ is a creation operator that creates a Λ in state α, b is an annihilation operator that destroys a neutron in the same state, and N is the number of neutrons in the target.

The analogue state is characterized by its permutation symmetry. According to the Pauli principle, the wave function for the target nucleus must be antisymmetric with respect to all neutrons and protons in the nucleus. For the strangeness exchange resonance this symmetry is retained only within one nuclear shell. No such symmetry principle exists between the hyperon and nucleon, and all possible permutation states are allowed. In the SAR attained by the (K^-, π^-) reaction, a neutron in the target is changed into Λ and the exact wave function remains unchanged, as does the permutation symmetry. The hypernuclear wave function is also antisymmetric with respect to the interchange of any neutron and Λ particle.

The description of recoilless Λ production with the SAR is elegant and thus very appealing. It neglects the differences between protons, neutrons, and lambdas, and can be classified according to the SU(3) symmetry. In fact, prior to the success of the octet model in hadron physics, Sakata (45) had already tried to create hadron states using the triplet basis of npΛ. This model is no longer interesting for hadron physics, but can be directly applied to hypernuclear physics. The SU(3) symmetry in hypernuclei is certainly not very good. The potential depth $D_\Lambda \approx 28$ MeV is half as small as the potential depth for nucleons; thus spacings between Λ and nucleon shells may differ so much that the strangeness exchange resonances corresponding to the particle-hole excitation in different nuclear shells do not mix at all. This in itself is not yet a sufficient reason for dropping such an elegant way of describing the hypernuclear states. It would still be very convenient to perceive the states produced in recoilless Λ production as eigenstates of a Hamiltonian with a high symmetry in order to classify the states according to their dominant features. And then, in the second step, the wave function could be corrected by including symmetry-breaking effects. The main objection to the use of SAR in recoilless Λ production comes from the consideration of the distortion in the (K^-, π^-) reaction. The reaction we are interested in takes place on the nuclear surface. Hence SAR is not a good description of the doorway state in recoilless Λ production, because in comparison with the nucleons in the last shell the nucleons from the inner shells are strongly suppressed on the surface. The doorway state is thus better represented by the strangeness exchange resonance on the last nuclear shell. If the particle-hole excitation of inner shells does not couple to the doorway state, there will also be little chance of observing it in experiments.

4.4   *Distortion in the (K^-, π^-) Reaction*

The use of the (K^-, π^-) reaction in hypernuclear spectroscopy makes sense only if we are sure that the reaction takes place on a single nucleon without any additional interaction of either incoming K^- or outgoing π^- with the rest of the nucleus. The kinematical conditions used in recoilless Λ production are such that they warrant clean spectra at the highest pion momenta observed. Any additional interaction would degrade the energy of pions and scatter them out of the 0° direction. The question is, however, how large is the momentum bite when only the contribution of a single interaction of the (K^-, π^-) reaction is expected. The first experiments on the (K^-, π^-) reaction have been done with rather small momentum acceptance, and there was some doubt (46) whether the well-separated recoilless and quasifree Λ production (Figure 9) was due to the experimental acceptance. In preliminary results the new experiment at CERN, performed with much larger momentum acceptance, reproduces the previous results well, and shows that the background, presumably coming from multi-interaction events, does not mask the energy region of the quasifree Λ production. Therefore we should briefly discuss why the one-step processes are separated from the rest of the events.

Let us first consider the cross section of the reaction. The cross sections at 0° summed over the recoilless and quasifree Λ production in all spectra are approximately the same and have a value of about 1.2 mb/sr. This cross section should be compared to the elementary cross section for the (K^-, π^-) reaction on a neutron, which is 2.3 mb/sr at 900 MeV/c and 0°. Only a fraction of this cross section is observed in the one-step strangeness exchange reaction on nuclei. The effective neutron number for the one-step strangeness exchange reaction on nuclei is about 0.5. This low a number is by no means surprising (37, 47). The mean free path of K^- and π^- in the nucleus is about 1 fm. Only a very thin ring of the nuclear surface contributes to the $0°$ single-step reaction. A very simple geometrical consideration shows that the volume of the ring that contributes to the reaction is independent of the nuclear radius, if the mean free path of the particle is kept constant. Therefore one does not expect a strong A dependence of the cross section for the one-step process (46).

In alternative calculations, however, Dalitz & Gal (32) and Epstein et al (48) found much larger cross sections than those observed experimentally, as well as a strong A dependence ($A^{0.3-0.5}$) for one-step processes. The effective neutron numbers of about 1.8 for a carbon target and 2.5–3.5 for a calcium target are found. As far as the absolute values of the cross section are concerned, we learned from the direct reaction on nuclei that distortion can be properly calculated only if the scattering of the incoming and outgoing particle on the nucleus at same energies is known. As soon as the results of the (K^-,π^-) reaction reach a sufficient degree of sophistication for us to worry about the absolute cross section, the scattering of kaons and pions on nuclei should be performed. Unfortunately, these measurements, at least those with kaons, are as time consuming as the (K^-,π^-) reaction itself. Less understandable, however, is the strong $A^{0.5}$ dependence found by Epstein et al, which is close to the $A^{\frac{2}{3}}$ value and signifies that the whole surface of the nucleus contributes to the one-step strangeness exchange reaction. This can hardly be reconciled with the physical picture of the process.

Let us finally turn to the most important question of why the (K^-,π^-) reaction is clear of background, at least in the region of the major contributions of the recoilless and quasifree Λ production, if measured at $0°$. In view of the strong distortion in the reaction, this is not self-evident. The most dangerous contamination is expected to come from the events where K^- and π^- involved in the strangeness exchange reaction scatter in the nucleus, losing a few MeV of energy. Inelastic scattering means that the nucleus becomes excited. In this process the quantum numbers of the nuclear state change, and for $0°$ inelastic scattering is strictly forbidden. Therefore the background can arise only from multiscattering events degrading the energy in each scattering, and can be energetically separated from the one-step events. The one-step events in the (K^-,π^-) reaction at $0°$ are selected by the requirement that K^- and π^- be collinear. Collinearity selects a very small part of phase space in which recoilless and quasifree Λ productions dominate.

Besides the strangeness exchange resonance, those states are of vital interest for hypernuclear spectroscopy in which the Λ particle produced on a valence neutron jumps to one of the lower-lying orbits. Pions leading to these states have higher energy than those leading to the recoilless Λ peak, and they are free of background.

5 Σ HYPERNUCLEI

The Σp interaction at low momenta has been studied, in a similar way to Λp, by means of the hydrogen bubble chamber used to generate the sigmas and to observe their interaction with hydrogen. The existing data on the Σp interaction are even more scarce than for the Λp interaction. This is mainly due to the shorter lifetime of the Σ hyperon, the ionization loss in the bubble chamber, and less distinct identification of the interaction events, which made counter experiments infeasible until now. The analysis is also more involved because of many competing channels in the Σ^-p interaction. The Σ hyperon has isospin $I = 1$ and can couple with a proton to $I = \frac{1}{2}$ and $\frac{3}{2}$. The Σ^+p interaction is pure in the $I = \frac{3}{2}$ state. The measured total cross section (49) at an incident Σ^+ momentum of 160 MeV/c is only about 110 mb, i.e. about a factor of 10 less than the unitary limit for the scattering cross section. This observation strongly suggests that the triplet cross section for ΣN in the $I = \frac{3}{2}$ state is very small, and thus the existence of a bound state or a low-lying resonance with these quantum numbers is unlikely. Furthermore, the experimental cross section does not even saturate the $I = \frac{3}{2}$ unitary limit for the scattering cross section in the singlet state.

At low Σ^- incident momenta the $\Sigma^- + p \rightarrow \Sigma^0 + n$ and $\Sigma^- + p \rightarrow \Lambda + n$ reactions compete with elastic scattering. Of particular interest is the $\Sigma^- + p \rightarrow \Lambda + n$ channel, which allows polarization measurements. Starting with polarized Σ^- hyperons, the Λ particles emerge at rest with a non-zero polarization even for the Σ^-p interaction, which presumably takes place in an s-wave (50). Since only the triplet state can contribute to the polarization, this information suggests that the Σ^-p interaction at low momenta takes place predominantly in the triplet state. Therefore the most elaborate search for a possible bound dibaryon state has been made for the $I = \frac{1}{2} \Sigma$N system in a triplet state. The best way to populate such a state is by means of the recoilless Σ production on a deuterium target. The kinematics for $K^- + p \rightarrow \Sigma^+ + \pi^-$ are very similar to the strangeness exchange reaction in which the Λ particle is produced, but with an optimal kaon momentum at 300 MeV/c. In the deuterium bubble chamber a strong enhancement in the $K^- + d \rightarrow \Lambda + p + \pi^-$ cross section is observed at the invariant mass close to the ΣN threshold. Model calculations using the relativistic deuteron form factor and kaon-nucleon amplitudes predict a dynamic cusp at the ΣN threshold owing to the strong final state $\Sigma N \to \Lambda N$ conversion. This effect alone is not sufficient to explain the narrow experimental peak, the energy-independent width, and the angular distributions of the reaction without assuming a rapidly varying $\Sigma + N \to \Lambda + N$ transition amplitude. In fact, the observed peak is satisfactorily described if, in addition to the dynamical cusp, a ΣN bound state of about 4 MeV below the threshold is assumed (51). Because of poor statistics and moderate resolution in the bubble-chamber experiments, there are still some ambiguities in separating the dynamic effects from the resonant cross section in the $K^- + d \to \Lambda + p + \pi^-$ reaction. The experiment should be redone with experimental techniques used in hypernuclear spectroscopy in order to determine the missing mass with adequate accuracy.

The strong indication of a "bound state" in the ΣN system triggered further search for strange dibaryon states, in particular those with strangeness $S = -1$ and $S = -2$. It will certainly be interesting to find whether more dibaryon states exist and whether they can be classified according to SU(3) multiplets, or six-quark states as predicted by the bag model (52).

As already pointed out, sigmas convert into lambdas in the nucleus via strong interaction. They may, however, live long enough for their interaction with the nucleus to be measured. If the width of the Σ hypernuclei is sufficiently narrow, as is the case for the strangeness -1 dibaryon in the $I = \frac{1}{2}$ and the $J^\pi = 1^+$ state, the production peak of Σ hypernuclei will show up from the background in the strangeness exchange reaction. But there are many reasons for believing that the width of heavy Σ hypernuclei will be larger than for the $\Sigma^+ n$ system. In heavy nuclei Σ hyperons are exposed to higher nuclear density even at the surface than they are in the two-body system, which certainly accelerates their conversion into the Λ particle. The states produced in recoilless Σ production lie in a continuum. The Q value for the $\Sigma + N \to \Lambda + N$ reaction is 80 MeV. To this energy should be added an excitation energy of 20–30 MeV of the strangeness exchange resonance, resulting in a total of about 100 MeV energy release in the Σ conversion. The chance to observe narrow states at 100-MeV excitation in the nucleus is not great, but one should try to discover experimentally if Σ hypernuclei can be detected.

6 CONCLUSIONS

Hypernuclear spectroscopy was first performed by means of a nuclear emulsion technique, which is generally applied to particle physics if experimental conditions are too difficult to allow for detection of particles by counters. The systematic knowledge of the light hypernuclear ground states stems from these experiments. The development of strong K^-

beams and new experimental methods permits counter experiments and allows the study of hypernuclear excited states, in particular those in the continuum. The hypernuclear ground states and the continuum states obtained in recoilless as well as quasifree Λ production are of central interest in hypernuclear physics, as they supply direct information about Λ-nucleus interaction. It has been demonstrated that all of these states can be detected in experiments that are free of background. The aim of the experiments in the immediate future is to collect systematic data on the ground states and the states populated by recoilless and quasifree Λ production in order to pin down the Λ-nucleus interaction. Analysis of these data will lead to the first direct determination of the quasiparticle properties in a strongly interacting medium.

Literature Cited

1. Alexander, G., Karshon, U., Shapira, A., Yekutieli, G., Engelmann, R., Filthuth, H., Lughofer, W. 1968. *Phys. Rev.* 173: 1452
2. Sechi-Zorn, B., Kehoe, B., Twitty, J., Burnstein, B. 1968. *Phys. Rev.* 175:1735
3. Londergan, J. T., Dalitz, R. H. 1972. *Phys. Rev. C* 6:76
4. Dalitz, R. H. 1969. In *Nuclear Physics*, ed. C. De Witt, V. Gillet, p. 701. New York: Gordon & Breach
5. Bamberger, A., Faessler, M. A., Lynen, U., Piekarz, H., Piekarz, J., Pniewski, J., Povh, B., Ritter, H. G., Soergel, V. 1973. *Nucl. Phys. B* 60:1
6. Bejidian, M., Filipkowski, A., Grossiord, J. Y., Guichard, A., Gusakow, M., Majewski, S., Piekarz, H., Piekarz, J., Pizzi, J. R. 1976. *Phys. Lett. B* 62:467
7. Bedjidian, M., Descroix, E., Grossiord, J. Y., Guichard, A., Gusakow, M., Jacquin, M., Kudta, M. J., Piekarz, H., Piekarz, J., Pizzi, J. R., Pniewski, J. 12–17 Sept. 1977. In *"Recent Development in Hypernuclear Spectroscopy Using the (K^-, π^-) Reaction."* Zvenigorod. In press
8. Gal, A. 1975. *Adv. Nucl. Phys.* 8:1
9. Herndon, R. C., Tang, Y. C. 1967. *Phys. Rev.* 153:1091
10. Herndon, R. C., Tang, Y. C. 1968. *Phys. Rev.* 165:1093
11. Dalitz, R. H., Herndon, R. C., Tang, Y. C. 1972. *Nucl. Phys. B* 47:109
12. Dalitz, R. H., von Hippel, F. 1964. *Phys. Lett.* 10:153
13. Brown, J. T., Downs, B. W., Iddings, C. K. 1970. *Ann. Phys.* 60:148
14. Kim, Y. E., Tubis, A. 1974. *Ann. Rev. Nucl. Sci.* 24:70–94
15. Cantwell, T., Davis, D. H., Kiełczewska, D., Zakrzewski, J., Jurić, M., Krecker, U., Coremans-Bertrand, G., Sacton, J., Tymieniecka, T., Montwill, A., Moriarty, P. 1974. *Nucl. Phys. A* 236:445
16. Jurić, M., Bohm, G., Klabuhn, J., Krecker, U., Wysotzki, F., Coremans-Bertrand, G., Sacton, J., Wilquet, G., Cantwell, T., Esmael, F., Montwill, A., Davis, D. H., Kiełczewska, D., Pniewski, T., Tymieniecka, T., Zakrzewski, J. 1973. *Nucl. Phys. B* 52:1
17. Lemonne, J., Mayeur, C., Sacton, J., Vilain, P., Wilquet, G., Stanley, D., Allen, P., Davis, D. H., Fletcher, E. R., Garbutt, D. A., Shaukat, M. A., Allen, J. E., Bull, V. A., Conway, A. P., March, P. V. 1965. *Phys. Lett.* 18:354
18. Rote, D. M., Bodmer, A. R. 1970. *Nucl. Phys. A* 148:97
19. Bodmer, A. R. 1973. *Proc. Summer Study Meet. Nucl. Hypernucl. Phys. Kaon Beams*, p. 64. Upton, NY: Brookhaven Natl. Lab.
20. Nield, K. J., Bowen, T., Cable, G. D., De Lise, D. A., Jenkins, E. W., Kalbach, R. M., Noggle, R. C. Pifer, A. E. 1976. *Phys. Rev. C* 13:1263
21. Kerman, A. K., Weiss, M. S. 1973. *Phys. Rev. C* 8:408
22. Dalitz, R. H. 1964. In *Proc. Int. Conf. Hyperfragments*, p. 147. St. Cergue: CERN Publication No. 64–1
23. Dalitz, R. H. 1963. *Nucl. Phys.* 41:78
24. Davis, D. H., Levi Setti, R., Raymund, M. 1963. *Nucl. Phys.* 41:73
25. Kiełczewska, D., Sacton, J., Cantwell, T., Montwill, A., Moriarty, P., Davis, D. H., Tymieniecka, T., Zakrzewski, J., Jurić, M., Krecker, U. 1975. *Nucl. Phys. A* 238:437

26. Zieminska, D., Dalitz, R. H. 1975. *Nucl. Phys. A* 238:453
27. Inglis, D. R. 1953. *Rev. Mod. Phys.* 25:390
28. Kurath, D. 1956. *Phys. Rev.* 101:216
29. Gal, A., Soper, J. M., Dalitz, R. H. 1971. *Ann. Phys.* 63:53
30. Gal, A., Soper, J. M., Dalitz, R. H. 1972. *Ann. Phys.* 72:445
31. Lane, A. M. 1960. *Rev. Mod. Phys.* 32:519
32. Dalitz, R. H., Gal, A. 1976. *Phys. Lett. B* 64:154
33. Bertini, R., Bing, O., Chaumeaux, A., Durand, J. M., Garreta, D., Kilian, K., Niewisch, J., Pietrzyk, B., Povh, B., Ritter, M. G., Schröder, H., Thirion, J. 1976. *Proposal CERN/EEC-76/9*
34. Aslanides, E., Bertini, R., Bing, O., Birien, P., Bricaud, B., Brochard, F., Catz, H., Durand, J. M., Faivre, J. C., Garreta, D., Gorodetzky, Ph., Hibou, F., Pain, J., Thirion, J. 29 Aug.–2 Sept. 1977. *Proc. Int. Conf. High-Energy Phys. Nucl. Struct., 7th, Zürich*, p. 376
35. Brückner, W., Granz, B., Ingham, D., Kilian, K., Lynen, U., Niewisch, J., Pietrzyk, B., Povh, B., Ritter, H. G., Schröder, H. 1976. *Phys. Lett. B* 62:481
36. Bonazzola, G. C., Bressani, T., Cester, R., Chiavassa, E., Dellacasa, G., Fainberg, A., Mirfakhrai, N., Musso, A., Rinaudo, G. 1974. *Phys. Lett. B* 53:297
37. Hüfner, J., Lee, S. Y., Weidenmüller, H. A. 1974. *Nucl. Phys. A* 234:429
38. Bouyssy, A., Hüfner, J. 1976. *Phys. Lett. B* 64:276
39. Schiffer, J. P., Lipkin, H. J. 1975. *Phys. Rev. Lett.* 35:708
40. Bouyssy, A. 1977. *Nucl. Phys. A* 290:324
41. Povh, B. 1976. *Z. Phys. A* 279:159
42. Auerbach, N., Nguyen van Giai, Lee, S. Y. 1978. *Phys. Lett. B* 68:255
43. Lipkin, H. J. 1965. *Phys. Rev. Lett.* 14:18
44. Feshbach, H., Kerman, A. K. 1966. *Preludes in Theoretical Physics*, p. 260. Amsterdam: North-Holland
45. Sakata, S. 1956. *Prog. Theor. Phys.* 16:782
46. Povh, B. 1976. *Rep. Prog. Phys.* 31:833
47. Deloff, A. 1973. *Nucl. Phys. B* 67:69
48. Epstein, G. N., Tabakin, F., Gal, A., Kisslinger, L. S. 1978. *Phys. Rev. C* 17:1501
49. Dosch, H. G., Filthuth, V. F., Hepp, V., Kluge, E. 1966. *Phys. Lett.* 21:236
50. Yamamoto, S. T., Stephens, D., Meisner, G. W., Kofler, R. R., Herzbach, S. S., Button-Shafer, J., Yamin, P., Berley, D. 5–7 May 1969. *Proc. Int. Conf. Hypernucl. Phys.*, ed. A. L. Brodmer, L. G. Hyman, p. 939. Argonne Natl. Lab., IL
51. Dosch, G., Hepp, V. 1977. *CERN Preprint TH 2310*
52. Jaffe, R. L. 1976. *Phys. Rev. Lett.* 38:195

Ann. Rev. Nucl. Part. Sci. 1978. 28:33–113

SYNCHROTRON RADIATION RESEARCH ✻5592

Herman Winick and Arthur Bienenstock

Stanford Synchrotron Radiation Laboratory, Stanford, California 94305

CONTENTS

1   INTRODUCTION

Synchrotron radiation, the electromagnetic radiation emitted by relativistic electrons traveling in curved paths, opens new possibilities in research and technology that have already produced important results and are

33

0066-4243/78/1201-0033$01.00

expected to have even greater impact in the future. With the development of high energy electron synchrotrons and storage rings for elementary particle research, very powerful sources of synchrotron radiation are now available and are extensively used for synchrotron radiation research, largely in a symbiotic manner with the high energy physics programs. In more than twenty laboratories throughout the world, the broad spectrum of intense, collimated, and polarized x rays and ultraviolet radiation is being used for experiments in structural biology, catalytic chemistry, surface physics, solid state physics, and many other areas, including applications to technology such as microstructure replication and x-ray topography.

Although synchrotron radiation research began in the early 1960s, most of the work until 1974 was concentrated in the ultraviolet part of the spectrum, and almost all the sources were synchrotrons. Starting in about 1974, a new generation of multi-GeV storage rings became available (SPEAR, DORIS, VEPP-3). These provide higher intensity, a more constant spectrum, and greater stability than synchrotrons.

The first major synchrotron radiation research facility utilizing a multi-GeV storage ring was the Stanford Synchrotron Radiation Laboratory (SSRL), which began operation in May 1974. The results from SSRL and from the other multi-GeV storage rings conclusively showed the enormous capability of synchrotron radiation from 150 eV to 35 keV as produced by a storage ring. Earlier results from smaller storage rings at Wisconsin and Orsay had already proven the case for lower photon energies.

As published work in this new energy region began to appear, an extremely large growth in user interest occurred. For example, at SSRL the number of active research proposals in January 1978 exceeded 200 and has approximately doubled each year for the last three years. Evaluation of national needs has prompted several countries (England, Germany, Japan, and the US) to construct new storage rings dedicated only to synchrotron radiation research and greatly to increase the programs on existing machines.

The unique properties of synchrotron radiation (markedly different both quantitatively and qualitatively from other radiation sources) are, of course, responsible for these developments. The most important characteristic of synchrotron radiation is the high and extremely stable intensity over a broad bandwidth. This frees researchers from working only at the discrete lines available from conventional sources in the ultraviolet (UV) and x-ray part of the spectrum. As a continuum x-ray source for example, synchrotron radiation makes available more than 100,000 times higher flux than the most powerful rotating anode x-ray generators. This has led, among other results, to the development of the technique known as Extended X-Ray Absorption Fine Structure (EXAFS) from an interesting laboratory curiosity to a powerful analytic tool for the determination of local atomic environment in complex systems. Spectra that took two weeks to acquire using powerful conventional sources are now routinely obtained in 20 minutes—and with improved signal-to-noise ratio. The widely applicable technique (it can be used on gases, liquids, and solids) has, for example, already contributed significantly to the understanding of amorphous materials, biological enzymes, industrial catalysts, phase transitions, and dilute impurities.

The high intensity also facilitates structural studies by large- and small-angle x-ray diffraction. A particularly exciting prospect is the ability to do dynamic structural studies to elucidate nerve and muscle functions, crystal growth, and mechanical failure processes on a milli-second time scale.

The extreme collimation (the radiation is emitted in a cone with an opening angle < 1 mrad for storage rings with energy > 1 GeV) means that highly monochromatic beams can be produced using gratings and crystals, permitting high resolution studies of absorption edges, core levels, etc. This feature also permits the use of soft x-ray lithography for the fabrication of microstructures such as memory devices, Josephson junctions, and integrated circuits. At soft x-ray wavelengths (10–50 Å), diffraction effects are much reduced compared to visible light. Thus, the naturally collimated, essentially parallel rays can be used to replicate the pattern of a mask (with features as small as 1000 Å or less) onto a silicon wafer without penumbral blurring. Future developments in this area are likely to have a major impact on very-large-scale integration (VLSI) technology.

Because electrons are not uniformly distributed around the perimeter of a storage ring, but rather are bunched, the radiation has a sharply pulsed time structure. In the SPEAR storage ring, for example, the pulses are of ~ 300 psec duration with a repetition rate of 1.28 MHz. This feature is used increasingly for lifetime measurements in atomic and molecular systems and may facilitate the pumping of an x-ray laser. In the future, with the increased intensity that will become available, it is likely that these timing properties will be used to extend dynamical structural studies to a submicrosecond time scale.

In storage rings, the high vacuum environment and small source size of the electron beam have enabled experimenters to image the source onto extremely small samples for studies of fresh surfaces prepared in situ in ultrahigh vacuum sample chambers. The measurement of the energy and angular distribution of photoelectrons produced by a tunable photon beam incident on clean surfaces and on surfaces covered with adsorbed gases is a very powerful technique for understanding electronic structure and surface processes such as oxidation, catalysis, and corrosion.

The very high polarization of synchrotron radiation (almost 100% linear polarization in the median plane, 75% integrated over all vertical angles) is an important feature that can be used to advantage in many measurements such as scattering and fluorescence experiments, including resonant nuclear (Mössbauer) scattering. For example, in trace element fluorescence measurements, increased sensitivity results from the decreased background scattering in the direction of polarization. Polarization has also made it possible to observe quantum beats in the fluorescence from magnetically oriented atoms.

Any one of the above-described characteristics would make synchrotron radiation an important experimental tool. The combination of all makes it an ideal source for an extremely broad, interdisciplinary range of spectroscopic and structural studies.

Almost all present sources are primarily operated for high energy physics purposes. Considerably higher intensity and brightness will be available when these machines become available as dedicated synchrotron radiation sources and when new dedicated synchrotron radiation sources, now in construction, become operational. Furthermore, all of the research to date has made use of the radiation produced in the ring bending magnets, where the magnetic field is typically $\lesssim 12$ kG. Over the next few years, it is likely that special insertions (such as high field wiggler magnets, multipole interference wigglers and possibly free electron lasers) will significantly extend experimental capabilities by enhancing and modifying the normal synchrotron radiation spectrum.

Similarly, in the experimental utilization of the radiation, it is likely that new designs for UV and x-ray optical systems, monochromators and detectors will result in instruments better matched to the properties of synchrotron radiation, and this will result in a further increase in experimental capability.

In this review, we describe the sources and properties of synchrotron radiation and cover some of the many scientific and technological applications and specific research results from laboratories throughout the world. In a review of this size, we cannot properly cover all of the important scientific accomplishments using synchrotron radiation. We have attempted to describe briefly a fairly large number of applications and cover a few in more detail. Most of the examples used will be taken from work at the Stanford Synchrotron Radiation Laboratory because of the association of the authors with this laboratory and because, at present, this is the laboratory with the broadest program of research in the UV and x-ray spectral regions. We apologize in advance for being unable to mention, because of space limitations, some of the excellent results obtained at SSRL and elsewhere.

2 SOURCES OF SYNCHROTRON RADIATION— SYNCHROTRONS AND STORAGE RINGS

With the development of electron synchrotrons (since the 1940s) and storage rings (since the 1960s) for high energy physics research, intense sources of synchrotron radiation have become available. For the high energy physicists this radiation is generally considered to be a nuisance causing background in detectors, requiring water cooling of vacuum chambers, and large rf systems to replenish the lost energy. However, as remarked by W. K. H. Panofsky, "The background for one experiment is the signal for another."

Most of the machines now used as synchrotron radiation sources were constructed and are still used primarily for high energy physics purposes. This situation is changing with the design and construction of several storage rings to be used as dedicated synchrotron radiation sources and the transition to dedicated synchrotron radiation operation on existing machines (the parasites are consuming the hosts).

The most capable source of synchrotron radiation is the storage ring. Figure 1 shows the basic features of an electron storage ring designed as a synchrotron radiation source. A closed continuous high vacuum chamber threads through various ring elements including: (a) bending magnets that bend the electrons in a circle and produce synchrotron radiation (only one magnet is shown); (b) special insertions (optional), such as the wiggler magnets shown in the drawing, to produce particularly intense or enhanced radiation; (c) an rf cavity and associated power supply, which replenishes the energy lost by the electron beam into synchrotron radiation; (d) vacuum pumps to evacuate the chamber; and (e) an inflector that permits electrons from a separate accelerator (not shown) to be injected.

In a properly designed storage ring, large currents of electrons (hundreds of milliamperes) can be accumulated within several minutes and adjusted within several additional minutes to any desired energy within range of the ring. Injection can be below the operating energy, with the storage ring then used briefly as a slow accelerator to increase the energy. Injection at the operating energy is quicker, provides the highest stored beam current, and maximizes experimental running time, but, of course, requires a more powerful injector accelerator.

With storage ring average pressure in the 10^{-9}-Torr range, electron beam encounters with residual gas molecules are sufficiently rare that the stored current decays with a time constant of two to twenty hours and in some cases even longer. The synchrotron light intensity as well as the spectrum and other source properties are, therefore, quite stable over long periods of time in a storage ring.

A synchrotron is quite similar in some ways; it consists of a roughly circular ring of magnets, vacuum chamber, and other components similar to the storage ring. The similarity is so close that a single ring could be made to operate as a synchrotron and as a storage ring as was done at the Cambridge Electron Accelerator (CEA) (1). However, a synchrotron is designed to accelerate rapidly (in ~ 10 msec) groups of $\sim 10^{11}$ electrons from a low injection energy to a maximum energy some 10–100 times higher. At the high energy the electrons strike an internal target or are extracted to strike an external target. The process of injection and acceleration of a new group of electrons then starts over again, repeating at 50–60 Hz.

Therefore, in synchrotrons the electron energy and consequently the synchrotron radiation spectrum are not constant. Also, the electron beam current, position, and cross-sectional area vary within an accelera-

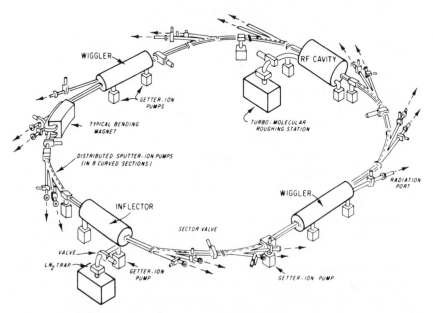

Figure 1   Artist conception of an electron storage ring designed as a source of synchrotron radiation. Not shown is the injector accelerator. Courtesy of J. Godell, Brookhaven National Laboratory.

tion cycle and from one cycle to the next. Furthermore, the large amount of high energy radiation near electron synchrotrons requires that synchrotron radiation experiments be totally enclosed in shielded areas, remotely controlled, and generally less accessible than those performed on a storage ring.

In spite of these difficulties, much synchrotron radiation research has been done with electron synchrotrons because they produce high photon flux extending to very high energy and, historically, they were available before storage rings.

Tangential ports permit the synchrotron radiation to leave the vacuum enclosure and travel to the experiments. One such port can accept sufficient radiation (10–50 mrad of arc) to serve more than one experiment simultaneously. Figure 2 shows how a single tangential port may be split and mirrors, gratings, and crystals used to serve several simultaneous experimental stations. Detailed descriptions of synchrotron radiation beam lines and research facilities are available, including descriptions of the facilities at Hamburg (2, 3), Orsay (4), and Stanford (5, 6).

3 HISTORY OF THEORETICAL AND EXPERIMENTAL WORK

Consideration of the radiation from accelerated charges goes back to the 19th century, and the work of Liénard (7) for example. Further work during the next fifty years was done by many authors (8–13). A useful bibliography (14) of works relating to synchrotron radiation gives additional references through 1974.

Among the first to be concerned about radiation effects in circular electron accelerators were Ivanenko & Pomeranchuk (10) and Blewett (12). Synchrotron radiation was first observed (accidentally) at the General Electric 70-MeV synchrotron in 1947.

Extensive theoretical work has been done by Sokolov and co-workers (15–18), Schwinger (13), and others. Quantum mechanical effects were shown to be negligible in practical cases. The theory is reviewed in the textbook by Jackson (19).

Experimental investigations on the properties of the radiation were carried out by Pollack and co-workers (20) using the General Electric 70-MeV synchrotron, by several groups (21–23) using the 250-MeV synchrotron at the Lebedev Institute in Moscow, by Corson (24) and Tomboulian and co-workers (25–27) using the Cornell 300-MeV synchrotron, by Codling & Madden (28) using the 180-MeV synchrotron at the National Bureau of Standards (NBS) in Washington, DC, and by

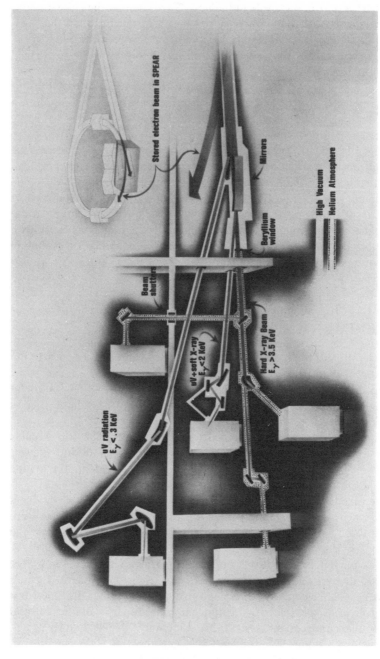

Figure 2   Artist conception of the first beam line implemented in 1974 at the Stanford Synchrotron Radiation Project. Courtesy of W. Zawojski, SLAC.

Haensel and co-workers (29) using the 6-GeV synchrotron in Hamburg. These studies verified the basic theoretical predictions regarding intensity, angular distribution, spectral distribution, polarization, etc, and also provided practical data and experience in the use of radiation.

A research program using synchrotron radiation was initiated in 1961 at the Frascati synchrotron, in 1963 at the NBS and Tokyo synchrotrons, and in 1966 at the Hamburg synchrotron. By 1972, programs were also underway at synchrotrons in Bonn, Yerevan, Glasgow, Lund, Moscow, and other locations. In most of these cases, the primary function of the synchrotron was high energy physics research.

The superior properties of the synchrotron radiation produced by a storage ring (e.g. constant spectral distribution, stable intensity) were first made available at the 240-MeV ring (30) at the University of Wisconsin, which began producing ultraviolet radiation in 1968 as a dedicated synchrotron radiation source. Larger storage rings extended these desirable characteristics to higher photon energy. In 1971, a synchrotron radiation program in the ultraviolet part of the spectrum was started on the 540-MeV ring (4) in Orsay, France. The x-ray part of the spectrum was opened in 1972 at the Cambridge Electron Accelerator (CEA) in Massachusetts, operating as a storage ring at 3 GeV (31). The Wisconsin and Orsay machines continue in operation with the French machine now also fully dedicated to synchrotron radiation research. The CEA ceased operation for high energy physics and synchrotron radiation research in 1973.

By 1972, the potential of synchrotron radiation as a multidisciplinary research tool had attracted the attention of many scientists. The proceedings of a symposium (32) on synchrotron radiation research held at Brookhaven in 1972 give an overview of the field and its prospects as seen at that time. In late 1972, a panel was formed by the National Science Foundation in the USA to study three proposals for synchrotron radiation research facilities. These proposals were for: (a) construction of a new 1.76-GeV electron storage ring at the University of Wisconsin to be dedicated to synchrotron radiation research; (b) dedication of the CEA storage ring at Harvard University to synchrotron radiation research; and (c) construction of the Stanford Synchrotron Radiation Project (SSRP) utilizing synchrotron radiation produced during high energy physics colliding-beam operation of the storage ring SPEAR at the Stanford Linear Accelerator Center (SLAC).

The panel concluded that synchrotron radiation offered interesting research prospects for radiation extending all the way into the short wavelength x-ray region. The third alternative, SSRP, (now SSRL) was authorized as the least costly way to evaluate these prospects.

SSRP (6, 33) was funded in June 1973 and began research operation on five experimental stations, each with a monochromator, in May 1974. This was the first time that synchrotron radiation from a multi-GeV storage ring became available to a large community of users. Within about a year, programs were also begun on the VEPP-3 storage ring in Novosibirsk, USSR, and the DORIS storage ring in Hamburg, Germany. The results obtained from the extended spectral range on these newly available large rings, plus continued results from smaller storage rings and synchrotrons throughout the world, have convinced scientists that synchrotron radiation is a powerful tool for basic and applied research in a large variety of fields such as atomic physics, structural biology, and microfabrication. A comprehensive report (5) on research with synchrotron radiation at SSRP and the prospects for future use of a multi-GeV storage ring was completed in 1976.

Compared with the cautious optimism about the usefulness of synchrotron radiation in 1973, by 1975 the capability was clearly proven, and the enormous increase in user demand created an urgent need to provide more experimental facilities. In 1975, the British, after completing a study of their national needs for synchrotron radiation facilities (34), authorized construction of the world's first multi-GeV storage ring as a dedicated source. Since then, additional proposals for dedicated sources have been made in Japan, the Netherlands, the Soviet Union, the United States, and elsewhere. Many of these have since been authorized and are now under construction. The present status is summarized in Tables 1–4.

Storage rings designed as dedicated sources of synchrotron radiation should offer advantages over machines designed for high energy physics purposes. These include ability to accommodate a larger number of synchrotron radiation beam lines and higher source point brightness due to reduced electron beam emittance. Design reports such as those produced at Brookhaven (35) and in the Netherlands (36) give more detailed considerations.

In early 1976, another panel was formed in the US to assess the national need for facilities dedicated to the production of synchrotron radiation. The panel projected the number of experimental stations that would be needed by 1986 to accommodate those scientists expected to be using synchrotron radiation in their research programs. The results of this projection are given in Table 5 taken from the panel report (37).

The major recommendation of the August 1976 panel report (37) was that "an immediate commitment be made to construct new dedicated national facilities and to expand existing facilities so that optimized XUV and x-ray capabilities are provided . . ." (p. 5).

The case was so compelling that the recommendations of the panel

Table 3 Synchrotron radiation sources—storage rings proposed as of February 1978

Machine, location	E (GeV)	I (mA)	R (m)	ε_c (keV)	Remarks
ERSINE Yerevan, USSR	2.5	500	5.21	6.67	Dedicated
PAMPUS Amsterdam, Netherlands	1.5	500	4.17	1.80	Dedicated (2 GeV operation possible)
IPP Moscow, USSR	1.35	100	2.5	2.2	Dedicated (2 GeV also considered)
CSRL Canada	1.2	100	3.1	1.33	Dedicated
Electrotechnical Lab Tokyo, Japan	0.6	100	2.0	0.24	Dedicated

Table 4 Synchrotron radiation sources—synchrotrons in operation—February 1978

Machine, location	E (GeV)	I (mA)	R (m)	ε_c (keV)
DESY Hamburg, Germany	7.5	10–30	31.7	29.5
ARUS Yerevan, USSR	4.5	1.5	24.6	8.22
BONN I Bonn, Germany	2.5	30	7.6	4.6
SIRIUS Tomsk, USSR	1.36	15	4.23	1.32
INS-ES Tokyo, Japan	1.3	30	4.0	1.22
PAKHRA Moscow, USSR	1.3	300	4.0	1.22
LUSY Lund, Sweden	1.2	40	3.6	1.06
FIAN, C-60 Moscow, USSR	0.68	10	1.6	0.44
BONN II Bonn, Germany	0.5	30	1.7	0.16

Table 5 Minimum predicted growth in synchrotron radiation facilities

	Present (Dec. 1976)		1986	
Use	Users	Stations	Users	Stations
X-Radiation	85	7	675	~60
XUV	120	16	480	~40

that "There will be a large discrepancy between the number of scientists who wish to use synchrotron radiation and the number of stations available on existing or proposed machines." The report stated (p. 1) that "A large effort to build a dedicated hard x-ray storage ring and appropriate advanced instrumentation with a design which goes beyond that of present day projects is recommended. This machine should be operational by 1985."

4   PROPERTIES OF SYNCHROTRON RADIATION

The most important properties of synchrotron radiation from an experimental point of view are the total power radiated, the angular distribution, the spectral distribution, and the polarization. Other properties of the radiation depend on the design of the particular machine. These include the pulsed time structure and the optical properties of the source point (i.e. electron beam size and divergence convoluted with the angular distribution of the emitted radiation).

We briefly discuss these topics in this section. More detailed treatments are given elsewhere (13, 19, 26, 41, 42). Particularly detailed tables and graphs describing the spectrum, angular distribution, and polarization functions are given by Green (43), who also gives an approximate treatment of the optical properties of the source. Elegant field plots showing the directionality of the synchrotron radiation from a charge in circular motion are given by Tsien (44).

4.1   Power Radiated

The Larmor formula (45) for power radiated by a single nonrelativistic accelerated charged particle is

$$P = \frac{2}{3} \frac{e^2}{c^3} \left| \frac{d\mathbf{v}}{dt} \right|^2 = \frac{2}{3} \frac{e^2}{m^2 c^3} \left| \frac{d\mathbf{p}}{dt} \right|^2. \qquad 1.$$

For a particle in circular motion with radius of curvature ρ the relativistic generalization of the Larmor formula is

$$P = \frac{2}{3} \frac{e^2 c}{\rho^2} \beta^4 \left[\frac{E}{mc^2} \right]^4.$$ 2.

The energy loss per turn can be readily obtained from Equation 2.

$$\Delta E = P \cdot \frac{2\pi\rho}{\beta c} = \frac{4\pi}{3} \frac{e^2}{\rho} \beta^3 \left[\frac{E}{mc^2} \right]^4.$$ 3.

In practical units (E in GeV, ρ in meters, I in amperes) for a highly relativistic electron ($\beta \approx 1$)

$$\Delta E \text{ (keV)} = 88.47 \, E^4/\rho$$ 4.

Multiplying Equation 4 by the current gives the total radiated power

$$P \text{ (kW)} = 88.47 \, E^4 I/\rho$$ 5.

or in terms of the magnetic field (B in kilogauss)

$$P \text{ (kW)} = 2.654 \, BE^3 I.$$ 6.

Equations 4, 5, and 6 are valid also when the orbit is not circular but consists, as it does in most machines, of arcs of bending radius ρ and field-free straight sections.

Reasonable design parameters for an intermediate energy synchrotron radiation source are $E = 1$ GeV, $B = 10$ kG (corresponding to a bending radius of 3.33 m) and $I = 0.5$ A. Such a machine would radiate 13.3 kW in a continuous spectrum that peaks at about 400 eV and extends to about 3 keV. Larger machines (2–4 GeV) such as SPEAR at Stanford and the National Synchrotron Light Source at Brookhaven National Laboratory (see Tables 1–3) radiate hundreds of kilowatts. At 4 GeV, the photon spectrum of SPEAR extends to about 50 keV. Even larger machines now in construction such as PEP at Stanford and PETRA at Hamburg will reach energies of 15–20 GeV and will radiate several megawatts of synchrotron radiation power with photon spectra extending above 200 keV. Proton storage rings produce negligible synchrotron radiation power because the fourth power dependence on the mass (see Equation 2) reduces the radiated power, compared to electrons, by about 13 orders of magnitude.

The most powerful rotating anode x-ray generators produce only about 10 W of x rays, even though electron beam powers of about 50 kW are used. Almost all of this radiation occurs within a narrow line at the characteristic fluorescence energy of the anode (e.g. 8 keV for copper).

Thus, these sources provide reasonably high intensity at particular photon energies, especially if a large fraction of the isotropically emitted radiation can be used. However, as a continuum x-ray source, synchrotron radiation provides five or more orders of magnitude higher intensity on small experimental samples.

We are assuming that the intensity and spectral distribution of the radiation from N electrons is simply N times that of a single electron i.e. the radiation emission is incoherent between electrons. This is a valid assumption for all cases of interest on present machines. It may be possible for a group of electrons to radiate coherently, producing much higher intensities, if the electron density is high enough and if a density variation with short wavelength components can be produced. This has been analyzed by Csonka (46).

4.2   Angular and Wavelength Distribution

For nonrelativistic electrons the angular distribution of the radiation is given by the familiar dipole radiation formula (19)

$$\frac{\mathrm{d}P}{\mathrm{d}\Omega} = \frac{e^2}{4\pi c^3} \left|\frac{\mathrm{d}^2\mathbf{r}}{\mathrm{d}t^2}\right|^2 \sin^2\Theta \qquad 7.$$

where Θ is measured relative to the direction of the acceleration. Integration of Equation 7 over all angles yields the Larmor formula, Equation 1. The pattern of the radiation described by Equation 7 is shown in Figure 3, Case I. This would also be the radiation pattern as seen in the instantaneous frame of a relativistic electron.

Light emitted at an angle θ relative to the electron direction of motion in the rest frame is viewed at an angle θ' in the lab frame. The transformation is given by (19)

$$\tan\theta' = \frac{\sin\theta}{\gamma[\beta + \cos\theta]}, \text{ where } \gamma = E/mc^2. \qquad 8.$$

At $\theta = 90°$, $\tan\theta' \approx \theta' \approx \gamma^{-1}$. Thus γ^{-1} is a typical opening half-angle of the radiation in the lab system. The concentration of virtually all of the radiation into such a small forward angle (< 1 mrad for machines > 1 GeV), as illustrated in Figure 3, Case II, results in extremely high flux densities on small experimental samples even at great distances from the source. For example, at SPEAR a flux of 2×10^{10} photons per second has been measured (47) in a 1-eV bandwidth at 7.1 keV on a 1-mm high by 20-mm wide sample located 24 m from the storage ring operating at 3.7 GeV and 20 mA. In a focusing system a flux of 1×10^{12} photons per second has been measured (47) on a smaller sample but with a 5-eV

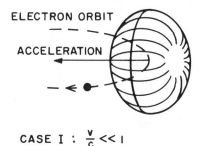

CASE I : $\frac{v}{c} \ll 1$

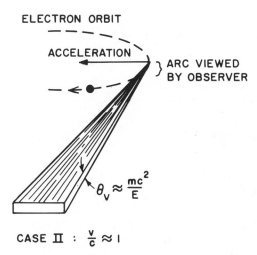

CASE II : $\frac{v}{c} \approx 1$

Figure 3 Pattern of radiation emitted by electrons in circular motion. *Case I*: non-relativistic electrons, $v/c \ll 1$. *Case II*: relativistic electrons, $v/c \approx 1$.

bandwidth (other conditions the same) corresponding to a factor of 150 increase in flux density. Even higher fluxes and flux densities will be possible in the future due to dedicated operation of machines like SPEAR at higher currents (39), the use of wigglers, and the operation of new dedicated storage rings designed for high brightness (35).

From Equation 8 it can be shown that the spectrum of the radiation varies as γ^3, as follows: Referring to Figure 4, an observer starts to see a pulse of light when the electron is at position 1 such that the angle, θ, between the electron's direction and the observer's direction is about γ^{-1}. The pulse ends when the electron reaches position 2, where the electron direction is again at an angle γ^{-1} from the observer's direction.

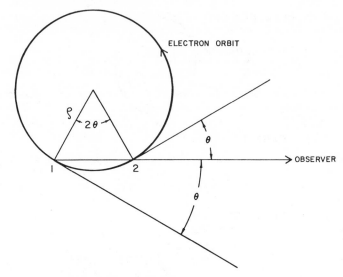

Figure 4   Diagram from which time duration of synchrotron radiation pulse as seen by a stationary observer can be calculated. See text.

For an electron travelling at a velocity β, the duration of the light pulse as seen by the observer is

$$\tau = \frac{\rho}{c}\left[\frac{\theta}{\beta} - 2\sin\left(\frac{\theta}{2}\right)\right] \approx \frac{\rho}{c}\theta^3 = \frac{\rho}{c}\gamma^{-3}.$$

A light pulse of this duration has frequency components up to about $\omega \approx \tau^{-1} = c\gamma^3\rho^{-1}$ corresponding to photon energies of $\varepsilon = \hbar\omega = \hbar c\gamma^3\rho^{-1}$. This can be put into a more convenient form:

$$\varepsilon = \left(\frac{\hbar c}{e^2}\right)\left(\frac{e^2}{mc^2}\right)\gamma^3\frac{mc^2}{\rho} = \frac{r_e}{\alpha}\gamma^3\frac{mc^2}{\rho}, \qquad\qquad 9.$$

where

$$\alpha = \frac{e^2}{\hbar c} = \frac{1}{137}, r_e = \frac{e^2}{mc^2} = 2.8 \times 10^{-15} \text{ m}, mc^2 = 0.51 \text{ MeV}.$$

Thus our earlier example of a 1-GeV storage ring with $\rho = 3.33$ m will produce a spectrum that extends to photons with an energy of about 500 eV. This qualitative analysis gives a photon energy, Equation 9, close to the critical energy, defined later in this section (Equation 11), which characterizes the exact spectral distribution. This analysis applies only when the electron is uniformly bent by an angle greater than about

$2\gamma^{-1}$. The interesting case of shorter bending magnets has been analyzed by Coïsson (48).

The results of Schwinger (13), expressed in terms of photon wavelength (26, 42) rather than frequency, give for the instantaneous power radiated per unit wavelength and per radian by a monoenergetic electron in circular orbit

$$I(\lambda\psi) = \frac{27}{32\pi^3} \frac{e^2 c}{\rho^3} \left(\frac{\lambda_c}{\lambda}\right)^4 \gamma^8 [1+X^2]^2 \left[K_{\frac{2}{3}}^2(\xi) + \frac{X^2}{1+X^2} K_{\frac{1}{3}}^2(\xi) \right], \qquad 10.$$

where $X = \gamma\psi$, $\xi = \lambda_c[1+X^2]^{\frac{3}{2}}/2\lambda$, $K_{\frac{1}{3}}$, and $K_{\frac{2}{3}}$ are modified Bessel functions of the second kind, and ψ is the angle between the direction of photon emission and the instantaneous orbital plane (ψ is not the polar angle). The critical wavelength, λ_c, and corresponding critical energy, ε_c are given by

$$\lambda_c = \frac{4\pi\rho}{3\gamma^3}, \quad \varepsilon_c = \frac{3\hbar c\gamma^3}{2\rho}. \qquad 11.$$

[Note: Some authors, including Jackson (19), define the critical wavelength to be twice that given by Equation 11.] In practical units (E in GeV, ρ in meters, B in kG)

$$\lambda_c(\text{Å}) = 5.59\ \rho/E^3 = 186.4/(BE^2)$$

$$\varepsilon_c(\text{keV}) = 2.218\ E^3/\rho = 0.06651\ BE^2. \qquad 12.$$

Half of the total power is radiated above the critical energy and half below. Storage rings and synchrotrons are designed with a fixed value of the bending radius, ρ, in the bending magnets, and hence the bending-magnet field is directly proportional to energy. For technical and economic reasons, the maximum magnetic field is generally chosen to be less than about 12 kG. It is possible to insert special devices (called wigglers) with much higher fields (up to 50 kG for superconducting wigglers) into ring straight sections. This is discussed in the next section.

It is also shown (13, 26, 42) that the angular distribution of radiated power integrated over all wavelengths is

$$I(\psi) = \frac{7}{16} \frac{e^2 c}{\rho^2} \gamma^5 \left(1+X^2\right)^{-\frac{5}{2}} \left(1+\frac{5}{7}\frac{X^2}{1+X^2}\right), \qquad 13.$$

and integrated over all emission angles the spectral distribution is

$$I(\lambda) = \frac{3^{\frac{5}{2}}}{16\pi^2} \frac{e^2 c}{\rho^3} \gamma^7 y^3 \int_y^\infty K_{\frac{5}{3}}(\Omega)\ d\Omega, \qquad 14.$$

where $y = \lambda_c/\lambda = \varepsilon/\varepsilon_c$.

The shape of the distribution given by Equation 14 is dependent only on λ_c and is given by

$$G(y) = y^3 \int_y^\infty K_{\frac{5}{3}}(\eta) \, d\eta$$

This function and other related synchrotron radiation functions have been plotted and tabulated (43, 49). Examples of spectral distributions are given in Figure 5.

At the critical wavelength, it may be shown that the total number of photons emitted per second within a 10% bandwidth per milliradian of angle in the orbital plane, integrated over the narrow range of angles out of the orbital plane, is given by

$$N_\gamma = 1.6 \times 10^{15} \, I(A) \, E(\text{GeV}).$$

The photon flux varies slowly for $\lambda > \lambda_c$ but drops exponentially for $\lambda < \lambda_c$. At $\lambda = \frac{1}{5}\lambda_c$ the flux is reduced by a factor of 25 compared to the flux at λ_c. The radiation from a single stored electron is easily detectable (50), including observation of the visible light by eye.

4.3   Polarization

The radiation is elliptically polarized with the intensities of the polarization components parallel and perpendicular to the electron orbit given

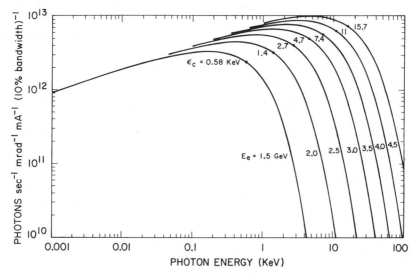

Figure 5   Spectrum of synchrotron radiation from the storage ring SPEAR; bending radius, $\rho = 12.7$ m.

by the two terms of Equation 10. In the plane of the orbit, the radiation is 100% linearly polarized with **E** parallel to the electron acceleration vector. Integration of Equation 10 over all angles and all wavelengths yields a 75% polarization parallel to the orbital plane independent of electron energy and bending radius.

Figure 6 is a plot of the parallel and perpendicular components of polarization as a function of the opening angle for several photon wavelengths. It is clear that polarization can be enhanced by the use of apertures.

Also, in the vacuum ultraviolet (VUV) part of the spectrum, polarization can be enhanced by use of the polarization dependence of reflectance from optical surfaces (51).

Vertically deflecting wiggler magnets (see Section 5) would produce radiation polarized perpendicular to the plane of the ring. Circularly polarized radiation is produced by electrons travelling helical paths as would be produced by special insertions [called helical wigglers (52)] that might be added to storage rings.

4.4 Pulsed Time Structure

The radiation is produced in a train of pulses that is different for each ring. The pulse duration, which is determined by the rf system, is typically about 10% of the rf period. The interval between pulses is determined by the pattern of filling of the number of rf wavelengths (or buckets) that can fit into the orbital circumference. For example, SPEAR operates at

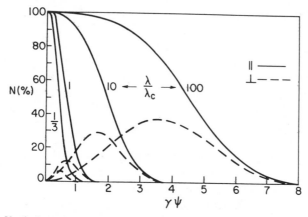

Figure 6 Vertical angular distribution of parallel and perpendicular polarization components. Courtesy of G. K. Green, Brookhaven National Laboratory.

a frequency of 358 MHz, the 280th harmonic of the 1.28 MHz orbital frequency. Thus the pulse length is typically 0.2–0.4 nsec and up to 280 buckets can be filled. By filling one or more desired buckets, a range of pulse intervals from 2.8 to 780 nsec may be obtained. By filling only one bucket, a long interval between pulses is obtained (780 nsec) with greater stability than is possible with multiple buckets each of which may be filled to a different level. This feature is particularly useful in measurements of fluorescence lifetimes.

4.5   Source Size and Divergence

As an optical source both the size and angular spread of the electron beam [due largely to betatron oscillations (53, 54)] must be considered. In a storage ring or synchrotron the position of an electron and its angle (both measured relative to the equilibrium orbit) are correlated (53, 54). Phase space plots of radial electron position vs angle in the median plane summarize this correlation in a so-called radial electron emittance ellipse. A similar plot of vertical position vs vertical angle gives the vertical electron emittance ellipse. The precise shape and orientation of these phase space ellipses change from point to point on the electron orbit, but their area (called the electron emittance) is invariant. Low emittance (or high source brightness) is generally desirable. To obtain the actual emittance of the synchrotron radiation the distribution of emission angles of synchrotron radiation must be convoluted with the electron emittances to obtain photon beam emittances.

These can then be used to determine the effect of slits, mirrors, crystals, etc on the spatial distribution and intensity of synchrotron radiation in a real experimental configuration. This use of phase space techniques is discussed further by several authors (43, 55–57).

The actual value of the phase space emittance of electrons is determined largely by the amplitude of the betatron oscillations. The emission of synchrotron radiation is a kind of frictional force that damps these oscillations. However, quantum fluctuations in the emission process excite these oscillations. Thus, the resultant oscillation amplitude and the electron emittance are determined by the balance between the damping and antidamping effects of synchrotron radiation. In a given machine, the electron emittance varies approximately linearly with electron energy. Thus, all machines have their smallest emittance, and therefore highest synchrotron radiation brightness, when operated at their lowest energy. Of course, at lower electron energy the synchrotron radiation spectrum cuts off at lower photon energy, so the above considerations are not the only factor determining the optimum operating conditions of the ring.

5 STANDARD WIGGLERS, INTERFERENCE WIGGLERS, AND THE FREE ELECTRON LASER

Until now, all research with synchrotron radiation has utilized the radiation produced by the ring bending magnets. Plans are now being implemented at several laboratories to utilize special magnetic structures called wigglers, which will enhance and modify the spectrum of synchrotron radiation compared to that produced by the bending magnets. This will considerably extend research possibilities.

The proceedings of a workshop on wiggler magnets (58) gives examples of various designs and analysis of their effects on storage ring performance as well as the properties of wiggler-produced radiation. We briefly summarize the main points here for the two general types of wiggler—the standard wiggler and the interference wiggler or undulator—and the related free electron laser.

5.1 Standard Wigglers

A standard wiggler (59–62) consists of several short sections of alternating-polarity transverse field magnets designed to produce no net displacement or deflection of the electron orbit. Thus, such a device could be placed in a field-free region (straight section) of a storage ring or synchrotron. The electron path through such a magnet would resemble a biased sine wave. The magnetic field of the wiggler could be higher (or lower) than that of the ring bending magnets. The synchrotron radiation spectrum would then have a critical energy (see Equation 12) shifted to higher (or lower) photon energy, and there would be an overall flux enhancement due to the superposition of radiation from the several oscillations of the orbit.

For example, an 18-kG wiggler magnet planned (63) for the 1.5-GeV storage ring ADONE would extend the critical energy from 1.5 keV in the bending magnets to 2.5 keV in the wiggler. Superconducting wigglers operating at 35–50 kG would produce even larger increase in critical energy and are being planned at Brookhaven (64), Daresbury (65), and Novosibirsk (G. Kulipanov, private communication). At SSRL, an 18-kG seven-pole wiggler magnet (66) is in construction to extend the synchrotron radiation spectrum when SPEAR operates for colliding-beam experiments with electron beam energies in the 2-GeV region.

Standard wiggler magnets could also be designed to deflect in the vertical plane, which would produce synchrotron radiation with polarization direction perpendicular to that produced by the ring bending magnets. This would be of advantage in certain experimental situations permitting, for example, convenient horizontal reflection by mirrors or crystals without loss of polarization.

5.2 Interference Wigglers

A device with many periods designed to produce quasimonochromatic synchrotron radiation is called an interference wiggler or undulator. Several authors have analyzed such devices in helical (52, 67, 68) and transverse geometries (69) or both (70, 71). Other examples of radiation produced by relativistic electrons in periodic fields include the radiation produced by high energy electrons in crystals (72) and Compton backscattering (73). Both of these produce quasimonochromatic peaks and are basically processes similar to interference effects in wigglers.

Interference wigglers can produce peaks in the spectrum of radiation corresponding to wavelengths given by

$$\lambda \approx \frac{\lambda_w}{2\gamma^2}\left[1+\gamma^2\theta^2+\frac{K^2}{2}\right]$$ 15.

where λ_w is the period or repeat length of the magnetic field, $\gamma = E/mc^2$, θ is the angle of observation relative to the average direction of the electron and K is a parameter that depends on the strength and geometry of the wiggler field (52, 70). For example, for a simple sinusoidal field $B_y = B_0 \sin(2\pi z/\lambda_w)$, $K = eB_0\lambda_w/2\pi mc^2 = \psi_0$, where ψ_0 is the maximum angle between the z axis and the approximately sinusoidal trajectory of the electron. The analysis is simplest for the "weak field" case, $K \ll 1$, and then the radiation contains only a single wavelength at each angle θ, given by Equation 15. For $K \gg 1$ (strong fields) larger angular deflections occur, the electron transverse motion becomes relativistic, and harmonics appear (52, 70, 71). The total power radiated at a wavelength given by Equation 15 is maximum for $K = 1$.

The fractional width of the peaks can be no smaller than $1/n$ where n is the number of wiggler oscillations. In a practical case, the width may be further broadened by inhomogeneity in electron energy or magnetic field or particularly if the angular divergence of the electron beam exceeds γ^{-1}. Partly for this reason, new storage ring designs (35) stress low electron beam emittance and provide locations for wiggler insertions where the angular divergence of the electron beam is particularly small.

Recently, measurements have been made on the spectrum and angular distribution produced by an interference wiggler installed in the Pakhra 1.3-GeV synchrotron (74). The ring and magnet parameters were chosen to produce the interference peak in the visible part of the spectrum for convenience.

A practical wiggler for a storage ring might have $\lambda_w \approx 10$ cm. Thus, in a 2-GeV storage ring ($\gamma = 4 \times 10^3$) the interference peak would occur

(Equation 15; $\theta = 0$, $K = 1$) at $\lambda \approx 45$ Å. Peaks at shorter wavelength can be obtained by reducing λ_w or increasing γ. Although some reduction in λ_w could be made, it becomes difficult to design a structure with λ_w less than the transverse aperture requirement for the circulating beam (typically 2–10 cm). Thus, interference peaks in the important region of $\lesssim 1$ Å will require higher energy storage rings such as PEP and PETRA. These machines will have stored beam energy of ≈ 16 GeV ($\gamma \approx 3.2 \times 10^4$) and be capable of producing interference peaks down to 0.7 Å with $\lambda_w = 10$ cm.

Coïsson (75) suggested that an interference wiggler could be used in a high energy ($\gtrsim 200$ GeV) proton machine to produce narrow band synchrotron radiation in the visible part of the spectrum. For example, he calculates (75) that an interference wiggler with a period of 13 cm, a field of 500 gauss, and a length of 5 m would produce more than 10^{11} visible photons per second per milliradian with a 6-A, 400-GeV proton beam. This opens the possibility of monitoring the intensity, cross-sectional area, and energy of such high energy proton beams as is now done on electron machines (76).

5.3   Free Electron Laser

This elegant and intriguing device is basically an interference wiggler structure (as described in the preceding section) plus a pair of mirrors forming a resonant cavity. Under the correct conditions, the mirrors provide feedback resulting in stimulated emission at a wavelength given by Equation 15. The device was first suggested by Madey (77), whose group was the first to demonstrate gain (78) and lasing action (79) in the device, using a linear accelerator as a source of electrons in both experiments. Several other authors have analyzed the device (80).

Although important in verifying the theory and establishing the practicality of the device, the experiments of Madey and co-workers (78, 79) produced only small average laser power (0.36 W, although the peak power was 7 kW) and at low efficiency (the total energy collected on the detector was 0.01% of the electron beam energy). By incorporating the device into an electron storage ring as shown schematically in Figure 7, much higher power at high efficiency may be attainable, thus providing the basis of a new class of tunable lasers.

A variation of the free electron laser, called the optical klystron, has been suggested by Vinokurov & Skrinskii (81). Their analysis shows that the threshold current for lasing can be reduced (perhaps by a factor of 10^3) in their two-stage device.

Neither the free electron laser nor the optical klystron has yet been tried in an electron storage ring. Studies for specially designed small

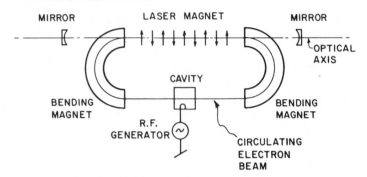

Figure 7   Schematic of free electron laser incorporated into an electron storage ring. Courtesy of J. M. J. Madey, Stanford University.

storage rings to incorporate such devices are in progress (private communications from C. Pellegrini and J. M. J. Madey).

6   INSTRUMENTATION FOR THE UTILIZATION OF SYNCHROTRON RADIATION

Much specialized instrumentation is required to collect, transport, monochromatize, and focus the synchrotron radiation onto an experimental sample and then to detect the transmitted, diffracted, or scattered photons or secondary electrons produced. It is clear that the development of instrumentation optimized for the characteristics of synchrotron radiation has already significantly expanded the research capabilities. Much remains to be done in this area, because many of the instruments now used were developed for other radiation sources and are not optimized for synchrotron radiation research. For example, advances are being made in the fabrication and polishing of mirrors (82) to reflect and focus synchrotron radiation in the VUV and x-ray regions using grazing incidence and total external reflection. Also multilayer reflective coatings (83) are being developed for normal incidence use in the VUV region.

Here, we give only a few examples of synchrotron radiation research instrumentation. More information can be found in the proceedings of several conferences and workshops on the subject (84–86) and in detailed descriptions of particular laboratories (2–5).

6.1   *Monochromators*

Perhaps the most important single instrument for research with synchrotron radiation is the monochromator. Specialized monochromators for the vacuum ultraviolet, soft x-ray, and hard x-ray regions of the spectrum are described in the above-mentioned proceedings and laboratory descriptions. A particularly successful vacuum ultraviolet and soft x-ray grazing incidence grating monochromator called the "Grasshopper" has been designed by Brown, Bachrach & Lien (87) and utilized since 1974 at SSRL. It covers an unusually large energy range (32 eV to $\gtrsim 1500$ eV). At DESY, Eberhardt and co-workers (88) have designed and built a grating monochromator called "Flipper," which covers a narrower energy range (20–300 eV) but with excellent order-sorting capabilities. More recently, instruments based on holographically made toroidal reflection gratings (89) appear to offer extremely high efficiency up to about 150 eV. Several authors (90, 91) suggest that holographically made transmission gratings can open up new possibilities for vacuum ultraviolet and soft x-ray monochromators. A review of monochromator designs for the VUV spectral region is given by Pruett (92). Sagawa (93) reviewed monochromators developed for synchrotron radiation use in Japan.

In the harder x-ray region, Bragg diffraction from crystals (e.g. Si, Ge, graphite) is the basis of monochromator designs. A review of the many possible configurations was given by Hastings (94) as well as by Beaumont & Hart (95). A very versatile yet simple monochromator utilizes a plane channel-cut crystal. Such a device was very successfully implemented by Kincaid (96) at SSRL in 1974 and is used extensively in many experiments, particularly on studies of Extended X-Ray Absorption Fine Structure (EXAFS) as described in Section 7 of this review. It provides rapid tunability, high transmission, narrow bandwidth ($\Delta E/E \approx 10^{-4}$), and almost constant exit beam position and direction.

In 1976, a major increase in capability was achieved with the implementation of a focusing version of this device by Hastings, Kincaid & Eisenberger (47) at SSRL. This powerful monochromator system employs a toroidal double-focusing mirror [first used by Horowitz (97)] in conjunction with a two-crystal monochromator. The properties of double-focusing grazing incidence mirrors were studied by Howell & Horowitz (98). The ability to adjust the crystals relative to one another facilitates compensation for thermal effects on the first crystal due to high power densities of focused radiation, and permits the suppression of harmonics. The mirror collects a large amount of synchrotron radiation and focuses it into a small spot (2 mm × 4 mm) resulting in flux densities 150 times higher than in unfocused beams. The platinum coated mirror is used at a grazing angle of incidence of 10 mrad, which reflects x rays up to ~9 keV with high efficiency but cuts off sharply at higher energy. This provides excellent rejection of harmonics above ~10 keV. The system is shown in Figure 8.

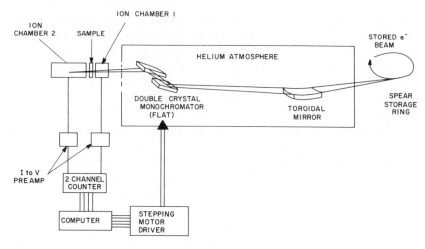

Figure 8   Schematic of separated function monochromator system. See discussion in text. Courtesy of J. Hastings, Brookhaven National Laboratory.

6.2   Detectors

A large variety of detectors have been used in synchrotron radiation research including scintillation counters, position sensitive detectors, photographic film, proportional counters, channeltrons, image intensifier systems, solid state detectors, and ionization chambers. A review of x-ray detectors for synchrotron radiation was given by Brown (99). Reviews of electronic area detectors of several types are given in chapters of a recently published book (100).

Improvements in many of the detectors have been made to meet the requirements of synchrotron radiation research (e.g. high counting rate, better angular and energy resolution). There is a particularly important need for the development of area detectors with good spatial resolution and high counting-rate capability to replace film as a recording medium, for example, in recording x-ray diffraction patterns. Developmental programs for such detectors are under way in several synchrotron radiation research laboratories (DESY, Novosibirsk, Orsay, SSRL).

A multiwire proportional chamber (MWPC) system has been developed by Xuong and co-workers (101) using a chamber developed by Perez-Mendez. The system has been used to solve protein structures by recording diffraction patterns generated by a conventional x-ray tube. A very promising MWPC system capable of higher counting rates and improved spatial resolution is the spherical drift chamber as developed by Charpak and co-workers (102). Several reports on position sensitive detectors were presented at the Orsay Instrumentation Conference (85) and at the Stanford Workshop on X-Ray Instrumentation for Synchrotron Radiation Research (86).

7   APPLICATIONS OF SYNCHROTRON RADIATION

In this main section of our review, we discuss several applications of synchrotron radiation to basic and applied research and technology. From the very large amount of work that has been done, we have tried to select important recent developments and those that best illustrate the special characteristics and capabilities of synchrotron radiation. We also include several suggested applications, some speculative, that may be developed in the future.

We have found it impossible, however, to be complete. The spectrum of scientific and technological activities affected by synchrotron radiation is so broad in scope that it is beyond our capabilities to describe all of them well. These activities range from determination of biological structures to analyses of electronic structures of metals and semiconductors at surfaces, using techniques that vary from the scattering and absorption of x rays to photoemission induced by ultraviolet radiation. Since one of us is close to the x-ray absorption and diffraction fields, and since synchrotron radiation activities in these fields are new and not yet described by such a review, we have concentrated more on these and less on the equally important, but more completely, previously described research lines in the ultraviolet portion of the spectrum.

Of necessity, several applications are not covered at all. These include: (*a*) the use of synchrotron radiation as a radiometric standard (42, 103); (*b*) the calibration and characterization of instrumentation such as gratings (104) and detectors; (*c*) the use of synchrotron radiation as a monitor and diagnostic tool in electron synchrotrons and storage rings (76) to measure the intensity, bunch dimensions, and energy of circulating electron beams; (*d*) radiation damage studies including the simulation of plasma conditions (105); (*e*) synchrotron radiation as an infrared source (106).

7.1   *X-Ray Diffraction*

7.1.1   INTRODUCTION   X-ray diffraction has been, for over fifty years, the basic tool for the determination of atomic arrangements in condensed matter. Synchrotron radiation offers the potential for major improvements in a number of experimental approaches because of its unique characteristics. In particular, as discussed in separate sections below, the capability of choosing the wavelength to be employed in an experiment, rather than being restricted to the characteristic radiations produced by x-ray tube anode materials, makes possible the exploitation of anomalous scattering in structure determinations. The intense white radiation makes possible experiments in which x-ray diffraction intensities can be measured extremely rapidly by sorting photon energy electronically (i.e. pulse height analysis) rather than by mechanical angular scans. In addition, it allows for the rapid production of x-ray topographs. Finally, the high degree of natural collimation of the beam leads to relatively simple small-angle scattering experiments as well as high resolution x-ray topographs. These applications are discussed in the sections that follow. Before proceeding, however, we review briefly some of the basics of x-ray diffraction.

The geometry of a typical experiment is shown in Figure 9. There, k_0 and k' (of magnitude $2\pi/\lambda$ where λ is the x-ray wavelength) are the wave vectors of the incident and diffracted beams, respectively. It is convenient to describe the scattering process in terms of the diffraction vector, which is defined by the relationship

$$\mathbf{k} = \mathbf{k}' - \mathbf{k}_0. \qquad\qquad 16.$$

It is easily shown that

$$|\mathbf{k}| = 4\pi \sin \theta/\lambda. \qquad\qquad 17.$$

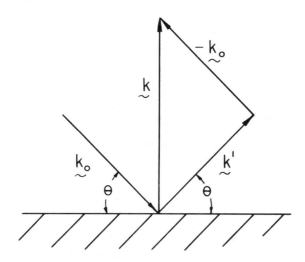

Figure 9   Geometry of Bragg scattering. k_0 and k' are the wave vectors of the incident and diffracted beams, respectively.

In most diffraction experiments, the incident photon energies are far from the absorption edge energies of the atoms in the sample and, in addition, multiple scattering of the x rays can be neglected. In this case, the diffracted intensities, $I(\mathbf{k})$, are related to the electron density, $\rho(\mathbf{r})$, in the sample through the following relationships.

$$I(\mathbf{k}) = C(\mathbf{k})\,|\,F(\mathbf{k})\,|^2, \qquad\qquad 18.$$

where

$$F(\mathbf{k}) = \int \rho(\mathbf{r})\exp(i\mathbf{k}\cdot\mathbf{r})\,d\mathbf{r} \qquad\qquad 19.$$

and $C(\mathbf{k})$ is a slowly varying function of $\mathbf{k}$ that depends on the geometry of the experiment. Thus, the scattering amplitude, $F(\mathbf{k})$, is the Fourier transform of the electron density. To further simplify, we define the atomic scattering factor f_i^0 of the ith atom through the relation

$$f_i^0(\mathbf{k}) = \int \rho_i(\mathbf{r})\exp(i\mathbf{k}\cdot\mathbf{r})\,d\mathbf{r}, \qquad\qquad 20.$$

where $\rho_i(\mathbf{r})$ is the electron density associated with the ith atom, but with $\mathbf{r}$ now measured relative to the center of that atom. Then, Equation 19 can be rewritten in its common form as the structure factor

$$F(\mathbf{k}) = \Sigma_i f_i^0 \exp(i\mathbf{k}\cdot\mathbf{r}_i), \qquad\qquad 21.$$

where the summation is over all the atoms in the sample. In a crystalline material, this may be expressed as N times the summation over one unit cell for Bragg reflections. Here, N is the number of unit cells. A unit cell is the basic repeating element in the crystal.

Equation 19 implies that if $F(\mathbf{k})$ were a measurable quantity, its Fourier transform would yield the electron density, which would be the maximum information obtainable about the atomic arrangement. Unfortunately, only the magnitude of $F(\mathbf{k})$ can, in general, be obtained through diffraction experiments. The phase associated with this complex quantity cannot be determined directly. This indeterminability leads to difficulties and ambiguities in determining crystal structures from x-ray diffraction data. For example, in protein crystal structure determinations, it is common to obtain these phases through the method of multiple isomorphous substitution. In this method, diffraction data are gathered from a number of similar crystals, which differ chemically only in that each crystal has a different heavy atom substituted into a specific known atomic site. It is assumed that the basic structure of the protein is undisturbed by this substitution. The differences in intensities are used to determine the phases. The method works only when suitable isomorphous derivatives can be prepared, a limited subset of all protein crystals. The anomalous scattering of x-rays offers a different approach, as explained below.

7.1.2 ANOMALOUS SCATTERING   When the incident photon energy is close to that of an absorption edge of an atom in the sample, Equations 19 and 20 are no longer valid. Instead, the atomic scattering factor, as used in Equation 21 must be replaced by

$$f_i = f_i^0 + \Delta f_i' + i\Delta f_i''.$$ 22.

Here, i, $\Delta f_i'$, and $\Delta f_i''$ are real functions of both photon energy and scattering vector magnitude, k. This phenomenon is reviewed by James (107) and by Ramaseshan & Abrahams (108). The forms of $\Delta f'$ and $\Delta f''$ near an absorption edge are shown in Figure 10. It is important to note that $\Delta f'$ becomes large and negative for photon energies just below the edge energy, so that large changes in f_i can be achieved through variation

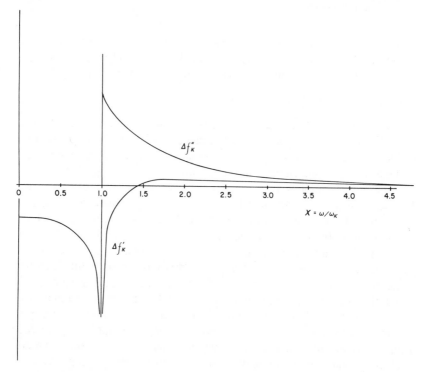

Figure 10   Real, f', and imaginary, f'', shifts of the atomic scattering factor as a function of angular frequency, ω, near an absorption edge, ω_K.

of the x-ray wavelength near the absorption edge. These changes have been employed by Phillips et al (109) in their synchrotron radiation study of the protein rubredoxin. This paper contains the details of the analysis as well as a large bibliography dealing with the theory of the use of anomalous scattering for this purpose. The great advantage of synchrotron radiation for this purpose is that λ can be tuned to achieve maximum changes in the f_i. It should also be noted, however, that Phillips et al (110) have also shown that the diffracted intensities observed in their synchrotron radiation experiments are a factor of at least 60 greater than those obtained with a sealed x-ray tube using the same crystal and instrumental parameters. In addition, in the few protein crystals studied with synchrotron radiation, it appears that there is essentially no loss of diffracting power during an exposure lasting several minutes. With conventional sources, exposures are often many hours, and radiation damage often causes reduced diffraction capability toward the end of the exposure. It appears that there are some radiation damage effects with a time constant measured in hours or days, since the crystals exposed to synchrotron radiation are badly damaged and diffract poorly several days later, even though they did not degrade measurably during the brief exposure. This is extremely important, since degradation of the protein crystals in the x-ray beam, coupled with the long exposure time and difficulties of producing crystalline samples, are major limitations in performing protein crystal structure analysis with conventional x-ray tubes.

Anomalous scattering has also been employed in the study of binary alloy order-disorder critical phenomena. In such phenomena, the alloy of composition AB is arranged in an ordered fashion at low temperatures. In CuZn, for example, the Cu atoms can be assigned to the corners of a cubic unit cell, while the Zn atoms occupy the body centers. As the temperature is elevated, the Cu and Zn atoms interchange positions until, at the critical temperature of approximately 460°C, the Cu and Zn atoms are equally distributed over the two sites.

The diffraction patterns from such alloys consist of two types of Bragg reflections plus a diffuse scattering, as discussed extensively by Guttman (111). One type of Bragg scattering comes from reflections whose intensities are independent of the state of order. These intensities are proportional to $|f_A + f_B|^2$. Those reflections that provide information about the state of long-range order in the material (i.e. the occupancies of the two different sites) have intensities proportional to $|f_A - f_B|^2$. Similarly, the diffuse scattering, which is related to the short-range order, or the probability that each A is surrounded by B atoms, is proportional to $|f_A - f_B|^2$.

It is usually difficult to measure the scattering that depends on the $|f_A - f_B|^2$, the order dependent scattering, because the alloys that show such transitions, (e.g. CuZn and CoFe) contain elements adjacent on the periodic table. As a result, there is little difference in their atomic scattering factors. The difference can be magnified greatly by tuning the photon energy to just below the absorption edge energy of either the A or B atom, so that the anomalous scattering is large. Synchrotron radiation experiments employing this approach have been performed by Sparks (unpublished).

It has also been proposed to use anomalous scattering of x rays to aid in the determination of atomic arrangements in amorphous materials. One approach, proposed by Keating (112) and discussed extensively by Bienenstock (113), involves the measurement of intensities diffracted by an amorphous binary (A-B) material at three different wavelengths. With such measurements, it should be possible to obtain separate A-A, A-B, and B-B atomic pair correlation functions. This method has been attempted using standard x-ray tubes by Waseda & Tamaki (114). Under such circumstances, when one cannot tune close to the edge, the method is extremely sensitive to experimental error, as discussed by Fuoss & Bienenstock (115). Since, however, the reliability of this approach can be increased markedly with increasing differences in the atomic scattering factors associated with the three different x-ray wavelengths, synchrotron radiation tunability offers the potential for further development of this technique as a reliable structural tool. However, a complete set of experiments has not yet been performed, although it has been initiated by Fuoss and co-workers.

The second approach, proposed by Shevchik (116), uses the strong variation with wavelength of the anomalous scattering factors near the absorption edge to obtain the coordination of individual atomic species in polyatomic amorphous materials. Because this variation is strong near the edge and weak everywhere else, the first and second derivatives of the intensity with respect to wavelength are determined by the coordination of the atomic species associated with the edge. Again, it is the tunability of the synchrotron radiation that makes this experiment feasible, although it has not, to our knowledge, been performed yet.

As discussed in Section 7.1.3, anomalous scattering may also be employed in small-angle x-ray scattering to increase the contrast between two phases.

The availability of tunable x-ray sources and the increased interest in the employment of anomalous dispersion has also lead to utilization of synchrotron radiation for the determination of $\Delta f'$ and $\Delta f''$. Such measurements are quite important because the methods described above depend rather sensitively on accurate knowledge of the wavelength dependence of these parameters. Bonse & Materlik (117) used an x-ray interferometric technique (118) to measure $\Delta f'$ for Ni. Fukamachi et al (119) studied the wavelength dependence of the relative intensities of centrosymmetric pairs of reflections to obtain $\Delta f'$ for Ga in GaP. Finally, Hodgson, Templeton, and co-workers (private communication) recently observed a $\Delta f'$ of over 20 for Cs, very close to its absorption edge. This important result implies that utilization of anomalous dispersion will increase markedly in the near future.

7.1.3 SMALL-ANGLE X-RAY SCATTERING   Small-angle x-ray scattering is the measurement of the scattering of x rays for small values of k. From Equations 17 and 19 it is apparent that such scattering is sensitive to long wavelength components of the electron density fluctuations. Consequently, it is commonly employed to obtain information about the size and shape of large (e.g. protein) molecules, the form of phase separation in a multicomponent alloy, or the form of the density fluctuations associated with critical phenomena.

Synchrotron radiation offers three advantages over x-ray tubes for the performance of this work. The first, and most important, is the natural collimation of the beam, which eliminates the necessity for the normally required sophisticated collimation systems to separate the diffracted from the transmitted beam. This natural collimation also leads to an extremely high usable intensity, relative to a normal x-ray tube. Finally, synchrotron radiation offers the possibility of obtaining increased contrast in special circumstances where a sample consists of two or more phases with almost the same electron density, but different chemical compositions. According to Equation 19, the intensity of normal scattering is rather small under such circumstances. By tuning the radiation close to the absorption edge of an element present in one phase, however, the effective contrast in electron density and the corresponding intensity can be made quite large.

A most dramatic display of the first two of these advantages is work carried out at Novosibirsk on the VEPP-3 storage ring as reported by Kulipanov & Skrinskii (50). The small-angle diffraction from frog muscle was recorded by means of a one-dimensional position-sensitive detector and accompanying multichannel analysis. Contraction of the muscle was produced by electrical excitation with a special stimulator triggered by pulses from a synchronizing generator. This generator synchronized the operation of the detecting apparatus with the phase of the muscle contraction. The muscle contraction cycle (~ 64 msec) was broken down into eight time intervals, and information from the detector in each of the intervals was recorded in one of the eight groups of the analyzer memory. This detection system permitted an eight-frame film to be made and permitted the change in the muscle structure to be observed in the various phases of the contraction. The first experiment was carried out by superposition of the information obtained from 100 muscle contractions. By contrast, many hours are required for the observation of usable small-angle diffraction data from muscle with a standard x-ray tube, and dynamic experiments are therefore difficult or impossible. With the increased intensity expected from synchrotron radiation sources in the future, it should be possible to obtain a complete diffraction pattern in a few milliseconds, and therefore a series of diffraction patterns could be recorded during a single muscle contraction.

Extensive studies, both static and dynamic, performed at DESY on muscle or muscle-related material have been described in the review of applications of synchrotron radiation to biological structure and chemical analysis by Barrington Leigh & Rosenbaum (120). Bordas et al (121) have reported work on rat tendon using the energy dispersive detector technique described below, while Kretzschmar et al (submitted for publication) have reported a detailed analysis of the shape and size of myosin subfragment-1. Blaurock (122) has shown that stacking disorder is characteristic of myelin in a freshly dissected nerve, using the small-angle mirror-monochromator system described by Webb (123).

7.1.4   ENERGY DISPERSIVE DETECTORS FOR POWDER DIFFRACTOMETRY AND SMALL-ANGLE SCATTERING   In most x-ray diffraction experiments employing a standard x-ray tube, it is customary to scan k-space by varying the scattering angle and using monochromatic radiation (see Equation 17). Such an approach takes advantage of the very large ratio of characteristic to white radiation intensities produced by an x-ray tube.

An alternative approach, employing the white synchrotron radiation and energy dispersive detectors, has been explored recently. In this approach, the detector is held at fixed scattering angle and has incident upon it radiation of all wavelengths scattered by the sample. The lithium-drifted silicon or germanium detector then sends to a multichannel analyzer a signal that is dependent on the x-ray photon energy. Thus k-space is covered by sorting diffracted photon energies by electronic means. The method has been used for powder diffractometry by Buras et al (124) and by Bordas et al (125). Small-angle scattering experiments on biological materials have also been performed by Bordas et al (121) in this manner.

The major advantage of this technique is the speed of data acquisition; diffraction patterns are obtained in a few seconds under optimum conditions. This speed should make possible the diffraction analysis of dynamic phenomena like phase transitions and biological changes.

7.1.5 TOPOGRAPHY   X-ray topography has been used for many years as a technique for imaging defects, cracks, voids, dislocations, etc in crystals. In this technique, the intensity of Bragg reflection as a function of position on the reflecting planes is measured carefully, yielding an image of that plane. Contrast is achieved in the image because of the difference in reflecting power between the perfect and distorted parts of the crystals.

Synchrotron radiation appears to be an ideal source for x-ray topography for the following reasons:

1. The high intensity reduces exposure times by two to four orders of magnitude.
2. The entire white spectrum can be used, or a narrow band can be selected with a monochromator. The use of white radiation offers two very great advantages. The first is that extremely simple crystal mounting techniques can be used since each Bragg reflection picks out the appropriate wavelength for diffraction. As a result, the topography measurements can be performed readily in combination with magnetic field or temperature variations. In addition, the high intensity means that very short exposures are required, so dynamic effects can be studied.
3. The extremely small angular divergence ($\sim 10^{-4}$ rad for multi-GeV machines) permits high spatial resolution ($\sim$ microns).

The general features of white radiation topography were explored by several groups including Tuomi et al (126) using the DESY synchrotron, Hart (127) using the Daresbury synchrotron, and more recently Parrish & Erickson (unpublished) using the SPEAR storage ring. The simple geometry used in these experiments is shown in Figure 11. Figure 12 gives some results. There have been many time dependent experiments,

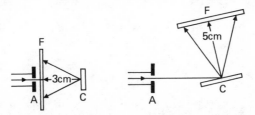

Figure 11   Geometry for x-ray topography with broadband synchrotron radiation (A = Aperture, F = Film, C = Crystal). (*Left*) Arrangement for back reflection. (*Right*) Arrangement for grazing incidence. Courtesy of W. Parrish, IBM, San Jose, CA.

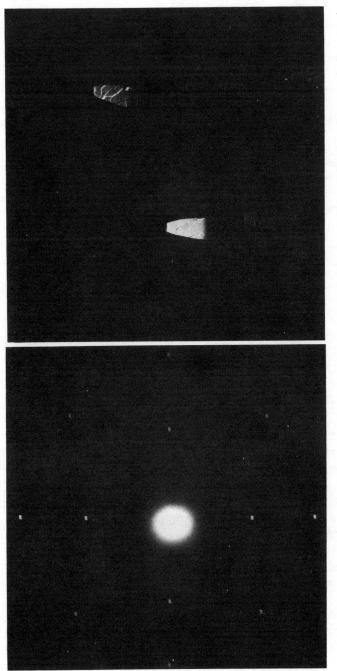

Figure 12   X-ray topographs of lithium fluoride cleavage surface. (*Bottom left*) Back reflection on Polaroid film 3 cm from sample. Exposure is 3 min with SPEAR at 1.85 GeV, 6.8 mA. (*Bottom right*) Grazing incidence on Kodak R single-coated film 5 cm from sample. Exposure is 5 min with SPEAR at 1.89 GeV, 7.2 mA. (*Top*) Optical enlargement of upper reflection spot in bottom right figure. Courtesy of W. Parrish, IBM, San Jose, CA.

including the study of phase transitions (128), domain wall motion (129), plastic deformation (130), and recrystallization (130). Composite crystals have been studied by several groups using synchrotron radiation including studies of polytypes in ZnS (131), bicrystals in germanium, and various heteroepitaxial systems (132).

It is clear that x-ray topography with synchrotron radiation will continue to attract more workers particularly on the large storage rings such as DORIS, DCI, SPEAR, and VEPP-3 that are ideally suited for these studies.

7.1.6 COHERENT NUCLEAR SCATTERING—MÖSSBAUER EFFECT   The high brightness, high polarization, and pulsed time structure of synchrotron radiation open the possibility of performing experiments on coherent nuclear scattering with improved intensities (compared to radioactive sources) and at any desired photon energy. Certain (Mössbauer) nuclei have, at an extremely well-defined energy, scattering amplitudes exceeding that of all their electrons. If crystals made only of such nuclei were available, it would be possible to Bragg-reflect photons of only that energy from the incident beam. For example, with ^{57}Fe, a 14-keV x-ray beam with energy defined to one part in 10^{11} can be obtained. This is five or six orders of magnitude better than can presently be achieved with conventional crystal monochromators. With present synchrotron sources such as SPEAR more than 10^4 photons per second are available within this narrow energy range. With dedicated operation and with wigglers, two or more orders of magnitude could be gained.

The primary experimental problem is to observe the narrow band radiation from the decay of the nuclear resonantly scattered radiation (lifetime ~ 100 nsec) in the presence of an intense, brief (~ 0.3 nsec) prompt pulse. This can be accomplished in several ways including the following:

1. Suppression of the wider band (~ 1 eV) atomic electronic Bragg-scattered radiation may be obtained using the fact that this radiation is electric dipole whereas the resonant nuclear scattering is magnetic dipole, i.e. the polarization selection rules differ.
2. The much larger nuclear cross section makes it possible to use very thin crystals, which are nearly transparent to electronic processes.
3. By using a multicomponent crystal, the electronic scattering from different types of atoms can be made to cancel approximately.
4. The larger Darwin width for nuclear reflections makes it possible to use successive reflections from two slightly misoriented crystals such that overlap of the nuclear scattering is large and that of the electronic scattering is small.

Efforts are under way by groups in Novosibirsk (V. A. Kabannik, private communication) and Stanford (P. A. Flinn, S. L. Ruby, private communication; and R. L. Cohen, G. L. Miller, private communication) to solve the technical problems involved such as obtaining sufficiently perfect, isotopically enriched single crystals containing [57]Fe and developing special gated detectors to register delayed emission of photons or conversion electrons.

If these efforts are successful, they will facilitate study of coherent nuclear scattering and open exciting possibilities in x-ray interferometry and holography, for which the large coherence length (~ 1 m) of these photons is a vital new property. Full exploitation of the potential of this highly monochromatic radiation (for x-ray holography, for example) will require the higher intensity that should be available in the future and also the development of higher resolution recording media. Meanwhile, it should be possible, for example, to duplicate the classical interference experiments, such as the Fresnel biprism, in the x-ray region.

More detailed treatments of coherent nuclear scattering with synchrotron radiation are given in the literature (50, 133, 134).

7.2   Extended X-Ray Absorption Fine Structure

7.2.1   INTRODUCTION   The development of Extended X-Ray Absorption Fine Structure (EXAFS) analysis as a means of determining the local coordination environment of individual atomic species in complex, polyatomic materials is one of the most important results of the availability of synchrotron radiation in the x-ray regime. Unlike the single-crystal diffraction techniques, this method may be applied to the determination of such coordinations for: (a) noncrystalline and poorly crystallized materials, as well as crystalline materials; and (b) dilute impurities in crystalline or amorphous materials. Here, by coordination we mean the number and type of atoms surrounding a specific atom and the interatomic distances.

These types of structural problems are often of considerable scientific and/or technological interest, since the determination of atomic arrangements is usually the first step in obtaining microscopic understanding of the physical and chemical properties of condensed matter. For example, the importance of these classes of problems becomes clear when one notes that most glasses used commercially or scientifically contain mixtures of SiO_2 or B_2O_3 with a number of metal oxides, which are added to control physical properties. Little is known about the coordination of the metal atoms. Similarly, small concentrations of impurities are used to control the electrical properties of crystalline and amorphous semiconductors. Their electrical action is intimately linked to their coordination. Finally, it is believed that the biological action of many large molecules is mediated by changes in the coordination geometry of metal atoms, which are often present in low concentrations.

While some techniques, such as nuclear magnetic resonance or electron spin resonance, might be applied to specific atomic species for the determination of coordination, there was no generally applicable approach prior to the development of EXAFS analysis.

The EXAFS is the fine structure on the x-ray absorption coefficient that appears for photon energies in the range of approximately 0–1500 eV above an absorption edge, as illustrated in Figure 13. Here, Hunter's (135) measurements of the relative absorption coefficient of crystalline CuAsSe$_2$ are shown as a function of photon energy in the region of the Cu K edge near 9 keV. For photon energies below the edge, the absorption coefficient, which is due here to lower energy processes, is a slowly varying and monotonically decreasing function of photon energy. At the edge, the photon has enough energy to excite the K-shell electron to the lowest energy empty states, yielding the sharp increase in the absorption coefficient. This change is shown more clearly in Figure 14, in which the absorption due to the lower energy processes has been subtracted out. Above the edge, the absorption coefficient consists of a slowly varying, monotonically decreasing part, superimposed upon which is a fine structure, the EXAFS.

As is discussed extensively below, this fine structure results from the backscattering of the photoelectron by the atoms surrounding that from which the electron was ejected. Consequently, it may be analyzed to obtain information about the average coordination of the atomic species

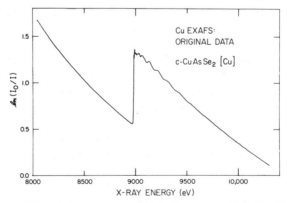

Figure 13   Relative x-ray absorption coefficient of crystalline CuAsSe$_2$ as a function of x-ray energy near the Cu absorption edge. Courtesy of S. Hunter, SSRL.

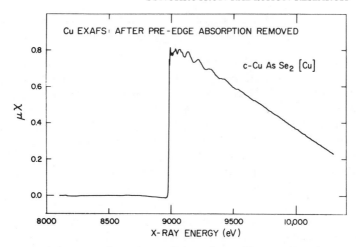

Figure 14   Relative x-ray absorption coefficient of crystalline CuAsSe$_2$ after removal of pre-edge absorption, as a function of x-ray energy near the Cu absorption edge. Courtesy of S. Hunter, SSRL.

whose absorption edge is being studied. Thus, if the absorption edges of all the atomic species in a sample being studied are sufficiently separated in energy, the average coordination of each may be determined separately through studies of the EXAFS associated with each edge. This feature makes EXAFS analysis a powerful structural tool in problems of the type discussed in this section.

Although EXAFS has been observed for nearly half a century, its importance as a structural tool became apparent through the pioneering work of Lytle (136, 137) and then Sayers et al (138, 139) in the mid-1960s and early 1970s. The availability of synchrotron radiation in the x-ray region during the past few years has led to dramatic progress in EXAFS data acquisition and analysis.

From the experimental point of view, the progress has been vastly accelerated because synchrotron radiation provides high intensity over a broad spectral range and is thus well suited for studies of the absorption coefficient as a function of photon energy over an extended energy range. In contrast, the classical x-ray tube source provides a high intensity of characteristic radiation at specific photon energies, but a very low intensity of the continuous Bremsstrahlung needed for such studies. As a result, a very large part of all the experimental EXAFS work published thus far has been performed at the Stanford Synchrotron Radiation Laboratory. It should also be noted that the polarization of synchrotron radiation makes it naturally useful for EXAFS studies of anisotropic materials. For such studies, radiation from a tube source must be polarized in a separate step, which further decreases the available intensity.

7.2.2 EXAFS—SOME BASIC CONSIDERATIONS   In order to provide a basis for the review of recent EXAFS research, the fundamental theory of Sayers et al (139) as well as of Ashley & Doniach (140) is reviewed here. As indicated above, the x-ray absorption process consists of the ejection of an electron from an atomic core state by a photon of energy E. Since the resulting photoelectron's energy is large relative to binding energies in the material, it may be considered as almost free, so that its wave-number magnitude, k, may be well defined by the relation

$$\hbar^2 k^2/2m = E - E_0,$$ 23.

where E and E_0 are the photon and the absorption edge energies, respectively. The EXAFS, $\chi(k)$, is defined by the relation

$$\chi(k) \equiv \{\mu(k) - \mu_0(k)\}/\mu_0(k).$$ 24.

Here, $\mu(k)$ is the experimentally observed absorption coefficient, while $\mu_0(k)$ is the smooth, monotonically decreasing portion of it. Figure 15 shows $\chi(k)$ for the previously illustrated $CuAsSe_2$ absorption, as obtained by Hunter (135).

Physically, the EXAFS arises in the following manner. The cross section for absorption of the photon is proportional to the square of the matrix element of the interaction, H_{int}, taken between the initial core

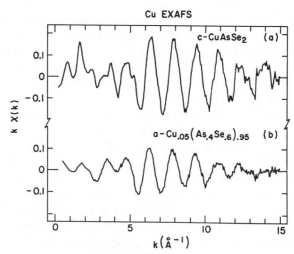

Figure 15 $k\chi(k)$ vs k for crystalline $CuAsSe_2$ and amorphous $Cu_{0.05}(As_{0.4}Se_{0.6})_{0.95}$ for the Cu absorption edge. Courtesy of S. Hunter, SSRL.

state, i, and the excited final state, f. The final state may be described as the sum of an outgoing wave with waves scattered by the atoms coordinating the photoexcited atom. Since the initial state, usually a K-shell state, is highly localized in the vicinity of a specific nucleus, this matrix element samples the form of the final state wave function in the vicinity of that nucleus. [For a discussion of EXAFS in which the initial state is an L state, see Lytle et al (141) and references contained therein.] This form is influenced by the interference between the outgoing and scattered waves. The interference depends strongly on the magnitude of the effective free electron wave vector of the photoemitted electron and the distance to the surrounding atoms. The change in k with changing photon energy, leading to the changing interference of outgoing and backscattered electrons, accounts for the oscillatory nature of the fine structure.

For K edges, the derivations of Sayers et al (139) as well as those of Ashley & Doniach (140) have led to a semiempirical equation for the $\chi(k)$ in terms of the atoms coordinating the specific species whose edge is being studied. It is

$$\chi(k) = -\sum_j (N_j/kR_j^2)|f_j(k)| \exp(-2k^2\langle u_j^2\rangle) \exp\{-R_j/\lambda(k)\}$$

$$\times \sin\{2kR_j + \alpha_j(k)\}. \quad 25.$$

Here, the summation is over coordination shells and atomic species within a shell, all denoted by the index, j. R_j is the average separation of atom j from the absorbing atom; $\langle u_j^2\rangle$ is the mean square deviation of that distance. The backscattering amplitude for the photoelectron from the neighbors is $|f_j(k)|$, N_j is the number of atoms in the shell at average distance R_j, $\lambda(k)$ is the energy dependent mean free path, and $\alpha_j(k)$ is twice the phase shift of the outgoing electron wave relative to the $1s$ core state together with the phase shift on backscattering.

Equation 25 implies that it is possible to determine the N_j, R_j, and $\langle u_j^2\rangle$ for each absorbing species in a complex, polyatomic material. These parameters constitute an almost complete description of the coordination of each species, and provide more information than is obtainable by any technique other than diffraction analysis of crystalline materials.

Since this equation is the basis of almost all attempts to use EXAFS to determine coordinations, it is worthwhile to review some of the important approximations associated with its derivation.

It is assumed, first of all, that the absorption process is dominated by a simple phenomenon in which a single core electron is promoted to an excited state, while all the other electronic states remain essentially unchanged. It has become apparent, through the work of Kincaid & Eisenberger (146), Lee & Beni (143), and others that this simplification leads to an overestimate, by something between 30% and 100% of the magnitude of the EXAFS. As discussed by Rehr et al (submitted for publication), the discrepancies arise from two phenomena. First, even in this simple process, there is a relaxation of the surrounding electrons caused by the presence of the core hole. In addition, as the photon energy is increased above the edge, additional channels for absorption open up as the simultaneous excitation of one or more electrons in addition to the core electron associated with the edge (referred to as shake-up and shake-off) becomes possible.

They estimate that the first effect leads to a reduction to 0.60, 0.64, and 0.64 ± 0.04 for F_2, Cl_2 and Br_2, respectively, of the EXAFS magnitude predicted by Equation 25. They estimate that the second effect leads to an additional reduction on the order of 10% for these molecules. In addition, their theory of these effects indicates that the reduction factors should be quite sensitive to the nature of the local chemical environment of the absorbing atom. The consequences of this are discussed below. We note here, however, that uncertainties in the magnitude of these reductions lead to uncertainties in coordination numbers, N_j, determined by EXAFS analysis and are therefore very important.

In addition, it is assumed that single scattering dominates the scattering processes. That is, multiple scattering processes in which the photo-excited electron is scattered by more than one neighboring atom before "returning" to the excited atom are neglected. This approximation has been examined carefully by Ashley & Donaich (140) and by Lee & Pendry (144) for metallic copper. In addition, an exactly solvable one-dimensional model has been studied by Rehr & Stern (145). All three works lead to the conclusion that such multiple scattering cannot be ignored when attempts are made to determine coordinations beyond the first coordination shell, but that they are unimportant for first coordination shell determinations.

An important early test of Equation 25 was performed by Kincaid & Eisenberger (146), who measured the EXAFS spectra of gaseous Br_2 and $GeCl_4$, and calculated the relevant backscattering amplitudes and phase shifts in the static-exchange Hartree-Fock approximation. They showed that the theory predicts EXAFS roughly a factor of two larger than what they observed experimentally. In addition, there were disagreements of peak positions that were well outside their stated experimental error limits. They noted, "If one adjusts the starting value of the interatomic spacing, which is the most sensitive parameter in the theory, to improve the peak-position agreement, the distance so determined is in error by 5%."

In associated work, Kincaid (147) presented two more results of particular importance in this context. The first was that the general shapes of the $|f_j(k)|$, and the values of the phase shifts, vary markedly with atomic number. The former have a peak as a function of k or $E - E_0$ approximately 100 eV above the edge for Cl, but 350 eV above the edge for Br. In general, the peak position moves to higher energy, and the peak itself becomes less pronounced, with increasing atomic number. This change of shape, which has also been noted by Lee & Beni (143), as well as the changes of phase imply that one should be able to identify the atomic constituents of each coordination shell observed if the discrepancies between theory and experiment can be resolved. Consequently, EXAFS analysis is a potential means of determining chemical ordering in amorphous materials and crystalline alloys.

In addition, Kincaid noted that the calculated backscattering amplitudes and phase shifts are quite sensitive to the many-electron approximation used. Consequently, some of the large difference between theory and experiment found by Kincaid & Eisenberger may be due to the inadequacies of the static-exchange Hartree-Fock approximation.

The absence of a complete theoretical basis for the analysis of EXAFS data coupled with clear experimental evidence of the insufficiencies of Equation 25, as well as the unavailability of reliable calculations of the backscattering amplitudes and phase shifts, mean that EXAFS analysis has been far from achieving the potential implied by Equation 25. In mature structure determination techniques, such as x-ray or neutron diffraction, there are a minimum of uncertainties in the theory and in important data analysis parameters like the scattering factors. Instead, the uncertainties are related to the atomic arrangements themselves. In EXAFS, on the other hand, the large uncertainties in the magnitude of the EXAFS, relative to the total absorption, and the smaller uncertainties in the phase shifts, have implied that structure analyses must necessarily lead to uncertain conclusions.

Consequently, two important lines of research have developed. One includes theoretical attempts to understand and calculate the spectra better. The other includes attempts to determine the backscattering amplitudes and phase shifts experimentally so that they may be used for structure determination. Included within this category are attempts to determine whether these parameters depend primarily on the atomic species being studied or whether, alternatively, they vary strongly from sample to sample even though the same elements are involved.

Initial emphasis in the latter line has been on the phase shifts, since these play an important role in the determination of interatomic distances. This importance arises from the fact that they may be written, to a very good approximation, as

$$\alpha_j(k) = \alpha_{j0} + \alpha_{j1}k + \alpha_{j2}k^2. \qquad\qquad 26.$$

Substitution of Equation 26 into the last factor of Equation 25 yields

$$\sin\left[2kR_j + \alpha_j(k)\right] = \sin\left[2kR_j + \alpha_{j0} + \alpha_{j1}k + \alpha_{j2}k^2\right] \qquad\qquad 27.$$

Consequently, α_{j1} looks like an interatomic distance. Fourier transformation of the EXAFS yields peaks that are shifted from R_j by $\frac{1}{2}\alpha_{j1}$. All analysis techniques produce interatomic distances that are as uncertain as are the linear parts of the phase shifts. It is therefore important to determine the phase shifts reliably.

In an extremely important work, Citrin et al (148) established the transferability of phase shifts. That is, they showed that phase shifts associated with the absorption of a photon by a specific atomic species, A, and the scattering by a neighbor, B, in a known structure may be used to analyze the EXAFS from a system with unknown structure, where A is again the absorber and B the scatterer. Thus, even though there have only recently been theoretical calculations of phase shifts that might allow for such accuracy, experimental determinations of interatomic distances relying in one way or another on experimentally determined phase shifts have proceeded and have been accepted with some degree of confidence. Some of these are discussed in a later section.

More sophisticated theoretical calculations of phase shifts were recently published by Lee et al (149). These will constitute a substantial step forward in our ability to analyze data if, as anticipated, they prove to be generally reliable.

Similarly, more sophisticated theoretical calculations of backscattering amplitudes have been published by Teo et al (150). These, with the phase shifts, should aid considerably in the use of EXAFS to identify the chemical species surrounding the scatterer. Their immediate use in the determination of coordination numbers is limited, however, by the factors affecting the EXAFS amplitudes discussed above.

No comparable test of the transferability of EXAFS magnitudes from one system to another has been published. In view of the comments of Rehr et al discussed above, on the potential sensitivity of these magnitudes to chemical environment, such studies seem particularly appropriate. Lacking them, experimental determinations of coordination numbers must be accepted with caution.

7.2.3 DATA ACQUISITION   Presented here is a brief discussion of the EXAFS measuring equipment presently being used by most experimentalists at SSRL. Far more detailed descriptions have been presented by Kincaid (96) and by Hunter (135).

As shown in Figure 16, the x-ray beam from SPEAR passes through a Be window and a slit before it is monochromatized by successive Bragg reflections from the parallel faces of a channel-cut crystal located about 20 m from the source point. The beam then passes through ion chamber No. 1 for a measurement of the intensity incident on the sample, through the sample, and into ion chamber No. 2 for a measurement of the transmitted intensity. As the diagram indicates, the experiment is computer controlled. Although a complete analysis of the experiment is inappropriate here, some details will interest the general reader.

The long wavelength cutoff of this arrangement, determined by the thickness of the Be window, is presently about 3.5 keV. Efforts are under way to produce thinner windows, which will be consistent with the need for reliable isolation from the storage ring high vacuum.

The channel-cut crystal, when set to the Bragg angle for a specific wavelength, λ_0, may also pass $\frac{1}{2}\lambda_0$ and/or $\frac{1}{3}\lambda_0$. As Hunter discusses, this harmonic contamination decreases the accuracy of coordination number determinations, but not determinations of interatomic distances. This problem may be dealt with by maintaining the stored electron energy at a sufficiently low value so that the intensity of these harmonics is negligible. It is also possible, at many wavelengths, to choose a mono-chromator configuration that does not strongly reflect harmonics, or to suppress harmonics with mirrors.

Finally, as discussed by Hunter (135), the use of ion chambers in the configuration shown does not allow for an absolute measurement of the absorption coefficient. Instead, the absorption coefficient plus a slowly varying function of photon energy is measured. This fact must be taken into account in some applications of the data.

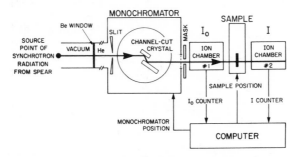

Figure 16 Block diagram of EXAFS experimental apparatus. Courtesy of S. Hunter, SSRL.

It should also be noted that the monochromatized beam emerges from the channel-cut crystal along a path parallel to, but displaced from, the original path. This displacement depends mostly on the crystal geometry and slightly on wavelength. Over the range of wavelengths in a typical EXAFS scan, the displacement of the monochromatic beam varies by about 1 mm or less. In some cases, this can be ignored. In other cases, the sample and detector system must be displaced to track the beam position.

7.2.4 DATA ANALYSIS Along with rapid developments in basic theory and data acquisition have come considerable advancements in data analysis, of which a very brief review is presented below.

As originally proposed by Sayers et al (139), Fourier transformation of the EXAFS data yields a radial structure function, $\phi(r)$, whose peak positions are shifted from the R_j as a result of the k dependence of the phase shifts, $\alpha_j(k)$, and whose peak heights and shapes are related in a complicated manner to the N_j and $\langle u_j^2 \rangle$. This complexity arises because of the shape of $|f_j(k)|$, the k dependence of the $\alpha_j(k)$, the presence of the Debye-Waller factor, and the finite region in k-space over which data is obtained, as discussed by Hayes et al (151, 152) as well as by Stern et al (153).

The Fourier transform, $\phi(r)$, of the Cu-edge EXAFS from crystalline $CuAsSe_2$ is shown in Figure 17. The first neighbor maximum is at 1.95 Å, whereas the interatomic distance is approximately 2.41 Å. Also shown

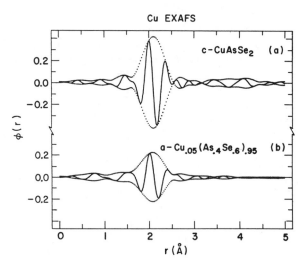

Figure 17 $\phi(r)$ vs r for crystalline $CuAsSe_2$ and amorphous $Cu_{0.05}(As_{0.4}Se_{0.6})_{0.95}$ from Cu-edge data. Courtesy of S. Hunter, SSRL.

is $\phi(r)$ obtained from the Cu-edge EXAFS from amorphous $Cu_{0.05}(As_{0.4}Se_{0.6})_{0.95}$, in which the Cu is also shown by Hunter et al (154) to be fourfold coordinated. Note that the peak height is much larger for the crystalline than the amorphous material. They show that this is because of the greater disorder in the amorphous material.

In the simplest analysis procedures aimed at determining coordination distances, the shifts in the peak positions due to absorber, A, and scatterer, S, are determined by using the $\phi(r)$ obtained from EXAFS studies of a sample in which the A-S interatomic distance is well known. Because of the complex factors that influence the peak shape, however, this approach may lead to small errors in interatomic distance. For this reason, more sophisticated methods have been developed.

These involve, in general, attempts to fit the contributions of the first or higher coordinations shells to the EXAFS or $\phi(r)$ with least squares or related analyses in which the r_j and $\langle u_j^2 \rangle$ are the parameters to be determined. Examples of these approaches are presented in References 142, 149, 150, 153, and 155.

7.2.5 EXAMPLES OF THE USE OF EXAFS FOR STRUCTURE DETERMINATIONS

Despite the brevity of the period in which EXAFS analysis and x-ray synchrotron radiation have been available, there are already more materials studied by this technique than can be described in such a review. To provide the reader with a slightly systematic summary, we concentrate, therefore, on four areas; (a) molecules of biological interest, (b) amorphous materials, (c) catalysts, and (d) surfaces.

7.2.5.1 *Molecules of biological interest* EXAFS analysis has been applied extensively to the determination of metal-ligand distances in molecules of biological interest. The applicability of EXAFS for studying crystalline materials is that it can be focused directly on the determination of the metal-ligand distances. In contrast, x-ray diffraction crystallographic structure determinations treat all the atoms in the unit cell equally, except for a weighting by the atomic number. Even the diffraction approach is unusable for many of these large molecules that have not yet been crystallized. In this case, EXAFS provides information unobtainable in any other way.

One of the most extensively studied proteins is rubredoxin, which has a molecular weight of approximately 6000 with one iron atom tetrahedrally coordinated by four cysteine sulphurs. Initial x-ray diffraction studies indicated that three Fe-S bonds were about 2.30 Å while the fourth was 2.05 Å (156). It was believed that appreciable strain energies were stored in the short bond.

EXAFS studies by Shulman et al (157), Bunker & Stern (158), Sayers et al (159), and Shulman et al (160) have shown that all four Fe-S distances are equal to within 0.10 Å. Consequently, the localized strain energy storage model does not seem to be applicable to this system.

Similar conclusions about the coordination of iron in different forms of hemoglobin have been drawn through the EXAFS studies of Eisenberger et al (161) as well as that of Eisenberger, Shulman, Kincaid, Brown & Ogawa available in preprint form. They have shown that the 3 kcal difference of oxygen binding energy between the high and low affinity forms of hemoglobin cannot be attributed to strain energy localized in the immediate vicinity of the iron atom. In addition, their determination of iron-nitrogen distances in oxy and deoxy complexes of hemoglobin also suggest that hemoglobin has a complex delocalized response to oxygen binding.

Several metal-containing molecules of biological relevance have been the subject of extensive EXAFS studies. Sayers et al (162) studied copper etioporphyrin I, Cramer et al (155) studied a number of iron porphyrins, while Kincaid et al (163) studied copper nickel tetraphenylporphyrin and methemoglobin. Hu et al (164) studied oxidized and reduced cytochrome c oxidase. Finally, the primary coordination environment of the "blue" copper iron in oxidized azurin has been determined by Tullius, Frank & Hodgson (preprint).

The coordination of Mo in various forms of nitrogenase has been investigated extensively by Cramer et al (165–167). This protein is important because of its role in bacterial nitrogen fixation. It is believed that the Mo is associated with the fixation site.

The reader may find a more detailed summary of the application of EXAFS to biological molecules in the review by Shulman, Eisenberger & Kincaid (preprint).

7.2.5.2 *Amorphous materials* Amorphous materials were among the very first studied by the Fourier analysis of EXAFS. Sayers, Stern & Lytle (139) presented an analysis of the data from crystalline amorphous Ge in 1971 and followed this by an analysis of amorphous Ge, GeO_2 and GeSe in 1972 (168) and of GeSe, $GeSe_2$, As_2Se_3, As_2A_3, and As_2Te_3 in 1973 (169). Among the contributions of this set of papers was the partial resolution of an important structural problem in the field of amorphous semiconductors, the coordinations of Ge and Se in amorphous GeSe. X-ray and electron diffraction radial distribution studies (170) of the amorphous germanium-monochalcogens had been unable to distinguish between two models for the atomic arrangements in these materials. In one, the germanium is fourfold coordinated while the chalcogen is twofold coordinated. In the other, each is threefold coordinated. The EXAFS work presented strong evidence for the first model. This model was rather important for understanding why many impurities have little effect on the electrical properties of amorphous semiconductors, as discussed by Mott (171).

More recently, Hunter et al (154, 172, 173) have studied an amorphous semiconducting system in which impurities have a drastic effect on the electrical conductivity, Cu in As_2Se_3. Using EXAFS, they were able to follow the changing coordination with copper concentration of the Cu, the As, and the Se separately. Similarly, Knights et al (142) have studied the coordination of arsenic impurities in amorphous silicon-hydrogen alloys. This work constitutes the first direct evidence for the occurrence of substitutional doping in an amorphous semiconductor.

Crozier et al (174) extended such studies of armorphous semiconductors into the liquid state in their analysis of the atomic arrangements in liquid As_2Se_3. The EXAFS data indicate that a major structural rearrangement does not occur in the nearest neighbor shell when As_2Se_3 is melted. These experiments, however, indicate small structural changes.

Finally, structural studies of the metallic glass $Pb_{78}Ge_{22}$ and sputtered amorphous $Pb_{80}Ge_{20}$ have been performed by Hayes, Allen, Tauc, Giesson & Hauser (preprint). The work indicates that the amorphous materials are chemically ordered, with each Ge being surrounded exclusively by Pb atoms. This result is important for understanding the formation and stability of metallic glasses.

7.2.5.3 *Catalysts* EXAFS has also been utilized extensively to determine the coordinations of metal atoms in catalysts. Its importance arises from the fact that the catalytic systems are often very finely dispersed, either on a support or in solution, so that normal x-ray diffraction procedures yield little information. EXAFS analysis yields the desired structural information, because it can be used to examine the average coordination of each metallic species in the catalyst. In addition, the valence state of the metals can often be determined.

In an early work, Lytle et al (175) used a standard x-ray tube to obtain EXAFS data from an alumina-supported CuCr catalyst. Cu is found to occupy both tetrahedral and octahedral sites in the supporting alumina lattice, while Cr was found only in octahedral sites. It was determined that the valence state was Cr^{+5} in the fresh catalyst, which changed to Cr^{+6} in the exhaust-cycled material. Confirmation of the valence state assignment was obtained by electron paramagnetic resonance.

Using similar experimental techniques, Bassi, Lytle & Parraveno (176) examined preparations of Au and Pt supported on Al_2O_3, MgO, and

SiO_2. The L_{III} x-ray absorption coefficient was normalized and Fourier-transformed to yield a radial structure function for each preparation. The analysis of the results showed that metals were present in two distinct phases: metallic with coordination (and metal-to-metal distance characteristic of bulk Au and Pt) as well as a highly dispersed one. It was concluded that the latter consisted of metal atomically dispersed or condensed in small, flat clusters of a few atoms. It was further estimated that the majority of Au and Pt was present in the highly dispersed form. The subsequent work described below was all performed at the Stanford Synchrotron Radiation Laboratory.

Lytle et al (177) examined the interaction of oxygen with a ruthenium-silica catalyst containing one weight percent ruthenium. Reed et al (178) studied the structure of the catalytic site of polymer-bound Wilkinson's catalyst. Reed & Eisenberger (179) studied the coordination of Rh and Pt in Bromo Tris(triphenylphosphine) rhodium(I). Reed, Eisenberger, Teo & Kincaid (preprint) studied the structural effects of cross linking in polymer-bound Bromo Tris(triphenylphosphine) rhodium(I) catalyst. Reed, Eisenberger & Hastings (preprint) studied the Ti-Cl bond distance in $TiCl_3$, a component in the Ziegler-Natta catalyst. Reed & Eisenberger (preprint) studied bond distances in $RhBr_3$ and Bromo Tris(triphenyl-phosphine) rhodium(I). Sinfelt et al (180) studied the coordination of platinum in two silica supported platinum catalysts as function of chemisorbed hydrogen.

In related work, Lytle (181) used the L_{III} x-ray absorption threshold to determine d-band occupancy in pure metals and supported catalysts, and to examine the correlation between percentage d character of the bulk metals and the catalytic activity of the supported metal.

7.2.5.4 *Surfaces*   Studies of atomic arrangements on surfaces have been performed in two different ways with EXAFS. The simpler, in terms of technique, is to examine the EXAFS above the absorption edge of an atomic species adsorbed on a substrate that does not contain that species. The analyses by Stern et al (182, 183) as well as Heald & Stern (preprint) of bromine adsorbed on graphite are examples of these.

A second approach, examined in detail by Lee (184), involves the measurement of the intensity of the Auger emission line as a function of photon energy. Since the number of Auger electrons emitted is proportional to the number of photons absorbed at the surface, the measurement is essentially the same as measuring the x-ray absorption coefficient of the atoms at the surface and therefore the EXAFS. Citrin et al (185) determined directly the bond length and position of I adsorbed on the (111) face of Ag in this manner. These authors claim that this approach, called surface-EXAFS (SEXAFS), offers three main advantages over low energy electron diffraction (LEED). It is easier to interpret, it is more accurate, and it does not require that the surface atoms be in a periodic array.

A related technique, described by Jaklevic et al (186) has been employed to measure the EXAFS spectrum from an atomic species present in very low concentrations in a material. Rather than measuring the transmitted radiation, the fluorescence radiation is counted as a function of incident photon energy. Since the fluorescent yield is expected to be a very slowly varying function of photon energy above the edge, the measurement is again an indirect measurement of the EXAFS associated with the fluorescing species. The gain in signal-to-noise is achieved by blocking out the fluorescent radiation from the other atomic species in the material. The authors demonstrate the power of the technique by showing EXAFS spectra from the Mn in a leaf, where it is present in concentrations of 10–50 ppm. They also show the As spectrum from a single crystal of Si doped with As at a concentration of 2–4×10^{19} cm^{-3}. This method was also employed by Shulman et al (160) in their study of rubredoxin. The concentration below which fluorescence gives better results than transmission depends on several factors, particularly the edge energy. For example, for energies below about 7 keV and concentrations below 10 millimolar in the metal atom, fluorescence is the preferred method.

7.3 X-Ray Fluorescence—Trace Element Analysis

The repopulation of an electron core level vacancy in an atom can result in the emission of characteristic fluorescence radiation, which can serve as a tag of the particular element and thus provide the basis of quantitative element identification. The creation of core level vacancies can be accomplished by exposing the sample to sufficiently energetic proton, electron, or electromagnetic radiation. All of these are used in trace element analysis work. The advantage of photons as the exciting radiation lies in their relatively high absorption cross section and relatively low cross section for scattering.

As a source of photons, synchrotron radiation has several favorable characteristics, particularly its high intensity, broad bandwith (which permits wavelength selection near absorption edges), and high collimation (facilitating the efficient use of focusing crystals to produce very small area beams of monochromatic radiation). Furthermore, the high linear polarization of the radiation, with its **E** vector in the plane of the electron acceleration means that the Rayleigh (elastic) and Compton (inelastic) scattering goes through a minimum in the direction of this **E** vector. Thus, a detector placed in this location will have much reduced background without decrease in the intensity of the isotropically emitted fluorescence radiation.

All of these properties have been very effectively used in an experimental search for super-heavy elements by Sparks et al (187, 188) at SSRL. Their experimental arrangement is shown in Figure 18, and typical fluorescence spectra are shown in Figure 19. The hot-pressed curved pyrolytic graphite (mosaic spread 0.4°) monochromator collected 2 mrad of synchrotron radiation in the horizontal plane and 0.14 mrad (about half of the total vertical divergence at 37 keV) in the vertical plane, which produced a focused flux of about 4×10^{10} photons per sec-mm^2 at 37.5 keV [300 eV full width half maximum (FWHM)] with SPEAR operating at 3.4 GeV and 20 mA. By tuning the monochromator, photon energies can be selected to bracket an absorption edge (in this case the L edge of element 126). Thus, a high degree of confidence can be obtained in any observed signal.

No evidence for the existence of super-heavy elements was found in a variety of samples, including monazite inclusions in mica crystals in which an earlier proton-induced fluorescence experiment (189) had reported possible evidence. The synchrotron radiation experiments had a sensitivity sufficient to observe about 5×10^8 super-heavy atoms per sample—about 55 times higher sensitivity than the proton experiment. Further refinements in the technique and higher incident fluxes should make it possible to detect concentrations as low as 10^6 atoms per sample (C. J. Sparks, private communication).

In the Soviet Union, trace element concentrations of $2–3 \times 10^{-8}$ g/g have been observed in samples of petroleum, powdered minerals, and salt solutions using the VEPP-3 storage ring in Novosibirsk (190). Concentrations of 10^{-7} g/g could be detected in 100 sec.

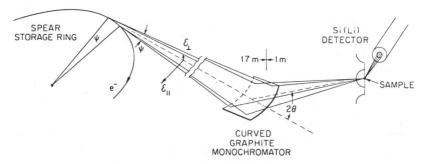

Figure 18 Schematic layout of Oak Ridge National Laboratory experiment used in the search for superheavy elements (188). Courtesy of C. J. Sparks, Oak Ridge National Laboratory.

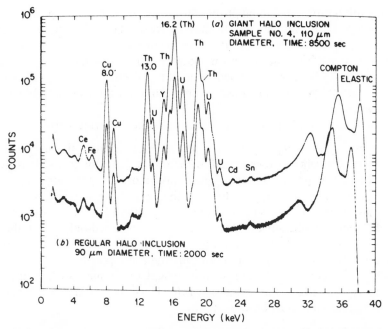

Figure 19   Typical fluorescence spectra obtained in the superheavy elements search (187). The L emission lines from superheavy elements, if present, would appear in the 22–31 keV region. The Cd and Sn peaks are due to trace quantities of these elements in the sample, sample chamber, or detector. The peak labeled Cd in the upper curve is due to the presence of about 5×10^9 atoms of cadmium in the sample. Courtesy of C. J. Sparks, Oak Ridge National Laboratory.

7.4   X-Ray Lithography—Microstructure Replication

The replication of patterns with micron- and submicron-sized features (such as integrated circuit patterns) by x-ray lithography is a new technique, begun in 1972 with the pioneering work of Spears & Smith (191) using conventional x-ray generators. The qualities of synchrotron radiation (e.g. high intensity, natural collimation) for this purpose were clearly demonstrated in the experiments of Spiller of IBM and collaborators in 1976 using the DESY synchrotron (192) and also by a group from the Thomson–C.S.F. Company and Orsay in 1976 using the ACO storage ring (193). A comprehensive review of x-ray lithography has recently been produced by Spiller & Feder (194).

The industrial interest in x-ray lithography using synchrotron radiation is increasing very rapidly. In both Germany and France, provisions have been made for private companies to use government-operated synchrotron radiation facilities for research and development work in x-ray lithography, including proprietary work. In some cases, production of devices is also anticipated. In the US, where there are greater restrictions on proprietary work at government-sponsored laboratories, a private company, the Electron Storage Ring Corporation in San Francisco (195) has been formed to build a privately owned storage ring for x-ray lithography and other commercial applications of synchrotron radiation.

The x-ray lithographic process consists of the following steps:

1. A mask containing a pattern is placed in close proximity to a wafer (normally silicon) overcoated with a thin layer of x-ray sensitive material called a photoresist—generally an organic polymer.
2. The photoresist is exposed through the mask to an x-ray beam. If the mask pattern consists of x-ray absorbing material (e.g. gold) on a thin backing film (e.g. mylar), then x rays are transmitted only where there is no absorber. The transmitted x rays cause chemical changes in the photoresist (e.g. they break bonds in long chain polymers).
3. The photoresist is developed in a solvent that removes exposed and unexposed regions at a different rate. Thus, the pattern of the mask is replicated in the photoresist.

Subsequent processes (e.g. plating, diffusion, ion implantation) are used to transform the pattern into an integrated circuit, a memory device, or other microstructure. Often successive exposures overlaying several different patterns with processing steps between exposures are required to make a single device. Precise registration of these different exposures is necessary to a tolerance of a fraction of the size of the smallest linewidth or feature.

The steps described above are basically the same whether x rays or more conventional radiation sources (visible or UV light, electrons) are used. However, x rays have several distinct advantages over the others, particularly in the replication of submicron structures. Briefly, these advantages are: (a) Short wavelength (10–50 Å) reduces diffraction effects. This makes possible the replication of structures with line-widths ∼0.1 μm. With visible and UV light, diffraction limits linewidth to ∼1 μm. (b) Mask need not be in direct contact with wafer. This increases mask life. (c) Relative insensitivity to dust. (d) Deep parallel grooves with large aspect ratios can be made (see Figure 20). (e) High intensity available with synchrotron radiation makes possible short exposure time (∼1 sec) even with high resolution, low sensitivity photoresists such as polymethylmethacrylate (PMMA).

In the Soviet Union, x-ray lithographic work has been done using the VEPP-3 storage ring in Novosibirsk (196). Patterns were recorded

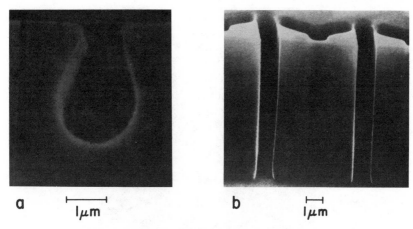

a ⊢——⊣ 1μm   b ⊢—⊣ 1μm

Figure 20 Comparison of grooves made by (*a*) electrons of 15-keV energy [Courtesy of M. Hatzakis, IBM, Yorktown Heights, NY (256)] and (*b*) x rays of 2.5-keV energy [Courtesy of E. Spiller, IBM, Yorktown Heights, NY (194)].

on chalcogenide glass films, producing a direct change in index of refraction without the need for development.

It appears likely that, with the increased availability of synchrotron radiation sources, particularly to industrial users, the technique may become developed and perfected to the extent that it will have a major impact on the semiconductor industry and other areas of microstructure fabrication. Examples of structures that have been fabricated using x-ray lithography are given in Figure 21.

7.5   X-Ray Microscopy

There is a growing interest in the application of synchrotron radiation to x-ray microscopy. Since 1972 three quite different x-ray microscopes have been built and operated using synchrotron radiation. Sayre et al (197) reviewed soft x-ray microscopy. Kirz et al (198) made a comparative analysis of x-ray emission microscopies for biological specimens using incident proton, electron, or x-ray beams focused to a small spot. They conclude that radiation damage limits the resolution for unstained biological specimens and that incident x-ray beams give the lowest dose for elements with atomic number above about 10.

We now discuss the three types of microscopy that have been demonstrated with synchrotron radiation.

7.5.1   SCANNING X-RAY MICROSCOPY   The principal of operation of this device, as implemented by Horowitz & Howell (97) in 1972, is shown in Figure 22. Synchrotron radiation from about 10 mrad of orbit from the

Figure 21   Examples of patterns replicated by x-ray lithography: (*a*) zone plate pattern and (*b*) magnetic bubble memory pattern. The individual features are 1 μm wide and 3 μm high for the bubble memory pattern. Courtesy of E. Spiller, IBM, Yorktown Heights, NY (194).

CEA storage ring was collected at grazing incidence, totally externally reflected, and focused by a quartz ellipsoidal mirror. Just upstream of the focus, the synchrotron radiation passed through a very thin (10 μm) beryllium foil, which terminated the high vacuum beam run. The focused x-ray beam was so intense that upon emerging from the beryllium window, it caused a readily visible air fluorescence as shown in Figure 23 (with the microscope removed). The focal spot was roughly 1 × 2 mm^2, but this was further collimated by a 2-μm pinhole in a 100-μm gold absorber foil.

The sample, which could be in an air or helium atmosphere, was scanned raster (line by line) fashion across the resulting 2-μm-sized beam. Fluorescent x rays, produced by the part of the sample being irradiated, were detected, energy analyzed, and the resulting signal used to intensify the beam of a cathode ray tube that was being swept synchronously with the sample. The main features of the microscope are the following (97): (a) The sample could be in air or other atmosphere, which facilitates the study of hydrated, even live, biological specimens. (b) A resolution of ~2 μm can be obtained, determined by the diameter of the pinhole. (c) The depth of field was about 1 mm, which facilitates stereoscopic viewing of pictures taken at different angles. However, the x rays penetrate 10–100 μm of most material. (d) It is possible to select a given element whose absorption edge is in the range of the instrument and determine the distribution of that element in the sample. The condensing mirror provides a cutoff above about 3.5 keV, and the beryllium window absorbs strongly below about 1 keV. (e) It takes about two minutes to form a picture.

The development of this microscope with synchrotron radiation

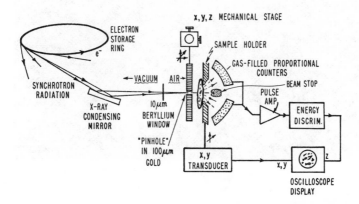

Figure 22   Schematic of scanning x-ray microscopy using synchrotron radiation. Courtesy of P. Horowitz, Harvard University.

Figure 23 Focused x-ray beam from CEA storage ring emerging from beryllium window and causing visible air fluorescence. Courtesy of P. Horowitz, Harvard University.

terminated in June 1973, when the CEA was shut down. The instrument has since been continued as a proton microprobe (199).

7.5.2 ZONE PLATE X-RAY MICROSCOPY Using synchrotron radiation from the DESY synchrotron, Niemann, Rudolf & Schmahl (200) operated the zone plate microscope shown schematically in Figure 24. The incident white synchrotron radiation is dispersed at grazing incidence by a reflection grating. A focal spot of monochromatic light is produced onto the sample by the first zone plate, which has a 5-mm diameter, 2600 zones, and a focal length of 522 mm for 46 Å radiation. A second zone plate with a diameter of 1 mm, 850 zones, and a focal length of 64.5 mm for 46 Å radiation forms a magnified image of the object. In their first tests, a resolution of 0.5 μm was obtained, and several biological objects were examined. Work is in progress to improve the resolution, reduce the exposure time (now some minutes long), and to study other samples (200).

7.5.3 CONTACT X-RAY MICROSCOPY Using a technique quite similar to x-ray lithography, a collaboration of scientists from IBM, DESY, NYU Medical School, and Yale University have obtained contact x-ray micrographs of biological objects with very high resolution (201). Both synchrotron radiation and carbon K_α x rays were used.

The specimen to be examined is placed in direct contact with a photoresist and exposed to the x rays (see Figure 25). The absorption profile of the specimen is recorded in the photoresist, which is then developed, coated with a metal film, and examined in a scanning electron microscope.

The highest resolution (~ 100 Å) was obtained using broadband synchrotron radiation with wavelengths in the range of about 30–44 Å. For

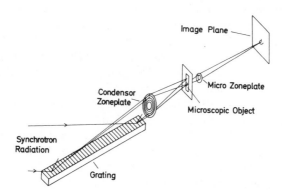

Figure 24 X-ray microscopy using zone plates. Courtesy of B. Niemann, Universitäts-Sternwarte, Göttingen (200).

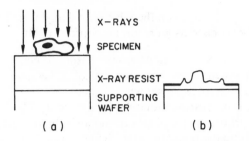

Figure 25   Schematic of contact x-ray microscopy technique. Courtesy of E. Spiller, IBM, Yorktown Heights, NY (201).

shorter wavelengths, the resolution is poorer because secondary electrons have a longer range in the photoresist. For longer wavelengths, diffraction limits the resolution. Some results are shown in Figure 26, which includes a scanning electron micrograph for comparison. The figure shows that the x-ray micrographs reveal the interior structure, while the scanning electron micrograph exhibits only surface structure.

A collaboration from the University of Paris and the Orsay Synchrotron Radiation Laboratory (202) performed similar measurements using monochromatized synchrotron radiation. By selecting incident photon energies corresponding to particular elemental absorption edges, they were able to do chemical microanalysis with a resolution of a few micrometers.

7.6   Ultraviolet and Soft X-Ray Studies of the Electronic Structure of Matter

7.6.1   INTRODUCTION   Compared to its utilization in the x-ray portion of the spectrum, vacuum ultraviolet (VUV) radiation research utilizing synchrotron radiation is a well-established field with a vast primary literature and several fine recent review articles (42, 203–215). Because this field has been reviewed so extensively and well, we present here a brief overview of the types of studies being performed, with references to recent reviews containing more detail.

The basic utility of VUV for exploring electronic states arises from the fact that the photon energies range from 10^1 to 10^3 eV. That is, the photon energies cover the range of binding energies from deep valence states to many core states. Consequently, the basic experimental approaches—reflection, absorption, fluorescence, and photoemission—provide valuable information about the nature of those electronic states and related collective excitations (plasmons, excitons, etc). At the same time, research in this field must necessarily be concerned with obtaining increased understanding of the interaction between the electromagnetic radiation and the substances being studied.

In performing VUV experiments, synchrotron radiation offers a number of advantages over other sources. As in most of the x-ray experiments, the most important feature is that high intensity is obtained over a continuous spectral range. That is, one need not be constrained to the use of characteristic radiation to obtain high intensity. It is quite apparent

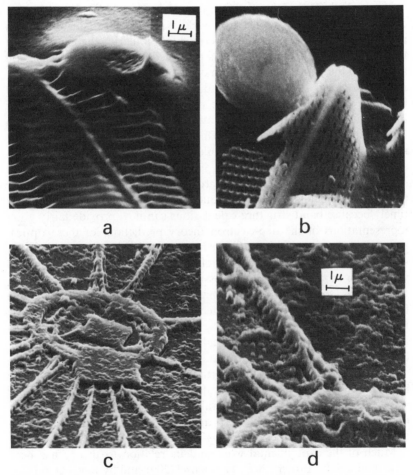

Figure 26 X-Ray replicas of diatoms (*a, c, d*) and scanning electron micrograph of a diatom (*b*). In (*a*) the x-ray source was carbon K_α radiation and the exposure time was 20 hr with a source-to-specimen distance of 15 cm. Part (*d*), a detail of (*c*), was taken with synchrotron radiation in 10 min with the specimen 40 m from the source. Courtesy of E. Spiller, IBM, Yorktown Heights, NY (201).

that this feature facilitates absorption and reflection measurements as a function of photon energy. Less obvious is that it allows the probing of electronic states as a function of depth in the photoemission experiment, as discussed below. This feature has led to the rapid growth of surface studies and studies of sorbed gases through photoemission in recent years, as discussed by Spicer (214).

Of equal importance to such studies is the fact that the electron storage ring is a high vacuum source, hence it does not contaminate the surfaces being studied. The high intensity also makes possible photoemission experiments in which the electron yield is sufficiently high to perform angle-resolved photoemission experiments. Similarly, there is the fact that a flat "pancake" of radiation is produced with very little dispersion away from the plane of the electron orbit. This lack of dispersion is very important because it allows the radiation to be collimated with much less loss than that given by more divergent sources. Finally, another important characteristic of the radiation is its natural linear polarization in the plane of the electron orbit. This can be particularly important for studies involving optical selection rules and/or angularly resolved photo-emission.

A particularly active area of research has been the study of the optical properties of metals and semiconductors. This research is particularly timely because band structure calculations can now provide fairly good representations of the one-electron theory predictions of these optical properties. Of particular importance in these analyses are the modulation spectroscopy techniques—electroreflectance, thermoreflectance, and wavelength modulation—made possible in the region above 10 eV by the intense, continuous, and stable synchrotron radiation available from electron storage rings. These techniques enhance weak structure in the optical spectra, emphasizing the contributions of localized regions in the Brillouin zone, such as critical points, and have been particularly effectively utilized at the Synchrotron Radiation Center in Wisconsin.

While data so obtained are important in the analysis of band structure and the testing of band structure calculations, it has been apparent for some time that not all of the spectral features can be treated within the one-electron, Bloch-type theory, so there is a strong interaction between these experimental studies and many-electron theory.

Much of the experimental work and its relation to theory has been described well by Brown (207), by Haensel (209), and by Koch et al (213). Lynch's review article (216) emphasizes the relationship between existing theory and experiment for metals, while that of Aspnes (217) deals with semiconductors. Mahan (218) has provided an excellent description of the many-body theoretical treatments of the VUV and x-ray optical properties of metals.

Another area that has advanced considerably as a result of synchrotron radiation is the use of VUV photoemission to probe the electronic structures of bulk solids, as well as surfaces. At the same time, the photoemission process itself has been the subject of considerable study. In these experiments, a monochromatic beam incident upon the samples yields photoemitted electrons. In many experiments, the total yield is measured as a function of photon energy. In energy-resolved photoemission, however, the energy distribution of the photoemitted electrons is determined as a function of photon energy. From such measurements, information about the densities of initial and final states can be obtained. If, in addition, angle-resolved photoemission studies, in which the direction of emission is also analyzed, are performed, information about the form of the initial and final states involved in the photoexcitation may be obtained. These energy- and angle-resolved studies, which involve a small portion of the total photoelectron yield, have benefited greatly from the high intensity of synchrotron radiation.

7.6.2 CROSS-SECTION EFFECTS   Examples of fundamental studies of the photoemission process are the studies of energy dependent cross sections, which take advantage of the tunability of the synchrotron radiation. Cooper & Fano (219, 220) predicted minima in the atomic photoemission cross section if the initial state wave function exhibits a radial node. This effect has been studied (215, 221, 222) for the d-band intensity in the noble metals Cu ($3d$, no radial node), Ag ($4d$, one radial node), and Au ($5d$, two radial nodes), as well as for In and Sb (223, 224). As shown in Figure 27, the energy dependence of the d-band intensity, which is proportional to the photoionization cross section (222), is significantly different for the three noble metals. Above threshold, the intensities first exhibit maxima that occur in the order Au $5d$ (< 40 eV), Ag $4d$ (~ 60 eV), and Cu $3d$ (~ 130 eV). This is due to an increasing overlap of the d-initial and f-final state wavefunction (the $d \rightarrow f$ channel dominates the $d \rightarrow p$ one above ~ 20 eV). This overlap is small above threshold where the f-wave is pushed out by the centrifugal barrier, but it increases at higher energies in the order $5d > 4d > 3d$, the order of decreasing radial expectation values of the d-orbitals. At still higher energies, the d-band intensity decreases steeply to a minimum for Ag and Au. These so-called Cooper minima arise from a change of sign in the radial dipole matrix element associated with the $d \rightarrow f$ channel. The photoexcitation cross section, which is proportional to the sum of the squares of the p and f channel radial matrix elements, will exhibit a minimum at the energy for which one of the matrix elements vanishes. As compared to the $4d$ function in Ag (one node), the $5d$ function in Au has two nodes and the Cooper minimum is smeared out. These minima have important consequences

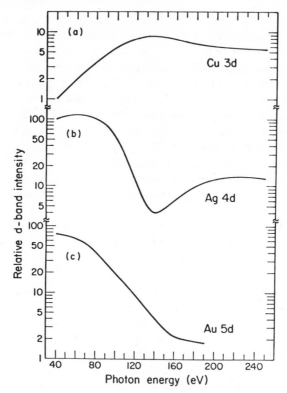

Figure 27 Relative *d*-band intensity vs photon energy for Cu 3*d*, Ag 4*d*, and Au 5*d* (221).

for the utilization of photoemission in the study of adsorbates on 4*d*- or 5*d*-type metals, as discussed below.

By tuning the photon energy to the Cooper minimum of the metal, a significant enhancement of the relative absorbate spectral intensity is achieved.

7.6.3 BAND STRUCTURE EFFECTS Photoemission has been used for many years to probe the band structure of solids. In these studies, the availability of synchrotron radiation has had two important consequences. First, the exciting photon energy may be varied continuously over a very large range, with a resultant variation in the relationship between the initial and final states involved in the process. This variation has allowed for more detailed determinations of the densities of states involved. Second, as discussed above, the high intensity has facilitated the development of energy- and angle-resolved photoemission. Thus the information obtained has become still more definitive. Recent work is reviewed by Koch et al (213), by Eastman (225), and by Shirley et al (215).

7.6.4 SURFACE AND ABSORBATE EFFECTS   An entire line of research that owes its existence very much to the availability of tunable synchrotron radiation, and that is also of considerable current theoretical and practical interest, is the study of structure and electronic states of surfaces and of absorbates on surfaces, as reviewed by Lindau (226), by Shirley et al (215), by Spicer (214), as well as by Plummer & Gustafsson (257).

The importance of a tunable source can be understood through reference to the plot (227) of the electron escape depth as a function of the electron energy above the Fermi level for a large number of materials shown in Figure 28. This depth has a minimum of the order of a few

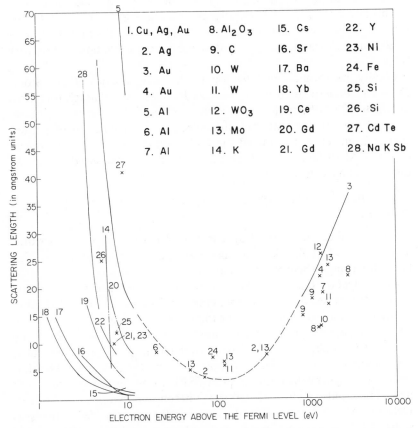

Figure 28 The escape depth, in Å, is shown as a function of the electron energy above the Fermi level, in eV, for a large number of materials (227).

angstroms at energies that range from ten to a couple of hundred eV, depending on the material. To study core electronic states of surfaces or of absorbates, one may tune the incident photon energy so that electrons photoemitted from a specific core level have energies close to the minimum. As a result, few electrons excited from the bulk of the material escape, and hence those detected are primarily from the surface or from the absorbate. Alternatively, in studies of absorbates on appropriate substrates the photon energy can be tuned to the Cooper minimum of the substrate, so absorbate emission dominates the detected electrons.

These studies are of considerable theoretical interest since methods for calculating the electronic states at the surface of solids are now being devised and employed actively. In addition, they are of considerable practical importance for the understanding of semiconductor surfaces, semiconductor passification, and catalytic activity. The reader interested in the latest work in this rapidly developing area is referred to the reports of the users' meetings of the Synchrotron Radiation Center at Wisconsin (228) and the Stanford Synchrotron Radiation Laboratory (229).

Finally, we should call attention to the very recent work of Stöhr, Denley & Perfetti (preprint) as well as Bianconi & Bachrach (preprint) in which SEXAFS studies have been performed through the measurement of inelastically scattered photoelectrons as a function of photon energy. Stöhr et al determined the O-Al near neighbor distance on an oxidized Al surface, while Bianconi & Bachrach measured the Al-Al distance on a clean Al (111) surface.

7.7   Time-Resolved Spectroscopy Using Synchrotron Radiation

The pulsed time structure of synchrotron radiation is very well suited to a variety of applications in time-resolved spectroscopy (TRS). The first TRS using synchrotron radiation excitation was done in 1974 by López-Delgado, Tramer & Munro at ACO in France (230). Several studies of time-resolved excitation, emission, and decay of excited states in organic and inorganic molecules have resulted from this work (231, 232). Nevertheless, the relatively short (73 nsec) interpulse period and rather broad pulse width (~ 1 nsec) available at ACO prevented those investigators from studying very short and very long lifetimes. The time structure of synchrotron radiation at SSRL from the SPEAR ring with its 300-psec pulse width and 780-nsec interpulse period, enabled López-Delgado, Miehé & Sipp (233) to measure fluorescence decay times as short as 0.2 nsec in tetracene crystals. A review of TRS techniques and results has been given by López-Delgado (234).

Several TRS experiments have since been done at SSRL (235). Matthias, Shirley, and co-workers have investigated fluorescence lifetimes in noble gases and simple molecules (236). Of special significance is their use of the polarization of synchrotron radiation to observe quantum beats in the fluorescence from magnetically oriented atoms (237). Simultaneously, Monahan & Rehn have been studying exciton-enhanced energy transfer in the pure and doped noble solids (238, 239). These investigators have demonstrated that TRS using synchrotron radiation can be used to determine the principal channels of energy transfer and their relative importance in these systems. This work has been highlighted by the discovery of previously unobserved electronic states, excimer (i.e. a diatomic molecule with one or more bound excited states and a repulsive ground state) species, and dissociation phenomena (239, 240).

The application of TRS techniques at SSRL and throughout the world are on the increase. A visible light and near-UV TRS port on the SPEAR ring was made operational in early 1978, and plans are being made to install a VUV port with large acceptance angle. Hahn, Jordan & Schwentner (241) have done some preliminary lifetime studies at the DORIS storage ring in Germany and are planning a number of experiments when DORIS is converted to single-bunch operation (~ 1-μsec period). In addition, storage rings in Novosibirsk have time structure suitable for TRS, and work in this area is expected in the near future. Many of the new storage rings now under construction as dedicated synchrotron radiation sources will be capable of operating in multibunch mode for maximum synchrotron radiation intensity and also single-bunch mode to facilitate timing studies.

7.8   Applications in Atomic and Molecular Physics

The continuous spectrum feature of synchrotron radiation, coupled with its high intensity, offers the opportunity, largely unexploited, for a vast array of experiments involving the interaction of electromagnetic radiation with atoms and molecules. One set of experiments, however, makes clear the powerful potential.

Ice & Crasemann (242) measured the differential cross sections for the total scattering of photons by H_2 at large momentum transfer. Such cross sections are of interest because total x-ray scattering is a two-electron expectation value, whereas coherent (elastic) intensity is a one-electron property (243). Thus, coherent scattering is routinely used to determine radial charge density distributions, i.e. the expectation of the electron-nucleus separations. The inelastic scattering intensity, on the other hand, is sensitive to electron correlation and depends upon the expectation of the inverse the electron-electron separation, $\langle r_{ij}^{-1} \rangle$. Of considerable interest in this regard are the two-electron systems H_2 and He. Because the cross sections are very small, of the order of hundreds of millibarns per steradian, they cannot be measured with adequate accuracy using ordinary x-ray sources.

In experiments now underway on neon, Ice, Parente & Crasemann (G. E. Ice, private communication) are utilizing the tunability of synchrotron radiation to perform a more sophisticated version of this experiment in which the elastic part of the scattering is experimentally removed. The motivation for this work is that the inelastic scattering is much more sensitive to the scattering model and the wave functions than is the elastic. Experimental removal of the elastic scattering provides a more direct and accurate measurement of the inelastic scattering.

This is achieved by using incident photons whose energies are a few eV higher than an absorption edge of an absorbing element placed between the scatterer and the detector. Consequently, the element absorbs heavily the elastically scattered radiation, but for most momentum transfers is relatively transparent for the inelastically scattered x rays, whose energies are below the edge. These then reach the detector and are counted. Such a technique is possible with a highly tunable x-ray source but is generally impossible with standard sources.

7.9   Some Future Applications

7.9.1   PUMPING OF X-RAY LASERS   Synchrotron radiation offers probably the most promising approach to achieve the high spatial and temporal energy densities required for pumping an x-ray laser (244). Furthermore, the ability to select any desired photon energy from the synchrotron radiation continuum is a major advantage. By selecting an energy just above the K edge, a much higher fraction of the incident energy results in pumping the desired level with corresponding reduction in unwanted ionization and heating of the medium. Since the lasing material remains relatively cool, and ionization is relatively low, homogeneous and inhomogeneous line broadening will be small, which thus enhances laser action. In addition, in a cool plasma, the outer electron structure of atoms can be rearranged by resonant transitions induced by low energy photons, and one is no longer restricted to configurations common in nature.

For example, a small volume of Li gas of 5×10^{18} cm^{-3} density irradiated with 1.8-eV photons from a conventional laser as well as with x rays just above the K edge of Li can be made to lase at 90 eV with about 100 times the photon pumping intensity now delivered by SPEAR in this energy range (244). The required intensity is likely to be reached with wigglers and with high current, low emittance dedicated operation of SPEAR and other storage rings.

7.9.2   X-RAY HOLOGRAPHY WITH SYNCHROTRON RADIATION : COHERENCE
The coherence properties of synchrotron radiation are superior to those of conventional x-ray sources, a feature that should greatly facilitate x-ray holography. For example, for VEPP-3, the power useful for holography has been calculated to be three orders of magnitude greater than the coherent radiation power from x-ray tubes (245).

Using such a source and a MWPC detector, it should be possible (50) to obtain x-ray microholograms with 1 Å incident radiation with a resolution of 50 Å transverse and 2500 Å longitudinal. The calculated exposure time is 1 sec.

Considerably higher coherent power would be available from a free electron laser (see Section 5.3) if such a device could be made to operate at the required photon energy and in a storage ring. It has also been suggested that synchrotron radiation could pump x-ray lasers (see previous section), which would be powerful sources of coherent x rays.

Lensless Fourier-transform x-ray holograms of simple one-dimensional objects have been obtained (246) with carbon K_α radiation (44.8 Å) and with synchrotron radiation (60 Å) (247). Two- and three-dimensional objects have also been studied (248) with aluminum K_α radiation (8.34 Å) with a resolution of 4 μm transverse and 4 mm longitudinal and with exposure times of ~ 1 hr.

By recording a hologram at one wavelength and reading it at another wavelength, a magnification corresponding to the ratio of the wavelengths may be obtained. Csonka (249, 250) recently analyzed the application of synchrotron radiation to x-ray holography.

7.9.3   RADIATION TECHNOLOGY   Synchrotron radiation opens possibilities in the field of radiation chemistry and technology that have not yet been evaluated. It is possible, for example, that techniques will be developed to modify the surface and bulk properties of materials by irradiation with synchrotron radiation, similar to radiation processes now in use commercially to manufacture "shrink-tubing" for electrical insulation.

With existing storage rings and synchrotrons, photon energies up to 50–100 keV are available for such work. With machines under construction and with wigglers, this could be extended to several hundred keV. Richter (251) has pointed out that a very large (100 GeV) electron storage ring would make available an enormous flux of x rays extending into the MeV range and has suggested that applications such as the sterilization of grain for storage in a worldwide "grain bank" should be considered. At these high photon energies, the walls of the ring vacuum chamber are transparent, so the radiation emerges from all parts of the orbit without the necessity of installing tangent beam pipes.

7.9.4   MEDICAL APPLICATIONS   Using conventional x-ray sources, it has been shown (252–254) that a significant improvement in the contrast of diagnostic x radiographs can be achieved with a technique called dichromatography. In this method, a pair of transmission radiographs are recorded, one below and one above an x-ray absorption edge. The difference between the signals obtained has high sensitivity for the detection of small concentrations of the particular element whose edge has been selected. This could be an aid to early detection of tumors that may have a higher concentration of a particular element than normal tissue. Other applications include improving the sensitivity and reducing the dose to the patient in thyroid dichromatography at the iodine K edge (33 keV).

Synchrotron radiation offers precise tunability and high intensity, which should greatly facilitate dichromatography. The use of improved electronic position-sensitive detectors and associated data storage systems should also be of great benefit in developing the technique. A system for using synchrotron radiation for dichromatography and other medical purposes is briefly described by Kulipanov & Skrinskii (50).

Combining dichromatography with computerized axial tomography (255) might result in contrast and sensitivity improvements beyond that available from either of these techniques alone. The possibility also exists that monochromatic radiation just above the K edge of a particular element might be useful in radiation therapy (50), if the concentration of that element was high enough in the tumor. Such selective absorption could also be the basis of molecular microsurgery (50), the cutting apart of individual molecules at particular locations.

8   CONCLUSION

The extraordinary properties of synchrotron radiation and its rapidly increasing availability throughout the world are having a profound, perhaps revolutionary, impact on a broad range of scientific and tech-nological disciplines. This is all the more remarkable when one realizes that the very substantial results to date, as reviewed in this article, have been accomplished largely symbiotically during high energy physics experimental runs.

With dedicated operation of existing sources and with the design and construction of dedicated storage rings in at least three countries, a significant increase in experimental capability will soon be realized. Furthermore, there is the exciting promise of enhanced and modified radiation produced by special insertions such as high field standard wiggler magnets, narrow-band interference wigglers, and even narrower-band, tunable, free electron lasers.

The breadth of application of intense synchrotron radiation, extending from the infrared through the visible and ultraviolet parts of the electromagnetic spectrum and deep into the x-ray and soon the γ-ray regions, brings into the same laboratory scientists from diverse specialties such as catalytic chemistry, surface physics, structural biology, and metallurgy. The interchange among these groups during experimental runs or at users' meetings provides added stimulation and enrichment to all, particularly to students just beginning their scientific careers.

It is clear to those of us close to the development of synchrotron radiation research over the past several years that man is now producing electromagnetic radiation in one of nature's most effective ways—witness the crab nebula. The simplicity and efficiency of producing electromagnetic radiation by using relativistic electrons following curved trajectories in an ultra-high vacuum and the scientific appropriateness of this radiation make synchrotron radiation a most elegant, aesthetically pleasing phenomenon.

ACKNOWLEDGMENTS

The task of summarizing scientific work over the vast domain covered here could not have been accomplished without considerable help from those more familiar with specific areas. For aid in writing portions of the manuscript, in finding the relevant literature, and in reviewing drafts, we are indebted to: R. Bachrach, B. Crasemann, P. Csonka, P. Eisenberger, P. Flinn, T. M. Hayes, K. Hodgson, I. Lindau, F. W. Lytle, K. Monahan, W. Parrish, E. M. Rowe, S. Ruby, M. Sauvage, D. Sayers, D. Shirley, W. E. Spicer, E. M. Stern, and J. Stöhr.

The typing and editorial help of Georgia Hathorne have been invaluable. To her we are most indebted.

Literature Cited

1. Averill, R. J., Colby, W. F., Dickenson, T. S., Hofmann, A., Little, R., Maddox, B. J., Mieras, H., Paterson, J. M., Strauch, K., Voss, G.-A., Winick, H. 1973. *IEEE Trans. NS* 20:813–15
2. Koch, E. E., Kunz, C., Weiner, E. W. 1976. *Optik* 45:395–410
3. Koch, E. E., Kunz, C. 1977. *Synchro-* tronstrahlung bei DESY Ein Handbuch für Benutzer. 420 pp.
4. Dagneaux, P., Depautex, C., Dhez, P., Durup, J., Farge, Y., Fourme, R., Guyon, P.-M., Jaeglé, P., Leach, S., López-Delgado, R., Morel, G., Pinchaux, R., Thiry, P., Vermeil, C., Wuilleumier, F. 1975. *Ann. Phys.* 9:9–65

5. Hodgson, K. O., Chu, G., Winick, H., eds. 1976. *SSRP Rep. 76/100*
6. Winick, H. 1974. *Proc. Int. Conf. High Energy Accel., 9th,* Stanford, CA, pp. 685–88; Winick, H. 1974. See Ref. 208, pp. 776–80
7. Liénard, A. 1898. *L'Eclairage Elect.* 16:5
8. Jassinsky, W. W. 1935. *J. Exp. Theor. Phys. (USSR)* 5:983; 1936. *Arch. Elektrotechnik.* 30:590
9. Kerst, D. W. 1941. *Phys. Rev.* 60:47–53
10. Ivanenko, D., Pomeranchuk, I. 1944. *Phys. Rev.* 65:343
11. Arzimovitch, L., Pomeranchuk, I. 1945. *J. Phys. (USSR)* 9:267; 1946. *J. Exp. Theor. Phys. (USSR)* 16:379
12. Blewett, J. P. 1946. *Phys. Rev.* 69:87–95
13. Schwinger, J. 1949. *Phys. Rev.* 75:1912–25
14. Marr, G. V., Munro, I. H., Sharp, J. C. 1972. *Synchrotron Radiat.: A Bibliography DNPL/R24.* Sci. Res. Counc.-Daresbury. 208 pp.; also 1974 *DL/Tm 127.* 95 pp.
15. Ivanenko, D., Sokolov, A. A. 1948. *Dokl. Akad. Nauk SSSR* 59:1551
16. Sokolov, A. A., Klepitov, N. P., Ternov, J. M. 1953. *Dokl. Akad. Nauk SSSR* 89:665
17. Sokolov, A. A., Ternov, I. M. 1955. *Sov. Phys. JETP* 1:227–30; 1957. *Sov. Phys. JETP* 4:396–400; 1964. *Sov. Phys. Dok.* 8:1203–5
18. Sokolov, A. A., Ternov, I. M. 1968. *Synchrotron Radiation.* Pergamon
19. Jackson, J. D. 1975. *Classical Electrodynamics,* Ch. 14. New York: Wiley. 848 pp.
20. Elder, F. R., Gurewitsch, A. M., Langmuir, R. V., Pollack, H. D. 1947. *Phys. Rev.* 71:829–30
21. Ado, I. M., Cherenkov, P. A. 1956. *Sov. Phys. Dokl.* 1:517–19
22. Korolev, F. A., Kulikov, O. F., Yarov, A. S. 1953. *Sov. Phys. JETP* 43:1653
23. Korolev, F. A., Kulikov, O. F. 1960. *Opt. Spectrosc. USSR* 8:1–3
24. Corson, D. A. 1952. *Phys. Rev.* 86:1052–53; 1953. *Phys. Rev.* 90:748–52
25. Tomboulian, D. H. 1955. *US AEC NP-5803*
26. Tomboulian, D. H., Hartman, P. L. 1956. *Phys. Rev.* 102:1423–47
27. Bedo, D. E., Tomboulian, D. H. 1958. *J. Appl. Phys.* 29:804–9
28. Codling, K., Madden, R. P. 1963. *Phys. Rev. Lett.* 10:516–18; 1964. *Phys. Rev. Lett.* 12:106–8; 1964. *J. Opt. Soc. Am.* 54:268; 1965. *J. Appl. Phys.* 36:830–37
29. Bathov, G., Freytag, E., Haensel, R. 1966. *J. Appl. Phys.* 37:3449
30. Rowe, E. M., Mills, F. E. 1973. *Part. Accel.* 4:211–27
31. Winick, H. 1973. *IEEE Trans. NS* 20(3):984–88
32. Watson, R. E., Perlman, M. L., eds. 1973. *Res. Appl. Synchrotron Radiat. BNL Rep. 50381.* 192 pp.
33. Lindau, I., Donaich, S., Spicer, W. E., Winick, H. 1975. *J. Vac. Sci. Technol.* (6) 12:1123–27
34. *The Scientific Case for Research with Synchrotron Radiation.* 1975. Daresbury *Rep. DL/SRF/R3.* 32 pp.
35. Blewett, J. P., ed. 1977. *Proposal for a National Synchrotron Light Source, BNL 50595,* Vol. I, pp. 1-1 to 10-1, Vol. II, pp. 1-1 to 4-28
36. *A Proposal for a 1.5 GeV Electron Storage Ring as a Dedicated Synchrotron Source.* 1976. *(Pampus) FOM Inst. Rep. 39760;* also *Univ. Technol. Rep. NK-235,* Oct.
37. Morse, R. 1976. *An Assessment of the National Need for Facilities Dedicated to the Production of Synchrotron Radiation 1976,* pp. 1–77. Washington, DC: Solid State Sci. Comm., Natl. Res. Counc.
38. Tigner, M. 1977. *IEEE Trans. NS* 24:1849–53
39. Cerino, J., Golde, A., Hastings, J., Lindau, I., Salsburg, B., Winick, H., Lee, M., Morton, P., Garren, A. 1977. *IEEE Trans. NS* 24:1003–5
40. *Synchrotron Radiation—A Perspective View for Europe.* 1977. Strasbourg, France: Eur. Sci. Found. 87 pp.
41. Haensel, R., Zimmerer, G. 1976. See Ref. 204, pp. 411–34; Basov, N. G., ed. 1975. *Synchrotron Radiation Proceedings (Trudy) of the PN Lebedev Phys. Inst.,* Vol. 80. Moscou Nauka Press. Transl. Consul. Bur., NY, 1976. 224 pp.
42. Godwin, R. P. 1968. *Springer Tracts in Modern Physics,* Vol. 51. Berlin/New York: Springer. 73 pp.
43. Green, G. K. 1977. *BNL Rep. 50522.* 90 pp.; see also *BNL Rep. 50595,* Vol. II
44. Tsien, R. Y. 1972. *Am. J. Phys.* 40:46–56
45. Larmor, J. 1897. *Philos. Mag.* 44:503
46. Csonka, P. 1978. *Part. Accel.* Vol. 8
47. Hastings, J. B., Kincaid, B., Eisenberger, P. 1977. *BNL Rep. 23353;* also 1978. *Nucl. Instrum. Methods.* 152:167–71
48. Coïsson, R. 1977. *Opt. Commun.* 22:135–37
49. Mack, R. 1966. *Cambridge Electron Accel. Rep. 1027.* 19 pp.
50. Kulipanov, G. N., Skrinskii, A. N. 1977. *Usp. Fiz. Nauk* 122:369–418 English translation *Sov. Phys. Usp.* 20(7):559–86
51. Rehn, V. 1976. See Ref. 5, pp. 32–34
52. Kincaid, B. 1977. *J. Appl. Phys.* 48:2684–91

53. Sands, M. 1971. *Proc. Int. Sch. Phys. "Enrico Fermi,"* Course 46, pp. 257–411. New York/London: Academic; also *SLAC Rep. 121*, Nov. 1970
54. Pellegrini, C. 1972. *Ann. Rev. Nucl. Sci.* 22: 1–24
55. Sabersky, A. P. 1973. *Part. Accel.* 5: 199–206
56. Lindau, I., Pianetta, P. 1976. See Ref. 204, pp. 372–87
57. Hastings, J. 1976. *J. Appl. Phys.* 48: 1576–84
58. Winick, H., Knight, T., eds. 1977. *Wiggler Magnets SSRP Rep. 77/05*. 138 pp.
59. Winick, H. 1978. *Nucl. Instrum. Methods.* 152: 9–15
60. Winick, H. 1976. See Ref. 204, pp. 27–42
61. Trzeciak, W. 1971. *IEEE Trans. NS* 18: 213–16
62. Brautti, G., Stagno, V. 1976. *Nucl. Instrum Methods* 135: 393
63. Bassetti, M., Cattoni, A., Luccio, A., Preger, M., Tazzari, S. 1977. *LNF Rep. 77/26 (R)*. Frascati, Italy. 24 pp.
64. Dahl, P. F., Sampson, W. B. 1975. *BNL Rep. 200/9*. 4 pp.
65. Clee, P., Canliffe, N., Simkin, J., Trowbridge, C. W., Watson, M., 1974. *SRS Rep. 74/64*. Daresbury, Engl.
66. Brunk, W. 1977. See Ref. 58, pp. II41–II44
67. Blewett, J. P., Chasman, R. 1977. *J. Appl. Phys.* 48: 2692–98
68. Chu, G. 1976. See Ref. 5, pp. 193–215
69. Motz, H., Thon, W., Whitehurst, R. N. 1953. *J. Appl. Phys.* 24: 826–33
70. Hofmann, A. 1978. *Nucl. Instrum. Methods.* 152: 17–21
71. Alferov, D. F., Bashmakov, Yu. A., Bessonov, E. G. 1974. *Sov. Phys. Tech. Phys.* 18: 1336–39
72. Diambrini Palazzi, G. 1968. *Rev. Mod. Phys.* 40: 611–31
73. Milburn, R. H. 1963. *Phys. Rev. Lett.* 10: 75–77
74. Alferov, D. F., Bashmakov, Yu. A., Belovintsev, K. A., Bessonov, E. G., Cherenkov, P. A. 1977. *Pisma v JETP.* 26: 525–29
75. Coïssou, R. 1977. *IEEE Trans. NS* 24: 1681–82; 1977. *Nucl. Instrum. Methods* 143: 241–43
76. Sabersky, A. P. 1973. *IEEE Trans. NS* 20: 638
77. Madey, J. M. J. 1971. *J. Appl. Phys.* 42: 1906–13
78. Elias, L. R., Fairbank, W. M., Madey, J. M. J., Schwettman, H. A., Smith, T. I. 1976. *Phys. Rev. Lett.* 36: 717–20
79. Deacon, D. A. G., Elias, L. R., Madey, J. M. J., Ramian, G. J., Schwettman, H. A., Smith, T. I. 1977. *Phys. Rev. Lett.* 38: 892–94
80. Colson, W. B. 1976. *Phys. Lett. A* 59: 187–90; Hopf, F. A., Meystre, P., Scully, M. O., Louisell, W. H. 1976. *Phys. Rev. Lett.* 37: 1342–45; Kwan, T., Dawson, J. M., Lin, A. T. 1977. *Phys. Fluids* 20: 581–88; Mayer, G. 1977. *Optics Commun.* 20: 200–4
81. Vinokurov, N. A., Skrinskii, A. N. 1977. *Inst. Nucl. Phys.* Novosibirsk, USSR. Preprint 77-59
82. Rehn, V., Jones, V. 1978. *Opt. Eng.* Sept./Oct.
83. Spiller, E. 1976. *Appl. Opt.* 15: 2333–38
84. McGowan, J. W., Rowe, E. M., eds. 1976. *Quebec Summer Work. Synchrotron Radiat. Facil., June 1976*
85. *Orsay Conf. Synchrotron Radiat. Instrum. Dev. Sept. 1977. Nucl. Instrum. Methods.* Vol. 152, June 1, 1978. 333 pp.
86. Brown, G., Winick, H., eds. 1978. *Proc. Work. X-Ray Instrum. Synchrotron Radiat. Res. SSRL Rep. 78/04*
87. Brown, F. C., Bachrach, R. A., Lien, N. 1978. *Nucl. Instrum. Methods.* 152: 73–79
88. Eberhardt, W., Kalkoffen, G., Kunz, C. 1977. See Ref. 3, pp. 112–19; *Nucl. Instrum. Methods.* 152: 81–83
89. Depautex, C., Thiry, P., Pinchaux, R., Pétroff, Y., Lepére, D., Passereau, G., Flamand, J. 1978. *Nucl. Instrum. Methods.* 152: 101–2
90. Bachrach, R. Z., Flodstrom, S. A., Schnopper, H. W., Delvaille, J. P. 1977. *Int. Conf. Vac. UV Radiat. Physics, 5th*, Montpellier, France
91. Källne, E., Schnopper, H. W., Delvaille, J. P., Van Speybroeck, L. P., Bachrach, R. Z. 1978. *Nucl. Instrum. Methods.* 152: 103–7
92. Pruett, C. 1976. See Ref. 84
93. Sagawa, T. 1976. See Ref. 204, pp. 125–46
94. Hastings, J. 1976. See Ref. 84
95. Beaumont, J. H., Hart, M. 1974. *J. Phys. E* 7: 823–29
96. Kincaid, B. 1975. *Synchrotron radiation studies of K-edge x-ray photoabsorption spectra: theory and experiment*, PhD thesis, Stanford Univ., *SSRL Rep. 75/03*. 151 pp.
97. Horowitz, P., Howell, J. 1972. *Science* 178: 608–11
98. Howell, J., Horowitz, P. 1975. *Nucl. Instrum. Methods* 125: 225–30
99. Brown, G. 1976. See Ref. 84
100. Arndt, U. W., Wonacott, A. J., eds. 1977. *The Rotation Method in Crystallography*, pp. 219–61. Amsterdam: North Holland
101. Cork, C., Fehr, D., Hamlin, R., Vernon, W., Xuong, Ng. H., Perez-Mendez, V. 1974. *J. Appl. Crystallogr.* 7: 319–23; Cork, D., Hamlin, R., Vernon, W.,

Xuong, Ng. H., Perez-Mendez, V. 1975. *Acta Crystallogr. A* 31:702–3
102. Charpak, G., Hajduk, A., Jeavons, A., Stubbs, R., Kahn, R. 1974. *Nucl. Instrum. Methods* 122:307–12; Charpak, G., Demierre, C., Kahn, R., Santiard, J. C., Sauli, F. 1977. *IEEE Trans. NS* 24(1):200–4
103. Ederer, D. L., Saloman, E. B., Ebner, S. C., Madden, R. P. 1975. *J. Res. Natl. Bur. Stand.* 79A:761
104. Schnopper, H. W., Van Speybroeck, L. P., Delvaille, J. P., Epstein, A., Källne, E., Bachrach, R. Z., Dijkstra, J. H., Lantward, L. 1977. *Appl. Opt.* 16:1088–91
105. Norem, J. H., Young, K. M. 1977. *Argonne Natl. Lab. ANL/FPP Tech. Memo. 79*
106. Stevenson, J. R., Ellis, H., Bartlett, R. 1973. *Appl. Opt.* 12:2884–89
107. James, R. W. 1962. *The Optical Principles of the Diffraction of X-rays*, Ch. 4. London: Bell
108. Ramaseshan, S., Abrahams, S. C., eds. 1975. *Anomalous Scattering.* Copenhagen: Munksgaard. 539 pp.
109. Phillips, J. C., Wlodawer, A., Goodfellow, J. M., Watenpaugh, K. D., Sieker, L. C., Jensen, L. H., Hodgson, K. O. 1977. *Acta Crystallogr. A* 33:445–55
110. Phillips, J. C., Wlodawer, A., Yevitz, M. M., Hodgson, K. O. 1976. *Proc. Natl. Acad. Sci. USA* 73:128–32
111. Guttman, L., 1956. *Solid State Phys.* 3:146–223
112. Keating, D. T. 1963. *J. Appl. Phys.* 34:923–25
113. Bienenstock, A. 1977. In *The Structure of Non-Crystalline Materials*, ed. P. H. Gaskell, pp. 5–11. London: Taylor & Francis. 262 pp.
114. Waseda, Y., Tamaki, S. 1975. *Philos. Mag.* 32:951; and 1976. *Z. Phys. B* 23
115. Fuoss, P., Bienenstock, A. 1976. *Bull. Am. Phys. Soc.* 22:405
116. Shevchik, N. 1977. *Philos. Mag.* 35:805–9, 1289–98
117. Bonse, U., Materlik, G. 1976. *Z. Phys. B* 24:189–91
118. Bonse, U., Materlik, G. 1975. See Ref. 108, pp. 107–9
119. Fukamachi, T., Hosoya, S., Kawamura, T., Hastings, J. 1977. *J. Appl. Crystallogr.* 10:321
120. Barrington Leigh, J., Rosenbaum, G. 1976. *Ann. Rev. Biophys. Bioeng.* 5:239–70
121. Bordas, J., Munro, I. H., Glazer, A. M. 1976. *Nature* 262:541–45
122. Blaurock, A. E. 1978. *Biochim. Biophys. Acta.* In press

123. Webb, N. G. 1976. *Rev. Sci. Instrum.* 47:545–47
124. Buras, B., Olsen, J. S., Gerward, L. 1976. *Nucl. Instrum. Methods* 135:193–95
125. Bordas, J., Glazer, A. M., Howard, C. J., Bourdillon, A. J. 1977. *Philos. Mag.* 35:311–23
126. Tuomi, T., Naukkarinen, K., Rabe, P. 1974. *Phys. Status Solidi A* 25:93
127. Hart, M. 1975. *J. Appl. Crystallogr.* 8:436
128. Bordas, J., Glazer, A. M., Hauser, H. 1975. *Philos. Mag.* 32:471
129. Tanner, B. K., Safa, M., Midgley, D., Bordas, J. 1976. *J. Magn. Mater.* 1:337; Tanner, B. K., Safa, M., Midgley, D. 1977. *J. Appl. Crystallogr.* 10:91
130. Miltat, J. 1978. *Nucl. Instrum. Methods.* 152:323–29
131. Steinberger, I. T., Bordas, J., Kalman, Z. H. 1977. *Philos. Mag.* 35:1257
132. Petroff, J. F., Sauvage, M. 1978. *J. Cryst. Growth.* In press
133. Ruby, S. L. 1974. *J. Phys. C* 6:209
134. Perlow, G. J., ed. 1977. *Workshop on New Directions in Mössbauer Spectroscopy, Argonne 1977.* Am. Inst. Phys. Conf. Proc. 38, p. 46, 91, 140
135. Hunter, S. H. 1977. *A structural study of amorphous copper-arsenic triselenide alloys using EXAFS*, PhD thesis, Stanford Univ.; *SSRL Rep.* 77/04. 136 pp.
136. Lytle, F. W. 1965. In *Physics of Non-Crystalline Solids*, ed. J. A. Prins, pp. 12–29. Amsterdam: North Holland
137. Lytle, F. W. 1966. In *Advances in X-ray Analysis*, ed. G. R. Mallett, M. J. Fay, W. M. Mueller, Vol. 9, pp. 398–409. New York: Plenum
138. Sayers, D. E., Lytle, F. W., Stern, E. A. 1970. In *Advances in X-ray Analysis*, ed. B. L. Henke, J. B. Newkirk, G. R. Mallett, Vol. 13, pp. 248–71. New York: Plenum
139. Sayers, D. E., Stern, E. A., Lytle, F. W. 1971. *Phys. Rev. Lett.* 27:1204–7
140. Ashley, C. A., Doniach, S. 1975. *Phys. Rev. B* 11:1279–88
141. Lytle, F. W., Sayers, D. E., Stern, E. A. 1977. *Phys. Rev. B* 15:2426–28
142. Knights, J. C., Hayes, T. M., Mikkelsen, J. C. Jr. 1977. *Phys. Rev. Lett.* 39:712–15
143. Lee, P. A., Beni, G. 1977. *Phys. Rev. B* 15:2862–83
144. Lee, P. A., Pendry, J. B. 1975. *Phys. Rev. B* 11:2795–2811
145. Rehr, J. J., Stern, E. A. 1976. *Phys. Rev. B* 14:4413–19
146. Kincaid, B. M., Eisenberger, P. 1975. *Phys. Rev. Lett.* 34:1361–64

147. Kincaid, B. M. 1975. See Ref. 96, Ch. 3
148. Citrin, P. H., Eisenberger, P., Kincaid, B. M. 1976. *Phys. Rev. Lett.* 36:1346–49
149. Lee, P. A., Teo, B. K., Simons, A. L. 1977. *J. Am. Chem. Soc.* 99:3856–57
150. Teo, B. K., Lee, P. A., Simons, A. L., Eisenberger, P., Kincaid, B. M. 1977. *J. Am. Chem. Soc.* 99:3854–55
151. Hayes, T. M., Sen, P. N., Hunter, S. H. 1976. *J. Phys. C* 4357
152. Hayes, T. M., Sen, P. N., Hunter, S. H. 1976. In *Structure and Excitations of Amorphous Solids*, ed. G. Lucovsky, F. L. Galeener, pp. 166–171. New York: AIP. 403 pp.
153. Stern, E. A., Sayers, D. E., Lytle, F. W. 1975. *Phys. Rev. B* 11:4836–46
154. Hunter, S. H., Bienenstock, A., Hayes, T. M. 1977. See Ref. 113, pp. 73–76
155. Cramer, S. P., Eccles, T. K., Kutzler, F., Hodgson, K. O., Doniach, S. 1976. *J. Am. Chem. Soc.* 98:8059–69
156. Watenpaugh, K. D., Sieker, L. C., Heriott, J. R., Jensen, L. H. 1973. *Acta Crystallogr. B* 29:943
157. Shulman, R. G., Eisenberger, P., Blumberg, W. E., Stombaugh, N. A. 1975. *Proc. Natl. Acad. Sci. USA* 72:4003–7
158. Bunker, B., Stern, E. A. 1977. *Biophys. J.* 19:253–64
159. Sayers, D. E., Stern, E. A., Herriot, J. R. 1976. *J. Chem. Phys.* 64:427
160. Shulman, R. G., Eisenberger, P., Teo, B. K., Kincaid, B. M., Brown, G. S. 1978. *J. Mol. Biol.* In press
161. Eisenberger, P., Shulman, R. G., Brown, G. S., Ogawa, S. 1976. *Proc. Natl. Acad. Sci. USA* 73:491–95
162. Sayers, D. E., Lytle, F. W., Weissbluth, M., Pianetta, P. 1975. *J. Chem. Phys.* 62:2514–15
163. Kincaid, B. M., Eisenberger, P., Hodgson, K. O., Doniach, S. 1975. *Proc. Natl. Acad. Sci. USA* 72:2340–42
164. Hu, V. W., Chan, S. I., Brown, G. S. 1977. *Proc. Natl. Acad. Sci. USA* 74:3821–25
165. Cramer, S. P., Hodgson, K. O., Stiefel, E. I., Newton, W. E. 1978. *J. Am. Chem. Soc.* 100:2748
166. Cramer, S. P., Hodgson, K. O., Gillum, W. O., Mortenson, L. E. 1978. *J. Am. Chem. Soc.* 100:11
167. Cramer, S. P., Gillum, W. O., Hodgson, K. O., Mortenson, L. E., Stiefel, E. I., Chisnell, J. R., Brill, W. J., Shah, V. K. 1978. *J. Am. Chem. Soc.* 100:3398–3407
168. Sayers, D. E., Lytle, F. W., Stern, E. A. 1972. *J. Non-Cryst. Solids* 8–10:401–7
169. Sayers, D. E., Lytle, F. W., Stern, E. A.

1974. In *Amorphous and Liquid Semiconductors*, ed. J. Stuke, W. Brenig, Vol. I, p. 403–12. London: Taylor & Francis. 735 pp.
170. Bienenstock, A., Mortyn, F., Narasimhan, S., Rowland, S. C. 1976. In *Frontiers in Materials Science—Distinguished Lectures*, ed. L. Murr, C. Stein, pp. 1–18. New York: Dekker. 590 pp.
171. Mott, N. F. 1967. *Adv. Phys.* 16:49–144
172. Hunter, S., Bienenstock, A. 1976. In *Structure and Properties of Non-Crystalline Semiconductors*, ed. B. T. Kolomiets, pp. 151–54. Leningrad: Ioffe Phys.-Tech. Inst. 546 pp.
173. Hunter, S. H., Bienenstock, A., Hayes, T. M. 1977. In *Amorphous and Liquid Semiconductors*, ed. W. E. Spear, pp. 78–82. Univ. Edinburgh. 891 pp.
174. Crozier, E. D., Lytle, F. W., Sayers, D. E., Stern, E. A. 1977. *Can. J. Chem.* 55:1968–74
175. Lytle, F. W., Sayers, D. E., Moore, E. B. Jr. 1974. *Appl. Phys. Lett.* 24:45–47
176. Bassi, I. W., Lytle, F. W., Parraveno, G. 1976. *J. Catal.* 42:139–47
177. Lytle, F. W., Via, G. H., Sinfelt, J. H. 1977. *J. Chem. Phys.* 67:3831–32
178. Reed, J., Eisenberger, P., Teo, B. K., Kincaid, B. M. 1977. *J. Am. Chem. Soc.* 99:5217–18
179. Reed, J., Eisenberger, P. 1977. *J. Chem. Soc. Chem Commun.* 18:628–30
180. Sinfelt, J. H., Via, G. H. Lytle, F. W. 1978. *J. Chem. Phys.* 68:2009
181. Lytle, F. W. 1976. *J. Catal.* 43:376–79
182. Stern, E. A., Sayers, D. E., Dash, J. G., Shechter, H., Bunker, B. 1977. *Phys. Rev. Lett.* 38:767–70
183. Stern, E. A. 1977. *J. Vac. Sci. Technol.* 14:461–65
184. Lee, P. A. 1976. *Phys. Rev. B* 13:5261–70
185. Citrin, P. H., Eisenberger, P., Hewitt, R. C., Schwartz, G. 1977. Presented at Ann. Stanford Synchrotron Lab. Users Group Meet., 4th, Stanford, CA
186. Jaklevic, J., Kirby, J. A., Klein, M. P., Robertson, A. S., Brown, G. S., Eisenberger, P. 1977. *Solid State Commun.* 23:679–82
187. Sparks, C. J., Raman, S., Yakel, H. L., Gentry, R. V., Krause, M. O. 1977. *Phys. Rev. Lett.* 38:205–8
188. Sparks, C. J., Raman, S., Ricci, E., Gentry, R. V., Krause, M. O. 1978. *Phys. Rev. Lett.* 40:507–11
189. Gentry, R. V. 1976. *Phys. Rev. Lett.* 37:11
190. Ilin, V. A., Kazakevich, G. M., Kulipanov, G. N., Mazalov, L. N., Matyushin, A. M., Skrinskii, A. N., Sheromov, M. A. 1977. *Nucl. Phys.*

Inst., Novosibirsk, USSR. Preprint 77-13
191. Spears, D. L., Smith, H. I. 1972. *Electron. Lett.* 8:102–4; *Solid State Technol.* 15:21–26
192. Spiller, E., Eastman, D. E., Feder, R., Grobman, W. D., Gudat, W., Topalian, J. 1976. *J. Appl. Phys.* 47:5450–59
193. Fay, B., Trotel, J., Petroff, Y., Pinchaux, R., Thiry, P. 1976. *Appl. Phys. Lett.* 29:370–72
194. Spiller, E., Feder, R. 1977. *Topics in Applied Physics*, ed. H. J. Queisser, Vol. 22, pp. 35–92. New York: Springer
195. Robinson, A. 1978. *Science* 199:411–13
196. Koronkevich, V. P., Kulipanov, G. N., Nalivaiko, V. I., Pindyurin, V. F., Skrinskii, A. N. 1977. *Nucl. Phys. Inst.*, Novosibirsk, USSR. Preprint 77-10
197. Sayre, D., Kirz, J., Feder, R., Kim, D. M., Spiller, E. 1977. *Science* 196: 1339–40
198. Kirz, J., Sayre, D., Dilger, J. 1977. See Ref. 197
199. Horowitz, P., Aronson, M., Grodzins, L., Ladd, W., Ryan, J., Merriam, G., Lechene, C. 1976. *Science* 194:1162–65
200. Niemann, B., Rudolf, D., Schmahl, G. 1976. *Appl. Opt.* 15:1883–84
201. Feder, R., Spiller, E., Topalian, J., Broers, A. N., Gudat, W., Panessa, B. J., Zadunaisky, Z. A. 1977. *Science* 197:259–60
202. Polack, F., Lowenthal, S., Petroff, Y., Farge, Y. 1977. *Appl. Phys. Lett.* 31:785–87
203. Haensel, R., Kunz, C. 1967. *Z. Angew. Phys.* 23:276
204. Mancini, A. N., Quercia, I. F., eds. 1976. *Int. Coll. Appl. Phys. Inst. Naz. Fis. Nucl. Course Synchrotron Radiation*, Vol. I, 481 pp., Vol. II, 84 pp.
205. Sokolov, A. A., Ternov, J. M. 1968. *Synchrotron Radiation*. Oxford: Pergamon. 207 pp. (from Russian)
206. Codling, K. 1973. *Rep. Progr. Phys.* 36:541
207. Brown, F. C. 1974. In *Solid State Physics*, ed. H. Ehrenreich, F. Seitz, D. Turnbull, 29:1–73. New York: Academic. 388 pp.
208. Koch, E. E., Haensel, R., Kunz, C., eds. 1974. *Vacuum Ultraviolet Radiation Physics*, Braunschweig: Vieweg/Pergamon. 832 pp.
209. Haensel, R. 1975. In *Festkörperprobleme, Advances in Solid State Physics*, ed. H. J. Queisser, 15:203–28. Braunschweig: Pergamon
210. Koch, E. E., Otto, A. 1976. *Int. J. Radiat. Phys. Chem.* 8:113
211. Kunz, C. 1976. In *Optical Properties of Solids—New Developments*, ed. B. O.

Seraphin, p. 473. Amsterdam: North-Holland
212. Sonntag, B. 1976. In *Rare Gas Solids*, ed. M. L. Klein, J. A. Venables, Vol. 2. New York: Academic
213. Koch, E. E., Kunz, C., Sonntag, B. 1977. *Phys. Rep.* 29:153–231
214. Spicer, W. E. 1978. In *Electron and Ion Spectroscopy of Solids*, pp. 54–89
215. Shirley, D. A., Stöhr, J., Wehner, P. S., Williams, R. S., Apai, G. 1978. *Phys. Scr.* 16:398–413
216. Lynch, D. W. 1976. See Ref. 204, pp. 26–85
217. Aspnes, D. A. See Ref. 204, pp. 286–97
218. Mahan, G. D. 1974. See Ref. 207, pp. 75–138
219. Cooper, J. W. 1964. *Phys. Rev. Lett.* 13:762–64
220. Fano, U., Cooper, J. W. 1968. *Rev. Mod. Phys.* 40:441–507
221. Wehner, P. S., Stöhr, J., Apai, G., McFeely, F. R., Williams, R. S., Shirley, D. A. 1976. *Phys. Rev. B* 14:2411–16
222. Stöhr, J., Apai, G., Wehner, P. S., McFeely, F. R., Williams, R. S., Shirley, D. A. 1967. *Phys. Rev. B* 14:5144–55
223. Lindau, I., Pianetta, P., Spicer, W. E. 1976. *Phys. Lett. A* 57:225–26
224. Lindau, I., Pianetta, P., Spicer, W. E. 1978. In *Proc. Int. Conf. Phys. X-ray Spectra, 1976*, ed. R. D. Deslattes, p. 78.
225. Eastman, D. E. 1974. See Ref. 208, p. 417
226. Lindau, I. 1976. See Ref. 204, pp. 319–71
227. Lindau, I., Spicer, W. E. 1974. *J. Electron. Spectrosc.* 3:409
228. Rowe, E. M., ed. 1977. *Notes on the Annual Synchrotron Radiation Users Group Conference*, The Synchrotron Radiation Center, Stoughton, WI
229. Brown, G. S., ed. 1977. *Ann. Stanford Synchrotron Radiat. Lab Users Group Meet., 4th, SSRL Rep. 77/11*, Stanford. 75 pp.
230. López-Delgado, R., Tramer, A., Munro, I. H. 1974. *Chem. Phys.* 5:72
231. Lindquist, L., López-Delgado, R., Martin, M. M., Tramer, A. 1974. *Opt. Commun.* 10:283–87
232. Benoist D'Azy, O., López-Delgado, R., Tramer, A. 1975. *Chem. Phys.* 9:327
233. López-Delgado, R., Miehé, J. A., Sipp, B. 1976. *SSRL Rep. 76/04*. 10 pp.
234. López-Delgado, R. 1976. See Ref. 204, pp. 63–124
235. Monahan, K. M., Rehn, V. 1978. *Nucl. Instrum. Methods.* 152:255–59
236. Matthias, E., Rosenberg, R. A., Poliakoff, E. D., White, M. G., Lee,

S.-T., Shirley, D. A. 1977. *Chem. Phys. Lett.* 52:239
237. Matthias, E., White, M. G., Poliakoff, E. D., Rosenberg, R. A., Lee, S.-T., Shirley, D. A. 1978. *Chem. Phys. Lett.* 54(1):30–34
238. Monahan, K. M., Rehn, V., Matthias, E., Poliakoff, E. 1977. *J. Chem. Phys.* 67:1784–86
239. Monahan, K. M., Rehn, V. 1977. *SSRL Rep. 77/10*
240. Monahan, K. M. 1978. *Bull. Am. Phys. Soc.* 23:61
241. Hahn, U., Jordan, B., Schwentner, N. 1977. *Extended Abstr. Int. Conf. Vac. Ultraviolet Radiat. Phys., 5th,* 3:202, Montpellier, France
242. Ice, G. E., Chen, M. H., Crasemann, B. 1978. *Phys. Rev. A* 17:650
243. Bentley, J. J., Stewart, R. F. 1975. *J. Chem. Phys.* 62:875–78
244. Csonka, P. 1977. *SSRL Rep. 77/03*; Csonka, P., Crasemann, B. 1975. *Phys. Rev. A* 12:611–14; Csonka, P. 1977. *Phys. Rev. A* 13:405–10; Univ. Ore. *Preprint ITS NT 065/77*
245. Kondratenko, A. M., Skrinskii, A. N.

1977. *Opt. Spectrosc.* 42:189–92
246. Kikuta, S., Aoki, S., Kosaki, S., Kohra, K. 1972. *Opt. Commun.* 5:86–89
247. Aoki, S., Ichihara, Y., Kikuta, S. 1972. *Jpn. J. Appl. Phys.* 11:1857
248. Aoki, S., Kikuta, S. 1974. *Jpn. J. Appl. Phys.* 13:1385–92
249. Csonka, P. 1977. Univ. Ore. Inst. Theor. Stud. *Preprint N.T.* 066/77
250. Csonka, P. 1978 *Part. Accel.* 8:161
251. Richter, B. 1976. *Nucl. Instrum. Methods* 136:47–60
252. Jacobson, B. 1964. *Am. J. Roentgenol. Radium Ther. Nucl. Med.* 91:202
253. Atkins, H. L., Hauser, W., Kraner, H. W. 1972. *Am. J. Roentgenol. Radium Ther. Nucl. Med.* 114:176
254. Mistretta, C. A., Ort, M. G., Kelcz, F., Cameron, J. R., Siedband, M. P., Crummy, A. B. 1973. *Invest. Radiat.* 8:402
255. Swindell, W., Barrett, H. H. 1977. *Phys. Today* 30:32–41
256. Hatzakis, M. 1971. *Appl. Phys. Lett.* 18:7–10
257. Plummer, E. W., Gustafsson, T. 1977. *Science* 198:165–70

Ann. Rev. Nucl. Part. Sci. 1978. 28:115–59
Copyright © 1978 by Annual Reviews Inc. All rights reserved

EXPERIMENTAL NUCLEAR ASTROPHYSICS[1]   ✕5593

C. Rolfs and H. P. Trautvetter

Institut für Kernphysik, Universität Münster, Münster, West Germany

CONTENTS

1   INTRODUCTION

To the casual observer the stars in the sky appear to be changeless. There are, however, dozens of stars that fade and brighten and whose behavior is clearly visible to the unaided eye. In fact, all stars evolve and have a finite lifetime, because their supply of energy is limited. Stars are born out of

[1] The survey of literature for this review was concluded in December 1977.

115

0066-4243/78/1201-0115$01.00

interstellar gas and dust (mainly hydrogen) and shine for long periods, during which the necessary energy is generated by a series of nuclear fusion reactions (1, 2). If the nuclear fuel is reduced to a critical limit, a star will die. The death throes of a star range in violence from the explosion of supernovae, in which stars nearly obliterate themselves and in which the debris can reach velocities of 15,000 km/sec, through the lesser outbursts of ordinary novae to the flares of more commonplace stars such as the sun. Nuclear reactions do not only provide the energy by which stars function but, as stars evolve (especially during the final stages), these reactions also provide the rich variety of the heavy nuclei out of the lighter ones (1, 2). The stellar debris therefore supply the interstellar medium with new elements, heavier than hydrogen. This transmuted material is mixed in interstellar space with the uncondensed hydrogen gas in the galaxy, and this mixed interstellar gas is available for condensation into second- and later-generation stars. This transfer of material between stars and the interstellar gas and dust is illustrated in Figure 1. It is the present understanding (1, 2), that most elemental nucleosynthesis (at least for elements $A \geq 12$) has taken place (and is still taking place) in successive generations of stars. Our solar system formed from such interstellar gas almost five billion years ago and inherited its heavy elements from the ashes of many previous stars.

The distribution of the elements and their isotopes, which can be observed on the earth and moon and in meteorites, as well as inferred spectroscopically from the solar and stellar atmospheres, lead to the so-called universal abundance curve (Figure 2). Although this curve may

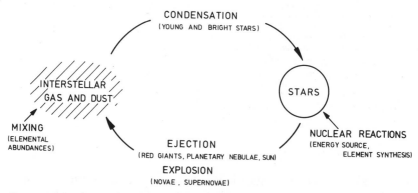

Figure 1   Transfer of material between stars and interstellar gas and dust. Synthesis of the elements occurs in the center of stars during their static burning stages and the supernovae, which end their lives. The space between the stars is the site of the mixing that results in the average elemental abundance distribution over fairly large astronomical regions. Mechanisms for the transfer as observed astronomically (1, 2) are indicated.

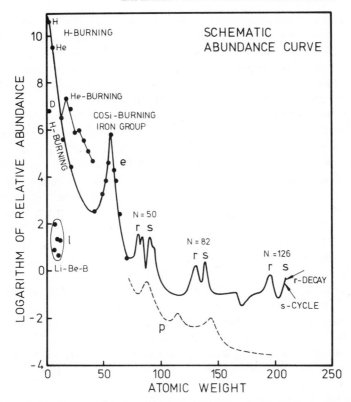

Figure 2   Schematic curve of atomic abundances as a function of atomic weight (adapted from 1). The nuclear processes involved in the synthesis of this abundance distribution are labeled and discussed in the text. The double peaks near $N = 50$, 82, and 126 are due to effects of closed nuclear shells.

not be universal, it at least represents the distribution for the great bulk of matter on which observations have been made (1–4). Since this abundance curve is a summary of all past events and activities in the galaxy and the universe, a theoretical reproduction of this curve will provide a picture of the history of the universe.

The close connection and interplay between astronomy and nuclear physics in this enterprise has led to the development of the interdisciplinary field of nuclear astrophysics, which tries to answer questions such as: why, where, when, and how has the abundance distribution been synthesized? These questions on the origin of the chemical elements and their isotopes are the modern expression of one of the most ancient problems in science. The attempt to answer these questions is comparable to the problem of solving a complex puzzle-game, where two major sets of building blocks must be provided: (a) The element-synthesizing nuclear reactions represent clearly one set of building blocks. These processes can be studied to a large extent in nuclear physics laboratories. (b) Theoretical studies (models) of stellar, galactic, and universal evolution represent the second set of building blocks. These models should determine the course of elemental nucleosynthesis.

The currently favored cosmological model of the universe is the "big-bang" hypothesis (5), i.e. the present state of the universe is the result of expansion from an extremely dense and hot singular origin. The most compelling evidence for this model is the observation (6) of the $2.7°K$ blackbody radiation, which remains from this hot singularity. In this model some nucleosynthesis occurs at the appropriate temperature during the expansion and continues until the rapidly falling temperature stops nuclear reactions. Calculations (7, 8) showed, however, that no significant nucleosynthesis past ^4He was possible due to the instability gaps at mass 5 and mass 8 as well as the constraints imposed by the present universal density and temperature.

Within about 10^9 years after the big bang, stars form (1, 2) and further nucleosynthetic activities (1) can then take place in these stars (i.e. "stellar nucleosynthesis"). In particular, the synthesis of the heavier elements $(A > 4)$ can now occur in the hot and dense cores of the stars. An examination of the elemental abundance curve and Coulomb barrier penetration considerations led Burbidge et al (1) to postulate a series of nucleosynthetic processes, which are, with slight modifications:

1. Hydrogen burning. The star first collapses gravitationally until its core heats sufficiently to ignite the lightest and most abundant nuclear fuel, namely hydrogen, where four hydrogen nuclei are converted into ^4He by a set of nuclear reactions (p-p chain and/or CNO cycles, see Section 4). The conversion of mass into energy in these fusion reactions keeps the stellar interior hot enough to prevent further gravitational collapse, i.e. the star is stabilized by nuclear reactions. This stability lasts (1, 2) for the relatively long periods necessary to consume the abundant hydrogen fuel (main sequence stage of stellar evolution). The hydrogen burning in the sun lasted for 4.5 billion years and will continue as long in the future. Stars can be considered therefore as gravitationally stabilized fusion reactors (1, 2). Gravitational energy is needed to raise the temperature to the ignition point for nuclear burning processes, but it cannot serve, of course, as the source of energy in a stabilized star. The nuclear reactions themselves do this.

While both sets of hydrogen-burning reactions produce ^4He ashes, the most important additional feature of the CNO cycles is the conversion of ^{12}C and ^{16}O (the dominant ashes from helium burning in earlier stars) mostly into ^{14}N. This is the origin of ^{14}N, which has been of critical importance in mankind's development (1).

2. Helium burning. When the hydrogen in the core of a star is exhausted, gravitational collapse will again increase the temperature (red giant stage of stellar evolution) until the helium ashes can be ignited. In order to bridge the mass 5 and mass 8 instability gaps, helium must burn (1, 2) first via a three-particle reaction 3^4He $\rightarrow$ ^{12}C (Section 5). The conversion of ^{4}He into heavier elements such as ^{12}C, ^{16}O, and ^{20}Ne is one of the most important features of helium burning, because many conclusions regarding stellar evolution and explosive nucleosynthesis hinge on these final abundances. If ^{14}N is present, helium burning synthesizes isotopes of oxygen and neon. Helium burning is also considered as the probable site of the s-process (see below).

It is perhaps interesting to note that in getting from ^{4}He to ^{12}C, the elements Li, Be, and B were skipped. The skipping of these elements by main-line stellar nucleosynthesis is quite consistent with the fact that these elements (l-elements circled in Figure 2) are much lower in abundance than their neighbors.

3. Carbon and oxygen burning. After helium exhaustion, the burning of the next available nuclear fuels, ^{12}C and ^{16}O, produces nuclides in the mass range $A = 16$–28 (9).

4. Silicon burning. The next nucleus regulating further abundance changes is ^{28}Si. Some of these nuclei are broken up by photodisintegration into alphas, and these alphas are captured by the remaining ^{28}Si nuclei to build heavier elements (10). In these equilibrium processes between synthesis and disintegration, matter is transmuted to the elements with the highest nuclear stability, i.e. the iron peak is produced (Figure 2). The iron core of a star will then collapse without being stopped by nuclear energy generation. The final stages of this collapse and the subsequent supernova (if it occurs) are of current interest in stellar evolution calculations and explosive nucleosynthesis (11).

5. The s- and r-processes. The fairly abrupt change to a small slope in the abundance curve above the iron peak (Figure 2) indicates that these elements are not synthesized by charged-particle reactions. It has been suggested therefore that the trans-iron elements are produced by neutron capture. These nuclei can be divided into two roughly equal groups—those due to the s-process and those due to the r-process (12). A third process called the p-process (Figure 2) produces a few very rare *proton*-rich nuclei (13). The term s-process is a mnemonic for *slow* neutron capture, slow being the rate relative to intervening β-decay times. The time scale for each neutron capture ranges from 10–10^5 years. Similarly, the r-process is a *rapid* neutron-capture process, where the neutrons are captured much

faster ($\sim$0.01–10 sec) than the appropriate β-decay rates. A possible site for this process is in the violent final stages of stellar evolution (11, 14).

6. The *l*-process. The production of the reactive *light* elements (*l* in Figure 2) is outside the main line of stellar nucleosynthesis (8, see also Section 7).

Even if the bulk of stellar nucleosynthesis should prove to be a product of the last events in a star's history [explosive nucleosynthesis (11)], studies of the earlier nucleosynthesis processes still remain extremely important. These processes represent the energy source for the greater part of the star's life and also provide the starting conditions for the final phases of stellar evolution. Furthermore, they will occur in appropriate concentric layers of the star during its final explosive death throes.

Much of the present framework of nuclear astrophysics is presented in the classical paper by Burbidge et al (1). The state of affairs up to 1970 of the s- and r-processes was reviewed by Allen et al (15) and that of the nucleosynthesis processes induced by charged particles was presented in the review by Barnes (16). The situation in this latter field is best described by the final remarks in Barnes' article (p. 97): "we have raised more questions than we have settled; conclusions are premature, to say the least." Major efforts have continued in the meantime to improve the theories of stellar and universal evolution as well as the knowledge of the nuclear reaction rates, the two major sets of building blocks requisite in the solution of the abundance puzzle-game.

The present review is restricted to charged-particle nucleosynthesis processes and concentrates on recent nuclear experiments of importance to astrophysics. The examples presented should illustrate both the spirit of the research problems and some established means of overcoming technical obstacles. It focuses attention also on those processes where new knowledge of nuclear physics may have important astrophysical consequences.

2   CHARGED-PARTICLE REACTION RATES UNDER STELLAR CONDITIONS

To determine what kind of experimental data must be obtained in the laboratory, one must first examine the temperature dependence expected for various types of nuclear reactions. For the most common case, in which two particles in the initial channel (1 and 2) form two particles in the final channel (3 and 4) and for stellar temperatures below 10^{10} °K, the reaction rate P_{12} is determined (1, 2) by

$$P_{12} = (n_1 n_2)(1 + \delta_{12})^{-1} \langle \sigma v \rangle, \qquad 1.$$

where n_1 and n_2 are the number densities of nuclei 1 and 2 and δ_{12} is the Kronecker symbol. The quantity

$$\langle \sigma v \rangle = \frac{(8/\pi)^{\frac{1}{2}}}{M^{\frac{1}{2}}(kT)^{\frac{3}{2}}} \int \sigma(E)E \exp\left(-E/kT\right) dE \qquad 2.$$

is the product of the energy-dependent reaction cross section $\sigma(E)$ and the velocity, averaged over the Maxwell-Boltzmann distribution of relative velocities. M is the reduced mass of particles 1 and 2.

The problem of calculating important quantities such as the nuclear energy production rate or the lifetime of a given nucleus under stellar conditions resolves itself (1, 2) into determining the cross-section factor $\langle \sigma v \rangle$. Since many nuclear reactions are involved in the nuclear networks (1, 2, 10–14) of stellar evolution calculations, these integrals $\langle \sigma v \rangle$ have to be evaluated and the results represented in manageable analytic expressions. The best method for evaluating $\langle \sigma v \rangle$ depends on the type of energy dependence exhibited by $\sigma(E)$, and a few commonly occurring cases are considered below. The more complex reaction types and their treatments in the $\langle \sigma v \rangle$ evaluation are described in References 1, 2, 14–17.

Analytic expressions for many astrophysically relevant nuclear reactions are given in the tabulations of Fowler et al (17).

2.1 Nonresonant Reactions

For nonresonant reactions the steepest energy dependence in $\sigma(E)$ is contained in the penetration factor for the Coulomb and angular momentum barrier. For low incident energies and for s-partial waves, this factor is approximately proportional to $\exp\left(-2\pi\eta\right)$, where η is the Sommerfeld parameter given (1, 2) by

$$2\pi\eta = 2\pi Z_1 Z_2 e^2/\hbar v = b/E^{\frac{1}{2}}. \qquad 3.$$

In this expression E is the interaction energy of the two particles, Z_1 and Z_2 their nuclear charges, and v their relative velocity. It is therefore convenient to factor out this well-known energy dependence, as well as an additional factor of $1/E$ (from the λ^2 factor that always appears in nuclear cross sections). The cross section can thus be written (1, 2) as

$$\sigma(E) = \frac{S(E)}{E} \exp\left(-2\pi\eta\right), \qquad 4.$$

where the remaining energy dependence from the effects of intrinsic nuclear properties is now absorbed in $S(E)$, the so-called astrophysical S-factor. For nonresonant reactions this S-factor is expected to have only a weak energy dependence. Inserting Equation 4 into Equation 2 yields

$$\langle \sigma v \rangle = \frac{(8/\pi)^{\frac{1}{2}}}{M^{\frac{1}{2}}(kT)^{\frac{3}{2}}} \int S(E) \exp\left(-\frac{E}{kT} - \frac{b}{E^{\frac{1}{2}}}\right) dE. \qquad 5.$$

The energy dependence of the integrand is determined predominantly by the product of the Maxwell-Boltzmann factor and the penetration factor (Figure 3). The nuclear reactions take place in stars at the relatively small energy region of $E_0 \pm \frac{1}{2}\Delta E_0$. For the case of $S(E) = S_0 = $ constant (for other cases, see 1, 2, 16, 17), these quantities are given by the expressions

$$E_0 = [\pi e^2 Z_1 Z_2 kT(Mc^2/2)^{\frac{1}{2}}/\hbar c]^{\frac{2}{3}} \qquad \text{and} \qquad 6.$$

$$\Delta E_0 = 4(E_0 kT/3)^{\frac{1}{2}}, \qquad 7.$$

and the cross section factor is given by

$$\langle \sigma v \rangle = (2/M)^{\frac{1}{2}} \Delta E_0 (kT)^{-\frac{3}{2}} S_0 \exp(-3E_0/kT). \qquad 8.$$

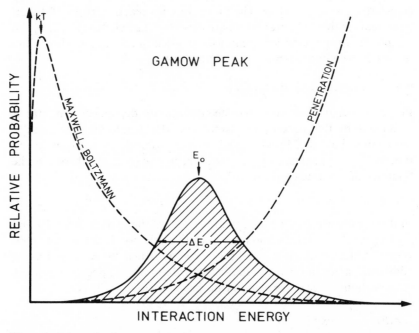

Figure 3 Schematic diagram of the Maxwell-Boltzmann distribution and the barrier penetration factor as a function of interaction energy. The product of these two quantities results in a peak (Gamow peak), which represents the most effective stellar energy region for nuclear burning processes. This energy region increases with nuclear charge and stellar temperature. The stellar rate of a nuclear reaction is proportional to the area of this peak.

2.2   Isolated Narrow Resonances

For an isolated narrow resonance, the energy dependence of the cross section is given by the well-known Breit-Wigner expression in the usual notation (1, 2):

$$\sigma(E) = \pi \lambdabar^2 \omega \frac{\Gamma_{12}\,\Gamma_{34}}{(E-E_R)^2 + \Gamma^2/4}. \qquad 9.$$

For the frequently occurring case $\Gamma \ll \Delta E_0$, Equation 2 can be integrated immediately to yield (1, 2, 16)

$$\langle \sigma v \rangle = (2\pi/MkT)^{\frac{3}{2}}\hbar^2(\omega\gamma)\exp(-E_R/kT). \qquad 10.$$

The quantity $\omega\gamma = \omega\Gamma_{12}\Gamma_{34}/\Gamma$ is called the resonance strength and can be determined experimentally (2, 16) from the thick-target yield. If several narrow resonances occur within ΔE_0, their contributions to $\langle \sigma v \rangle$ are simply summed (16).

2.3   Broad Overlapping Resonances

For broad resonances it is necessary to include the energy dependence of the partial widths and the total width in the Breit-Wigner expression for the determination of the $S(E)$-factor curve. The energy dependence of a width is given by the well-known relation (2, 33)

$$\Gamma_l(E) = (2\hbar^2/\lambdabar M R_0)\,P_l(E,R_0)\,\Theta_l^2, \qquad 11.$$

where l represents the orbital angular momentum of the channel, R_0 is the nuclear radius, Θ_l^2 is the energy-independent reduced particle width (in the Wigner limit, $\Theta_l^2 = 1$), and $P_l(E,R_0)$ is the well-known penetrability function (2). The energy dependence of the width of a γ-ray transition of multipolarity L is given by $\Gamma_\gamma(E) \propto E_\gamma^{2L+1}$ (2, 16, 17). In cases where there are overlapping resonances of the same spin and parity, interference effects between the resonant amplitudes occur in the total cross section. It is therefore necessary, when integrating Equation 2 for broad resonances, to know whether the interference is constructive or destructive.

3   EXPERIMENTAL PROCEDURES AND SOME PRACTICAL CONSIDERATIONS

The cross sections of the nuclear processes involved in hydrostatic stellar burning are very small at the relevant stellar energies, and they cannot be measured directly in the laboratory using any currently known techniques. Rather, the usual procedure (1, 2, 16), is to study nuclear reactions over a wide, yet higher range of beam energies, starting with the lowest practical energy. These low-energy cross sections are among the smallest measured in the nuclear laboratory and often require long integration times with painstaking attention to background. Such data are then extrapolated down into the stellar energy region with the guidance of theoretical and other considerations. In the Kellogg Radiation Laboratory at the California Institute of Technology (CIT) much of the pioneering work in this research field has been carried out under the guidance of W. A. Fowler and his colleagues. It should be pointed out, however, that with the high beam current accelerators and the improved target- and detector-technologies available today, direct measurements of the reaction cross sections can be extended to extremely low energies, whereby the uncertainties in the extrapolation procedures can be reduced significantly.

An illustrative example of the above procedure is the reaction $^{18}O(p,\alpha_o)^{15}N$, which is one of the reactions involved in the CNO cycles (Section 4.2). This reaction has been studied by the Münster group (19) over a wide range of beam energies (Figure 4). The cross section could be studied over ten orders of magnitude from $\sigma = 70$ pb at the lowest point of measurements ($E_p = 72$ keV). Solid-state targets as well as a window-less and recirculating gas target system and detectors in close geometry were employed. In the measurements at low beam energies, proton beam currents of up to 300 μA were used. The gross energy dependence in the data is largely due to the effects of the Coulomb barrier, as can be seen from the conversion of the data into the $S(E)$-factor: this quantity varies more gently with energy. A reliable estimate of the stellar rates at still lower beam energies can be obtained by extrapolating $S(E)$, as long as the reaction mechanism does not change over the small energy region of $E_p \leq 72$ keV. The $^{18}O(p,\alpha_o)^{15}N$ reaction is a good example of the danger of extrapolating high-energy data over a wide energy gap far down to stellar energies. The previously available data at $E_p \geq 500$ keV indicated a constructive (destructive) interference between the two broad $J^\pi = \frac{1}{2}^+$ resonances at $E_R = 680$ and 846 keV at beam energies between (outside) the two resonance energies, which led (20) to an extrapolated value of $S(0) = 1600$ keV-b. However, the new data below $E_p = 500$ keV show—aside from the new and narrow resonances—a nonresonant yield, which is not decreasing nearly as fast as expected (20) from the high-energy data. The slower decrease in $S(E)$ together with the pronounced anisotropic angular distributions (asymmetric around 90°) indicate clearly the presence of another nonresonant reaction mechanism at low beam energies. The observed anisotropies (19) are also in contrast to the general assumption that at energies far below the Coulomb barrier only s-partial waves in the incoming channel (and therefore isotropic angular distributions) dominate the picture. Clearly, the interpretation of these new data requires higher partial waves. It is not understood yet (19) whether the observed nonresonant features are due to interference effects of broad and distant resonance amplitudes of different J^π values or whether a direct (p,α) reaction mechanism (e.g. "heavy-particle" pick-up) is present here. Experimental guidance to the solution of this question requires additional (p,α) studies on other light-target nuclei at energies far below the Coulomb barrier. At least one such experiment is already under way.

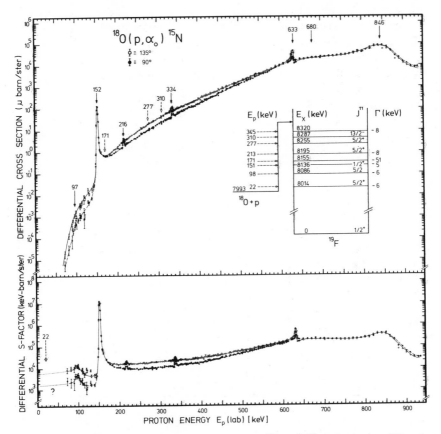

Figure 4 Differential cross sections and S-factor for $^{18}O(p,\alpha_0)^{15}N$, obtained at 90° and 135° and at $E_p = 72$–930 keV. The measurements (19) were carried out at accelerators in Münster, CIT, Mainz, and Stuttgart. No data were available previously below $E_p = 500$ keV. Four of the seven expected resonances (*insert*) at low energies could be found. The expected $E_p = 22$-keV resonance near the proton threshold must be investigated by other means (Sections 3 and 4.2). The lines through the data points are to guide the eye.

If the observed data for $^{18}O(p,\alpha_o)^{15}N$ are reproduced theoretically over the entire energy region, the extrapolation into the remaining small energy gap should give reliable stellar reaction rates. Such extrapolated rates represent, however, only lower limits, if there exist low-energy resonances near the particle threshold, which can increase significantly the extrapolated rates (Figure 5). These resonances will usually be very narrow (unless they involve a neutron channel or a high-energy charged-particle channel), but their existence cannot be detected in the laboratory as an anomaly even at the lowest energy points of the direct yield measurements. Furthermore, since at the low energies $\Gamma_{12} \ll \Gamma \sim \Gamma_{34}$, the resonance strength is given by $(\omega\gamma) \sim \omega\Gamma_{12}$, and this strength decreases rapidly at low energies because of the barrier penetration factor contained in Γ_{12} (Equation 11). Thus, these low-energy resonances are inaccessible to the experimenter by a direct strength measurement even at the energy of the resonance. In the search for such potential stellar resonances, one must

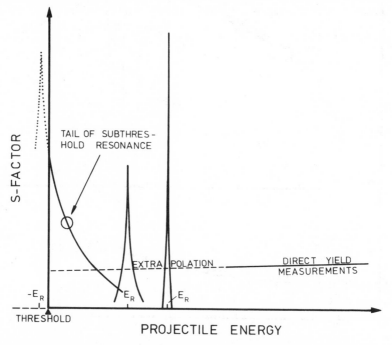

Figure 5 Schematic diagram illustrating the problem of low-energy resonances in the determination of stellar reaction rates. Possible resonances below the lowest energy point of direct yield measurements can increase significantly the rates extrapolated from the high-energy data alone. Also indicated is a compound state below the particle threshold, which could as a subthreshold resonance also increase the stellar reaction rates through its high-energy tail.

investigate the number of compound states near the particle threshold with other nuclear reactions. In order to prevent a biased or incomplete selection of these states in the compound nucleus, they should be studied with a set of different nuclear reactions. If the number of states has been established, their contributions to the stellar rate can be calculated with the use of Equation 10, provided the resonance energy, spin, and widths are known. Often, only a single width needs to be determined. For example, in (p,γ) reactions it frequently happens that these are the only two channels open and that because of the effects of the Coulomb barrier at low energies, $\Gamma_p \ll \Gamma_\gamma \approx \Gamma$. In this case, the stellar rate depends only on the proton width i.e. $(\omega\gamma) \sim \omega\Gamma_p$. The barrier effects on Γ_p are computable (Equation 11), so that the major data needed to calculate the effects of low-energy resonances are in these cases the resonance energy, the spin, and the reduced proton width Θ_p^2. Measurements of these quantities frequently demand nuclear studies that bear only subtle connections to the stellar reaction of interest. In the case of the ^{18}O(p,α)^{15}N reaction, studies with other reactions revealed (21) a potential resonance at $E_R = 22$ keV having a total width of $\Gamma \sim 6$ keV (Figure 4). Although a third channel, the γ channel, is open in this reaction, its influence can be neglected in the calculation of the (p,α) rate, since $\Gamma_\gamma \lesssim$ few eV and consequently $\Gamma_\alpha \approx \Gamma$. The (p,$\alpha$) rate is therefore determined entirely by the proton width $\omega\gamma = \omega\Gamma_p\Gamma_\alpha/\Gamma \sim \Gamma_p$. Studies of the direct capture (DC) process into this state via ^{18}O(p,γ)^{19}F reveal (22) a finite reduced proton width Θ_p^2, which implies a definite increase in the extrapolated rates from the high-energy data.

It is obvious that precise measurements of the particle threshold and the excitation energies of states above the threshold are needed to clarify their astrophysical importance. But the exact energy is also important if the state is located in the negative-energy domain (Figure 5), because such a state may—due to its natural width—provide a resonant wing in the positive-energy domain (subthreshold resonance). In a few cases where direct measurements of $\omega\gamma$ have not been feasible either because of unstable target nuclei or because of a location of a state near the threshold, it has been possible to measure the important quantities such as the reduced particle widths, the ratios Γ_{12}/Γ or Γ_{34}/Γ, and the total width Γ (or $\tau = \hbar/\Gamma$) by various novel experimental techniques. Subsequently, their stellar rates can be calculated from these quantities. There are, however, many energy levels for which neither kind of measurement has been accomplished, and these potentially important levels constitute a major challenge to the experimenter. Further illustrating examples of such cases are discussed in the following sections.

Because of the need for absolute total cross section measurements, particularly at energies far below the Coulomb barrier, special requirements and precautions must be employed:

1. The use of high beam currents facilitates the measurement of small cross sections. Improvements in accelerator technology allow traditional ion sources as well as several new types of heavy-ion sources [e.g. the sputtering source of Middleton (24)] to provide extremely high beam currents (e.g. hydrogen and helium beams of up to 4-mA and carbon beams of up to 15-μA particle currents) (23, 24, 33). A beam-handling system particularly designed for such extreme currents is described by Hammer et al (25).

2. The energy resolution, stability, and absolute value of the beam energy must be known rather accurately, particularly at low energies due to the steep drop in cross section (23, 24, 26, 33).

3. Special attention must also be focused toward a "clean" vacuum system, especially near the target (23, 27, 33), in order to minimize any deposition of contaminants (like carbon) on the targets. Such a deposition can not only produce background radiation but will degrade, in an uncontrolled manner, the energy of the incident beam before it can strike the true target surface.

4. In fabricating solid-state targets (23, 28, 33), special considerations must be given for possible contaminants in the target and/or the backing material, particularly those with smaller nuclear charges than the target nucleus of interest. Contaminant reactions can otherwise extinguish completely the signal-to-noise ratio. These targets must also be able to withstand the high beam loads usually employed without significant deterioration. Many of these problems can be reduced appreciably by the use of a windowless and differentially pumped gas target. When the gas can also be recirculated, studies on rare and expensive isotopes (like ^{21}Ne) can be carried out (29, 30). A novel type of cryo-pumped gas target has been described recently (31).

5. For the determination of absolute cross sections, it is necessary to have an accurate measurement of the number of incident particles. This is normally obtained by measuring the total accumulated charge. Special precautions must be taken (23, 33) for the effects of secondary electrons as well as for changes in the charge state of the incident ion (on the way from the energy analyzer to the target). In the case of gas targets, balanced calorimeters on the beam stop (23) or measurements of the Rutherford scattering of the beam from the target nuclei (30, 32) have been employed as "charge-measurement" devices.

6. In order to improve the signal-to-noise ratio in measurements of small cross sections, detectors must be employed with good energy resolution and high efficiency. A review on these detectors has been given recently (33).

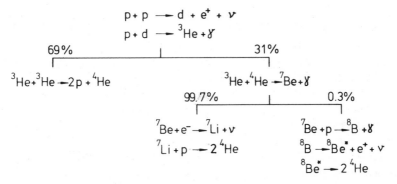

Figure 6 Hydrogen burning in the p-p chain (1, 2). There are three possible routes that lead to the conversion of $4p \rightarrow {}^4$He. The numbers shown (38) indicate the estimates of the relative importance of each branch in finishing the chain.

4 HYDROGEN BURNING

4.1 *The Proton-Proton Chain and the Solar Neutrino Problem*

The major energy production in first-generation stars, as well as later-generation stars (like our sun) with $M \leq M_\odot$, proceeds through a sequence of nuclear reactions, called the proton-proton (p-p) chain (Figure 6). The first reaction in the chain is by far the slowest because it involves the positron decay of an unbound state of two protons. Thus, the overall rate of the energy production is mainly determined by this reaction (1, 2).

The sun, the nearest and best studied of all stellar objects, is believed (2) to burn hydrogen in its core predominantly through this p-p chain. The neutrinos produced in this chain easily escape the sun and reach the earth within eight minutes. Thus, the detection of these solar neutrinos offers a unique possibility of "looking" into the central hot core of the sun. The neutrino-detector of Davis (34) is particularly sensitive to the high-energy neutrinos from the decay of ^{8}B, the weakest ending of the p-p chain. Since the flux of these neutrinos is very sensitive to the central temperature of the sun, this detector represents a solar thermometer (35). The first results of the still ongoing Davis experiment have set an upper limit (34) of 1.0 SNU[2] on the neutrino flux from the sun, which is in sharp disagreement with the theoretical prediction of 9 SNU, calculated (36) on the basis of accepted reaction cross sections and solar model parameters. This disagreement, the so-called solar neutrino problem, has

___

[2] Mnemonic for Solar Neutrino Unit, where 1 SNU = 10^{-36} captures per target atom per second.

caused considerable consternation among nuclear astrophysicists and has led to dramatic possible explanations (35) such as: (a) neutrinos may decay in flight, (b) the temperature in the solar interior may have decreased substantially in the past 10^7 years, (c) the sun may oscillate in shape, and (d) the sun contains a black hole in its center.

In view of the solar neutrino problem, there have been attempts to find ways of calibrating the efficiency of the neutrino detector. A sufficient accelerator production of the ^{8}B neutrinos through the reactions ^{9}Be(p,2n)^{8}B or ^{6}Li(^{3}He,n)^{8}B requires (37) projectiles of energy $\geqq 25$ MeV and beam currents $\geqq 30$ mA. These requirements cannot be fulfilled with available accelerators. Alvarez (37) suggested alternatively a reactor production of a ^{65}Zn source, which could provide a sufficiently intense neutrino flux to test the Davis detector.

Because the solar neutrino problem may also involve nuclear physics through the reaction cross sections, the data pertaining to the p-p chain have been reexamined (38) for possible errors. Further experimental checks[3] on these reactions are very valuable, especially if different approaches or novel techniques are employed. Comments on some recent developments are given here.

4.1.1 THE PROTON-PROTON REACTION   Because of the weak interaction involved in the $p+p \rightarrow d+e^+ +\nu$ reaction, its cross section has to be derived from theoretical calculations, which involve (2) well-established aspects of nuclear scattering and conventional β-decay theory. It is believed that the calculated rate is unlikely to be in error by more than 10%. The recent report of Slobodrian et al (40) of the observation of deuterons produced by the reaction ^{3}He + ^{3}He $\rightarrow$ d+e$^+$ +ν+ ^{4}He caused a flurry of astonishment and cast doubt on the calculated p+p rate, since the two reactions are quite similar. It was suggested (40) that the p+p rate in the sun might have been underestimated by a factor of 10^8, and that consequently ^{8}B is not produced in the sun, which thereby solves the solar neutrino problem. It was then shown by Newman & Fowler (41) that a small increase in the p+p rate would indeed explain the absence of high-energy solar neutrinos but that an increase by 10^8 in the p+p rate was completely incompatible with current models of solar structure and with other empirical determinations of the strength of the weak interaction. The flurry subsided, when Davies et al (42) were unable to confirm the quoted results.

---

[3] The absolute cross section of ^{7}Be(p,γ)^{8}B has been checked recently (39) at $E_p = 360$ keV. The result is in agreement with previous work (38).

4.1.2 THE ^{3}He(^{3}He,2p)^{4}He REACTION   In the search for possible nuclear causes of the solar neutrino problem, the ^{3}He(^{3}He,2p)^{4}He reaction seemed to offer some hope. Fowler and independently Fetisov & Kopysov (43) pointed out that a hitherto undetected and narrow resonance near the particle threshold (Section 3) could solve the problem. If this resonance had suitable properties, it could increase the extrapolated reaction rates significantly, such that this reaction would short circuit the high-energy neutrino-producing reactions. Such a narrow resonance just above the threshold of ^{3}He + ^{3}He would correspond to a new state in ^{6}Be at $E \approx 11.50$ MeV. Since this state is highly unstable against several different particle decays, it must have an almost pure ^{3}He + ^{3}He configuration in order to be narrow. Consequently, it may be hard to excite this state in any reaction other than through the ^{3}He + ^{3}He reaction channel itself. In spite of these problems of interpretation, several nuclear reactions have been applied (44) in the search for this hypothetical state. No evidence for such a state has been found, however, in all these experiments, neither in ^{6}Be nor in the ^{6}Li analogue nucleus. Because of the expected pure ^{3}He + ^{3}He configuration of this hypothetical state, a search in the direct capture process of ^{3}He(^{3}He,γ)^{6}Be at beam energies of $E \geqq 1$ MeV may represent a more sensitive method (50). Although this reaction lacks the above problems of interpretation, the expected DC yield into this state is [even for $\Theta^2(^3$He$) = 1$] in the nanobarn range, and the observation of this process will therefore be difficult.

The most direct test at present for a possible enhancement of the ^{3}He(^{3}He,2p)^{4}He reaction rate at low energies comes from the measurements of Dwarakanath (45), who extended the previous measurements on this reaction down to $E_{cm} = 30$ keV ($\sigma = 0.06$ nb). The results eliminate the possibility of all but an extremely narrow resonance ($\Gamma \sim 100$ eV) at still lower beam energies. A search for such a narrow resonance was also carried out (45) by measuring thick-target yields. From the upper yield limits obtained at $E_{cm} = 14$–28 keV, Dwarakanath (45) could not exclude the possibility of such a resonance, but he could put severe constraints on the properties of a resonance that could bring about the required reduction of the high-energy solar neutrino flux. It therefore appears unlikely that this resonance exists at all.

The results of the Davis experiment over the period 1970–1977 are (35) 1.6 ± 0.3 SNU, while the standard solar model predicts (46) 4.7 ± 1.6 SNU using the best nuclear and atomic input data presently available. The solar neutrino problem is therefore still with us and it may tell us— as Fowler suggests (35)—that there is something "neu" under the sun after all.

4.2 The CNO Tri-Cycle

In stars somewhat more massive than the sun, hydrogen is converted into helium predominantly by the operation of the CNO bi-cycle (1, 2). The sequence of reactions involved in the main CN cycle is

$$^{12}C(p,\gamma)^{13}N(e^+ \nu)^{13}C(p,\gamma)^{14}N(p,\gamma)^{15}O(e^+ \nu)^{15}N(p,\alpha)^{12}C.$$

The loss of CN catalyst through the reaction $^{15}N(p,\gamma)^{16}O$ is replenished by the sequence of reactions $^{16}O(p,\gamma)^{17}F(e^+ \nu)^{17}O(p,\alpha)^{14}N$. Most of the relevant cross-section measurements (16, 17, 23) were made some years ago, particularly at CIT. Because of the importance of this cycle as a power source as well as in the nucleosynthesis of the carbon, nitrogen, and oxygen isotopes (48), there has been a continuing effort in recent years to improve the knowledge on the stellar reaction rates of the above processes. A few illustrative examples of these recent activities are described below.

4.2.1 PROTON CAPTURE BY ^{15}N In order to probe the capture mechanisms involved in $^{15}N(p,\gamma)^{16}O$, this reaction has been reexamined by Rolfs & Rodney (47) over a wide range of beam energies. The data for the dominant ground-state transition (Figure 7) demonstrate that the capture process is primarily controlled by the two known $J^\pi = 1^-$ resonances at $E_p = 338$ and 1028 keV with a small contribution from the $J^\pi = 1^+$ resonance at $E_p = 1640$ keV. The data were first analyzed in terms of the two $J^\pi = 1^-$ resonances. If the usual energy dependence of the partial and total widths as well as an interference term between the two resonances are taken into account, a curve indicated by the dashed line in Figure 7 is obtained. The deduced $S(0)$-factor of 22 keV-b is in fair agreement with previous work (49). The calculated yield deviates significantly, however, from the data at the tails of both resonances. These discrepancies can be removed, if an expected DC process (47, 50) is included in the capture mechanism, which results in an extrapolated $S(0)$-value higher by a factor of 2.5 than the previous value. This result implies that for every 880 cycles of the main CN cycle (rather than every 2200 cycles) CN catalyst is lost through the $^{15}N(p,\gamma)^{16}O$ reaction.

This reaction illustrates the problems involved in broad overlapping resonances (Section 2.3), which can interfere not only with each other but also in an important way with an additional nonresonant background process. It is perhaps interesting to note that a clarification of the reaction mechanisms involved in this capture process required yield measurements over a wide and continuous range of beam energies and that a reliable extrapolation of the data to stellar energies hinged sensitively on

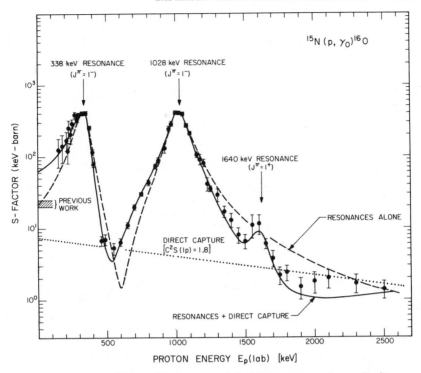

Figure 7 S-factor for $^{15}N(p,\gamma_0)^{16}O$ (47). The dashed line represents the results from an analysis solely in terms of the two $J^\pi = 1^-$ resonances. The solid line includes the results when interfering effects of a direct capture component (*dotted line*) are included in the capture process (47).

the features observed on the tails of both $J^\pi = 1^-$ resonances, especially at energies between the two resonances. An extrapolation of the very-low-energy data alone with their relatively large errors would have resulted in a large uncertainty in $S(0)$.

4.2.2 HYDROGEN BURNING OF ^{17}O It has been assumed previously (1, 2, 51) that hydrogen burning of ^{17}O proceeds entirely through the $^{17}O(p,\alpha)^{14}N$ reaction ($Q = 1.19$ MeV). The low-energy tails of all observed resonances lead (20, 21) to an extrapolated S-factor of $S_{p\alpha}(0) = 1.0$ keV-b (Figure 8). Because of the existence of compound states in ^{18}F near the proton threshold (21, see also Section 3), Brown (51) suggested that the stellar rate of this reaction was dominated by a resonance at $E_p = 66$ keV through the $E(J^\pi) = 5668$ (1^-)-keV compound state. Interference effects arising from 5604 (1^-)-keV compound state (bound by 2 keV) were also taken into account. In this rate calculation

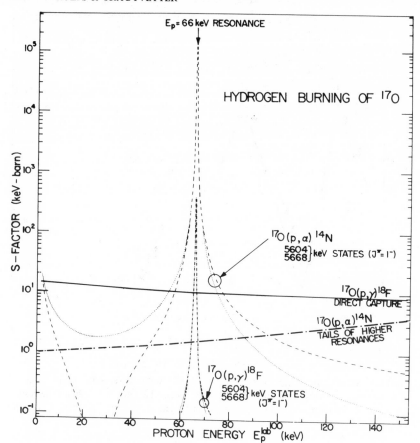

Figure 8   Contributions to the total *S*-factor for the reactions $^{17}O(p,\alpha)^{14}N$ and $^{17}O(p,\gamma)^{18}F$. The calculated *S*-factors for both reactions through the $J^{\pi} = 1^{-}$ compound states at $E = 5604$ and 5668 keV are upper limits, where the dotted (dashed) lines represent the results for constructive (destructive) interference between the two states.

(Section 3) Brown used a two-level formula, where each $J^{\pi} = 1^{-}$ resonance was described by a Breit-Wigner shape. All necessary resonance parameters for this calculation, except the proton partial widths, can be deduced from the available data (21) of $^{14}N(\alpha,\alpha)^{14}N$ and $^{14}N(\alpha,\gamma)^{18}F$. Reduced proton widths of $\Theta_p^2(l=1) = 0.007$ have been assumed (51) for both $J^{\pi} = 1^{-}$ resonances. With this value and Equation 11, the cross sections were calculated for both constructive and destructive interference between the two states. The calculated rate for the assumption of destructive interference has since been used in many astrophysical calculations.

Investigations by the Toronto group on the resonant (52) and direct capture (50) mechanisms in $^{17}O(p,\gamma)^{18}F$ ($Q = 5.61$ MeV) resulted in extrapolated S-factors (20) of $S_{res}^{p\gamma}(0) = 0.01$ keV-b and $S_{DC}^{p\gamma}(0) = 14$ keV-b (Figure 8). In view of these new data for $^{17}O(p,\gamma)^{18}F$ and because of the strong dependence of the $^{17}O(p,\alpha)^{14}N$ rate on the assumed values of Θ_p^2 for the two $J^\pi = 1^-$ threshold states, the above picture for hydrogen burning of ^{17}O was doubtful. In order to elucidate this situation, Rolfs & Rodney (20) used the DC process in $^{17}O(p,\gamma)^{18}F$ to obtain experimental values for these Θ_p^2 widths of the threshold states. This information can be deduced (50) from the absolute cross sections of DC-γ-ray transitions into these states. A search for these DC transitions in the nonresonant beam energy region of $^{17}O(p,\gamma)^{18}F$ yielded (20) only upper limits on the DC cross sections. However, the deduced upper limits on the reduced proton widths were significantly lower than the values assumed previously (51). With these upper limits, the stellar reaction rates of the threshold states were calculated (Figure 8).

As a result of these studies, the $^{17}O(p,\alpha)^{14}N$ reaction rate is reduced by at least a factor of 60 compared to the previous estimates (51). All the available data (Figure 8) demonstrate, that the $^{17}O(p,\gamma)^{18}F$ reaction cannot be neglected in the hydrogen burning of ^{17}O. If the subsequent hydrogen burning of ^{18}O (made after e^+ decay of ^{18}F) proceeds predominantly through the $^{18}O(p,\alpha)^{15}N$ reaction (see below), one must conclude (20) that the CNO cycle is tri-cycling (Figure 9). These results do not significantly change the energy release in the CNO cycle but affect principally the ^{17}O and ^{18}O abundances (20, 48). In particular, the isotope ^{17}O survives hydrogen burning in greater abundance than

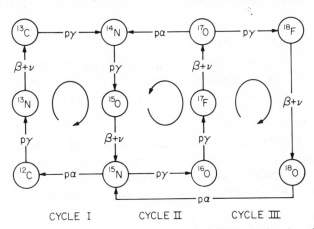

Figure 9 Illustration of the three CNO cycles involved (20) in the burning of hydrogen into helium.

heretofore thought and thus can serve as a source of neutrons in helium burning via $^{17}O(\alpha,n)^{20}Ne$.

It should be pointed out, that the DC process can be used as an accurate method (50) for determining reduced particle widths (or limits) of states near the particle threshold. These widths are crucial (Section 3) for the calculation of overall stellar reaction rates. For threshold states with high orbital angular momenta ($l \geqq 3$), the expected DC cross sections (even for $\Theta^2 = 1$) are significantly reduced, and it will therefore be difficult to determine their reduced particle widths. However, such states have—because of their high centrifugal barriers—often negligible importance for the overall stellar reaction rates.

4.2.3 THE $^{18}O(p,\alpha)^{15}N$ REACTION In order to elucidate the "ending" of the CNO tri-cycle (Figure 9), the $^{18}O(p,\alpha_o)^{15}N$ reaction has been investigated by the Münster group (19). The results are shown in Figure 4. The features reported previously (21) at $E_p \geqq 500$ keV were reproduced, and four new resonances were found[4] at low energies. The results of these measurements (Section 3) together with studies (22) on the stellar rate for the competing $^{18}O(p,\gamma)^{19}F$ reaction indicate that the picture of the ending of the CNO tri-cycle (Figure 9) is most likely correct.

If the CNO cycles operate on a rapid time scale (e.g. in the explosive phases of stellar evolution), β-unstable nuclei like ^{13}N will not have enough time to decay but rather will be burned by nuclear reactions. In this case, the CNO cycles develop into a many-cycled monster rivaling the many-headed Hydra (48). A determination of the stellar rates of the reactions involved in this monster (mostly reactions on unstable target nuclei) represents a challenge to the experimentalist (Section 3).

4.3 The NeNa Cycle

At high stellar temperatures, additional hydrogen-burning cycles such as the NeNa cycle, suggested in 1957 by Marion & Fowler (53), are operative. The sequence of reactions involved in this cycle is

$$^{20}Ne(p,\gamma)^{21}Na(e^+v)^{21}Ne(p,\gamma)^{22}Na(e^+v)^{22}Ne(p,\gamma)^{23}Na(p,\alpha)^{20}Ne.$$

Because of the higher Coulomb barriers involved in these reactions, when compared with the CNO reactions, the NeNa cycle is expected to be relatively unimportant as an additional energy source in stars. However, it will be important in the nucleosynthesis of the Ne and Na

[4] The intense and narrow ($\Gamma \leqq 0.5$ keV) resonance at $E_p = 152$ keV (Figure 4) may be of interest as a nuclear probe in oxygen depth-profile measurements (e.g. investigation of silicon oxide layers in silicon technology).

isotopes. The ^{21}Ne produced is also of interest for subsequent helium burning in stars, since it can act as a source of neutrons via ^{21}Ne$(\alpha,n)^{24}$Mg for the s-process synthesis.

Experimental data have become available only in recent years on the stellar rates of these NeNa reactions. Studies at CIT (54) reveal that the stellar rate of the ^{23}Na$(p,\alpha)^{20}$Ne reaction is large enough, compared to the competing reaction ^{23}Na$(p,\gamma)^{24}$Mg, to guarantee cycling. Data from studies of the proton-induced reactions on the Ne isotopes have been published (30), but the results of ongoing experiments at CIT and Münster are necessary before astrophysically relevant conclusions can be drawn.

The investigations of the ^{20}Ne$(p,\gamma)^{21}$Na reaction (30, 55) are completed and some of the results are discussed here, because they illustrate convincingly the importance of subthreshold resonances (Section 3) to overall stellar reaction rates. The study of this reaction over a wide range of energies (30) seems to indicate that its stellar rate is dominated by the DC process to the 332- and 2425-keV bound states in ^{21}Na (Figure 10). The 2425-keV state, 7 keV below the proton threshold (30, 56), could, however, contribute significantly to the overall stellar rate (Section 3). From the quoted lifetime (56) of $\tau \leq 1$ fs and for the assumption of an M1 transition strength of ≤ 10 Weisskopf units for the 100% decay to the ground state (56), the γ-ray width of this state is restricted to values of $\Gamma_\gamma = 0.7$–3.0 eV. Furthermore, the DC studies revealed (30) a reduced proton width of $\Theta_p^2(l = 0) = 0.9$ for this state. All this information made it possible to calculate (Section 3) the high-energy tail of this subthreshold state, where a single-level Breit-Wigner shape was assumed, since no other interfering $J^\pi = \frac{1}{2}^+$ states are reported below 5-MeV excitation energy (56). The calculations (55) revealed (Figure 10) that this high-energy tail dominates the stellar rates and should be detectable in an experiment. The tail should be found by searching for the ground-state transition from the 2425-keV state at $E_p \geq 500$ keV, where the unique identifying feature would be the predicted energy dependence of this γ-ray transition (i.e. decreasing S-factor with increasing beam energy). The experimental results of the measurements by Rolfs & Winkler (55) confirmed (Figure 10) these predictions, where the fit shown (dashed-dotted line) was obtained (55) for a γ width "at resonance" of $\Gamma_\gamma = 0.31$ eV.

The importance of a bound state near threshold for stellar burning rates was noted long ago. In the case of ^{20}Ne$(p,\gamma)^{21}$Na, Marion & Fowler (53) previously estimated the contribution of the 2425-keV state to stellar rates. However, the information on the nuclear properties of this state was imprecise, so the calculated rates were quite uncertain. A first experimental confirmation of such cases was obtained by Dyer &

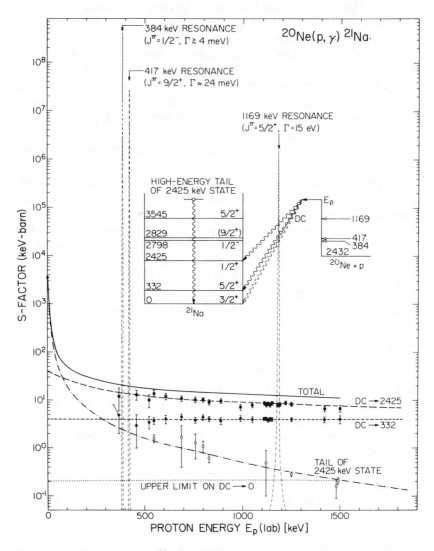

Figure 10 S-factors for the ^{20}Ne(p,γ)^{21}Na reaction. The dashed lines through the DC → 332- and DC → 2425-keV data points (30) are DC-model predictions. The dashed-dotted line is the prediction of the high-energy tail of the 2425-keV bound state, fitted with the free parameter Γ_γ to the observed data (55). A possible DC → 0 process would have almost the same energy dependence as the DC → 332-keV transition and is indicated as a dotted line. Also shown for comparison are the S-factor curves for the observed resonances (30).

Barnes (57) for the reaction $^{12}C(\alpha,\gamma)^{16}O$, in which a bound state interferes with a higher-lying unbound state (Section 5.2). It seems, however, that the tail of a subthreshold resonance alone has never been observed in charged-particle capture reactions.

It should be pointed out that the laboratory observation of the high-energy tail of the 2425-keV bound state is favored by approximately maximum partial widths (γ width: M1 = 1.0 W.u.; reduced proton width: $\Theta_p^2 = 0.9$) as well as the s-wave formation. The exact position of this state relative to the threshold is important for the stellar burning rate, but is not very important for its laboratory observation (Figure 10). If the state had required, for example, d-wave protons to form it from ^{20}Ne, the yield of the high-energy tail would be reduced by two orders of magnitude. In this case, the subthreshold state would not have significantly affected the stellar rate, and it would have been impossible to detect the high-energy tail by the experiment.

5 HELIUM BURNING

Because of the crucial importance of the $3\alpha \rightarrow {}^{12}C$ and $^{12}C(\alpha,\gamma)^{16}O$ helium-burning reactions to nuclear astrophysics (Section 1), a large number of experiments have been carried out recently in order to improve the accuracy of their stellar rates. In addition to these two reactions, there have been also major experimental efforts to elucidate the nature and rate of other basic helium-burning reactions.

5.1 The Triple-α Process

The fusion of three alpha particles to form ^{12}C was originally suggested by Öpik and Salpeter (58) as a method of bypassing the $A = 5$ and $A = 8$ gaps in the chain of stable nuclides. The fusion proceeds (2) via a two-step process: (a) the formation of a small concentration of ^8Be in thermal equilibrium with ^4He [$^4He + {}^4He \rightleftarrows {}^8Be$] and (b) the radiative capture of an additional alpha particle to form ^{12}C [$^8Be + {}^4He \rightleftarrows {}^{12}C^* \rightarrow {}^{12}C + \gamma$ (or an e^+e^- pair)]. The more condensed notation $3\alpha \rightarrow {}^{12}C$ is referred to as (1, 2) as the "triple-α process." It was pointed out many years ago by Hoyle (59) that the large abundance of ^{12}C in the universe implies the enhancement of the $^8Be(\alpha,\gamma)^{12}C$ reaction by a fortuitously located resonance in the stellar energy region. Subsequent experiments have established this "astrophysically predicted" nuclear state, now known (21) as the $J^\pi = 0^+$ second excited state of ^{12}C at $E = 7.65$ MeV (Figure 11). The stellar rate of this resonance cannot be determined, however, by a direct yield measurement, because unstable ^8Be target nuclei would have to be involved. This resonant state in ^{12}C represents therefore another example of a difficult-to-measure level (Section 3), and many nuclear techniques had to be involved in determining just this one astrophysical reaction rate.

Since the reactions $^4\text{He} + {}^4\text{He} \rightleftarrows {}^8\text{Be}$ and $^8\text{Be} + {}^4\text{He} \rightleftarrows {}^{12}\text{C}^*(7.65 \text{ MeV})$ proceed much more rapidly (a factor $> 10^3$) than the radiative decay of $^{12}\text{C}^*(7.65)$, the concentration of $^{12}\text{C}^*(7.65)$ nuclei can be calculated (1, 2) from the equilibrium abundance ratio $^{12}\text{C}^*/^4\text{He}$ (using equations from statistical mechanics). The ^{12}C ground state is produced then as the precipitate from this equilibrium between three α particles and the excited state $^{12}\text{C}^*(7.65)$ owing to a small but irreversible leakage out of the equilibrium, namely by radiative decay from the 7.65-MeV state. This decay proceeds (Figure 11) by γ-ray (Γ_γ) and E0 internal-pair (Γ_π) emission, hence $\Gamma_{\text{rad}} = \Gamma_\gamma + \Gamma_\pi$. It is readily seen from the above comments and Section 3, that the rate of the triple-α process $P_{3\alpha}$ can be calculated from Equation 10 (Section 2) to give $P_{3\alpha} \propto \Gamma_{\text{rad}} \exp(-Q/kT)$, where Q is the mass difference between $^{12}\text{C}^*(7.65)$ and three alpha particles. The measurement of the quantities Q and Γ_{rad} has become the subject of intensive experimental work and provides a fascinating example of the

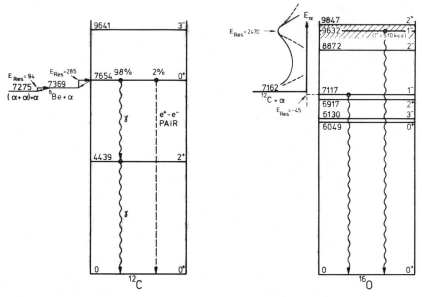

Figure 11 Energy-level diagrams (21) relevant for the $3\alpha \to {}^{12}\text{C}$ and $^{12}\text{C}(\alpha,\gamma)^{16}\text{O}$ helium-burning reactions. The 7.65-MeV state in ^{12}C serves as a resonance in the $3\alpha \to {}^{12}\text{C}$ process. The $J^\pi = 1^-$, 7.12-MeV subthreshold state in ^{16}O can interfere at this high-energy tail constructively (*solid line*) or destructively (*dashed line*) with the $J^\pi = 1^-$ broad state at 9.63 MeV.

way in which nuclear astrophysics must draw upon many facets of nuclear knowledge.

The Q value must be known rather accurately, since $P_{3\alpha}$ depends exponentially on it. Here again a precise knowledge of nuclear energies is often of crucial importance in nuclear astrophysics. The results of different experimental groups have been reviewed by Nolen & Austin (60). In most of the experiments, the excitation energy of the 7.65-MeV state was determined with high precision, from which Q could be obtained by subtracting $3M_\alpha c^2$. A direct measurement of Q, via the energy released in the 3α breakup from the 7.65-MeV state (Figure 11), was carried out by Barnes & Nichols (61). The Q values derived from the more recent measurements are in good agreement and lead to a recommended weighted average of $Q = 379.38 \pm 0.20$ keV (60). This error in Q implies a negligible uncertainty (2.5%) in $P_{3\alpha}$ at $T = 10^8$ °K.

The Γ_{rad} width of the 7.65-MeV state is determined indirectly (2) as the product of three experimentally measured quantities, $\Gamma_{rad} = (\Gamma_{rad}/\Gamma)$ $(\Gamma/\Gamma_\pi)\Gamma_\pi$, where Γ is the total width of the 7.65-MeV state.

Inelastic scattering of high-energy electrons was used to determine the E0 matrix element connecting the ground state and 7.65-MeV state in ^{12}C, which resulted (62) in a e^+e^- pair emission width of $\Gamma_\pi = 60.5 \pm 3.9$ μeV.

The branching ratio Γ_{rad}/Γ has been the subject of several recent experimental studies, especially motivated by the results of Chamberlin et al (63), which gave a value nearly 45% larger than the older accepted value (64). The basic experimental techniques used in these measurements are the associated-particle technique and the particle-γ-γ triple-coincidence technique. The 7.65-MeV state is populated in both methods through a particle channel $P_{7.65}$ of a selected nuclear reaction. In the first technique, the branching ratio Γ_{rad}/Γ for the radiative decay of the 7.65-MeV state was measured by detecting the ^{12}C(0.0) recoil ions (remaining after radiation emission) in coincidence with the particles populating the 7.65-MeV state. Thus, of all the ^{12}C* nuclei formed in the 7.65-MeV state, only those decaying to the ^{12}C ground state are selected by means of time-of-flight, energy, and kinematic constraints. The branching ratio is then given by the number of ^{12}C(0.0) recoil coincidence events divided by the total number of single events $P_{7.65}$, suitably corrected for counting losses and for the recoil detection efficiency of the system. In the particle-γ-γ technique, the radiative decay of the 7.65-MeV state is observed through its γ-γ cascade via the 4.44-MeV state (Figure 11). Since this procedure yields only a value for Γ_γ/Γ, a small correction (1.7%) for the E0 pair emission width has to be incorporated to obtain Γ_{rad}/Γ. The results reported by a number of experimental groups have been reviewed recently by Markham et al (65), which leads to a recommended weighted average of $\Gamma_{\mathrm{rad}}/\Gamma = (4.13 \pm 0.11) \times 10^{-4}$ and confirms the results of Chamberlin et al (63).

The ratio Γ_π/Γ was determined by Alburger (66) with the use of a pair-spectrometer and by Robertson et al (67) with a method similar to the second technique discussed above. The results are in excellent agreement and yield a weighted average (66) of $\Gamma_\pi/\Gamma = (6.8 \pm 0.7) \times 10^{-6}$.

The experimental results of the three sets of experiments determine (66) the radiative width of the 7.65-MeV state to $\Gamma_{\mathrm{rad}} = 3.67 \pm 0.46$ meV.

As a summary, the stellar reaction rate of the triple-α process is known now with a precision of 13% which is sufficient for all practical purposes.

5.2   The $^{12}C(\alpha,\gamma)^{16}O$ Reaction

The state in ^{16}O at $E_x(J^\pi) = 7.12(1^-)$ MeV, which is 45 keV below the ^{12}C$+\alpha$ threshold (Figure 11), is expected to dominate the stellar reaction rate of $^{12}C(\alpha,\gamma)^{16}O$ by its high-energy wing at the relevant stellar energy $E_0 = 0.3$ MeV (Equation 6). Furthermore, coherent interference effects with the broad $J^\pi = 1^-$ state at 9.63 MeV ($E_{\mathrm{res}} = 2.47$ MeV) will also affect to some extent the capture yield at the energy E_0. The determination of the contribution of this subthreshold state to stellar helium burning represents therefore another example of a difficult-to-measure level (Section 3). Since resonance fluorescence techniques determined (21) the γ width of the 7.12-MeV state to $\Gamma_\gamma = \Gamma = 62 \pm 5$ meV, the contribution of this subthreshold resonance to the rate of $^{12}C(\alpha,\gamma)^{16}O$ can be calculated (Section 3) if its reduced α-particle width $\Theta_\alpha^2(7.12)$ is available. With the known nuclear properties of the 9.63-MeV state, the influence of interference effects between the two $J^\pi = 1^-$ states to the stellar rate at E_0 can be incorporated. For the determination of this stellar rate, two different approaches have been explored.

5.2.1   INDIRECT DETERMINATION   From the analyses of α-transfer reactions, such as $^{12}C(^6Li,d)^{16}O$ and $^{12}C(^7Li,t)^{16}O$, values of $\Theta_\alpha^2(7.12) = 0.06$–$0.14$ and ~ 0.25 have been deduced (68). The spectrum of α particles (69) emitted from $^{16}O^*$ (produced through β decay of ^{16}N) has also been analyzed (70) yielding $\Theta_\alpha^2(7.12) = 0.013$–$0.24$. The considerable scatter in these values implies a large uncertainty in the calculated reaction rates.

5.2.2   DIRECT DETERMINATION   A more reliable way to obtain the $^{12}C(\alpha,\gamma)^{16}O$ stellar rate is to measure the capture cross section over a wide range of energies in order to observe directly the interference between the two $J^\pi = 1^-$ states. It is essential in this approach to have precise measurements of the capture cross sections at energies as far below the $E_{res} = 2.47$-MeV resonance (Figure 11) as is technically feasible. Since the capture cross section at the peak of this resonance is only 40 nb, observation of interference effects arising from the subthreshold state requires the measurement of cross sections considerably smaller than 1 nb. The formidable problems encountered in these measurements arise from the combination of a low γ-ray capture yield and a high γ-ray background, induced by the sensitivity of the γ-ray detector to neutrons. The $^{13}C(\alpha,n)^{16}O$ reaction is a prolific source of neutrons. Any ^{13}C contained in the target is therefore undesirable, as is the deposition of any natural carbon on the target or the beam-defining collimators. In the measurements of Jaszczak et al (71) and Dyer & Barnes (57), targets of high ^{12}C isotopic purity $[^{13}C/^{12}C \leq 0.005\% \ (57)]$ were used therefore in combination with the neutron-gamma time-of-flight difference to separate the effects of the neutrons and capture γ rays in the NaI(Tl) detectors as well as to suppress the γ-ray room background. The measurements of Jaszczak et al (71) have been carried out at $E_{cm} = 1.2$–3.1 MeV, but because of the large statistical errors the cross section extrapolated down to 0.3 MeV is essentially undetermined (200% error). Significantly improved accuracy has been obtained in the work of Dyer & Barnes (57), where the yield measurements have been carried out in the energy region $E_{cm} = 1.41$–2.94 MeV (Figure 12) with $\sigma(1.41$ MeV$) = 0.3$ nb. These new capture data together with the p-wave phase

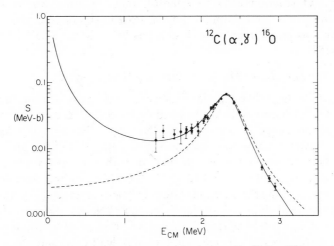

Figure 12 S-factor for the measured E1 contributions to the $^{12}C(\alpha,\gamma)^{16}O$ reaction (57) and the best fit (*solid line*) of the hybrid model (73). The dashed curve is a fit for $\Theta_\alpha^2(7.12$-MeV state) $= 0$, i.e. no interference effects with the subthreshold state. The theoretical curves for the R- and K-matrix models (57, 74) are qualitatively similar.

shifts (72) for $^{12}C(\alpha,\alpha)^{12}C$ have been parameterized in terms of a three-level R-matrix model (57), which includes the subthreshold state at 7.12 MeV, the 9.63-MeV state, and a (fictitious) state at high excitation energy to represent any other $J^\pi = 1^-$ "background" in the energy region below $E_{cm} = 2.94$ MeV. The fit to the data yields an S-factor for the E1 portion of the $^{12}C(\alpha,\gamma)^{16}O$ cross section of $S(0.3$ MeV) = $0.14^{+0.14}_{-0.12}$ MeV-b. A "hybrid" R-matrix optical-model parameterization (73) results in $S(0.3$ MeV) = $0.08^{+0.05}_{-0.04}$ MeV-b, and a K-matrix parameterization (74) yields $S(0.3$ MeV) = $0.09^{+0.13}_{-0.08}$ MeV-b. All methods of parameterizing the available data give consistent results (i.e. constructive interference at energies between the two $J^\pi = 1^-$ states). Although these models add reliability to the order of magnitude of the S-factor, the remaining uncertainty in this extrapolated value is still significant. Any improvement of this situation requires higher accuracy of the $^{12}C(\alpha,\gamma)^{16}O$ data, which is extremely difficult to achieve. Humblet et al (74) suggested that some improvement might already be obtained from better data on the $^{12}C+\alpha$ elastic scattering. It should be pointed out also that E2 resonant capture through the high-energy tail of the $J^\pi = 2^+$ state at 6.92 MeV (Figure 11) may be significant (57) in the stellar region, if the reduced α-particle width of this state should prove to be several times larger than that of the 7.12-MeV state.

Robertson et al (67) used the helium-burning nucleosynthesis calculations of Arnett (75) together with the new results of the $3\alpha \rightarrow {}^{12}C$ and $^{12}C(\alpha,\gamma)^{16}O$ reaction rates to establish that ^{12}C is likely to be the dominant product of helium burning, at least in stars lighter than about 15 solar masses. A similar conclusion has been drawn also by Dyer & Barnes (57).

5.3   Further Basic Helium-Burning Reactions

Since the $^{16}O(\alpha,\gamma)^{20}Ne$ reaction has a very low cross section, the ratio of ^{12}C to ^{16}O is essentially determined by the $3\alpha \rightarrow {}^{12}C$ and $^{12}C(\alpha,\gamma)^{16}O$ rates. A knowledge of the $^{16}O(\alpha,\gamma)^{20}Ne$ and $^{20}Ne(\alpha,\gamma)^{24}Mg$ reaction rates is, however, of interest for the nucleosynthesis of ^{20}Ne and ^{24}Mg. The $^{16}O(\alpha,\gamma)^{20}Ne$ rate is dominated at higher temperatures by the resonances observed at $E_\alpha \geq 1.12$ MeV (21), while at lower temperatures a DC process to the three bound states of ^{20}Ne is assumed (17). Although no experimental data are available for the $^{20}Ne(\alpha,\gamma)^{24}Mg$ reaction at energies of $E_\alpha < 1.6$ MeV (56), the potential presence of several favorably located resonances in this reaction (56) may lead eventually to a much faster burning rate of ^{20}Ne than of ^{16}O. This would yield then much larger amounts of ^{24}Mg than ^{20}Ne, while the observed abundances clearly favor neon over magnesium. In order to elucidate the constraints and possible sites of the stellar nucleosynthesis of ^{20}Ne and ^{24}Mg, further experimental investigations of both (α,γ) reactions appear necessary.

In hydrogen-depleted regions of second-generation stars, a significant concentration of ^{14}N is also expected from the operation of the CNO tri-cycle. Helium burning of ^{14}N can produce the isotope ^{18}O by the reaction ^{14}N$(\alpha,\gamma)^{18}$F$(e^+ \nu)^{18}$O, and the subsequent reaction ^{18}O$(\alpha,\gamma)^{22}$Ne is an important link in one of the reaction chains by which heavier elements can be produced.

In the case of ^{14}N$(\alpha,\gamma)^{18}$F, there was the possibility that this reaction was considerably faster than the triple-α process, thus helium burning could be ignited in most second-generation stars by this reaction before the onset of $3\alpha \rightarrow {}^{12}$C. The measurements of Couch et al (76) led, however, to a decrease of this rate by several orders of magnitude, thereby removing the above possibility.

A knowledge of the ^{18}O$(\alpha,\gamma)^{22}$Ne rate is desirable for two reasons: (a) Black (77) and others found an almost pure ^{22}Ne gas in meteorites, which has been given the name Ne-E. In the attempt to understand the origin of the solar system, the nucleosynthesis of this neon gas represents a major mystery. In order to solve this mystery, ^{22}Ne-synthesizing reactions have to be known, e.g. ^{18}O$(\alpha,\gamma)^{22}$Ne. (b) Recent stellar evolution calculations indicated (12), that the reaction ^{22}Ne$(\alpha,n)^{25}$Mg could be the major source of neutrons for the s-process synthesis and that ^{22}Ne could be provided in the star through the sequence of reactions ^{14}N(α,γ) ^{18}F$(e^+ \nu)^{18}$O$(\alpha,\gamma)^{22}$Ne. In order to determine the conditions of such calculations, the ^{18}O$(\alpha,\gamma)^{22}$Ne reaction must be studied and compared with the concurring reaction ^{18}O$(\alpha,n)^{21}$Ne.

One of the principal difficulties encountered in measuring the ^{18}O$(\alpha,\gamma)^{22}$Ne cross sections arises from the inevitable neutron flux due to the concurring reaction ^{18}O$(\alpha,n)^{21}$Ne. At beam energies above $E_\alpha = 0.86$ MeV, the number of neutron-induced γ-ray events is far greater than the number of capture γ rays. With the use of a high-resolution Ge(Li) detector in conjunction with high beam currents and improved target-technologies, the Münster group (79) extended the previous measurements at E_α(lab) ≥ 2.15 MeV down to energies as low as $E_\alpha = 0.6$ MeV. The results (Figure 13) illustrate that the reaction proceeds through a relatively small number of isolated resonances. The rates of the two concurring helium-burning reactions of ^{18}O are compared in Figure 14. The observation of strong resonances in the ^{18}O$(\alpha,\gamma)^{22}$Ne reaction below the neutron threshold and the consequent dominance of the capture reaction at $T_9(10^9$ °K$) \leq 0.6$ may help to untangle the mystery of Ne-E. If the ^{22}Ne$(\alpha,n)^{25}$Mg reaction is indeed the

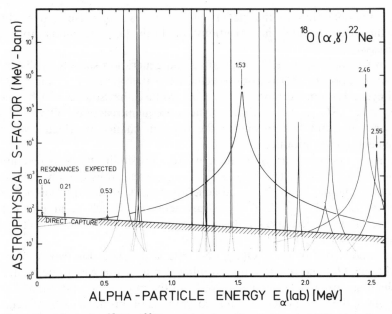

Figure 13   S-factor for $^{18}O(\alpha,\gamma)^{22}Ne$ as taken from Trautvetter et al (79). Also shown is the expected location of possible low-energy resonances, which can eventually increase (Section 3) the stellar reaction rates significantly.

key reaction for producing the s-process neutrons[5] (12), the conversion of ^{14}N into ^{22}Ne by the above sequence of reactions must take place at stellar temperatures of $T_9 \leqq 0.6$.

6   CARBON AND OXYGEN BURNING

At the end of helium burning, a core remains with predominantly ^{12}C ashes. The energy output from carbon burning, now setting in (9), will once again stabilize the star. Chandrasekar (80) pointed out, however, that degenerate electrons can support up to 1.4 $M_\odot$ of a carbon core. Such stars will therefore neither collapse nor ignite their carbon fuel. Stars, which form carbon cores slightly above the 1.4 $M_\odot$ limit, ignite their carbon fuel under conditions where the pressure is dominated by degeneracy effects. Under these conditions the star blows up in a so-called carbon detonation supernova (9). Stars above a certain critical total mass ($\sim 7\ M_\odot$) are able to burn their carbon fuel under non-

[5] Major experimental efforts have been focused recently on the determination of the rates of the neutron-producing (α,n) reactions on the carbon, oxygen, and neon isotopes (78).

degenerate conditions. These stars do not detonate but evolve to more advanced evolutionary stages. The critical stellar mass separating these two different paths of stellar evolution is determined (9) by the rate of the $^{12}C + ^{12}C$ reaction at energies as low as $E_{cm} = 1$ MeV. The rate of this reaction is therefore one of the basic quantities needed for the study of stellar evolution and nucleosynthesis.

After carbon burning, the core of a more massive star is primarily ^{20}Ne, however there is also some ^{23}Na and ^{23}Mg present. As the ^{20}Ne core collapses and heats up, the neon photodisintegrates into ^{16}O. If there are unburned ^{12}C nuclei left, the $^{12}C + ^{16}O$ fusion reaction will

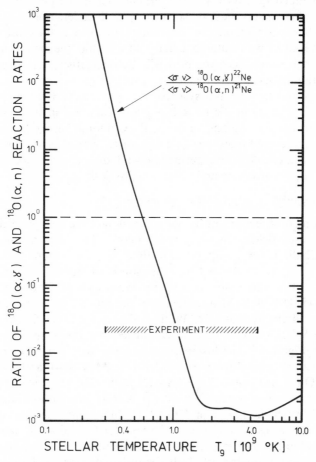

Figure 14   Ratio of the rates $\langle \sigma v \rangle$ for the reactions $^{18}O(\alpha,\gamma)^{22}Ne$ and $^{18}O(\alpha,n)^{21}Ne$ is plotted as a function of stellar temperature (79). Also indicated is the temperature region, for which the reaction rates of $^{18}O(\alpha,\gamma)^{22}Ne$ are determined by observation.

take place at this stage of stellar evolution. When the dominant ^{16}O core collapses and heats to high enough temperatures, oxygen burning ($^{16}O + ^{16}O$) will produce primarily ^{28}Si and will be followed eventually by the silicon and explosive nucleosynthesis processes.

The heavy-ion reactions $^{12}C + ^{12}C$, $^{12}C + ^{16}O$, and $^{16}O + ^{16}O$ play an important role in the behavior of highly developed stars. These reactions determine how stars evolve and produce many of the chemical elements, how massive stars explode as supernovae, and what sort of remnants these stars leave behind (black hole, neutron star, or white dwarf).

In order to extrapolate the heavy-ion reaction yields, obtained at higher beam energies, down to the relevant stellar energies, the involved reaction mechanisms must be sufficiently well understood. For a well-behaved heavy-ion interaction, the dominant feature of the cross sections should be given by Coulomb barrier effects, and the behavior of the exit-channels should be given primarily by the statistical averages of an evaporation model. The S-factor is expected therefore to be rather featureless and to vary in a systematic way as the heavy-ion species is changed. The study of the three heavy-ion reactions, involved in astrophysics, at energies near and far below the Coulomb barrier has revealed, however, richer and more puzzling structures, which have subsequently triggered a great deal of experimental and theoretical research in the areas of nuclear structure and heavy-ion physics.

6.1   *Absorption Under The Barrier*

The fusion cross sections of the three reactions have been studied either by charged-particle or by γ-ray spectroscopy. The early measurements of the Chalk River group (81) for the $^{12}C + ^{12}C$ reaction at $E_{cm} = 5.0-12.5$ MeV were extended by Patterson et al (82) to energies $E_{cm} = 3.23-8.75$ MeV and more recently by Mazarakis & Stephens (83) to $E_{cm} = 2.45-4.91$ MeV. Measurements of the $^{12}C + ^{12}C$ gamma-ray yield were carried out by Spinka & Winkler (32) at $E_{cm} = 3.6-7.5$ MeV. The resulting S-factors from the charged-particle measurements (82, 83) are shown in Figure 15. The fusion cross sections for $^{12}C + ^{16}O$ have been determined (Figure 16) at $E_{cm} = 3.9-12.0$ MeV (84) and those for $^{16}O + ^{16}O$ (Figure 16) at $E_{cm} = 6.8-11.9$ MeV (32). The $^{12}C + ^{12}C$ reaction (and to some extent $^{12}C + ^{16}O$) reveals pronounced resonance structure, which is absent in $^{16}O + ^{16}O$. In addition to the observed narrow resonances ($\Gamma \sim 100$ keV), the S-factor for $^{12}C + ^{12}C$ shows also an anomalously steep rise with decreasing beam energy at $E_{cm} \leq 3.5$ MeV, in contrast to the relatively smooth behavior in the other two reactions. The theoretical interpretation of this steep rise at energies far below the Coulomb barrier has led to two rival models, which differ in

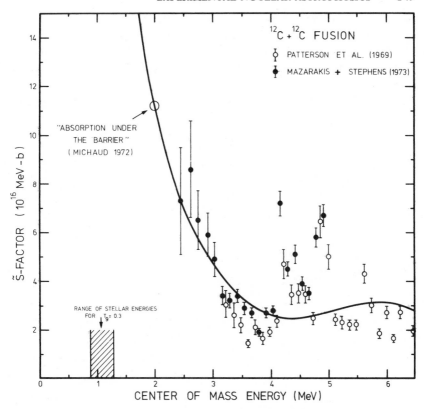

Figure 15   Fusion rate of $^{12}C + ^{12}C$ as observed by Patterson et al (82) and Mazarakis & Stephens (83). The steep rise at low energies is cited by Michaud (86) as evidence for the phenomenon of absorption under the barrier, which gives rise to a large extrapolated fusion rate at stellar energies. The rate scale is presented here in form of the $\tilde{S}(E)$-factor, defined (82) by $\tilde{S}(E) = S(E)\, exp\,(-gE)$ with $g = 0.46$.

their predictions of the extrapolated stellar rates by more than two orders of magnitude. The essential difference between these models is related to the question of whether the fusion of two nuclei can be initiated far out in the tenuous tail of the nuclear mass distribution (i.e. several times the radii of the nuclei) in what might be called the "nuclear stratosphere." In the standard model of Coulomb barrier penetration (17) extensive use has been made of the conventional optical model to fit the gross features of the available data (i.e. averages over the observed resonance structures) by varying the parameters specifying the real and imaginary potentials. The steep rise in the S-factor at low energies cannot be explained with this model and has been interpreted (17) as the high-energy wing of a broad resonance, so that the S-factor may be again a decreasing function of energy around stellar energies of 1 MeV. From a detailed study of the optical-model potentials, Michaud et al (85) focused attention on an interesting new physical concept, known as

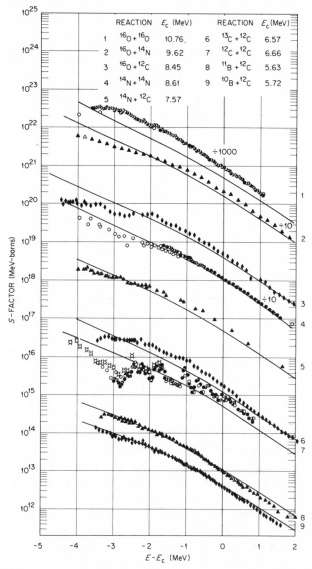

Figure 16 S-factor curves for nine heavy-ion systems as taken from Stokstad et al (96). The solid curves are the results of optical-model calculations. The energy scales have—for a better comparison—been normalized to the heights of the respective Coulomb barriers.

"absorption under the barrier." If the absorbing imaginary part of the optical potential has a surface thickness large enough (i.e. if the nuclear density extends to large radii), the nuclear fusion process at energies far below the Coulomb barrier will take place at large internuclear separations before a significant penetration of the Coulomb barrier has occurred. In this case the reaction cross section does not drop as rapidly with decreasing energy, since the effect of the Coulomb barrier is partially compensated by shifting the maximum of the interaction integrand outwards. Consequently, a significant rise in the S-factor at very low energies can be expected (85). Since (a) the shape of the imaginary potential far in the tail is not known, and (b) some reaction rates at stellar energies may very well be dominated by the tail of this imaginary part, it was argued (85) that the standard optical model may lose in such cases its utility in extrapolating high-energy data down to stellar energies. The steep rise in the S-factor for $^{12}C + ^{12}C$ fusion at low energies (Figure 15) has been cited by Michaud (86) as evidence for this new phenomenon of absorption under the barrier. In this case, the rise in the S-factor will continue resulting in a very different extrapolated rate than that obtained from the standard model. Because of the widely differing extrapolations of the two rival models and since experimental studies of other ion systems in the mass range $A = 9$–16 have revealed no additional evidence for this phenomenon, a more precise remeasurement of the $^{12}C + ^{12}C$ reaction rate far below the Coulomb barrier was desirable.

Using the intense ^{12}C beam (up to 15 particle μA) from the Bochum Dynamitron Tandem Accelerator, the $^{12}C + ^{12}C$ process has been studied by the Münster group (87) at $E_{cm} = 2.5$–6.2 MeV via γ-ray spectroscopy. The ubiquitous hydrogen contamination in the targets seriously hampered the measurements at low energies due to an intense γ-ray background created by interactions of the ^{12}C beam with the ^{1}H and ^{2}H contaminating nuclei. The use of targets with low hydrogen content was therefore of crucial importance for this experiment. The target thickness (9–55 $\mu g/cm^2$) and its profile were determined via the $^{12}C(p,\gamma)^{13}N$ reaction and stayed constant to better than 1 $\mu g/cm^2$ for a 20-particle Coulomb deposition. The energy calibration of the accelerator was checked and found to be correct and constant in time within 0.1%. All these considerations and precautions are crucial for reliable measurements of fusion cross sections at extreme low energies (Section 3). The resulting yields of the two most intense γ-ray lines (representing a somewhat varying but sizable fraction of the total fusion yield) are shown in Figure 17. The substantial differences[6] from the Pennsylvania data (83) particularly

---

[6] The differences could largely be removed if the energy scale of the Pennsylvania data were actually higher by $\Delta E_{cm} \sim 100$ keV.

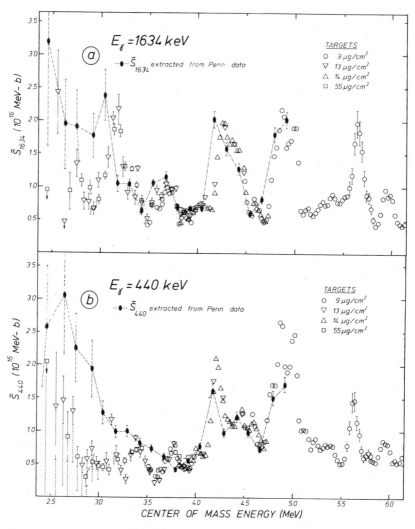

Figure 17   Fusion rate of $^{12}C + ^{12}C$ for the (*a*) alpha and (*b*) proton exit channels, as obtained by Kettner et al (87) is compared with previous work (83). The results for different targets (87) are indicated by different symbols. If the γ_{440} data points at $E_{cm} = 2.45$ and 2.55 MeV for the 55- and 14-μg/cm^2 targets, respectively, are identified entirely with Coulomb excitation of a ^{23}Na contamination in the targets (*dotted error bars*), only the values in the neighborhood will be uncertain to a limited amount. (For definition of $\tilde{S}(E)$, see Figure 15.)

below 3.5 MeV is apparent. The steep increase of the rates at low energies has not been confirmed, but the pronounced resonance structure continues down to the lowest energies, in that narrow peaks are common to both the α- and p-channels. These results obviate the need for absorption under the barrier down to at least 2.4 MeV[7] and suggest the use of the standard model (17) for extrapolations.

Since the direct population of the ground states of ^{20}Ne and ^{23}Na cannot be observed in the γ-ray spectroscopy method, additional charged-particle measurements are required to determine the total fusion cross sections. Such measurements are under way (89).

6.2 Intermediate Structure

Since the ^{12}C + ^{12}C resonant structures can continue into the stellar energy region and since such pronounced structures can also exist at these energies for the other two reactions, a significant uncertainty remains in the extrapolation procedure. This uncertainty has stimulated experimental studies of other systems in the $A = 9\text{--}16$ mass range (carried out primarily at CIT), in order to probe the nature of these structures. It was found (Figure 16) that such structures are absent in the energy regions studied. The physical explanation of the presence or absence of such structures is a long-standing problem in heavy-ion physics and has been the subject of continuing interest, speculations, and many contrary opinions. A review of the state of affairs was recently presented by Feshbach (90).

The narrow width of the structures does not allow us to identify them as compound nuclear resonances, and an explanation in terms of statistical fluctuations fails because of the strong correlations observed among all exit channels. It was realized therefore (90) that some new type of intermediate structure was present. The earliest data from Chalk River were interpreted (81, 90) as resonances in a ^{12}C $-$ ^{12}C single-particle shallow potential well and were termed "quasimolecular" resonances. These molecular states are considered to be an intermediate step (doorway) in the reaction mechanism lying between entrance channel and compound nucleus formation and being able to decay either into the compound nucleus or back into the entrance channel. The resonance-like creation of the doorway state then results in cross section resonances for the particles evaporated from the compound nucleus. Thus, the resonances appear as "entrance-channel" effects. This hypothesis is supported by experimental data (90), since other

[7] A similar conclusion has been drawn by High & Cujek (88).

heavy-ion reactions leading to the same compound nucleus do not show evidence for intermediate structure. In the model of Imanishi (91), the incident channel is coupled to a doorway state where one of the carbon nuclei is in its first excited state. Although this model and its extensions (90) generate resonance structures that can be linked with observation at low energies, they do not provide an explanation for the absence of structures in other systems. It has also been suggested that the density of levels of the compound nucleus as well as the number of open channels may enter importantly into the presence or absence of resonances (90) in a given system, since both quantities are small for $^{12}C + ^{12}C$ and increase significantly going to $^{12}C + ^{16}O$ and $^{16}O + ^{16}O$. However, there are other systems such as $^9Be + ^{12}C$ and $^{12}C + ^{13}C$ with small numbers for both quantities, and neither of these systems (92) exhibits resonance structures. The above constraints seem therefore to be inconclusive. Michaud & Vogt (85, 86) suggested a mechanism in which the resonances corresponded to intermediate α-cluster states. This model considers (90) the excitation of one carbon nucleus to its $J_\pi = 0^+$, 7.65-MeV state, which has a good 3α-particle structure (Section 5.1), and this doorway state is trapped in the effective potential of both colliding nuclei. Although some of these models (90) have met with success in providing partial explanation of the experimental features, none are able to give an entirely satisfactory answer. Obviously, a great deal of experimental and theoretical work remains to be done.

The experimental determination of J^π assignments for the observed resonances will help to test the proposed models (90, 93). Since the α-particle model suggests (85) the presence of intermediate structure in the $^{12}C + ^{20}Ne$ system and since this system may also have astrophysical importance similar to the $^{16}O + ^{16}O$ system (because of the presence of ^{12}C and ^{20}Ne nuclei in the core of a star and because of a similar height in the Coulomb barrier), an experimental investigation of the $^{12}C + ^{20}Ne$ fusion reaction is desirable. Such studies are currently underway by the Münster group. Feldman et al (94) have shown that heavy-ion capture reactions, leading directly to the lowest states of the compound nucleus, contribute very little to the total stellar reaction rate. It has been pointed out, however, by Vogt (95) that—because of the intermediate structure of the resonances—the capture process may proceed predominantly by γ-γ cascades through higher excited states. This possibility and the difference in the energy dependence between a γ-ray and a charged-particle exit channel make the importance of heavy-ion capture reactions at stellar energies an open question. A search for heavy-ion capture represents a novel challenge to experimentalists.

6.3 *Gross Energy Dependence*

The recent cross-section measurements with ions in the mass region of $A = 9$–16 point (96) to nonsystematic trends at subbarrier energies (Figure 16). The gross energy dependence varies markedly from one system to the next, even though differences of only one or two nucleons are involved. This observation is in sharp contrast to the situation usually encountered at energies above the barrier and in systems of heavier ions. The observed features for different ion pairs at subbarrier energies cannot be reproduced with a consistent set of optical-model parameters (except for $^{10,11}B + ^{12}C$) and suggest that nuclear structure interactions on a microscopic level may sensitively determine the reaction mechanism between heavy ions in the above mass region. These features may also reflect interesting variations in the ion-ion interaction potential. All these data provide a rich source of information on the macroscopic and microscopic physics involved in heavy-ion reactions. The explanation of these data is an important and interesting subject for further experimental and theoretical work.

Once the mechanisms involved in these features, as well as in the intermediate structures, are completely clarified, one may have also learned how two deformed nuclei interact when they are in close contact[8] and under what circumstances nuclear molecule systems may exist.

7 OTHER ASPECTS OF NUCLEAR ASTROPHYSICS

For the final stages of stellar evolution (silicon burning and explosive nucleosynthesis) a large number of nuclear reactions are involved in the medium mass region. Many of these reactions have not been experimentally measured and represent an open field for future activities in the nuclear laboratory (see 14 and 23 for tables of astrophysically important reactions). Considerable theoretical effort is currently being devoted (99) to produce a program that will predict reaction rates for the large number of cases where laboratory data are unobtainable because of unstable target nuclei or the importance of excited states of the target nuclei to the overall reaction rate in stellar environments. It is, however, imperative to check—wherever possible—theoretically predicted reaction rates, in part as an incentive for further improvements

---

[8] Studies of Arnould & Howard (97) indicate, that the intrinsic deformation of ^{12}C nuclei and their relative orientation during the collision $^{12}C + ^{12}C$ may play a critical role in determining low-energy fusion cross sections.

in theory. For example, large Wigner cusps (18) have been predicted (99) in the limit of strong isospin mixing for several (p,γ) reactions on medium mass nuclei. Substantial Wigner cusps have been observed (98) recently on ^{54}Cr and ^{64}Ni target nuclei. They will be useful eventually as a technique to measure the extent of isospin mixing in the continuum.

Nuclear astrophysics requires not only activities at extremely low and medium energies, but measurements of nuclear cross sections at very high energies are also needed to provide spallation yields from the interactions of cosmic rays and of flare-accelerated particles with interstellar atmospheres and stellar objects. These measurements are crucial for the understanding of the rare light elements (Group *l* in Figure 2). In recent years there has been significant progress in understanding the processes that created these light elements (8). An unexpected byproduct of this understanding is the most convincing answer to date of the cosmological question: Will the presently observed expansion of the universe continue forever or will the universe eventually collapse again to a hot dense singularity. From the measurements of the spallation yields and other arguments, there is good evidence (8) that the present abundances of deuterium and ^{7}Li are remnants from the primeval fireball phase of the universe and that they have been largely preserved until the present epoch. These abundances can be used therefore to place limits on the mean baryon density of the universe. In the simplest cosmological model, these limits imply (8) that the expansion of the universe will continue forever (open universe).

Nuclear astrophysics involves most of astrophysics, because there are few important events in astronomy, cosmogony, and cosmology that have not left nuclear clues, which in many cases can be studied in nuclear laboratories. This article discussed only a few examples of these nuclear clues. It is obvious, that much remains to be worked out and proven on the nuclear physics involved in the many aspects of astrophysics. From a purely nuclear point of view, the nuclear reactions studied are often of comparatively little interest. The intellectual motivation is to be found more in wresting from nature a hard-won number, that she has presumably used herself, and in evaluating its astrophysical consequences. However, experience reveals many occasions when the evaluation and interpretation of the reaction rates present unsuspected intellectual rewards in nuclear physics itself. Some of the painstaking data presented in this review would probably never have been initiated from a pure nuclear point of view, yet they have presented in turn a rich source of new and interesting nuclear phenomena.

The quantitative understanding of the nuclear physics involved in astrophysics is one of the vital ingredients in our attempts to explain the structure and evolution of stars, galaxies, and the universe, as well as the origin of the chemical elements. It has been man's most enduring challenge to understand the stars, and the association of the research in nuclear astrophysics with attaining this goal should give all groups engaged in this enterprise an additional appreciation of the significance of their work to applications in other sciences.

ACKNOWLEDGMENTS

The authors appreciate the numerous collaborations and discussions with A. E. Litherland, R. E. Azuma, K. P. Jackson, W. S. Rodney, and H. Winkler. C.R. gratefully acknowledges the fruitful and enjoyable stay at the Kellogg Radiation Laboratory at CIT, where many ideas were born.

Literature Cited

1. Burbidge, E. M., Burbidge, G. R., Fowler, W. A., Hoyle, F. 1957. *Rev. Mod. Phys.* 29:547
2. Clayton, D. D. 1968. *Principles of Stellar Evolution and Nucleosynthesis.* New York: McGraw-Hill. 606 pp.
3. Suess, H. E., Urey, H. C. 1956. *Rev. Mod. Phys.* 28:53
4. Cameron, A. G. W. 1972. *Explosive Nucleosynthesis.* Austin: Univ. Texas. 301 pp.
5. Alpher, R. A., Bethe, H., Gamow, G. 1948. *Phys. Rev.* 73:803
6. Penzias, A. A., Wilson, R. W., 1965. *Astrophys. J.* 142:419; Dicke, R. H., Peebles, P. J. E., Roll, P. G., Wilkinson, D. T. 1965. *Astrophys. J.* 142:414
7. Wagoner, R. V., Fowler, W. A., Hoyle, F. 1967. *Astrophys. J.* 148:3; Wagoner, R. V. 1973. *Astrophys. J.* 179:343; Reeves, H., Audouze, J., Fowler, W. A., Schramm, D. N. 1973. *Astrophys. J.* 179:909
8. Reeves, H., Fowler, W. A., Hoyle, F. 1970. *Nature* 226:727; Schramm, D. N., Wagoner, R. V. 1974. *Phys. Today* 27:40; Gott, J. R., Gunn, J. E., Schramm, D. N., Tinsley, B. M. 1974. *Astrophys. J.* 194:543; Epstein, R. I., Arnett, W. D., Schramm, D. N. 1976. *Astrophys. J.* 31:Suppl. 1, p. 111; Austin, S. M. 1977. *Preprint MSUCL-254*
9. Reeves, H., Salpeter, E. E. 1959. *Phys. Rev.* 116:1505; Fowler, W. A., Hoyle, F. 1964. *Astrophys. J.* 9:Suppl., p. 201; Arnett, W. D. 1969. *Astrophys. Space Sci.* 5:180; Arnett, W. D. 1970. *Astro-*

phys. J. 162:349; Arnett, W. D., Truran, J. W., Woosley, S. 1971. *Astrophys. J.* 165:87; Arnett, W. D. 1973. *Astrophys. J.* 179:249
10. Bodansky, D., Clayton, D. D., Fowler, W. A. 1968. *Astrophys. J.* 16:Suppl. 148, p. 299; Arnett, W. D. 1977. *Astrophys. J.* 35:Suppl. 2, p. 145
11. Arnett, W. D. 1973. *Ann. Rev. Astron. Astrophys.* 11:73; Schramm, D. N., Arnett, W. D. 1973. See Ref. 4
12. Couch, R. G., Arnett, W. D. 1972. *Astrophys. J.* 178:771; Lamb, S. A., Howard, W. M., Truran, J. W., Iben, I. 1977. *Astrophys. J.* 217:213
13. Truran, J. W. 1973. See Ref. 4
14. Clayton, D. D., Woosley, S. E. 1973. *Proc. Int. Conf. Nucl. Phys., Munich* 2:718. Amsterdam: North-Holland
15. Allen, B. J., Gibbons, J. H., Macklin, R. L. 1971. *Adv. Nucl. Phys.* 4:205
16. Barnes, C. A. 1971. *Adv. Nucl. Phys.* 4:133
17. Fowler, W. A., Caughlan, G. R., Zimmerman, B. A. 1975. *Ann. Rev. Astron. Astrophys.* 13:69
18. Wigner, E. P. 1948. *Phys. Rev.* 73:1002
19. Lorenz-Wirzba, H., Rolfs, C. 1977. *Ann. Rep. Inst. Kernphys. Münster.* 98 pp.
20. Rolfs, C., Rodney, W. S. 1974. *Astrophys. J. Lett.* 194:63; 1975. *Nucl. Phys. A* 250:295
21. Ajzenberg-Selove, F. 1972. *Nucl. Phys. A* 190:1; 1975. 248:1; 1977. 281:1
22. Wiescher, M., Trautvetter, H. P., Kettner, K. U., Rolfs, C. 1977. See Ref. 19
23. Fowler, W. A. 1968. *New Uses of Low*

Energy Accelerators. Washington, DC: Natl. Acad. Sci.
24. Iss. Large Electr. Accel. 1974. *Nucl. Instrum. Methods* 122:1; Aitken, D. 1976. *Nucl. Instrum. Methods* 139:125
25. Hammer, J. W., Schüpferling, H. M., Bergandt, E., Pflaum, Th. 1975. *Nucl. Instrum. Methods* 128:409
26. Freye, T., Lorenz-Wirzba, H., Cleff, B., Trautvetter, H. P., Rolfs, C. 1977. *Z. Phys. A* 281:211; Andersen, H. H., Horenshoj, P., Hojsholt-Poulsen, L., Knudsen, H., Nielsen, B. R., Stensgaard, R. 1976. *Nucl. Instrum. Methods* 136:119
27. Trautvetter, H. P., Rolfs, C. 1975. *Nucl. Phys. A* 242:519
28. Proc. Int. Symp. Res. Mater. Nucl. Meas., 3rd. 1972. *Nucl. Instrum. Methods* 102:373
29. Parks, P. B., Beard, P. M., Bilpuch, E. G., Newson, H. W. 1964. *Rev. Sci. Instrum.* 35:549; Bloch, R., Pixley, R. W., Winkler, H. 1967. *Helv. Phys. Acta* 40:832; Litherland, A. E., Ollerhead, R. W., Smulders, P. J., Alexander, T. K., Broude, C., Ferguson, A. J., Kuehner, J. A. 1967. *Can. J. Phys.* 45:1901; Bussière, J., Robson, J. M. 1971. *Nucl. Instrum. Methods* 91:103; Dwarakanath, M. R., Winkler, H. 1971. *Phys. Rev. C* 4:1532; Rolfs, C. 1977. See Ref. 19
30. Rolfs, C., Rodney, W. S., Shapiro, M. H., Winkler, H. 1975. *Nucl. Phys. A* 241:460; Berg, H. L., Hietzke, W., Rolfs, C., Winkler, H. 1977. *Nucl. Phys. A* 276:168
31. Allen, K. W., Dolan, S. P., Holmes, A. R., Symons, T. J. M., Watt, F., Zimmerman, C. H., Litherland, A. E., Sandorfi, A. M. J. 1976. *Nucl. Instrum. Methods* 134:1
32. Spinka, H., Winkler, H. 1974. *Nucl. Phys. A* 233:456
33. Cerny, J. 1974. *Nuclear Spectroscopy and Reactions,* Part A, pp. 3, 290, Part C, p. 143. New York: Academic
34. Davis, R. Jr., Harmer, D. S., Hoffman, K. C. 1968. *Phys. Rev. Lett.* 20:1205; Davis, R. Jr., Evans, J. C. 1976. *Brookhaven Natl. Lab. Rep.*
35. Fowler, W. A. 1977. *Cal. Tech. Preprint OAP-507*
36. Bahcall, J. N., Bahcall, N. A., Ulrich, R. K. 1969. *Astrophys. J.* 156:559
37. Marrs, R. E., Bodansky, D., Adelberger, E. G. 1973. *Phys. Rev. C* 8:427; Davison, N. E., Canty, M. S., Dohan, D. A., McDonald, A. 1974. *Phys. Rev. C* 10:50; Alvarez, L. W. 1973. *LRL Phys. Note 767*
38. Kavanagh, R. W. 1972. *Cosmology, Fusion and other Matters.* Denver:

Colorado Univ. Press. 169 pp.
39. Wiezorek, C., Kräwinkel, H., Santo, R., Wallek, L. 1977. *Z. Phys. A* 282:121
40. Slobodrian, R. J., Pigeon, R., Irshad, M. 1975. *Phys. Rev. Lett.* 35:19
41. Newman, M. J., Fowler, W. A. 1976. *Phys. Rev. Lett.* 36:895
42. Davies, W. G., Ball, G. C., Ferguson, A. J., Forster, J. S., Horn, D., Warner, R. E. 1977. *Phys. Rev. Lett.* 38:1119
43. Fowler, W. A. 1972. *Nature* 238:24; Fetisov, V. N., Kopysov, Y. S. 1972. *Phys. Lett. B* 40:602
44. McDonald, A. B., Alexander, T. K., Bune, J. R., Mak, H. B. 1977. *Nucl. Phys. A* 288:529
45. Dwarakanath, M. R. 1974. *Phys. Rev. C* 9:805
46. Bahcall, J. N. 1977. *Astrophys. J. Lett.* 216:115
47. Rolfs, C., Rodney, W. S. 1974. *Nucl. Phys. A* 235:450
48. Caughlan, G. R. 1977. *CNO-Isotopes in Astrophysics.* Netherlands: Reidel. 121 pp.
49. Hebbard, D. F. 1960. *Nucl. Phys.* 15:289
50. Rolfs, C. 1973. *Nucl. Phys. A* 217:29
51. Brown, R. E. 1962. *Phys. Rev.* 125:347
52. Rolfs, C., Charlesworth, A. M., Azuma, R. E. 1973. *Nucl. Phys. A* 199:257
53. Marion, J. B., Fowler, W. A. 1957. *Astrophys. J.* 125:221
54. Zyskind, J. L., Rolfs, C. Unpublished.
55. Rolfs, C., Winkler, H. 1974. *Phys. Lett. B* 52:317
56. Endt, P. M., van der Leun, C. 1973. *Nucl. Phys. A* 214:1
57. Dyer, P., Barnes, C. A. 1974. *Nucl. Phys. A* 233:495
58. Öpik, E. J. 1951. *Proc. R. Ir. Assoc. A* 54:49; Salpeter, E. E. 1952. *Phys. Rev.* 88:547; 1952. *Astrophys. J.* 115:326
59. Hoyle, F. 1954. *Astrophys. J. 1*: Suppl., p. 121
60. Nolen, J. A., Austin, S. M. 1976. *Phys. Rev. C* 13:1773
61. Barnes, C. A., Nichols, D. B. 1973. *Nucl. Phys. A* 217:125
62. Crannell, H., Griffy, T. A., Suelzle, L. R., Yearian, M. R. 1967. *Nucl. Phys. A* 90:152; Strehl, P., Schucan, Th. H. 1968. *Phys. Lett. B* 27:641; Strehl, P. 1970. *Z. Phys.* 234:416
63. Chamberlin, D., Bodansky, D., Jacobs, W. W., Oberg, D. L. 1974. *Phys. Rev. C* 9:69
64. Seeger, P. A., Kavanagh, R. W. 1963. *Astrophys. J.* 137:704; 1963. *Nucl. Phys.* 46:577
65. Markham, R. G., Austin, S. M., Shahabuddin, M. A. M. 1976. *Nucl. Phys. A* 270:489
66. Alburger, D. E. 1960. *Phys. Rev.* 118:

235; 1977. *Phys. Rev. C* 16:2394
67. Robertson, R. G. H., Warner, R. A., Austin, S. M. 1977. *Phys. Rev. C* 15:1072
68. Loebenstein, H. M., Mingay, D. W., Winkler, H., Zaidins, C. S. 1967. *Nucl. Phys.* 91:481; Puehlhofer, F., Ritter, H. G., Brommunat, G., Schmidt, H., Bethge, K. 1970. *Nucl. Phys. A* 147:258
69. Hättig, H., Hünchen, K., Roth, P., Wäffler, H. 1969. *Nucl. Phys. A* 137:144; Neubeck, K., Schober, H., Wäffler, H. 1974. *Phys. Rev. C* 10:320
70. Barker, F. C. 1971. *Aust. J. Phys.* 24:777; Werntz, C. 1971. *Phys. Rev. C* 4:1591
71. Jaszczak, R. J., Gibbons, J. H., Macklin, R. L. 1970. *Phys. Rev. C* 2:63, 2452
72. Jones, C. M., Phillips, G. C., Harris, R. W., Beckner, E. H. 1962. *Nucl. Phys.* 37:1; Clark, G. J., Sullivan, D. J., Treacy, P. B. 1968. *Nucl. Phys. A* 110:481
73. Koonin, S. E., Tombrello, T. A., Fox, G. 1974. *Nucl. Phys. A* 220:221
74. Humblet, J., Dyer, P., Zimmerman, B. A. 1976. *Nucl. Phys. A* 271:210
75. Arnett, W. D. 1972. *Astrophys. J.* 176:681
76. Couch, R. G., Spinka, H., Tombrello, T. A., Weaver, T. A. 1971. *Nucl. Phys. A* 175:300
77. Black, D. 1972. *Geochim. Cosmochim. Acta* 36:377
78. Bair, J. K., Haas, F. X. 1973. *Phys. Rev. C* 7:1356; Haas, F. X., Bair, J. K. 1973. *Phys. Rev. C* 7:2432; Mak, H. B., Ashery, D., Barnes, C. A. 1974. *Nucl. Phys. A* 226:493
79. Trautvetter, H. P., Wiescher, M., Kettner, K. U., Rolfs, C., Hammer, J. W. 1978. *Nucl. Phys. A* 297:489
80. Chandrasekar, S. 1931. *Astrophys. J.* 74:81
81. Almquist, E., Bromley, D. A., Kuehner, J. A. 1960. *Phys. Rev. Lett.* 4:515; 1963. *Phys. Rev.* 130:1140; 1964. 136:B84; Davis, R. H. 1960. *Phys. Rev. Lett.* 4:521
82. Patterson, J. R., Winkler, H., Zaidins, C. S. 1969. *Astrophys. J.* 157:367
83. Mazarakis, M. G., Stephens, W. E. 1973. *Phys. Rev. C* 7:1280
84. Cujek, B., Barnes, C. A. 1976. *Nucl. Phys. A* 266:461; Christensen, P. R., Switkowski, Z. E., Dayras, R. A. 1977. *Nucl. Phys. A* 280:189
85. Michaud, G., Scherk, L., Vogt, E. W. 1970. *Phys. Rev. C* 1:864; Michaud, G., Vogt, E. W. 1969. *Phys. Lett. B* 30:85; 1972. *Phys. Rev. C* 5:350
86. Michaud, G. 1972. *Astrophys. J.* 175:751; 1973. *Phys. Rev. C* 8:525
87. Kettner, K. U., Lorenz-Wirzba, H., Rolfs, C., Winkler, H. 1977. *Phys. Rev. Lett.* 38:337; Kettner, K. U. 1977. *Diplomarbeit.* Münster: Inst. Kernphys.
88. High, M. D., Cujek, B. 1977. *Nucl. Phys. A* 282:181
89. Becker, H. W. 1978. *Diplomarbeit.* Münster: Inst. Kernphys.; Marquardt, N. 1977. Univ. Bochum. Private communication
90. Feshbach, H. 1976. *J. Phys. C* 5: Suppl., p. 177 and ref. therein
91. Imanishi, B. 1969. *Nucl. Phys. A* 125:33; Park, J. Y., Greiner, W., Scheid, W. 1977. *Phys. Rev. C* 16:2276
92. Dayras, R. A., Stokstad, R. G., Switkowski, Z. E., Wieland, R. M. 1976. *Nucl. Phys. A* 265:153; Switkowski, Z. E., Wu, S. C., Overlay, J. C., Barnes, C. A. 1977. *Nucl. Phys. A* 289:236
93. Galster, W., Treu, W., Dück, P., Fröhlich, H., Voit, H. 1977. *Phys. Rev. C* 15:950
94. Feldman, W., Heikkinen, D. W. 1969. *Nucl. Phys. A* 133:177
95. Vogt, E. W. 1977. Private communication with A. E. Litherland, Univ. Toronto
96. Stokstad, R. G., Switkowski, Z. E., Dayras, R. A., Wieland, R. M. 1976. *Phys. Rev. Lett.* 37:888
97. Arnould, M., Howard, W. M. 1976. *Nucl. Phys. A* 274:295
98. Mann, F. M., Dayras, R. A., Switkowski, Z. E. 1975. *Phys. Lett. B* 58:420; Zyskind, J. L. Davidson, J. M., Esat, M. T., Shapiro, M. H., Spear, R. H. 1977. *Cal. Tech. Preprint OAP-512*
99. Woosley, S. E., Fowler, W. A., Holmes, J. A., Zimmerman, B. A. 1975. *Cal. Tech. Preprint OAP-422*

Ann. Rev. Nucl. Part. Sci. 1978. 28:161–205

HIGH ENERGY INTERACTIONS OF NUCLEI

✖5594

Alfred S. Goldhaber[1]

Institute for Theoretical Physics, State University of New York, Stony Brook, New York 11794

Harry H. Heckman[2]

Lawrence Berkeley Laboratory, University of California, Berkeley, California 94720

CONTENTS

[1] On sabbatical leave, 1977–78.
[2] Alexander von Humboldt Award, 1977–78.

161

0066-4243/78/1201-0161$01.00

1 INTRODUCTION

Nuclear physics and nuclear chemistry developed during the past half century as fields of research concerned with the nature of nuclear matter at or near equilibrium. The emergence in recent years of heavy ion accelerators with energies surpassing 100 A MeV (where A is the nuclear mass number) has opened the possibility of studying in the laboratory the effects of compressing nuclear matter to densities two or more times greater than normal while simultaneously "heating" the matter to temperatures in the 100 MeV range. This article is devoted to an assay of the present status of high energy nuclear collision studies and of the important questions that remain to be addressed.

A collision of two nuclei may be labeled by the impact parameter b. At one extreme is the case of maximum b, a peripheral or grazing collision, in which only one or a few nucleons leave the projectile or the target, and large, slightly excited projectile and target fragments remain. Such a process is widely supposed to be predictable in terms of conventional nuclear structure and collision theory. Indeed, plausible model approximations in this framework give a fair account of experimental data. Thus, from the viewpoint of a nuclear theorist, peripheral processes might be thought of mainly as a testing ground for conventional concepts. However, nuclear theory does not always attain that degree of precision and reliability required to make theoretical predictions useful for applications. For this reason, if no other, it has proved important to study such processes, especially projectile fragmentation, which plays a critical role in determining the composition of cosmic rays in proximity to the earth, and relating that composition to the nature of cosmic-ray sources.

At the opposite extreme in impact parameter is a central collision, in which the projectile and target overlap completely. Such collisions could produce the high densities and temperatures that might lead to important and perhaps surprising transformations of nuclear matter. It is obvious that theoretical predictions of such phenomena must be difficult and unreliable. However, the experimental analysis of central collisions is also a formidable challenge. First of all, there is not even an established criterion to identify a central collision. Second, assuming unusual effects occur during a stage of high compression, it is not clear how such effects will be manifested in the final distribution of reaction products.

The study of high energy nuclear collisions is now at a natural "vista point" from which to appraise the progress already made and to set a future course. Fragmentation processes are well in hand, especially on the experimental side. Central processes have great potential, if we can find the right way to look at them.

The following section contains a brief historical review, starting with the early observations of cosmic-ray nuclear interactions. Also included are some kinematic terms, with definitions and motivations for their use, along with a description of accelerator facilities and a cursory outline of experimental techniques.

Section 3 is a description of the most mature branch of the subject, processes that leave significant near-projectile-velocity and near-target-velocity fragments.

Section 4 describes experiments not restricted to projectile or target fragments, and contains comparisons of data with conventional theoretical models representing the most conservative extrapolation of traditional nuclear and particle physics concepts. There is also a brief description of theoretical speculations about possible new phases of matter at higher-than-normal baryon density. In both Sections 3 and 4, some theoretical comments are included in the experimental discussion, while others are put in a separate subsection.

Finally, in Section 5, we present some conclusions and suggestions for the future, hoping that these will stimulate other and better proposals from readers. To help compensate for omissions imposed by the brevity of this article, we list several other recent reviews on experimental or theoretical aspects of our subject.

2   PRIMER

2.1   History: Cosmic-Ray Period

Two independent works thirty years ago were forerunners of current approaches to high energy interactions between nuclei. Freier et al (1, 2) discovered $Z \geqq 2$ components in the primary cosmic radiation, which verified a prediction by Alfvén (3) and opened the way to experimental studies long before high energy heavy ion beams became available at laboratory accelerators. Meanwhile Feenberg & Primakoff (4) were led by the properties of the nucleon-nucleon tensor force to speculate that the familiar saturation of nuclear density might be overcome, and tightly bound "collapsed" nuclei might occur in nature. In the past decade similar speculations based on various hypothetical mechanisms have appeared.

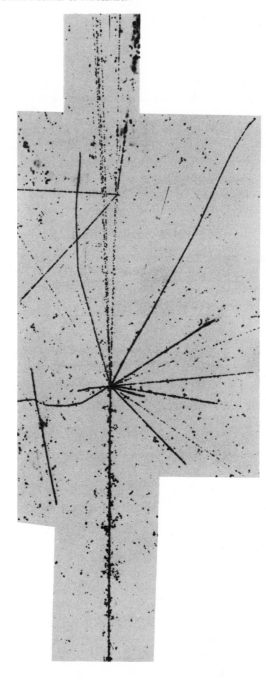

Figure 1   Interaction of an ^{40}Ar projectile, $T = 1.8$ A GeV, with a Ag(Br) nucleus in Ilford G.5 emulsion that shows characteristics of both projectile and target fragmentation.

Today the intriguing problem of high nuclear densities and the possible role of high energy nuclei in achieving these densities has become the most dramatic and difficult topic for experimental and theoretical investigation in this field.

Early studies of heavy ion interactions in matter were primarily concerned with interaction mean free paths and the production of nuclear fragments, as such data are most pertinent to the physics of cosmic rays. Here, the principal goal is to learn from the chemical and, eventually, isotopic composition of the observed heavy cosmic-ray nuclei, the conditions of their origin, acceleration mechanisms, and subsequent propagation. The astrophysically important information derived from heavy cosmic-ray nuclei is described in comprehensive review articles by Shapiro & Silberberg (5) and Waddington (6), and the classic monograph of Powell et al (7), with its wealth of information and pictorial examples of particle interactions in nuclear track emulsions.

2.1.1   QUALITATIVE CLASSIFICATION OF RELATIVISTIC HEAVY ION COLLISIONS   The pioneering cosmic-ray work that pertains to the interactions of nuclei is given in a series of papers by Bradt & Peters (8–11), where one is introduced to the concepts of peripheral and central collisions between nuclei (9). In a peripheral collision, part of the projectile that overlaps the target nucleus may be sheared off. The remaining nuclear matter proceeds with its original velocity as fragment nuclei of reduced charge and/or mass, alpha particles, and nucleons (10–12). Figure 1 is an example of a collision between an ^{40}Ar nucleus, kinetic energy $T = 72$ GeV (1.8 A GeV), with a Ag or Br nucleus in Ilford G.5 emulsion. This event shows the characteristics of both projectile and target fragmentation. The narrow forward cone of five He fragments indicates that a peripheral collision has taken place. The velocities of the He nuclei are approximately equal to the velocity of the incident ^{40}Ar (an example of persistence of velocities), and hence, correspond to low velocities (energies) in the projectile frame. In the laboratory, or target frame, the counterparts of such fragments are emitted as low energy, heavily ionizing tracks, several of which are seen in this event. Note that a π^- meson is produced, comes to rest, and forms a three-prong star.

Figure 2 illustrates a central collision between a 1.8 A GeV-^{40}Ar projectile and a Pb target nucleus in a streamer chamber. In this case both nuclei are destroyed. Such an event involves high levels of excitation and the emission of a large number of secondary fragments. At least 63 particles, predominantly nucleons, light fragments, and pions (including six π^-) are produced in this particular event.

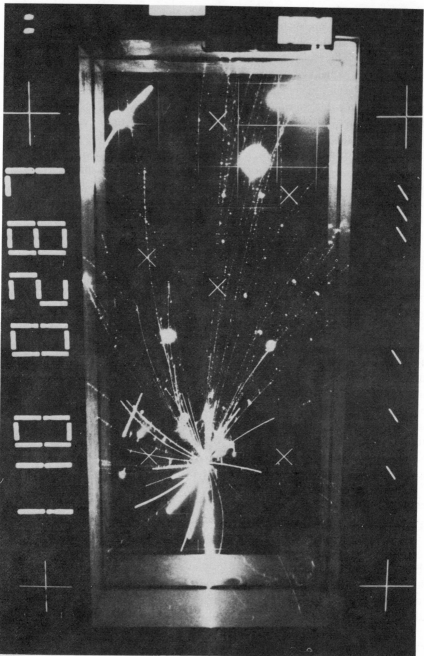

Figure 2 Interaction of an ^{40}Ar projectile, $T = 1.8$ A GeV, with a Pb nucleus (Pb$_3$O$_4$ target) in the LBL streamer chamber leading to the catastrophic destruction of target and projectile nuclei. Negative particles have counter-clockwise trajectories in the magnetic field of the chamber (courtesy of UC Riverside/LBL collaboration).

2.1.2 REACTION CROSS SECTIONS A semiempirical "black sphere" expression for the reaction cross section of beam and target nuclei, with mass numbers A_B and A_T, respectively, introduced by Bradt & Peters (10) and extensively tested with nuclear emulsion detectors using comic ray nuclei $2 \leq Z \leq 26$ is

$$\sigma = \pi r_0^2 (A_B^{\frac{1}{3}} + A_T^{\frac{1}{3}} - \delta)^2. \qquad 2.1$$

The "overlap parameter" δ is meant to represent the diffuseness and partial transparency of the nuclear surfaces.

An assessment of the mean-free-path data for heavy cosmic-ray nuclei in emulsion by Cleghorn (13, 14) resulted in the following conclusions: (a) The overlap model accounts for the measured mean-free-path data for $2 \leq Z \leq 26$ to an accuracy of about 10–15% for $r_0 = 1.2$ fm and $\delta = 0.5$. Because r_0 and δ are coupled in Equation 2.1, these values are not unique. (b) The reaction cross sections of nuclei are essentially independent of energy from 0.1 to 30 A GeV. (c) The fragmentation parameters p_{ij}, the average number of i-type nuclei produced in the fragmentation of a j-type projectile, are also consistent with energy independence over this energy range. The errors in p_{ij} are typically 10–20% (5, 15).

Eisenberg (16) refined the Bradt-Peters approach, proposing a "grey sphere" model. Using a mean-free path $\lambda = 5$ fm of a nucleon in nuclear matter, of uniform density, he found a 25% reduction of the geometric cross section due to transparency for $A_B = 25$ and $A_T = 64$; for $A_T = 207$, the reduction was 16%.

In the same paper, Eisenberg introduced what is now generally known as the abrasion-ablation model of nuclear collisions (17). In this model fragment production occurs in collisions where part of the incident projectile is "cut"; nucleons in the overlapping region are removed (abrasion), and nucleons subsequently evaporate from the residual, excited prefragment (ablation).

Alexander & Yekutieli (18) and Aizu et al (19), noting that Equation 2.1 is independent of the nucleon-nucleon cross section in nuclear matter as well as the nucleon-density distributions of the colliding nuclei, calculated the reaction cross section σ_r for the nucleus-nucleus interaction at high energies from the optical model (20, 21). Alexander & Yekutieli presented their results as a function of σ_0, the effective nucleon-nucleon cross section in nuclear matter, and concluded that both the calculated nucleon-nucleus and nucleus-nucleus cross sections were in general agreement for values of $\sigma_0 = 30$–40 mb. Their calculations showed that σ_r is not sensitive to σ_0. For example, a 10% change in σ_0 results in only a 2% change in the σ_r for carbon projectiles. Thus, any energy dependence of σ_0 will be suppressed in σ_r.

2.1.3 REACTION PRODUCTS   Although encumbered by low intensities and uncertainties in charge, mass, and energy determinations, experimental studies of nucleus-nucleus interactions using cosmic rays revealed important aspects of the high energy interactions of nuclei. Early experiments were concerned with meson production by α particles and heavier nuclei. Investigations by Jain et al (22) [α particles and nuclei $Z \geqq 6$, $7 \leqq T \leqq 100$ A GeV], Alexander & Yekutieli (18) [$3 \leqq Z \leqq 8$, $T \geqq 1.5$ A GeV], Tsuzuki (23) [$6 \leqq Z \leqq 8$, $T \approx 40$ A GeV], Rybicki (24) [$Z \geqq 3$, $T \geqq 100$ A GeV], Abraham et al (25) [$Z \geqq 2$, $T \geqq 1000$ A GeV], and Andersson et al (26) [$3 \leqq Z \leqq 26$, $T \geqq 1.7$ A GeV] are representative studies of shower-particle, i.e. fast meson and nucleon, production in cosmic-ray, heavy ion collisions over the energy range $1-10^3$ A GeV.

Typical angular distributions of shower particles produced by cosmic-ray nuclei, $3 \leqq Z \leqq 26$, at kinetic energies greater than 1.7 A GeV (mean energy $= 7$ A GeV) are shown in Figure 3 (26). The principal feature of the angular distribution of shower particles is that they are composed of two main components: first, protons from the fragmenting projectile confined to a narrow cone, and second, pions and protons that have a

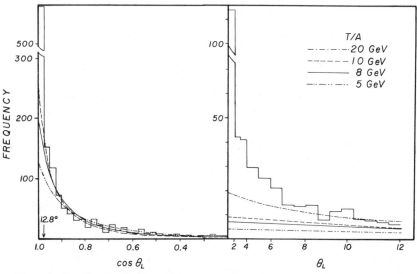

Figure 3 Angular distributions of shower particles from the interactions of cosmic-ray heavy ions, $3 \leqq Z \leqq 26$ and $T \geqq 1.7$ A GeV, with emulsion nuclei (a) $0° \leqq \theta_{lab} \leqq 80°$ and (b) $0° \leqq \theta_{lab} \leqq 12°$. Curves are calculated distributions of pions for projectile energies 5–20 A GeV (26).

wider distribution. Figure 3a shows the angular distribution of shower particles $0 \leq \theta_{lab} \leq 80°$, and Figure 3b gives a more detailed presentation at small angles $0 \leq \theta_{lab} \leq 12°$. The curves show calculated angular distributions for pions, assumed isotropic in the nucleon-nucleon center of mass. The results of the work described above as well as other emulsion experiments by the University of Lund group (27–34) were in qualitative agreement with the hypothesis that the nucleus-nucleus interaction could be described as a superposition of independent nucleon-nucleon (18, 22, 24, 25), nucleon-nucleus (26), or α-α (23) collisions. However, notable discrepancies remained:

1. The apparent recoil velocity, $\beta = v/c \approx 0.02$, of the residual target nucleus, as evident from the forward:backward ratios ≈ 1.4 for low energy ($\lesssim 30$ A MeV) fragments (black prongs, in emulsion terminology), is virtually independent of the mass of the projectile (18, 26, 28, 35, 36).

2. The spectra of energies, angles, and average multiplicities of black tracks, $\langle n_b \rangle$, are nearly the same for incident protons and heavy nuclei (28); $\langle n_b \rangle$ is usually considered to be a measure of the excitation energy transferred to the target nucleus.

3. For $T > 1.7$ A GeV, the mean multiplicities of pions $\langle n_\pi \rangle$ increase with energy [as $T^{0.4 \pm 0.08}$ between 10 and 40 A GeV (22)] and mass of the projectile, whereas the mean multiplicity of heavy prongs (nonpions) with energies below 400 A MeV, $\langle n_h \rangle$, is nearly independent of these quantities (28).

4. Departures of the nucleus-nucleus interaction from the characteristics of the nucleon-nucleus interaction increase markedly as n_h increases, becoming most evident for AgBr interactions with $n_h \geq 28$ and for "central" collisions with $7 \leq n_h \leq 28$ and no $Z \geq 2$ projectile fragments produced (34, 37).

Considering the limited control over experimental conditions and the low statistics, precise understanding of the complex interactions of high energy nuclei could not be expected from cosmic-ray experiments. Nonetheless, questions of basic importance were raised. For example, how is momentum transferred to nucleons and, in particular, helium, even when kinematically forbidden by models of quasielastic nucleon-nucleon or nucleon-alpha scattering (29)? Why does the structure of the projectile affect the charge (mass) distributions of its fragments (38, 39)? And finally what are the possibilities for making and observing new modes of collective interaction in nuclear matter, only hinted at in the cosmic-ray data on production of shower particles in highly central and disruptive collisions?

Cosmic-ray research has given us a preview of the physics of interactions between relativistic nuclei. We now focus our attention on this area of research, which came into existence with the development of relativistic, heavy ion beams at Princeton, Berkeley, Saclay, and Dubna in the early 1970s.

2.2 Glossary

As part of this primer we present a few definitions, mainly of kinematic quantities. For this discussion, we have drawn from Reference 40.

Momentum, p (GeV/c): $p = \beta\gamma E_0/c$, with $\beta = v/c$, $\gamma = (1-\beta^2)^{-\frac{1}{2}}$, and E_0 the rest energy of the particle in GeV.

Rigidity (i), R (gauss-cm): $R = B\rho$, where ρ is the radius of curvature of a particle's trajectory in a magnetic field of strength B (the component normal to particle's momentum).

Rigidity (ii), R (GV): $R = pc/Ze$, with $pc = \beta\gamma E_0$ (GeV), and Z the number of units of charge carried by the ion. R is numerically equal to momentum per unit charge p/Z (GeV/c). R (GV) $= 0.02998$ R (kilogauss-meter).

Kinetic energy per nucleon, T/A (GeV): $T/A = (\gamma-1)E_0/A$, with A the mass number of the nucleus. Note that the definition implies trivially $T = (T/A) A$ GeV; in other words, the energy of a nucleus with mass number A is the energy per nucleon, multiplied by A. Thus, the common usage giving energy per nucleon in GeV/A is incorrect; rather, the proper unit is GeV. If atomic mass units are used, T/u is given in GeV, with $u = E_0/0.9315016(26)$ GeV.

Rapidity, y: For a particle in any system, $y = \frac{1}{2} \ln \left[(E+p_{\parallel})(E-p_{\parallel})^{-1} \right] = \tanh^{-1} \beta_{\parallel}$, where E is the total energy, and $p_{\parallel}$ and $\beta_{\parallel}$ are the longitudinal momentum and velocity respectively ($c \equiv 1$). Use of the rapidity variable in place of velocity has the important property of relating two longitudinally moving frames by a simple translation along the rapidity axis. Therefore, distributions expressed in y have Lorentz-invariant shapes. At nonrelativistic energies, $\beta_{\parallel} \ll 1$, $\beta_{\perp} \ll 1$, one obtains $y = \tanh^{-1} \beta_{\parallel} \approx \beta_{\parallel}$ and $p_{\perp}/E_0 \approx \beta_{\perp}$. For an isotropic distribution in a system with $y = 0$, a contour plot of $p_{\perp}/E_0$ vs y at low velocities will form concentric semicircles about the origin. If the isotropic system is moving at β_c in the laboratory, the distribution will be the same, but centered on $y' = \tanh^{-1} \beta_c$.

At highly relativistic energies, one has $E \approx p$ and

$$y \approx \frac{1}{2} \ln \frac{p+p_{\parallel}}{p-p_{\parallel}} = \frac{1}{2} \ln \frac{1+\cos\theta}{1-\cos\theta} = -\ln\left(\tan\frac{\theta}{2}\right).$$

Thus, the shape of the distribution of particles in high energy events, plotted in the variable $-\ln\left[\tan\left(\theta/2\right)\right]$, will be nearly independent of the velocity of the observation frame along the beam direction.

2.3   Heavy Ion Facilities and Experimental Techniques

We give a brief overview of high energy heavy ion accelerators and the methods and techniques used in the experimental programs of these facilities. A full discussion is not only impractical but unwarranted, considering the recent review articles by Grunder & Selph (41) on heavy ion accelerators, Goulding & Harvey (42) on the electronic identification of nuclear particles, and Price & Fleischer (43) on particle tracks in solids, and the texts by Fleischer, Price & Walker (44) on tracks in solids, and Barkas (45) and Powell et al (7) on nuclear emulsions. An exposition of experimental techniques has been given by Stock (46) in his review of nuclear reactions between relativistic heavy ions.

2.3.1   HIGH ENERGY HEAVY ION ACCELERATORS: $T/A > 0.1$ GeV, $A \geq 4$
By 1970–71 it had become evident that proton accelerators in the few GeV region such as the Princeton Particle Accelerator (PPA), and the Bevatron, had largely completed their missions in high energy particle physics. There was also an increased awareness of new and beneficial applications of high energy accelerators to fields other than high energy physics (47). Long-standing interest in the use of high energy heavy ions for biomedical research and the realization that acceleration of heavy nuclei to relativistic energies would open a new field of nuclear research were the principal scientific justifications for acceleration of heavy ions (48, 49). The first successful acceleration of relativistic heavy ions in the US was achieved in 1971 at the PPA, a feat shortly followed at the Bevatron (50). Similar developments were taking place at Dubna (JINR), where acceleration of alpha particles in the synchrophasotron began in 1970. The possibility of using the ISR at CERN for heavy ion experiments was also proposed (51).

Table 1 summarizes the performance characteristics of synchrotrons that are accelerating beams of heavy ions, $Z \geq 2$, to energies $T/A \geq 0.1$ GeV. The closing of the PPA in 1972 made the Bevatron/Bevalac the only high energy heavy ion facility in the US.

Major heavy ion facilities under construction or in the planning-proposal stages are: (a) Project GANIL, Caen, France. The maximum energy for this multiaccelerator system will be 8 A MeV for ^{238}U increasing to 0.1 A GeV for ^{12}C (53). (b) Numatron Project, Japan. A synchrotron and storage ring with energies from ≈ 0.7 A GeV (U) to ≈ 1.5 A GeV (Ne) (54). (c) Nuklotron Project, Dubna, USSR. A super-conducting synchrotron with beam energies from 15–20 A GeV (55). (d) GSI, Darmstadt, Germany. First stage: A synchrotron with energies from 0.8 A GeV (U) to 2.0 A GeV (light ions). Second stage: 8–10 A GeV intersecting storage rings (56).

Table 1   Heavy ion facilities, $Z \geqq 2$ and $T/A \geqq 0.1$ GeV

Location	Accelerator	Ion	T/A (GeV)	Intensity (extracted, ppp)	Pulse rate
Saclay	Saturne[a]	^{4}He	$\leqq 1.2$	2×10^{10}	15/min
Dubna (JINR) (52)	Synchrophasotron	^{4}He	$\leqq 5.0$	—	—
		^{12}C	$\leqq 5.0$	—	—
		^{20}Ne	$\leqq 5.0$	—	—
Berkeley (LBL)	Bevalac[b]	^{6}Li	0.1–2.1	—	
		^{12}C	0.1–2.1	2×10^9	
		^{14}N	0.1–2.1	2×10^9	
		^{16}O	0.1–2.1	2×10^9	10/min, 15/min
		^{20}Ne	0.1–2.1	8×10^8	at $T/A \leqq 0.4$ GeV
		^{40}Ar	1.8	5×10^8	
		^{56}Fe	1.9	1–2×10^5	
Berkeley (LBL)	Bevatron[c]	^{4}He	0.1–2.1	1–2×10^{10}	
		^{12}C	0.1–2.1	1×10^8	

[a] By end of 1978, Saturne will accelerate $Z/A = 1/2$ nuclei through ^{20}Ne; $T/A = 0.10$–1.1 GeV.
[b] SuperHILAC injector: $T/A = 8.5$ MeV.
[c] Local linear accelerator for injector: $T/A = 5$ MeV.

The Bevalac, proposed by Ghiorso in 1971 (49, 57, 58), employs an 8.5 A MeV heavy ion linear accelerator, the SuperHILAC, as an injector of heavy ions for the Bevatron, which then continues the acceleration of the ions to a maximum energy of 2.6 A GeV (for particles with $Z/A = 1/2$).

The SuperHILAC delivers beam intensities up to 10^{10}–10^{12} particles per pulse at a maximum pulse rate of 36 sec^{-1}. Because the Bevatron accepts particles for 500 μsec each 4 to 6 seconds, the SuperHILAC needs to divert only a small fraction, $\approx 3\%$, of its duty cycle to the Bevatron. With two sources available the SuperHILAC can accelerate two different beams on a pulse-to-pulse basis, delivering one beam to the Bevatron while accelerating the other for low energy heavy ion experiments. By use of time sharing, programs at the SuperHILAC and Bevatron can proceed largely independently of each other. Essential for operation of the Bevalac is complete computer control of all aspects of the acceleration process.

The requirement that the Bevatron accept only fully stripped ions for acceleration means that the beam from the SuperHILAC must pass through a stripper foil at the maximum energy of 8.5 A MeV in order to produce the highest intensity of stripped ions. Both the stripping efficiency and the losses due to charge changing of the ions in matter increase as the charge of the ion increases, placing a practical limit on the heaviest element that can be accelerated. At the moment, this limit appears to be Fe. With a vacuum of 10^{-9}–10^{-10} Torr, and an upgraded SuperHILAC, the Bevalac would be able to accelerate partially stripped ions. This is the basis for a proposal to install a vacuum liner in the Bevalac and a third injector in the SuperHILAC to produce high intensity uranium beams at 1.1 A GeV (U^{+72}) (59). With such an improved system, the use of partially stripped ions would permit efficient acceleration and extraction of particle beams at energies as low as 35 A MeV, and would give the Bevalac an extraordinary range of beam masses and energies.

2.3.2 EXPERIMENTAL TECHNIQUES   Experiments in high energy heavy ion physics, because of their wide ranges in particle mass, charge, and energy, necessarily involve the application of particle detection methods from both low energy nuclear and high energy particle physics. The added dimension of high multiplicities of highly charged nuclear fragments, nucleons, and mesons that often occur in high energy collisions between nuclei offers formidable technical and interpretative problems— problems that are now only in embryonic stages of solution.

The techniques used in any particular experiment correlate quite well with the rapidity variable. Low values of rapidity in the laboratory are characteristically target related, so that traditional methods of nuclear physics are applicable, e.g. particle identification by ΔE-E and time-of-flight (TOF) (60), with measurements over practically 4π steradians. Experiments in the realm of high rapidity, associated with projectile fragmentation where the beam-velocity fragments are emitted within a few degrees of the incident projectile, require the application of high energy techniques, magnetic spectrometers, and large beam-transport systems (61, 62). Such experiments therefore may incorporate measurements of dE/dx, rigidity, TOF for slow particles, Čerenkov radiation (for relativistic particles) and particle trajectories, as well as use of streamer chamber (63) or other multiple-track, electronically triggered detectors. Visual methods of detection by emulsions, plastics, and AgCl monocrystals (64, 65) are also important in heavy ion experiments because of their wide range of sensitivities, 4π geometry, versatility, and small demands for beam time.

To date, experiments have been designed largely to emphasize the low and high rapidity regions, with only limited capabilities for excursions into the central regions and to high transverse momenta. Because the measurements of total kinetic energy by range are rapidly curtailed by the finite size of the detectors and losses due to nuclear interactions, experiments that pertain to particles of intermediate energy, $\gtrsim 0.2\ A$ GeV, will tend to use dE/dx, rigidity, TOF and/or Čerenkov radiation for particle identification. The minimal requirement here is to design magnetic spectrometers of modest size to measure particle rigidities

(p/Z) and, by rotation about the target, to measure particle production over a wide range of angles (66).

The principal feature that distinguishes heavy ion from typical high energy physics experiments is the presence of a large range of particle charges, Z. Although a measurement of $dE/dx \approx Z^2 f(\beta)$ alone is sufficient to distinguish elements at relativistic velocities, technical problems arise when charge identification and accurate time measurements must be made over a broad range of nuclear charges and velocities. In such cases, an amplifier-discriminator system is required that first samples a signal to find whether it is large or small, and then selects the appropriate sensitivity for accurate signal measurement.

3   NUCLEAR FRAGMENTATION

Fragments observed to have low velocities ($\beta \lesssim 0.3$) in the rest frame of a projectile or target nucleus may be considered "spectators" of a high energy collision. The high rapidity of projectile fragments in the laboratory is important because it permits measurements of fragment momenta that correspond to very low velocities in the projectile frame (even zero)— measurements not possible with traditional target fragmentation experiments because of the inability of the fragments to escape targets of finite thickness.

3.1   *Projectile Fragmentation*

Single-particle inclusive experiments form the basis of our information on projectile and target fragmentation. In these experiments, the reaction investigated is $B + T \rightarrow F + X$, where B and T represent the beam and target nuclei, F is the (single) detected fragment nucleus, and X refers to all other (undetected) reaction products.

Two general terms, discussed in both high energy particle physics and low energy nuclear physics, have wide, though not universal, application to these data. In the high energy language, the first feature is called "limiting fragmentation" (67) or "scaling" (68). This means that a distribution of products with finite energies in the rest frame of projectile or target approaches a limiting form as the bombardment energy increases. Practical tests of limiting fragmentation demonstrate that, in a given range of bombarding energies, a particular distribution shows negligible change. The second term is "factorization" (69). This means that the cross section for production of a particular projectile fragment may be written as the product of a factor γ_T depending only on the target and a factor γ_B^F depending only on the beam and fragment. Obviously the roles of projectile and target may be interchanged to describe target fragmentation. In the language of nuclear physics, both factorization and limiting fragmentation are examples of Bohr's independence hypothesis for decay products of a "compound nucleus" (70). This amounts to the statement that the object emitting a fragment keeps little or no memory of the formation or excitation mechanism that produced it. Of course, these concepts are not automatically valid, but must be tested by experiment. It appears that fragment-velocity distributions do assume a limiting form over successively larger domains of velocity as collision energy increases. On the other hand, as common sense would suggest, factorization seems to hold only for products of grazing collisions, and in particular is untrue for total cross sections.

3.1.1 NUCLEUS-NUCLEUS CROSS SECTIONS   Measurements of the total nucleus-nucleus cross sections, σ_{tot}, using "good geometry" techniques, have been made by Jaros (71) for each of the 32 possible target/projectile/energy combinations of the light nuclei p, d, ^{4}He, and ^{12}C at 0.87 and 2.1 A GeV. The total cross sections σ_{tot} (AA) at 2.1 A GeV for identical target and projectile masses are shown in Figure 4. The

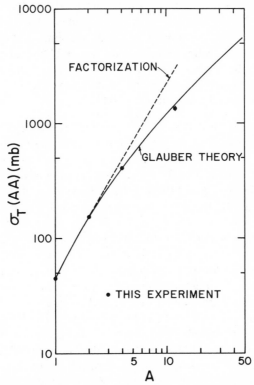

Figure 4 Measured total cross sections $\sigma_{tot}(AA)$ vs A compared with predictions by Glauber theory and from factorization (71).

data are well accounted for by Glauber theory. Another possible A dependence of σ_{tot} would be the factorization form, σ_{tot} $(AA) = \sigma^2_{tot}(pA)/\sigma_{tot}(pp)$, which obviously disagrees with the data.

The inelastic cross sections σ_{in} for reactions initiated by d, ^{4}He, and ^{12}C are found to be independent of beam energy, with σ_{in} (0.87 A GeV) = (1.00 ± 0.01) σ_{in} (2.1 A GeV) (71).

Reaction, or more precisely, transmutation, cross sections σ_r for projectile nuclei including ^{4}He, ^{12}C, ^{14}N, ^{16}O, and ^{40}Ar have been measured at ≈ 2.0 A GeV (72–75), and at 0.15–0.2 A GeV (76) in emulsion, and for a variety of target materials, H through U, with a tungsten-scintillator calorimeter (77) and an elemental transmission detector-telescope system (78). In general, these techniques are sensitive only to interactions that involve changes in the charge Z of the projectile, i.e. $\Delta Z \geq 1$. When supplemented by an estimate of neutron stripping, the resultant cross section is σ_r ($\Delta A \geq 1$), hence the term transmutation. Within their respective errors, the mean-free-path lengths in emulsion are consistent with energy independence down to 0.15 A GeV (76). The σ_r values measured by Lindstrom et al (78) for ^{12}C, ^{16}O, and ^{40}Ar, $T/A \approx 2$ GeV, gave the first experimental evidence that the Bradt-Peters overlap parameter δ is not constant, but depends upon $A_{min} = \min(A_T, A_B)$. They found that for $r_0 = 1.29$ fm and $\delta = 1.0 - 0.028$ A_{min} (with $\delta = 0$ for $A_{min} \gtrsim 30$), Equation 2.1 fits σ_r for all beam and target combinations to 10%. Optical model estimates for σ_r (79, 80) show the same pattern as the experimental data (72).

3.1.2 ISOTOPE PRODUCTION   The major feature of the isotope production cross sections from fragmentation of relativistic projectile nuclei, $T = 1.05$–2.1 A GeV, is that they are independent of energy, and are factorable (62, 81–84). The 0-degree fragmentation cross sections of ^{12}C and ^{16}O projectiles, $T = 1.05$ and 2.1 A GeV, have been measured by Lindstrom et al (82) for all nuclear fragments, $1/3 \leq Z/A \leq 1$ for targets H to Pb. Some 470 cross sections for 35 isotopes were determined. An analysis of the systematics of the measured cross sections leads to the following conclusions: (a) σ^F_{BT} is energy independent: σ^F_{BT} (2.10)/σ^F_{BT} (1.05) = 1.01 $\pm$ 0.01 for all fragments (F) of ^{12}C (B). Limiting fragmentation is satisfied. (b) Factorization, i.e. $\sigma^F_{BT} = \gamma^F_B \gamma_T$, is valid to a high degree. Exceptions are apparent for H and possibly He (85) targets and one-nucleon-loss cross sections in high-Z targets. The latter is explained qualitatively as Coulomb dissociation of the projectile in the target's electric field: the projectile is excited to the giant-dipole resonance and decays by particle emission (86). (c) A target factor $\gamma_T \propto A_T^{\frac{1}{4}}$ accounts for the data to $\approx 10\%$, as does the form $A_T^{\frac{1}{3}} + $ constant.

The functional forms of γ_T are essentially geometrical in character, and imply that projectile fragmentation results from peripheral, i.e. large impact-parameter interactions. Other significant features of the fragmentation cross sections are: (*a*) no nucleon-pickup isotopes are observed ($\sigma \leqq 10\ \mu$b); (*b*) from 30% (Pb target) to 90% (H target) of the beam charge is accounted for by summing the fragmentation cross sections; and (*c*) nuclear structure effects of both projectile and fragment are evident, often dominant, in establishing the isotope production cross sections.

3.1.3 MOMENTUM DISTRIBUTIONS The isotope production cross sections at beam energies $\geqq 1\ A$ GeV are consistent with factorization and energy independence. This suggests that the momentum distributions

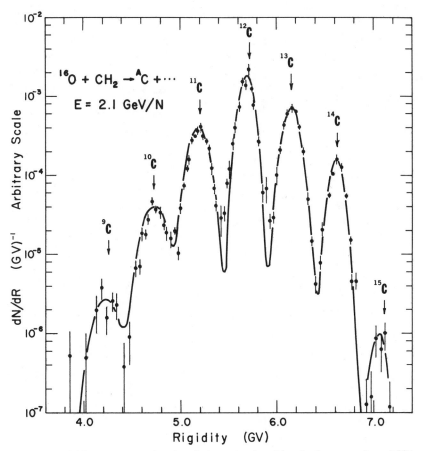

Figure 5 Rigidity spectrum of carbon isotopes produced by the fragmentation of ^{16}O projectiles at 2.1 A GeV. Arrows indicate the rigidities for each isotope evaluated at beam velocity (87).

of fragments in the projectile rest-frame should also exhibit independence of target structure and beam energy. These limiting conditions are, in fact, met to better than 10% accuracy for fragmentation products of ^{12}C and ^{16}O (87) and ^{4}He (62, 83), for fragment momenta $p \lesssim 400$ MeV/c (projectile frame), at beam energies $T \geqq 1.05\ A$ GeV. Evidence for energy-dependent changes in the fragment-momentum spectra from ^{4}He at 0.4 A GeV (62) indicates that the fragment distributions become limiting somewhere between beam energies 0.4 and 1.05 A GeV.

The rigidity ($p_{\parallel}c/Ze$) spectrum shown in Figure 5 is typical of all such isotopic spectra from the 0-degree fragmentation of ^{12}C at 1.05 and 2.1 A GeV and ^{16}O at 2.1 A GeV. The rigidity distribution of each isotope is peaked near the beam velocity, and the widths of all isotope peaks are about the same. In the rest frame of the projectile, the longitudinal momentum distributions show a Gaussian dependence on $p_{\parallel}$, Figure 6. Irrespective of projectile, beam energy $\geqq 1.05\ A$ GeV, and target nucleus, the $p_{\parallel}$ distributions for all fragments from ^{12}C and ^{16}O, with the exception of protons, are characterized by: (a) a Gaussian shape, with rms widths $\sigma(p_{\parallel}) \approx 50$ to 200 MeV/c and values of $\langle p_{\parallel} \rangle \approx -10$ to -130 MeV/c. The latter show that the mean velocities of the fragments are

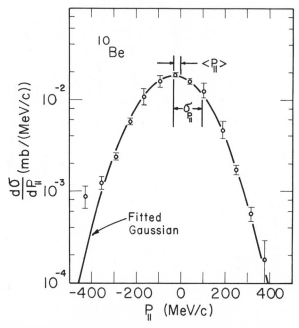

Figure 6 The projectile frame longitudinal momentum distribution for ^{10}Be fragments from 2.1 A GeV ^{12}C on a Be target (87).

less than that of the projectile, evidence of nuclear "friction"; (b) rms widths $\sigma(p_{\parallel})$ and $\sigma(p_{\perp})$ that are equal to an accuracy of 10%, consistent with isotropic production of fragments in a frame moving at $\langle \beta_{\parallel} \rangle \approx - \langle p_{\parallel} \rangle / E_0$ in the projectile frame; (c) $\sigma(p_{\parallel})$ and $\langle p_{\parallel} \rangle$ that are independent of target mass and beam energy but dependent on the masses of the beam B and fragment F. These observations satisfy limiting fragmentation and factorization. The general trend of the measured $\sigma(p_{\parallel})$ is reproduced by the expression $\sigma(p_{\parallel}) = 2\sigma_0 [x(1-x)]^{\frac{1}{2}}$, where $x = A_F/A_B$. The parabolic dependence of $\sigma(p_{\parallel})$ on fragment mass arises from a variety of theoretical approaches, all dependent on simple postulates including conservation of momentum (70, 88, 89). Although the parabolic shape reproduces the general trend of the data, deviations indicate the importance of fragment structure and final-state binding energy effects on $\sigma(p_{\parallel})$ (90).

Fragmentation of ^{4}He has been studied in the reactions ^{4}He+(C, CH$_2$, Pb) → (p, d, ^{3}He, ^{4}He)+X at beam energies 0.4, 1.05, and 2.1 A GeV. Extending the work of Papp et al (83), Anderson (62) measured fragment momenta over the intervals $0.5 \leqq p \leqq 11.5$ GeV/c and $0 \leqq \theta_{\text{lab}} \leqq 12°$, giving $p_{\perp}$ up to 600 A MeV/c.

The $p_{\perp}$ distribution of beam-velocity protons from ^{4}He on a ^{12}C target are presented in Figure 7 for beam momenta 0.93, 1.75, and 2.88 A GeV/c. They show no dependence on beam momenta to an accuracy $\lesssim 10\%$. The momentum distributions for all fragments, p through ^{3}He, produced with $p_{\perp} = 0$, are asymmetric in the projectile rest frame. This asymmetry is observed at each beam momentum, and is due in part to longitudinal momentum transfer between the projectile and target. Angular distributions in the projectile frame also show forward-transverse asymmetries. Distributions of $p_{\perp}$ are broader than those of $p_{\parallel}$, which suggests contributions by hadronic scattering processes. The same conclusion comes from an experiment on the inclusive production of He in the reaction ^{4}He+p → ^{3}He+X measured with incident ^{4}He at 6.85 GeV/c (1.02 A GeV) (91), where the kinematics of ^{3}He production are dominated by the quasielastic scattering of the target proton on the ^{3}He component in the ^{4}He projectile, the remaining neutron acting as a spectator.

Momentum spectra of fast pions, viewed as projectile fragments, also show limiting fragmentation down to beam energies of 1 A GeV (55, 83). This is shown in Figure 8 (83), where the invariant cross sections for π^- production at $\theta_{\text{lab}} = 2.5°$ are plotted as a function of the scaling parameter $x' = k_{\parallel}^* / k_{\parallel}^*$ (max), with $k_{\parallel}^*$ the longitudinal momentum in the center of mass, and $k_{\parallel}^*$ (max) the maximum value allowed by energy and momentum conservation. The scaling property is demonstrated for each projectile type by the fact that the pion yield depends on x' only, and not on the energy of the projectile. The data show excellent agreement with a model of Schmidt & Blankenbecler (92).

Frankel et al (93, 94) observed fast backward fragments from heavy targets bombarded by protons. Their results, as well as those from fast forward projectile fragments, indicate the presence of short-range structure in the nuclear wave function. There have been several semiphenomeno-

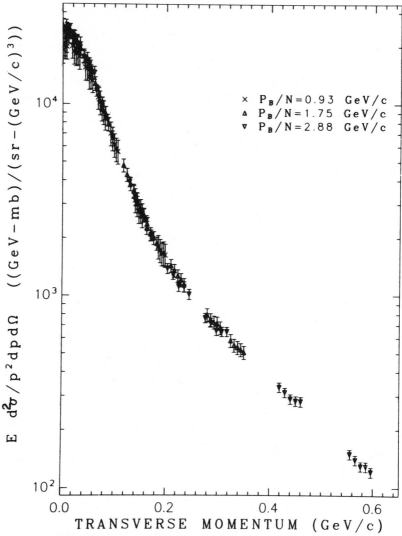

Figure 7   Transverse momentum distributions of beam-velocity protons produced in the reaction α + C for three different beam momenta (62).

logical approaches to parameterize these data, each based on a different picture of the collision process. At one extreme is the notion of the "cumulative effect," in which the effective projectile is supposed to be a compact object of mass larger than a nucleon (95–97). By kinematics alone, this permits fast forward fragment emission. A second approach is to assume a spectrum of nucleons with high virtual three-momentum inside the projectile. The scattering process releases this momentum. For an exponentially falling virtual momentum spectrum, good agreement with target fragmentation data is obtained (94). Finally, a relativistic analogue of this notion, the parton model, relates the distribution of observed x' values to a vertex function, that is, a generalization of a momentum-space wave function. Simple powers of $(1 - x')$ give good fits to light projectile fragmentation (92). The different approaches actually give similar parameterizations of data and are probably closer conceptually than their presentations suggest.

3.1.4   FRAGMENTATION REACTIONS AT NONRELATIVISTIC ENERGIES   Cross sections for production of projectile fragments He through N in ^{16}O-emulsion nucleus interactions at 0.15–0.2 A GeV were measured and compared with those measured at 2.1 A GeV (76). The cross sections for Be, B, C, and N were the same within experimental errors (10–25%), but production of Li and He were found to be larger at 0.2 than at 2.1 A GeV,

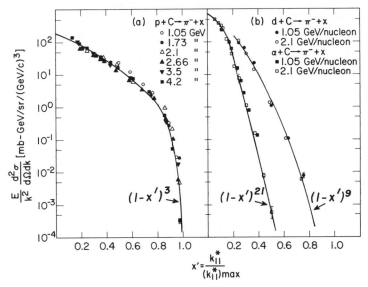

Figure 8   Invariant cross sections for π^- production at $\theta_{lab} = 2.5°$ by (*a*) incident protons, (*b*) incident deuterons and alphas, on ^{12}C (83). Curves are from Reference 92.

up to a factor of 2.8 ± 0.4 for Li. Differences in the topology of the ^{16}O interactions leading to the production of Li and He fragments at these energies were also observed. For example, the production of Li via $^{16}O \rightarrow Li + Li + X$ occurs in about 25% of the Li-producing reactions at 0.2 A GeV, but is unobserved (in approximately 2800 interactions) at 2.1 A GeV (72, 74, 75). This is a consequence of increased excitation energies at 2.1 A GeV beam energy, which suppresses multiple emission of light nuclei.

Unexpectedly, the cross sections and kinematics of peripheral collisions between ^{16}O at 0.315 GeV (0.02 A GeV) on target nuclei $94 \leq A_T \leq 232$ show remarkable similarities to the fragmentation of ^{16}O projectiles at 33.6 GeV (2.1 A GeV) (98–101). The relative isotope yields are essentially target independent, and within experimental errors, the relative yields from ^{16}O on ^{208}Pb are identical to those measured at 2.1 A GeV (98). Differences in the isotopic production cross sections are observed, however, since the trends are toward lower production cross sections for neutron-deficient isotopes at the lower energy. The angular distributions and energy spectra in the excited projectile frame are similar to those seen at Bevalac energies (100, 101).

3.2 Target Fragmentation

Cross sections for fragmentation of light nuclear targets ($A \leq 27$) by proton beams (see 102) and by alpha beams (see 85) are in generally good agreement with projectile fragmentation data for comparable energies. The fragmentation of heavy target nuclei ($A \geq 67$) leads to a wide range of fragment residues, the products of spallation and nuclear fission. The large number of identifiable radionuclides (up to 76 for ^{238}U) gives a good statistical basis for comparing the mass and charge distributions for different target/projectile/energy combinations, as a means for revealing new interaction mechanisms that may arise.

Radioanalytical studies have been done on the energy dependence of ^4He-induced spallation of niobium, $0.08 \leq T \leq 0.22$ A GeV (103). Heavy ion irradiations of Cu have included ^4He at 0.18 A GeV (104), ^{12}C at 2.1 A GeV (105), ^{14}N at 0.28 A GeV (106), and ^{40}Ar at 1.8 A GeV (107). Target-dependent data are available from radioanalytical experiments on the reactions of 2.1 A GeV ^{12}C nuclei with Ag (108), Au, Pb (109), and U (110). Track detectors have been used to study fission reactions induced by ^{14}N bombardment of U, Bi, Au, and Ag at energies 0.14, 0.28, and 2.1 A GeV (111).

The principal result of these spallation experiments is, except for some nonfission products of U (110), the remarkable similarity of the mass-yield and charge-dispersion (at each fixed mass) curves for all projectiles, including protons (105–108). Cumming et al found no detectable difference in the shapes of the charge-dispersion curves for fragment masses $37 \leqq A_F \leqq 61$ from the spallation of Cu by 25 GeV ^{12}C (105) and by ^{14}N and protons, both at 3.9 GeV (106). The 3.9 GeV ^{14}N and proton data

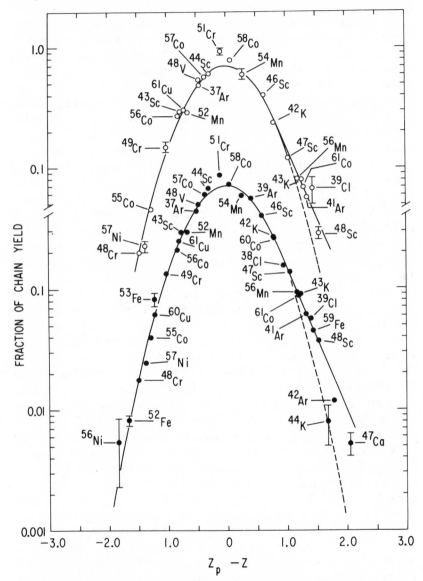

Figure 9   Charge dispersion curves for products in the mass range $37 \leqq A \leqq 61$ from the spallation of copper by 3.9 GeV ^{14}N ions and by protons. The lower curves and filled points are for protons; the upper curves and open points are for ^{14}N. The curves are displaced vertically by a factor of 10 for display purposes (106).

are compared in Figure 9. Illustrated here is evidence for factorization and, because of the identity of the 3.9 and 29 GeV charge-dispersion curves, limiting fragmentation. This implies that peripheral reactions are the dominant source for these (mid-mass) spallation products of Cu.

Because the observed spallation products result from all impact parameters, central collisions also contribute to the mass distributions. For central collisions, however, factorization fails and the yields from such collisions will depend on the mass of the projectile. This may account for the observation that some light fragment yields (^{3}He, ^{7}Be) do not factor (107, 108). Loveland et al (110) observed that the interaction of 2.1 A GeV ^{12}C with U produces a particularly large enhancement in the mass-yield curve in the region $160 \leqq A_F \leqq 190$, a feature completely absent in the 28 GeV p+U mass-yield curve. Such an effect must be a consequence of nonperipheral collisions. There is more evidence for contributions of nonperipheral collisions to target fragmentation. Westfall et al (102) remarked on the appearance of a "high temperature," roughly 15 MeV, Maxwellian energy distribution of fragments from nuclei bombarded by protons. This high temperature component becomes increasingly conspicuous (relative to the ≈ 8 MeV Maxwellian characteristic of light nucleus fragmentation) as target mass increases. The effective temperature rises above 15 MeV as projectile mass increases (60, 112), which shows a breakdown of factorization.

Heavily ionizing black tracks (momenta $p \lesssim 0.25$ A GeV/c) from relativistic interactions in nuclear emulsion are attributable to target fragmentation (34, 73, 113, 114). The rapidity distributions of such fragments (assumed to be protons) from AgBr under bombardment of ^{4}He, ^{16}O, and ^{40}Ar projectiles at ≈ 2 A GeV are found to be Maxwellian (114). To within the errors of the experiment, the values of $\langle y \rangle$ and the standard deviation σ_y are independent of projectile mass, the mean of all distributions being consistent with $\langle y \rangle = 0.014 \pm 0.002$ and $\sigma_y = 0.082 \pm 0.001$, the latter corresponding to a temperature $\tau = 6.3 \pm 0.2$ MeV. These results are impressively close to values of $\langle \beta_{\parallel} \rangle$ and τ obtained from the momentum spectra of projectile fragments (87).

The near constancy of the c.m. velocity $\beta_{\parallel}$ of the emitting system has been demonstrated for a variety of projectiles, π^-, p, and heavy ions, when the projectiles are relativistic, and fragment ranges are in the interval 0–4 mm (tabulations are given in 28, 36, and 114). The velocity $\beta_{\parallel}$ is not an invariant feature of heavy ion collisions, however, since it has been found to increase when (a) the fragment range (energy) is increased (114); (b) the amount of charge, ΔZ, lost by the target nucleus, i.e. the degree of dissociation, increases (74); and (c) the projectile energy decreases (60, 100, 101, 115–117). Similar phenomena are observed in

^{14}N-induced fission reactions where the binary fission and single (energetic spallation) tracks show increased angular asymmetries as either the beam energy, or mass of the target, decreases (111).

A two-step kinematic model (118) is applicable to these observations (100, 101). In this model, target fragmentation proceeds via (a) $B + T \rightarrow B^* + T^*$ and (b) $T^* \rightarrow F + X$. From the conservation of energy and momentum, the recoil velocity of the target is given by the expression, valid for recoil ≪ excitation ≪ nuclear rest energies,

$$\langle \beta_T \rangle \approx (\gamma E_T^* + E_B^*)/M_T \beta \gamma, \qquad\qquad 3.1$$

with β the beam velocity, $\gamma = (1 - \beta^2)^{-\frac{1}{2}}$, E_T^* and E_B^* the respective excitations of target and beam nuclei, and M_T the target mass. Equation 3.1 explains the following qualitative features of the cited data: (a) $\langle \beta_T \rangle$ is independent of M_B (factorization); (b) for given values of E_T^* and E_B^*, asymmetry anticorrelates with target mass; (c) given $E_T^* \ll E_B^*$, i.e. low target excitation, asymmetry anticorrelates with beam momentum; and (d) asymmetry correlates with energy deposition in the collision.

3.3   Theoretical Aspects

The phenomena of nuclear fragmentation have been analyzed by a number of theoretical techniques well tested in familiar domains of nuclear and particle physics. Such analyses have given successful explanations of available experimental data, ranging in precision from percent-level accuracy for total cross sections to semiquantitative or qualitative descriptions of fragment production rates and momentum distributions.

3.3.1   TOTAL CROSS SECTIONS   The high energy model of Glauber (21), which has been remarkably successful in describing nucleon-nucleus cross sections (119), supplies a physically reasonable and computationally tractable framework for deducing nucleus-nucleus cross sections also (120). The fundamental approximation of the model is that during the collision any projectile-nucleon (P) target-nucleon (T) pair wave function suffers a complex phase shift depending only on the nucleon-nucleon relative impact parameters. Therefore, the nucleus-nucleus wave function is multiplied by a phase factor $\exp\left[i\sum \chi_{jk}(\mathbf{b}_{Pj} - \mathbf{b}_{Tk})\right]$ where the sum goes over all pairs of nucleons. For a given nucleus-nucleus impact parameter $\mathbf{b}$, the above phase factor averaged over projectile and target ground-state nucleon distributions yields $\exp\left[i\chi(\mathbf{b})\right]$, the survival amplitude at impact parameter $\mathbf{b}$. Huyghens' construction gives the small q elastic scattering amplitude (q = momentum transfer) as proportional to the Fourier transform of $\{1 - \exp\left[i\chi(\mathbf{b})\right]\}$, and hence by the optical theorem the total and reaction cross-section formulas,

$$\sigma_{tot} = 2Re \int d^2b[1 - \exp(i\chi(\mathbf{b}))], \quad \text{and} \qquad 3.2$$

$$\sigma_r = \int d^2b\{1 - |\exp(i\chi(\mathbf{b}))|^2\}. \qquad 3.3$$

Thus, given sufficiently accurate nucleon distributions inside each nucleus, and given $\chi_{jk}(\mathbf{b}_{Pj} - \mathbf{b}_{Tk})$ (which may be obtained by fitting nucleon-nucleon scattering data), one may compute the cross sections by integration. As Figure 4 shows, the results for light nuclei are in good agreement with experiment. The actual calculations require inclusion of Coulomb phase shifts, since the Coulomb parameter $\alpha Z_1 Z_2$ can be quite large. The model is physically reasonable, since results are sensitive mainly to interactions of surface nucleons. For these nucleons the approximations (a) that their internal motion during the collision can be neglected and (b) that they interact as if free are most plausible. Under some additional approximations, Glauber's model becomes equivalent to the optical model, which therefore gives the same results when valid (120).

3.3.2 FRAGMENTATION High energy reactions that deposit small amounts of energy in nuclear residues are naturally described as occurring in two stages—a fast collision that may lead to knock-out of one or more nucleons, followed by deexcitation on a time scale characteristic of internal nuclear motion (70). Calculational approaches involving high energy approximations, such as the Glauber model or the intranuclear cascade model (121), may be used to describe the first stage. In the cascade model, individual projectile and target nucleons move at uniform velocities between collisions with other nucleons (different particles, like pions, may also be included). The position and outcome of each collision in a given cascade are determined by random sampling of probabilities derived from the elementary cross sections. Thus, even for a fixed nucleus-nucleus impact parameter it may be necessary to follow many cascades in order to obtain a probability distribution for configurations at the end of the fast stage. The energy deposited by the cascade in a "spectator" residue gives the input for the second stage, in which fragments "evaporate" from the residue, assumed to be in rough thermal equilibrium. While there is an extensive literature on cascade calculations for protons incident on nuclei, attempts to deal with complex-nuclear projectiles have begun only recently, and with an emphasis on products other than projectile or target fragments (122).

The cascade calculations, once they become available, should be most appropriate for describing the high temperature component of fragment distributions, as well as examples of isotope production that do not exhibit factorization. That is, processes that involve large overlap of projectile and target are likely to be insensitive to quantum interference effects, so that a minimally quantum-mechanical scheme may be used to describe them. The proton data (102) indicate that these nonfactorizing processes exhibit limiting fragmentation for bombardment energies greater than a few A GeV.

In contrast to the above, the properties of the low temperature component and of fission (which also involves very small energy deposit), namely factorization and limiting fragmentation at about 20 A MeV bombardment energy, invite a different, more quantum-mechanical approach. Hüfner and collaborators (123) have given the most systematic treatment to date, using Glauber theory to describe the fast stage. In this approach, the survival factor $\exp[i\sum\chi_{jk}(\mathbf{b}_{Pj}-\mathbf{b}_{Tk})]$ is treated as an operator on the projectile and target nuclear wave functions. As nuclear overlap increases with decreasing impact parameter, the survival factor differs more from unity, so that higher states are excited. Even at a given impact parameter, the excitations of projectile and target are likely to be correlated. If the exact nuclear Hamiltonian were known, the result of the fast stage could be used to give the initial condition for the slow evolution of the excited nucleus. Lacking this, one must make further approximations to describe the second stage, and improvement in these approximations may be the most fruitful area for further developments in the theory of low energy processes leading to nuclear fragment emission.

In its domain of applicability, $T \gtrsim 1$ A GeV, the Glauber approach automatically implies limiting fragmentation at least to the extent that the nucleon-nucleon cross section is constant. Approximate factorization for light projectile fragmentation is also a natural consequence, since highly peripheral interactions primarily involve direct contact of one or a few nucleon pairs in a localized overlap region, and hence the resulting excitations are insensitive to the shape of the other nucleus. It should be mentioned that another aspect of factorization, namely noncorrelation of projectile and target excitations, has been little studied in theory or experiment.

Projectile fragment momentum spectra and relative isotope frequencies both indicate an effective temperature of 8–10 MeV (87, 100). The Gaussian momentum spectra could be due to a combination of factors, the approximate validity of harmonic oscillator shell model wave functions, and the fact that several nucleon momenta are being added randomly, resulting in almost instant emission of the fragment. On the other hand, the appearance of a Boltzmann factor in the frequency distribution of modes of fragmentation (124) suggests a slower process, in which the mean energy deposited is distributed over many degrees of freedom before fragment emission occurs. From this point of view, it would be interesting to study fragmentation of heavier projectiles by unexcited targets, to see if the (presumably similar) local energy deposit is spread out more, resulting in lower effective temperatures.

4   GENERAL PROCESSES

In the previous section we discussed reactions in which only particles with velocities near that of the target and/or projectile are detected. Of course, many kinds of processes besides highly peripheral interactions should produce such particles, but general processes will also lead to population of intermediate rapidities and large transverse momenta. In this section we consider single-fragment production in the near-target region, where target emission remains distinctive, but where effects of the projectile nucleus are evident, and then go on to the intermediate region, where fragment emission can no longer be identified with either target or projectile.

4.1   Near-Target Rapidities

Information on fragment production in the near-target regions has come from studies on the emission of fast, light nuclei, $Z \leqq 2$, $T \lesssim 250$ A MeV, from interactions of heavy ions in emulsions (73, 74, 76, 113, 114, 116) and AgCl monocrystalline detectors (115, 117) at beam energies $0.1 \lesssim T \lesssim 4.2$ A GeV, from measurements of the energy and angular distributions for fragments produced in the reaction 2.1 A GeV $^{12}C + Au \rightarrow Z$, with $5 \leqq Z \leqq 9$ (125), and from a comprehensive study of the single-fragment inclusive reactions $^{Ne}_{He} + ^{U}_{Al} \rightarrow F + X$ at selected beam energies $T = 0.25, 0.40,$ and 2.1 A GeV (60). In the experiments of Gosset et al (60), the detected fragments were protons through nitrogen with energies $30 \leqq T \leqq 150$ A MeV and angles $25° \leqq \theta_{lab} \leqq 150°$.

Some qualitative features of fragment emission in the near-target region are: (a) The angular distributions for the hydrogen and helium isotopes are smooth, and exhibit forward peaking that increases with mass and energy of the fragment (60, 73, 113, 114, 116, 125), and with decreasing beam energy (60, 113, 115–117). (b) Proton production is predominant, especially for $\theta_{lab} > 90°$, since other fragment energy spectra fall more steeply with increasing angle. The cross sections for ^{3}He are enhanced at high fragment energies, characteristic of neutron-deficient isotopes (60). (c) The spectra of all fragments deviate from exponential (evaporation-like) spectra, tending to power-law distributions at high energies (60, 74, 113, 125, 126). (*d*) The production of light fragments, e.g. ^{3}He, is dependent on the target. The angular distributions are nearly independent of the projectile (60). (*e*) The emission of fast He nuclei, $10 \leqq T \leqq 250$ *A* MeV, is associated with high excitation energies (60, 74). Up to seven target-related He nuclei per interaction have been observed in emulsion events, with combined kinetic energies often exceeding 1 GeV (33).

Figure 10 shows contours of the invariant cross sections $p^{-1}d^2\sigma/dEd\Omega$,

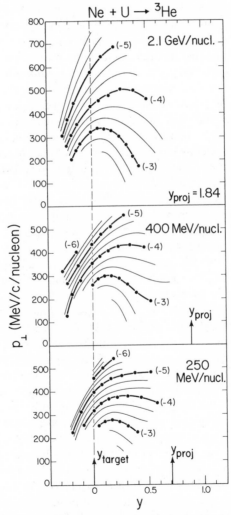

Figure 10   Contours of constant invariant cross section in the $(y, p_\perp)$ plane for ^{3}He fragments from ^{20}Ne on U at different bombarding energies (60).

from Gosset et al (60) for the reaction ^{20}Ne + U → ^{3}He + X, plotted in the rapidity versus $p_\perp/A_F$ plane for each bombarding energy indicated. The heavy contours are identified by the $\log_{10}$ of the invariant cross section. The spacing between contours corresponds to a constant factor in the cross sections. Note that at 2.1 A GeV, the contours are well separated from $y_{proj} = 1.84$, and the identification of target-related production of ^{3}He is quite clear. At 250 A MeV, the smaller rapidity gap between the target and projectile ($y_{proj} = 0.71$) does not allow such clean separation between the target and projectile regions. To interpret these patterns, recall that for fragments emitted isotropically from a unique moving source, the contours will center about the rapidity of that source. The data show that no such unique source exists. At the lowest value of $p_\perp$, the apparent sources have rapidities $y \lesssim 0.1$. As $p_\perp$ increases, the shifts in the centroids of the contours toward higher, intermediate rapidities indicate increasing source velocities.

To account for the apparent spectrum of velocities of the particle-emitting sources, Westfall et al invoked the concept of the nuclear fireball (127). They assumed that in a collision between nuclei, the "participating" nucleons in the overlapping volumes of the projectile and target stick together, forming an entity called a fireball. The fireball thus moves at rapidities intermediate between target and projectile, and is assumed to be an equilibrated, nonrotating ideal gas that expands isotropically in its rest frame with a Maxwellian distribution in energy.

Because of the unequal projectile and target masses in the Ne + U collision, the kinematics of fireball production depend on the impact parameter b, and a spectrum of b-dependent fireball (source) velocities naturally arises from the model. The key assumption of a fireball system characterized only by b of the collision cannot be directly verified by these data alone. For this, equal target and projectile masses are required.

Light fragment (cluster) emission comprises a significant fraction of the total baryonic cross section (about 50% at energies 30–50 A MeV from U) (60). The cross section in velocity space for emitting a light fragment consisting of A nucleons is roughly proportional to the Ath power of the cross section for emitting single nucleons of the same velocity. Models that correlate light fragments with nucleon production in velocity space in this manner are the coalescence (60, 128) and thermodynamic (129, 130) models. In the coalescence model, nucleons are assumed to form a fragment nucleus when a total of A neutrons and protons corresponding to that nucleus are emitted with vector momenta all terminating within a coalescence sphere of radius p_0. Figure 11 shows the impressive results of fitting the production cross sections to the coalescence model for d, t, ^{3}He, and ^{4}He from the reaction Ne on U

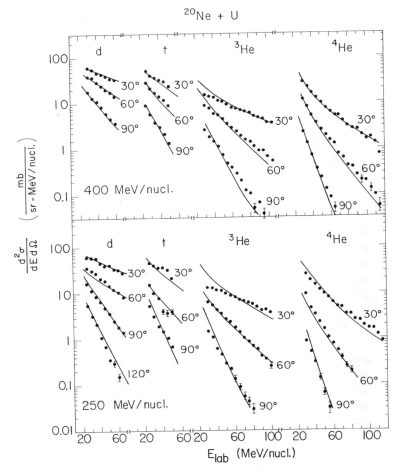

Figure 11   Double differential cross sections for hydrogen and helium isotopes from ^{20}Ne on U compared with calculations (*curves*) using the coalescence formalism (60).

at 250 and 400 A MeV, using the measured proton cross sections. The fitted values of p_0 for these eight sets of data are in the range $126 \leqq p_0 \leqq 147$ MeV/c, typical of Fermi momenta.

4.2   *Intermediate Rapidities*

The experiment of Nagamiya et al (66) on inclusive proton spectra from the reactions 800 A MeV ^{20}Ne$+^{\text{Pb}}_{\text{NaF}} \to$ p$+$X employs a target-centered, rotating magnetic spectrometer to obtain data at high $p_\perp$ for production angles $15° \leqq \theta_{\text{lab}} \leqq 145°$. Proton momenta are measured in the interval $0.4 \leqq p \leqq 2.4$ GeV/c, the upper limit being 1.6 times the momentum per nucleon of the beam.

Figure 12 is a contour plot of the invariant cross sections $\sigma_I = Ed^2\sigma/p^2dpd\Omega$ in the $(y, p_\perp/m_p c)$ plane for proton emission from Ne on Pb. The centroids of the contours shift from $y \approx 0$ as $p_\perp/m_p c \equiv \eta_\perp$ increases. At the highest values of $\eta_\perp$ the contours approach symmetry about the average of the target and projectile rapidities $\frac{1}{2}(y_P + y_T)$.

The centroid shifts may be linear for small $\eta_\perp$. For nonrelativistic, isotropically produced protons a linear relationship, in terms of velocities, is $\beta_0 = a + b\beta'$, where β_0 is the velocity of the emitting source for protons with velocity $\beta' \approx \eta_\perp$ (max) in the moving frame. Stevenson et al (126) were the first to note that such a relation was a good approximation for all fragments $A \leqq 16$. This led them to find an exponential σ_I versus momentum distribution for light fragments, independent of the fragment mass.

The reaction 800 A MeV ^{20}Ne + NaF $\rightarrow$ p + X closely approximates

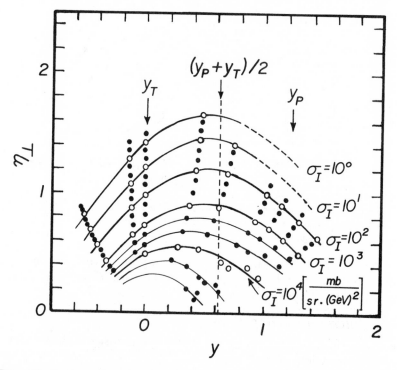

Figure 12 Contours of constant invariant cross section, σ_I, in the $(y, \eta_\perp)$ plane for the reaction 0.8 A GeV Ne + Pb $\rightarrow$ p + X (66).

collisions between nuclei of equal masses. The contours of σ_I for this reaction (66) show symmetry about $\bar{y} = \frac{1}{2}(y_P + y_T) = 0.62$, the rapidity corresponding to 90° in the center-of-mass frame. Proton emission is not isotropic in the c.m. frame, but peaks forward and backward, particularly at small $\eta_\perp$. To reproduce the observed shape of the contours of σ_I requires a superposition of a spectrum of effective sources over a range of rapidities $\Delta y \approx \pm 0.2$ centered about $\bar{y}$. The elemental concept of the production of a unique fireball is not supported and, at the least, refinements of the fireball model that allows a continuum of source velocities are necessary.

Other features of the inclusive proton spectra are: (a) The proton energy distributions at $p_\perp$ (max) for the reactions (all at 800 A MeV) C + C, Ne + NaF, C + Pb, and Ne + Pb are of the form $\sigma_I \propto \exp(-T/\tau)$, with τ in the range 68–78 MeV. (b) The target dependence of the proton yield ($\propto A_T^n$) from 800 A MeV ^{20}Ne, expressed in terms of the exponent n, approaches $n = \frac{1}{3}$ at small angles (projectile fragmentation), and increases with angle, becoming $n \approx 1$ (or slightly greater) in the backward hemisphere.

4.3   Nuclear Shock Phenomena

Searches for compression phenomena in nucleus-nucleus collisions have been carried out by Schopper and co-workers (115, 117, 131), based on studies of the angular distributions of fragments emitted from large-prong-number events in AgCl monocrystalline detectors irradiated by ^{12}C and ^{16}O projectiles, $0.25 \leq T \leq 4.2$ A GeV. Because of the limited sensitivity of AgCl detectors, only fragments with ionization rates $I \geq 8.5$ I_{min} are observed. This restricts the detection of hydrogen nuclei to $T \leq 28$ A MeV, and He nuclei to $T \leq 200$ A MeV. The excess of nonevaporation He fragments, $28 \leq T \leq 200$ A MeV, in the forward hemisphere and the dependence of the peak angle of this excess on beam energy observed in the AgCl experiment have been attributed to nuclear density effects (132, 133). Other experimental searches for nuclear shock waves have not yielded positive results (73, 74, 113, 114, 116, 134, 135).

To account for the excess of fragments observed at forward angles, Schopper et al (117) interpreted the $d\sigma/d\theta$ distributions observed in AgCl detectors as superpositions of (predominantly) two spectra: (a) a "background" spectrum of low energy fragments, mainly H and He, with energies $\lesssim 28$ MeV; and (b) a spectrum of He nuclei with energies $\lesssim 200$ A MeV, whose angular distribution is strongly peaked in the forward hemisphere owing to their larger mass and high energies (60, 74, 116, 134). Using the measured cross sections for $Z = 1$ and $Z = 2$ fragments (60), and the angular distribution from intranuclear cascade calculations for protons with $T \leqq 250$ A MeV (122), Schopper et al were able to reproduce their observed distributions.

Thus, there is no experimental evidence for the existence of shock phenomena. It is not clear, though, that experiments performed thus far were capable of establishing the existence of such effects, in that they were predominantly single-particle inclusive measurements and/or lacked essential information on the multiplicities, energies, and isotopic identification of fragments, including pions.

4.4 Pion Production

4.4.1 INCLUSIVE SPECTRA

Rudimentary information on the angular distributions of pions produced in high energy, heavy ion collisions is available from cosmic-ray studies on shower production and from preliminary results of Bevalac experiments on the reactions 0.8 A GeV $^{20}\text{Ne} + ^{\text{NaF}}_{\text{Pb}} \rightarrow \pi^{+} + X$ (66, 136). Known features of the angular distributions in the c.m. system of equal-mass particles are: (a) At high p, i.e. $\gtrsim 2\, m_{\pi}c$, pion emission is nearly isotropic (22, 34, 66). (b) At $p_{\perp} \approx m_{\pi}c$ (66) and c.m. energies $T_{\pi} \approx 100\text{--}150$ MeV (136) the angular distribution shows a forward/backward peaking. (c) At c.m. energies 20–50 MeV the pions are more nearly isotropic, with a possible broad peaking at $\theta_{\text{c.m.}} \approx 90°$ (136).

The fact that the production of pions does not depend sensitively on the structure of the projectile/target suggests independent-particle production. However, under the assumption that the cross sections are given by $d\sigma(\text{B} + \text{T} \rightarrow \pi) \approx Z_{\text{B}} d\sigma(\text{p} + \text{T} \rightarrow \pi) + N_{\text{B}} d\sigma(\text{n} + \text{T} \rightarrow \pi)$, where Z_{B} and N_{B} are the proton and neutron numbers of the projectile, it is found that the calculated cross sections agree qualitatively with experiment for $20 \lesssim T_{\pi}^{\text{lab}} \lesssim 100$ MeV and $\theta_{\text{lab}} < 60°$, but are overestimates by factors of two or more for $\theta_{\text{lab}} \geqq 90°$ (136).

The target factor index ($\sigma \propto A_{\text{T}}^{n}$) for pion production varies from about $\frac{1}{3}$ to $\frac{2}{3}$ under the following conditions: (a) for high pion momenta, $p_{\pi} > 1$ GeV/c, near $\theta_{\text{lab}} = 0°$, n is about $\frac{1}{3}$ (83, 137), indicative of production in peripheral collisions; (b) n increases to about $\frac{1}{2}$–$\frac{2}{3}$ as p_{π} decreases (83). This suggests that low momentum pions at $\theta_{\text{lab}} \approx 0°$ are associated with more central collisions, as are pions produced at high transverse momenta, $1 < p_{\perp}/m_{p}c < 5$, which also have $n \approx \frac{2}{3}$ (66).

4.4.2 MULTIPLICITIES

The multiplicity distributions of pions $P(n)$, the probability for the emission of n pions per collision, have been measured for π^{-} for a variety of targets and projectiles in a streamer chamber experiment (138) and for $\pi^{\pm}$ for ^{12}C and ^{16}O in emulsions (139) at beam energies ≈ 2 A GeV. The principal results are: (a) copious production of high multiplicity events does not occur and (b) the measured P(n) can be fitted to a sum of impact-parameter-dependent Poisson distributions P(n, b) that are found in a large class of dynamical models, independent of whether equilibrium is reached, under the condition that multipion correlations can be neglected (140).

The π^- multiplicity distribution for the reaction 1.8 A GeV ^{40}Ar + Pb$_3$O$_4 \rightarrow n\pi^- + X$ is shown in Figure 13a. The data are to be compared with curve 2, the calculated P(n) averaged for the Pb and O components of the target appropriate for the particular "trigger"-mode used to select events (i.e. the number of charged tracks per event > 5). Figure 13b shows the model calculations of P(n, b) as a function of impact parameter for Ar and Pb. Gyulassy & Kauffmann (140) stress that any deviation of P(n, b = 0) from a Poisson distribution would be direct evidence for unusual, coherent pion production in heavy ion collisions.

4.5 Theoretical Aspects

A full description of central collisions between nuclei (or, for that matter, between nucleons) would require a quantum theory capable of dealing with infinite degrees of freedom, since both real and virtual excitations may occur. For very light nuclei a phenomenological treatment based on fits to hadron collisions is possible. For intermediate mass nuclei there may not be a very good calculational scheme. If at least one and

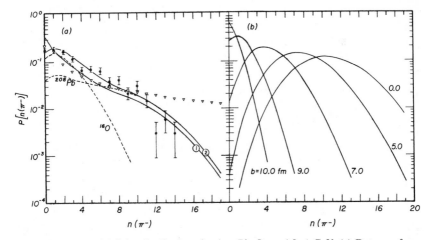

Figure 13 π^- multiplicity distributions for Ar + Pb$_3$O$_4$ at 1.8 A GeV. (a) Data are from Reference 138; curves 1 and 2 are from Reference 140. Curve 1: no impact parameter b cutoff; curve 2: with $b_{\max}$(Pb) = 9.6 fm, $b_{\max}$(O) = 5.0 fm. Dashed curves give contributions to curve 2 from Pb and O targets. Points denoted by ∇ are from Reference 141. (b) Calculated distributions plotted as a function of impact parameter b (140).

preferably both partners in a collision are very heavy, then one may hope once more to find useful models, because errors arising from crude approximations to the full theory may have little effect when averaged over almost macroscopic regions. Many different models have been compared with data on single-particle-inclusive reactions (142–144).

4.5.1 LONGITUDINAL MODELS   One group of models might be called "longitudinal" or "one-dimensional": the collision is divided into different regions in the dimensions transverse to the beam, and communication among these regions is ignored, at least for the initial phase of the collision. The simplest of these is the fireball model, already mentioned (127). The "firestreak" model assumes sticking together and equilibration, not of the whole collection of overlapping projectile and target particles as a unit, but rather of the matter in individual, narrow, imaginary tubes drawn parallel to the beam direction (145). Each tube then acts as a separate emitter of product particles. In comparison to the fireball, the firestreak has the advantage that the nuclear surface diffuseness is easily accommodated, but this is not so important for large nuclei. At moderate energies (100–400 A MeV) the fireball assumption of complete equilibration in the whole overlap region seems more reasonable. At very high energies (well above Bevalac) complete sticking together even in tubes seems unlikely. However, at energies near the top of the Bevalac range, the firestreak seems more promising, especially for large projectile and target.

A more quantum-mechanical example is the one-dimensional "row-on-row" model using Glauber theory for collisions of particles in a tube (146). This is really a way of computing and checking, rather than assuming the initial conditions of the firestreak, that is, stopping of projectile and target material in the center-of-mass frame for a given tube.

4.5.2. CLASSICAL MICROSCOPIC MODELS   In these models, trajectories of individual particles are followed throughout the collision, but quantum effects are put in "by hand." The first group are cascade models of the type described in Section 3. Here quantum mechanics is represented by the probabilistic method of determining locations and results of collisions. The simplest approach is to represent the A nucleons of a projectile by following A individual nucleon-nucleus cascades and adding the results (142). This clearly neglects all interactions of projectile nucleons among themselves, as well as most interactions of struck target nucleons with each other. A modified cascade model, in which baryon number is conserved only on the average and individual particles have probability distributions to be in various places at a given time, is computationally easier than the standard technique when applied to nucleus-nucleus collisions (122). Full-fledged standard cascades, applicable in the moderate energy domain, are nearing completion at this writing (Z. Fraenkel & Y. Yariv, unpublished; 144).

Another kind of calculation proceeds by numerical solution of Newtonian equations of motion for nucleons interacting by two-body forces (147, 148). The forces can be made velocity dependent in order to imitate the effect of the Pauli principle. Such models, like the cascade, are not easily extended to relativistic energies, where particle degrees of freedom proliferate.

There is even a sort of cross between the cascade and equation-of-motion approaches, in which the impact parameter of a nucleon-nucleon collision determines the scattering angle. For classical hard spheres, this gives an exact solution of the equations of motion, while for more realistic nucleon-nucleon potentials it may still be a useful approximation (142, 149). This method can be made formally relativistic, but then it is no longer exact for any kind of interaction.

4.5.3 CLASSICAL MACROSCOPIC MODELS   Since the proper degrees of freedom for nuclear matter at high density are not well determined from first principles, there are fundamental as well as practical arguments for describing nuclear collisions by fluid dynamics, in which the matter is treated as a fluid, always in equilibrium, with an equation of state perhaps containing phenomenological parameters to be fixed by fitting to experiment (142, 150). Two major numerical investigations were made, one using conventional relativistic nonviscous fluid dynamics (150), and the other treating colliding nuclei as drops of two different fluids coupled by a frictional drag (151). At low bombarding velocities, less than the Fermi velocity $\beta_F \approx 0.2$, the one-fluid formulation makes sense, but at much higher speeds it is less reasonable, since nucleon-nucleon collisions cannot instantly share energy and momentum between projectile and target matter.

The great advantage of fluid dynamics is its power to relate experimental results to the equation of state for matter with high baryon and energy densities. The corresponding disadvantage is the assumption of local equilibrium, even in the two-fluid model. At energies much higher than Bevalac this assumption almost certainly breaks down; at least one more fluid associated with mesons must be introduced, as is done in hydrodynamic models of nucleon-nucleon collisions (152).

4.5.4 SPECIAL EFFECTS   Of course, any macroscopic approach sacrifices description of singular processes such as one-step nucleon knockout, as well as of quantum correlations among emitted particles. The latter are clearly beyond any classical framework.

Spectral shapes for protons with high transverse momenta have been computed on the assumption of a single hard nucleon-nucleon scattering. Fitting to experimental data gave acceptable values for the average number of such scatterings in a collision, but could not confirm the single-step mechanism (153).

While copious data on correlations among produced particles may be expected from experiments in the near future, the main information available so far is on the multiplicity distribution for negative pions, obtained from streamer chamber photos (138). The results have been explained in both a thermal model, where a fireball decoupling density was assumed (140), and in an approximate cascade calculation (141). The point is that minimal correlations giving a Poisson distribution of multiplicities fit well, so that at least the impact-parameter-averaged distributions carry little information.

Potentially more interesting correlations have been discussed. If some feature of the outgoing particles could signal that the collision was central with respect to both projectile and target, then single and multiple particle distributions for events with this central signal should be especially informative. It seems to us that the absence of large projectile-velocity and/or target-velocity fragments, from collisions of similar mass heavy nuclei, should be a central signal. The assumption of similar masses is important because a light projectile could be destroyed in a peripheral collision with a heavy target. On the other hand, for equal small masses the nuclei are mainly "surface," so that again destruction could occur in collisions that were far from head-on.

Two-particle momentum correlations might be used to test a single-step knockout mechanism (153), or to search for wave coherence effects in the correlation functions of identical particles, which could indicate the spatial and temporal dimensions of the effective source (154). However, such results would be interesting only if the source turned out to be surprisingly big or small.

4.5.5 APPRAISAL   Comparison of model predictions for single-particle-inclusive cross sections with each other and with experiment indicate generally better than order-of-magnitude agreement in absolute rate, and good agreement on trends with energy and angle (144), at least for bombarding energies in the range of 100–500 A MeV. This suggests that these cross sections are largely determined by geometry and conservation laws, and that the large number of degrees of freedom in the colliding system makes it possible to get the same "statistical" results with widely different approximation schemes. The rough agreement with experiment means that we have yet to find evidence for surprising or unusual phenomena in such collisions. While this might be disappointing, it suggests that existing theoretical concepts may be applicable at least to the stage of compression in central collisions of large nuclei. If so, then the crucial question becomes how, and to what degree, experimental results are affected by the nature of the evolution from that stage.

4.6   New Phases

The most dramatic suggestion for the later stages of compression is the possible formation of new phases of matter with high baryon density. Such a phase could well be more stable than ordinary nuclear matter at that density, or perhaps even more stable than matter at normal nuclear density. In the latter case, a very high barrier evidently separates ordinary matter from the new phase, since spontaneous transitions do not occur at a perceptible rate.

Proposed mechanisms leading to a new phase include the original notion of the nucleon-nucleon tensor force (4, 155); the not-so-different possibility of a very short-range attraction between nucleons (155, 156); "quark" matter, in which these hypothetical constituents of strongly interacting particles (hadrons) would not be confined to individual nucleons, but instead could move separately throughout the nucleus (155, 157); pion condensation, in which a field operator that can create and destroy pions acquires a nonzero expectation value in matter, varying from point to point in space (158); fundamental chiral symmetry, causing nucleons to become massless at sufficiently high density (159); or tensor meson exchange between nucleons (160).

More than one kind of new phase is likely to occur. A phase that is unstable with respect to ordinary matter might still produce a resonance-like energy dependence in the total cross section or in special channels, at an energy appropriate for its excitation (161).

If a new phase were absolutely stable, it probably would have much higher baryon density than ordinary nuclei. Otherwise it would be hard to understand why light nuclei do not collapse spontaneously at an appreciable rate. If a stable phase had high density, then surface effects could well make light nuclei stable in the familiar phase, while for heavy nuclei the greater barrier and greater numbers of nucleons needing to pass it would inhibit spontaneous transitions (A. K. Kerman, unpublished). In addition, a first order phase transition at relatively low density would be likely to have a great effect on neutron stars, generally agreed to be visible as pulsars, whose properties are constrained by astronomical observations (160).

Therefore, if a new stable phase can be made at all in heavy ion collisions—a proposition that has been questioned (155)—it is likely to require densities at least an order of magnitude higher than normal, which at best would be attained with near head-on collisions of the largest nuclei at center-of-mass energies of a few A GeV, well above the Bevalac range (162). Thus, somewhat more subtle signals of new phases, which are not actually stable, may be the most exciting prey to pursue in the immediate future. A possible source for such a signal would be some unusual thermodynamic property of matter at high baryon density. Perhaps the most dramatic proposal up to now extends speculations about hadron structure (163) to this new domain, raising the possibility that dense matter might exhibit a "limiting temperature" $kT \approx m_\pi \approx 140$ MeV (164). That is to say, the specific heat would diverge as this temperature was approached, because more and more different states could be excited. As with other new kinds of phase, the biggest question is what effect this phenomenon would have on the particles remaining after expansion had lowered the density (164).

5 CONCLUSIONS

V. F. Weisskopf (unpublished) has compared collisions between hadrons to collisions between fine Swiss watches. His point is that the products of watch collisions are usually present before impact while the products of hadron collisions are usually formed after impact.[3] Precisely for that reason, one should not try to learn about the structure of matter by smashing watches on each other, or even smashing atoms on each other. There are far better ways to probe these systems, so that the best to be hoped from such experiments is confirmation of insights derived in other ways. However, the case of nuclei is quite different. High energy nuclear collisions are the only way to explore the frontier of high baryon density (165) and that alone justifies, indeed demands, further pursuit of this field.

Up to now, studies of nuclear fragmentation have revealed a rich structure, with a hierarchy of different particle energies in the rest frame of a fragmenting nucleus. Low energy (< 10 MeV temperature) fragments are associated with peripheral interactions that "tickle" the nucleus. Slightly higher energy fragments (< 20 MeV temperature) arise from

---

[3] The word hadron is used here in the sense of a strongly interacting "elementary" particle of the type listed in Reference 40.

greater overlap of the colliding partners, and quite possibly are indications of the nuclear response to moderate compression and heating. In the limit of very high bombardment energies, particles with hundreds of MeV, or even GeV, energies should doubtless be considered as projectile or target fragments, just as they are in hadron collisions. Again, these very energetic fragments should indicate the response to compression and heating.

Experiments indicate that at energies in the several hundred A MeV region head-on collisions lead to temporary pressing together of projectile and target matter, if not full thermal equilibration. It is important to find the highest energies at which this phenomenon persists, since these will give the highest compressions and largest departures from familiar domains. In such experiments the "central" trigger coming from absence of large fragments with the velocity of the projectile, assumed equal in mass to the large target, should be a valuable tool.

In single-particle studies, searches for K^+ should give a direct measure of strangeness production, since strangeness is not easily destroyed once created. This would give a valuable clue about the nature of compressed matter.

Looking over the results reviewed here, we see a beginning has been made in the exploration of high baryon density, a frontier which is outside traditional nuclear and particle physics. As at any frontier, there is still plenty of room for imagination and adventure, which ought to attract increasing bands of able explorers.

We wish to acknowledge valuable help from many colleagues, especially at the Lawrence Berkeley Laboratory (LBL). This work was supported in part by the US Department of Energy and in part by the US National Science Foundation. A. S. G. thanks the Nuclear Science Division of Lawrence Berkeley Laboratory and H. H. H. thanks the Institut für Kernphysik, Universität Frankfurt, for hospitality during the writing of this review.

Several related reviews have appeared recently or will appear soon. Experimental: Stock (46, 166), Schroeder (63), and Steiner (84). Theoretical: Bertsch (167), Nix (144), and Gyulassy (143, 168).

Literature Cited

1. Freier, P., Lofgren, E. J., Ney, E. P., Oppenheimer, F., Bradt, H. L., Peters, B. 1948. *Phys. Rev.* 74:213–17
2. Freier, P., Lofgren, E. J., Ney, E. P., Oppenheimer, F. 1948. *Phys. Rev.* 74:1818–27
3. Alfvén, H. 1939. *Nature* 143:435
4. Feenberg, E., Primakoff, H. 1946. *Phys. Rev.* 70:980–81
5. Shapiro, M. M., Silberberg, R. 1970. *Ann. Rev. Nucl. Sci.* 20:323–92
6. Waddington, C. J. 1960. *Prog. Nucl. Phys.* 8:1–45
7. Powell, C. F., Fowler, P. H., Perkins,

D. H. 1959. *The Study of Elementary Particles by the Photographic Method.* London: Pergamon. 669 pp.

8. Bradt, H. L., Peters, B. 1948. *Phys. Rev.* 74:1828–37

9. Bradt, H. L., Peters, B. 1949. *Phys. Rev.* 75:1779–80

10. Bradt, H. L., Peters, B. 1950. *Phys. Rev.* 77:54–70

11. Bradt, H. L., Peters, B. 1950. *Phys. Rev.* 80:943–53

12. Kaplon, M. F., Peters, B., Reynolds, H. L. Ritson, D. M. 1952. *Phys. Rev.* 85:295–309

13. Cleghorn, T. F. 1967. *The energy dependence of the fragmentation parameters and interaction mean free paths in nuclear emulsion for heavy cosmic ray nuclei.* MS thesis and *Rep. CR-104*, Univ. Minn. 75 pp.

14. Cleghorn, T. F., Freier, P. S., Waddington, C. J. 1968. *Can. J. Phys.* 46:572–77

15. Lohrmann, E., Teucher, M. W. 1959. *Phys. Rev.* 115:636–42

16. Eisenberg, Y. 1954. *Phys. Rev.* 96:1378–82

17. Bowman, J. D., Swiatecki, W. J., Tsang, C. F. 1973. *LBL Rep. 2908.* 22 pp.

18. Alexander, G., Yekutieli, G. 1961. *Nuovo Cimento* 19:103–17; Alexander, G., Avidan, J., Avni, A., Yekutieli, G. 1961. *Nuovo Cimento* 20:648–61

19. Aizu, H., Fujimoto, Y., Hasegawa, S., Koshiba, M., Mito, I., Nishimura, J., Yokoi, K. 1961. *Prog. Theor. Phys. Suppl.* 16:54–168

20. Fernbach, S., Serber, R., Taylor, T. B. 1949. *Phys. Rev.* 75:1352–55

21. Glauber, R. J. 1959. *Lectures in Theoretical Physics,* ed. W. B. Britton et al, Vol. 1, pp. 315–414. New York: Wiley-Interscience

22. Jain, P. L., Lohrmann, E., Teucher, M. W. 1959. *Phys. Rev.* 115:643–54

23. Tsuzuki, Y. 1961. *J. Phys. Soc. Jpn.* 16:2131–39

24. Rybicki, K. 1963. *Nuovo Cimento* 28:1437–54

25. Abraham, F., Gierula, J., Levi-Setti, R., Rybicki, K., Tsao, C. H. 1967. *Phys. Rev.* 159:1110–23

26. Andersson, B., Otterlund, I., Kristiansson, K. 1966. *Ark. Fys.* 31:527–48

27. Otterlund, I., Andersson, B. 1967. *Ark. Fys.* 35:133–47

28. Otterlund, I. 1968. *Ark. Fys.* 38:467–87

29. Otterlund, I., Resman, R. 1969. *Ark. Fys.* 39:265–93

30. Resman, R., Otterlund, I. 1971. *Phys. Scr.* 4:183–89

31. Kullberg, R., Otterlund, I., Resman, R. 1972. *Phys. Scr.* 5:5–12

32. Kullberg, R., Otterlund, I. 1973. *Z. Phys.* 259:245–62

33. Jakobsson, B., Kullberg, R., Otterlund, I. 1974. *Z. Phys.* 268:1–9

34. Jakobsson, B., Kullberg, R., Otterlund, I. 1975. *Z. Phys. A* 272:159–68

35. Skjeggestad, O., Sörensen, S. O. 1959. *Phys. Rev.* 113:1115–24

36. Hyde, E. K., Butler, G. W., Poskanzer, A. M. 1971. *Phys. Rev. C* 4:1759–78

37. Gagarin, Yu. F., Ivanova, N. S., Kulikov, V. N. 1970. *Sov. J. Nucl. Phys.* 11:698–703

38. Gottstein, K. 1954. *Philos. Mag.* 45:347–59

39. Cester, R., DeBenedetti, A., Garelli, C. M., Quassiati, B., Tallone, L., Vigone, M. 1958. *Nuovo Cimento* 7:371–99

40. Particle Data Group. 1976. *Rev. Mod. Phys.* No. 2, Pt. II 48:S1–245

41. Grunder, H. A., Selph, F. B. 1977. *Ann. Rev. Nucl. Sci.* 27:353–92

42. Goulding, F. S., Harvey, B. G. 1975. *Ann. Rev. Nucl. Sci.* 25:167–240

43. Price, P. B., Fleischer, R. L. 1971. *Ann. Rev. Nucl. Sci.* 21:295–330

44. Fleischer, R. L., Price, P. B., Walker, R. M. 1975. *Nuclear Tracks in Solids.* Univ. Calif. Press. 605 pp.

45. Barkas, W. H. 1963. *Nuclear Research Emulsions.* New York: Academic. 518 pp.

46. Stock, R. 1978. *Heavy Ion Collisions,* ed. R. Bock, Vol. 1. Amsterdam: North-Holland. In press

47. Rosen, L. 1971. *IEEE Trans. Nucl. Sci.* NS-18 (3):29–35

48. White, M. G. 1971. *IEEE Trans. Nucl. Sci.* 174:1121–23

49. Ghiorso, A., Grunder, H., Hartsough, W., Lambertson, G., Lofgren, E., Lou, K., Main, R., Mobley, R., Morgado, R., Salsig, W., Selph, F. 1973. *IEEE Trans. Nucl. Sci.* NS-20 (3):155–58

50. White, M. G., Isaila, M., Prelec, K., Allen, H. L. 1971. *Science* 174:1121–23; Grunder, H. A., Hartsough, W. D., Lofgren, E. J. 1971. *Science* 174:1128–29

51. Farley, F. J. M. 1970. *Speculations on Nucleus-Nucleus Collisions with the ISR,* CERN NP Intern. Rep. 70-26. 6 pp.

52. Baldin, A. M. 1977. *Sov. J. Part. Nucl.* 8:430–77

53. GANIL Study Group. 1975. *IEEE Trans. Nucl. Sci.* NS-22 (3):1651–54

54. Study Group of Numatron Project. 1977. *Proposal for Numatron.* Inst. Nucl. Study: Univ. Tokyo. 354 pp.

55. Baldin, A. M. 1975. *Proc. High-Energy Phys. Nucl. Struct. 1975,* pp. 621–41. New York: AIP

56. Study Report. 1977. *Ges. Schwerionenforsch.* GSI-P-2-77. 39 pp.

57. *Nuclear Chemistry Annual Report*. 1972. 444 pp. *LBL-1666*; see also Ref. 49
58. Grunder, H. A. 1975. *IEEE Trans. Nucl. Sci.* NS-22 (3): 1621–25
59. Morgado, R. E., Poskanzer, A. M., Myers, W. D., eds. 1975. *High Intensity Uranium Beams from the SuperHILAC and the Bevatron.* LBL Proposal-32. 110 pp.
60. Gosset, J., Gutbrod, H. H., Meyer, W. G., Poskanzer, A. M., Sandoval, A., Stock, R., Westfall, G. D. 1977. *Phys. Rev. C* 16: 629–57
61. Greiner, D. E., Lindstrom, P. J., Bieser, F. S., Heckman, H. H. 1974. *Nucl. Instrum. Methods* 116: 21–24
62. Anderson, L. 1977. *Fragmentation of relativistic light nuclei: longitudinal and transverse momentum distributions.* PhD thesis. Univ. Calif., Berkeley; also *LBL-6769.* 133 pp.
63. Schroeder, L. S. 1977. *Acta Phys. Pol. B* 8: 355–87
64. Childs, C. B., Slifkin, L. 1963. *Rev. Sci. Instrum.* 34: 101–4
65. Haase, G., Schopper, E., Granzer, F. 1973. *Photogr. Sci. Eng.* 17: 409–12
66. Nagamiya, S., Tanihata, I., Schnetzer, S., Anderson, L., Brückner, W., Chamberlain, O., Shapiro, G., Steiner, H. 1977. *J. Phys. Soc. Jpn.* 44: Suppl. 378–85
67. Benecke, J., Chou, T. T., Yang, C. N., Yen, E. 1969. *Phys. Rev.* 188: 2159–69
68. Feynman, R. P. 1969. *Phys. Rev. Lett.* 23: 1415–17
69. Gell-Mann, M. 1962. *Phys. Rev. Lett.* 8: 263–64; Gribov, V. N., Pomeranchuk, I. Ya. 1962. *Phys. Rev. Lett.* 8: 343–45
70. Feshbach, H., Huang, K. 1973. *Phys. Lett. B* 47: 300–2
71. Jaros, J. 1975. *Nucleus-nucleus total cross sections for light nuclei at 1.55 and 2.89 GeV/c/nucleon.* PhD thesis; also *LBL Rep. 3849.* 116 pp.
72. Heckman, H. H., Greiner, D. E., Lindstrom, P. J., Shwe, H. 1978. *Phys. Rev. C* 17: 1735–47
73. Chernov, G. M., Gulamov, K. G., Gulyamov, U. G., Nasyrov, S. Z., Svechnikova, L. N. 1977. *Nucl. Phys. A* 280: 478–90
74. Jakobsson, B., Kullberg, R. 1976. *Phys. Scr.* 13: 327–38
75. Judek, B. 1975. *Proc. Int. Conf. Cosmic Rays, 14th,* München, pp. 2342–47
76. Kullberg, R., Kristiansson, K., Lindkvist, B., Otterlund, I. 1977. *Nucl. Phys. A* 280: 491–97
77. Cheshire, D. L., Huggett, R. W., Johnson, D. P., Jones, W. V., Rountree, S. P., Verma, S. D., Schmidt, W. K. H., Kurz,

R. J., Bowen, T., Krider, E. P. 1974. *Phys. Rev. D* 10: 25–31
78. Lindstrom, P. J., Greiner, D. E., Heckman, H. H., Cork, B., Bieser, F. S. 1974. See Ref. 75, pp. 2315–18
79. Barshay, S., Dover, C. B., Vary, J. P. 1975. *Phys. Rev. C* 11: 360–69
80. Karol, P. J. 1975. *Phys. Rev C* 11: 1203–9
81. Heckman, H. H., Greiner, D. E., Lindstrom, P. J., Bieser, F. S. 1972. *Phys. Rev. Lett.* 28: 926–29
82. Lindstrom, P. J., Greiner, D. E., Heckman, H. H., Cork, B., Bieser, F. S. 1975. *LBL Rep. 3650.* 10 pp.
83. Papp, J., Jaros, J., Schroeder, L., Staples, J., Steiner, H., Wagner, A., Wiss, J. 1975. *Phys. Rev. Lett.* 10: 601–4
84. Steiner, H. L. 1977. *Proc. Int. Conf. High-Energy Phys. Nucl. Struct., 7th,* Zurich, 29 Aug–2 Sept., ed. M. Locher, pp. 261–86
85. Raisbeck, G. M., Yiou, F. 1975. *Phys. Rev. Lett.* 35: 155–59
86. Heckman, H. H., Lindstrom, P. J. 1976. *Phys. Rev. Lett.* 37: 56–59
87. Greiner, D. E., Lindstrom, P. J., Heckman, H. H., Cork, B., Bieser, F. S. 1975 *Phys. Rev. Lett.* 35: 152–55
88. Lepore, J. V., Riddell, R. J. Jr. 1974. *LBL Rep. 3086.* 24 pp.
89. Goldhaber, A. S. 1974. *Phys. Lett. B* 53: 306–8
90. Masuda, N., Uchiyama, F. 1975. *LBL Rep. 4263.* 24 pp.
91. Bizard, G., LeBrun, C., Berger, J., Duflo, J., Goldzahl, L., Plouin, F., Oostens, J., Van Den Bossche, M., Vu Hai, L., Fabbri, F. L., Picozza, P., Satta, L. 1977. *Nucl. Phys. A* 285: 461–68
92. Schmidt, I. A., Blankenbecler, R. 1977. *Phys. Rev. D* 15: 3321–31
93. Frankel, S., Frati, W., Van Dyck, O., Werbeck, R., Highland, V. 1976. *Phys. Rev. Lett.* 36: 642–45
94. Frankel, S. 1977. *Phys. Rev.* 38: 1338–41
95. Dar, A. 1977. *Proc. Meet. Nucl. Prod. 1977 Very High Energies,* ed. G. Bellini, L. Bertocchi, P. G. Rancoita, p. 591. Trieste: ITCP
96. Burov, V. V., Lukyanov, V. K., Titov, A. I. 1977. *Phys. Lett. B* 67: 46–48
97. Fujita, T. 1977. *Phys. Rev. Lett.* 39: 174–76
98. Buenerd, M., Gelbke, C. K., Harvey, B. G., Hendrie, D. L., Mahoney, J., Menchaca-Rocha, A., Olmer, C., Scott, D. K. 1976. *Phys. Rev. Lett.* 37: 1191–94
99. Gelbke, C. K., Buenerd, M., Hendrie, D. L., Mahoney, J., Mermaz, M. C., Olmer, C., Scott, D. K. 1976. *Phys.*

Rev. Lett. 37:1191–94
100. Gelbke, C. K., Olmer, C., Buenerd, M., Hendrie, D. L., Mahoney, J., Mermaz, M. C., Scott, D. K. 1977. *LBL Rep.* *5826.* 116 pp.
101. Gelbke, C. K., Scott, D. K., Bini, M., Hendrie, D. L., Laville, J. L., Mahoney, J., Mermaz, M. C., Olmer, C. 1977. *Phys. Lett. B* 70:415–17
102. Westfall, G. D., Sextro, R. G., Poskanzer, A. M., Zebelman, A. M., Butler, G. W., Hyde, E. K. 1978. *Phys. Rev. C* 17:1368–81
103. Korteling, R. G., Hyde, E. K. 1964. *Phys. Rev. B* 136:425–36
104. Karol, P. J. 1974. *Phys. Rev. C* 10:150–55
105. Cumming, J. B., Stoenner, R. W., Haustein, P. E. 1976. *Phys. Rev. C* 14:1554–63
106. Cumming, J. B., Haustein, P. E., Stoenner, R. W., Mausner, L., Naumann, R. A. 1974. *Phys. Rev. C* 10:739–55
107. Cumming, J. B., Haustein, P. E., Ruth, T. J., Virtes, G. J. 1978. *Phys. Rev. C* 17:1632–41
108. Rudy, C. R., Porile, N. T. 1975. *Phys. Lett. B* 59:240–43
109. Loveland, W., Otto, R. J., Morrissey, D. J., Seaborg, G. T. 1977. *Phys. Lett. B* 69:284–86
110. Loveland, W., Otto, R. J., Morrissey, D. J., Seaborg, G. T. 1977. *Phys. Rev. Lett* 39:320–22
111. Katcoff, S., Hudis, J. 1976. *Phys. Rev. C* 14:628–34
112. Zebelman, A. M., Poskanzer, A. M., Bowman, J. D., Sextro, R. G., Viola, V. E. Jr. 1975. *Phys. Rev. C* 11:1280–86
113. Jakobsson, B., Kullberg, R., Otterlund, I. 1977. *Nucl. Phys. A* 276:523–32
114. Heckman, H. H., Crawford, H. J., Greiner, D. E., Lindstrom, P. J., Wilson, L. W. 1978. *Phys. Rev. C* 17:1651–64
· 115. Baumgardt, H. G., Schott, J. U., Sakamoto, Y., Schopper, E., Stöcker, H., Hofmann, J., Scheid, W., Greiner, W. 1975. *Z. Phys. A* 237:359–71
116. Heckman, H. H., Crawford, H. J., Greiner, D. E., Lindstrom, P. J., Wilson, L. W. 1977. *Proc. Meet. Heavy Ion Collisions,* Fall Creek Falls, TN, June 13–17; also *Oak Ridge Natl. Lab. Rep. CONF-770602,* pp. 411–32
117. Schopper, E., Baumgardt, H. G., Obst, E. 1977. See Ref. 116, pp. 398–410
118. Masuda, N., Uchiyama, F. 1977. *Phys. Rev. C* 15:1598–1600
119. Glauber, R. J., Matthiae, G. 1970. *Nucl. Phys. B* 21:135–57; Blieden, H. R., Finocchiaro, G., Goldhaber, A. S.,

Grannis, P. D., Green, D., Hietarinta, J., Hochman, D., Kephart, R., Kirz, J., Lee, Y. Y., Nef, C., Thun, R., Faissler, W., Tang, Y. W. 1975. *Phys. Rev. D* 11:14–28
120. Franco, V., Varma, G. K. 1977. *Phys. Rev. C* 15:1375–78; Franco, V., Tekou, A. 1977. *Phys. Rev. C* 16:658–64; Franco, V., Nutt, W. T. 1978. *Phys. Rev. C* 17:1347–58
121. Bertini, H. W., Santoro, R. T., Hermann, O. W. 1976. *Phys. Rev. C* 14:590–95
122. Smith, R. K., Danos, M. 1977. See Ref. 116, pp. 363–80
123. Hüfner, J., Schäfer, K., Schürmann, B. 1975. *Phys. Rev. C* 12:1888–98; Abul-Magd, A., Hüfner, J. 1976. *Z. Phys. A* 277:379–84; Abul-Magd, A., Hüfner, J., Schürmann, B. 1976. *Phys. Lett. B* 60:327–30; Celenza, L. S., Hüfner, J., Sander, C. 1977. *Nucl. Phys. A* 276:509–22
124. Lukyanov, V. K., Titov, A. I. 1975. *Phys. Lett. B* 57:10–12
125. Crawford, H. J., Price, P. B., Stevenson, J., Wilson, L. W. 1975. *Phys. Rev. Lett.* 34:329–31
126. Stevenson, J., Price, P. B., Frankel, K. 1977. *Phys. Rev. Lett.* 38:1125–29
127. Westfall, G. D., Gosset, J., Johansen, P. J., Poskanzer, A. M., Meyer, W. G., Gutbrod, H. H., Sandoval, A., Stock, R. 1976. *Phys. Rev. Lett.* 37:1202–5
128. Gutbrod, H. H., Sandoval, A., Johansen, P. J., Poskanzer, A. M., Gosset, J., Meyer, W. G., Westfall, G. D., Stock, R. 1976. *Phys. Rev. Lett.* 37:667–69
129. Mekjian, A. 1977. *Phys. Rev. Lett.* 38:640–43; Mekjian, A., Bond, R., Johansen, P. J., Koonin, S. E., Garpman, S. I. A. 1977. *Phys. Lett. B* 71:43–47
130. Alard, J. P., Baldit, A., Brun, R., Costilhes, J. P., Dhermain, J., Fargeix, J., Fraysse, L., Pellet, J., Roche, G., Tamain, J. C. 1975. *Nuovo Cimento* 30:320–44
131. Baumgardt, H. G., Schopper, E., Schott, J. U., Kocherov, N. P., Vorohov, A. V., Issinsky, I. D., Markov, L. G. 1976. *Proc. Int. Work. Gross Prop. Nucl., 4th,* Hirschegg, Austria. AED Conf. 76-015-000, pp. 105–10
132. Hofmann, J., Stöcker, H., Gyulassy, M., Scheid, W., Greiner, W., Baumgardt, H. G., Schott, J. U., Schopper, E. 1976. *Proc. Int. Conf. Sel. Topics Nucl. Struct.,* Dubna, USSR. D-9920, Vol. 2, p. 370
133. Hofmann, J., Stöcker, H., Heinz, U., Scheid, W., Greiner, W. 1976. *Phys. Rev. Lett.* 36:88–91; Stöcker, H. 1977. *Proc. Int. Symp. Nucl. Collisions Their*

Microsc. Descr., Bled, Yugoslavia, 26 Sept–1 Oct. FIZIKA 9 (Suppl. 4): 671–706

134. Poskanzer, A. M., Sextro, R. G., Zebelman, A. M., Gutbrod, H. H., Sandoval, A., Stock, R. 1975. Phys. Rev. Lett. 35:1701–4

135. Basova, E. S., Bondarenko, A. I., Gulamov, K. G., Gulyamov, U. G., Nasyrov, Sh. Z., Svechnikova, L. N., Chernov, G. M. 1976. JETP Lett. 24:229–32

136. Nakai, K., Chiba, J., Tanihata, I., Nagamiya, S., Bowman, H., Ioannou, J., Rasmussen, J. O. 1977. See Ref. 66

137. Schimmerling, W., Vosburgh, K. G., Koepke, K. 1974. Phys. Rev. Lett. 33:1170–73

138. Fung, S. Y., Gorn, W., Kiernan, G. P., Liu, F. F., Lu, J. J., Oh, Y. T., Ozawa, J., Poe, R. T., Schroeder, L., Steiner, H. 1978. Phys. Rev. Lett. 40:292–95

139. Jakobsson, B., Kullberg, R., Otterlund, I., Ruiz, A., Bolta, J. M., Higón, E. 1978. Nucl. Phys. A 300:397–410

140. Gyulassy, M., Kauffmann, S. K. 1978. Phys. Rev. Lett. 40:298–302

141. Vary, J. P. 1978. Phys. Rev. Lett. 40:295–98

142. Amsden, A. A., Ginocchio, J. N., Harlow, F. H., Nix, J. R., Danos, M., Halbert, E. C., Smith, R. K. Jr. 1977. Phys. Rev. Lett. 38:1055–58

143. Gyulassy, M. 1977. See Ref. 133, pp. 623–70

144. Nix, J. R. 1978. Prog. Part. Nucl. Phys. ed. D. H. Wilkinson, Vol. 1. London: Pergamon; Los Alamos Sci. Lab. Rep. LA-UR-77-2952. 101 pp.

145. Myers, W. D. 1978. Nucl. Phys. A 296:177–88

146. Hüfner, J., Knoll, J. 1977. Nucl. Phys. A 290:460–92

147. Bodmer, A. R., Panos, C. N. 1977. Phys. Rev. C 15:1342–58; Bodmer, A. R. 1977. See Ref. 116, pp. 309–62

148. Wilets, L., Henley, E. M., Kraft, M., MacKellar, A. D. 1977. Nucl. Phys. A 282:341–50; Wilets, L., MacKellar, A. D., Rinker, G. A. Jr. 1976. See Ref. 131, pp. 111–14; Wilets, L., Yariv, Y., Chestnut, R. 1978. Nucl. Phys. A 301:359–64

149. Bondorf, J. P., Feldmeier, H. T., Garpman, S. I. A., Halbert, E. C. 1976. Phys. Lett. B 65:217–20

150. Amsden, A. A., Bertsch, G. F., Harlow, F. H., Nix, J. R. 1975. Phys. Rev. Lett. 35:905–8; Amsden, A. A., Harlow, F. H., Nix, J. R. 1977. Phys. Rev. C 15:2059–71

151. Amsden, A. A., Goldhaber, A. S.,

Harlow, F. H., Nix, J. R. 1978. Phys. Rev. C 17:2080–96

152. Landau, L. D. 1953. Isv. Akad. Nauk, SSRR, Ser. Fiz. 17:51; Cooper, F. 1975. Particles and Fields-1974. AIP Conf. Proc. 23, ed. C. E. Carlson. New York: AIP. 499 pp.

153. Koonin, S. E. 1977. Phys. Rev. Lett. 39:680–84

154. Goldhaber, G., Goldhaber, S., Lee, W., Pais, A. 1960. Phys. Rev. 120: 300–12; Kopylov, G. I. 1974. Phys. Lett. B 50:472–74; Koonin, S. E. 1977. Phys. Lett. B 70:43–47

155. Bodmer, A. R. 1971. Phys. Rev. D 4:1601–6; The Nuclear Many-Body Problem, ed. F. Calogero, C. Ciofi Degli Atti, Vol. 2, pp. 505–34. Bologna: Editrice Compositori

156. Ne'eman, Y. 1968. Proc. Conf. Symmetry Principles High Energy, 5th, pp. 149–51. New York: Benjamin; 1974. Proc. Int. Astron. Union Symp. No. 53 Physics Dense Matter. Dordrecht, Netherlands: Reidel. 111 pp.

157. Collins, J. C., Perry, M. J. 1975. Phys. Rev. Lett. 34:1353–56

158. Migdal, A. B. 1971. Zh. Eksp. Teor. Fiz. 61:2209; Migdal, A. B. 1972. Sov. Phys. JETP 34:1184; Sawyer, R. F. 1972. Phys. Rev. Lett. 29:382–85; Scalapino, D. J. 1972. Phys. Rev. Lett. 29:386–88; Brown, G. E., Weise, W. 1976. Phys. Rep. 27:1–34; Ruck, V., Gyulassy, M., Greiner, W. 1976. Z. Phys. A 277:391–94

159. Lee, T. D., Wick, G. C. 1974. Phys. Rev. D 9:2291–303; Lee, T. D. 1975. Rev. Mod. Phys. 47:267–75; Lee, T. D. 1978. Mesons in Nuclei, ed. M. Rho, D. H. Wilkinson. Amsterdam: North-Holland

160. Canuto, V., Datta, B., Kalman, G. 1978. Astrophys. J. 221:274–83

161. Chapline, G. F., Kerman, A. K. 1978. MIT Rep. CTP 695. 11 pp.

162. Goldhaber, A. S. 1978. Nature. In press

163. Hagedorn, R. 1973. Cargése Lectures in Physics, ed. E. Schatzman, Vol. 6, pp. 643–716. New York: Gordon & Breach

164. Glendenning, N. K., Karant, Y. J. 1977. Phys. Rev. Lett. 40:374–77

165. Chapline, G. F., Johnson, M. H., Teller, E., Weiss, M. S. 1973. Phys. Rev. D 8:4302–8

166. Stock, R. 1979. Phys. Rep. In press

167. Bertsch, G. F. 1978. Proc. Summer Sch. Heavy Ions Mesons Nucl. Phys., Les Houches, France, ed. M. Rho, G. Ripka. Amsterdam: North-Holland. In press

168. Gyulassy, M. 1979. Phys. Rep. In press

Ann. Rev. Nucl. Part Sci. 1978. 28: 207–37

INTENSE SOURCES OF FAST NEUTRONS

×5595

H. H. Barschall

Department of Nuclear Engineering, University of Wisconsin, Madison, Wisconsin 53706

CONTENTS

0066-4243/78/1201-0207$01.00

In recent years there has been a growing interest in intense sources of neutrons of energy higher than that of fission neutrons. Such sources are needed for the testing of materials under consideration for fusion reactors, and for medical applications.

The present review is limited to accelerator-based neutron sources and does not include plasma devices, such as fusion test reactors and dense plasma focus sources.

1   REQUIREMENTS FOR INTENSE SOURCES OF FAST NEUTRONS

1.1   Radiation Damage Studies

Present designs of fusion reactors are based on the interaction of deuterons and tritons. Each interaction produces a 14-MeV neutron. In designs in which the reactions occur in a magnetically confined plasma, the plasma is contained in a large vacuum vessel. The inner wall of this vessel will be subjected to bombardment by 14-MeV neutrons with a flux density of about 10^{14} cm^{-2} sec^{-1}, or a fluence of 3×10^{21} cm^{-2} in a year (1).

Neutron bombardment will damage the walls in two distinct ways: It will produce bulk radiation damage, i.e. deterioration of the mechanical properties of the wall material, and it will produce surface effects, such as the emission of recoiling atoms and the release of surface atoms caused by radiation damage near the surface. Bulk radiation damage has again two distinct causes, displacement of atoms caused by charged particles produced in neutron interactions, and the presence of atoms generated in transmutations induced by neutrons.

The cross sections for transmutations induced by 14-MeV neutrons are higher than for less energetic neutrons, such as fission neutrons. The 14-MeV neutrons are likely to produce hydrogen and helium inside materials through (n,p) and (n,α) reactions. In addition most transmutations produce, either directly or after radioactive decay, atoms of heavy elements not originally present in the material. For example, neutron reactions with Fe produce Mn and Cr. The dominating effect of transmutations caused by 14-MeV neutrons is expected to be that of helium production, especially embrittlement of metals at the high temperatures in an operating reactor.

Although radiation damage caused by fast neutrons has been investigated extensively for the development of fast breeder reactors, results obtained for fission neutrons cannot be directly applied to fusion reactors because of the much larger effect of transmutations for fusion neutrons. If one takes the helium production as a measure of the effect of transmutations and the displacements per atom (dpa) as a measure of recoil effects, the ratio of helium production to dpa in stainless steel is estimated to be about 0.1 ppm He/dpa for a fast breeder, and 20 ppm He/dpa for a Tokamak-type fusion reactor (2).

Radiation damage studies are important to fusion technology because of the difficulty of replacing the vacuum vessel in Tokamak reactors as presently conceived. Not only is the vacuum vessel surrounded by a thick blanket and magnets, but it will become highly radioactive. In some current designs the surface area of the vacuum vessel is of the order of 3000 m^2 and the mass is of the order of 100 tons. Replacement of the vacuum vessel is time consuming and costly, and the problem of the disposal of the material, which may have activities of the order of 10^9 Ci, is difficult. The feasibility and economics of fusion reactors are influenced by the lifetime of the vacuum vessel, presently estimated to be about two years. Hence development of materials that would extend the lifetime of the vacuum vessel is of greatest importance.

Studies of radiation damage for the design of fusion reactors require a neutron source that produces a recoil spectrum and a helium-production-to-dpa ratio similar to those produced by fusion neutrons. The test source should produce a higher flux density than is present in the fusion reactor so that the effect of the neutrons produced over the reactor lifetime can be studied within a reasonable time. Hence a flux density several times that expected on the wall of the fusion reactor, i.e. several times 10^{14} cm^{-2} sec^{-1}, would be desirable.

Accelerator-based sources produce neutrons in a small volume, and the flux falls off according to the inverse square law. The source strength required to attain a certain flux density depends therefore both on the size of the source and of the sample to be irradiated, and on the uniformity of flux density required over the test sample. For example, if the average distance between source and sample is 3 mm, an isotropic source producing 10^{14} neutrons/sec would produce a flux density of the order of 10^{14} cm^{-2} sec^{-1}. On the other hand, if either the neutron source or the sample have to be larger so that the average source-to-sample distance is, say, 3 cm, the source strength to achieve a flux density of 10^{14} cm^{-2} sec^{-1} would have to be 10^{16} sec^{-1}.

The difficulty of the neutron source problem may be appreciated by recalling that typical laboratory 14-MeV neutron sources have source strengths of 10^{10} sec^{-1} and that in 1977 the most intense steady state 14-MeV neutron source had a strength of a few times 10^{12} sec^{-1}.

Because of the difficulty of obtaining the desired source strength from the reaction of deuterons with tritons, other source reactions that produce a broad neutron spectrum are promising for obtaining adequate intensities, even though the effects of these neutrons differ somewhat from those of fusion neutrons.

The first 14-MeV neutron source specifically designed for radiation damage studies for fusion reactors is expected to become operational in 1978. It can produce, however, a flux density of only about 10^{13} cm^{-2} sec^{-1} in a very small volume. A much more intense 14-MeV neutron source has been designed, but at this writing the question of its actual construction is still under discussion. The design of a very intense broad-spectrum neutron source for radiation damage studies is underway, but when construction will begin has not been determined.

There is evidence that radiation damage induced by neutrons depends not only on fluence but also on the flux density, i.e. that the effect differs depending on whether a given fluence is obtained from a steady source or a pulsed source. Some proposed neutron sources produce a steady flux; others, such as linear accelerators, produce short bursts of neutrons. The same applies to fusion reactors. Some proposed reactors have long burn cycles, such as Tokamak reactors, while others, especially inertially confined plasmas, produce very short bursts of neutrons. Neutron sources designed for radiation damage studies should preferably have a duty cycle similar to that of the reactor for which the study is made.

1.2   Radiotherapy

The use of neutrons for cancer therapy was proposed by Lawrence (3) in 1936, soon after he had developed the cyclotron. Between 1938 and 1943 over 200 cancer patients were treated with neutrons from the Berkeley cyclotron. Although there appeared to be some benefits from this treatment, many patients suffered severe side effects, and the physicians who had directed the neutron treatments concluded that these complications made neutron treatment an undesirable procedure (4). A more recent evaluation (5) of the early Berkeley treatments has shown, however, that, on the basis of present knowledge of the biological effects of neutrons, the doses of radiation the patients received were too large.

The more recent interest in the use of neutrons in therapy arose from radiobiological studies that showed the importance of what is called the oxygen effect. This term describes the observation that cells deprived of oxygen (hypoxic cells) show greater resistance to radiation than well-oxygenated cells. Cells at the center of a tumor are often hypoxic, hence radiation resistant. This oxygen effect is much smaller for neutrons than for x rays. This observation led to the hope that neutrons might cure some malignant diseases that do not respond to conventional x-ray treatment.

The clinical use of neutrons resumed in 1966 at the Hammersmith Hospital in London. The reports from Hammersmith by Catterall (6) and her associates have been so encouraging that radiotherapists in other European countries, in Japan, and in the US have started to use neutrons for treating cancer patients.

Neutron sources for radiotherapy must satisfy minimum requirements of both neutron energy and source strength (7). The energy requirement is based on the need for the neutrons to penetrate to the depth at which the tumor is located without too much absorption in the healthy tissue through which they pass. If the neutron energy is too low, the radiation dose to the healthy tissue is so high that the healthy tissue may not recover. Radiotherapists, who have much experience with the γ rays from ^{60}Co sources and find their penetration satisfactory, would like a neutron source that gives comparable or better penetration in tissue. The attenuation of the neutrons is determined largely by the hydrogen scattering cross section, which decreases with neutron energy and becomes sufficiently small around 10-MeV neutron energy.

The decrease of neutron intensity as the neutrons penetrate into tissue depends not only on the nuclear interactions, but also on the distance from the source because of the effect of the inverse square law. In order to provide adequate depth dose and in order to have space for shielding, the source-to-skin distance should be at least 100 cm, preferably 125 cm. This requirement combined with the limitation of a practical treatment time to about five minutes determines the needed source strength for an isotropic neutron source as about 10^{13} sec^{-1}. If the source is anisotropic, the total source strength may be lower; only the number of neutrons emitted per unit solid angle is important, and this should be at least 10^{12} sr^{-1} sec^{-1}.

In the application in radiotherapy there is also a limitation on acceptable source diameter. In order to limit the treatment volume, a collimator must be used. If the source is too large, the design of the collimator is difficult and an undesirable penumbra is introduced. The collimator will increase the flux density at a given distance from the source over that observed without the collimator because the collimator scatters neutrons into the collimated beam. How much increase occurs depends on the field size and the design of the collimator. If the neutron source has a large diameter, the collimator could reduce the flux density rather than increase it. For most practical sources and collimators the collimator increases the flux density by 10–20%.

Sources that produce neutrons with an average energy above 10 MeV and an intensity of 10^{12} sr^{-1} sec^{-1} are relatively easy to construct. There are, however, some requirements for the clinical use that are more difficult to satisfy. If neutron therapy is to be used in many hospitals, the cost of neutron sources should be not much more than that of presently used x-ray sources, such as electron linear accelerators. Furthermore, it is important that the neutron source can be rotated around the patient as is the practice with electron accelerators and radioactive sources. Such a facility is called isocentric.

Several approaches to the neutron source problem are being pursued. The most widely used sources use the reaction of deuterons on beryllium, the reaction that was used in the early Berkeley work and at Hammersmith. At the Fermi National Accelerator Laboratory 60-MeV protons from a linear accelerator bombard a beryllium target. A third reaction, which produces relatively energetic neutrons with deuterons from a small cyclotron, is the D-D reaction.

The difficulty with cyclotrons and proton or deuteron linear accelerators is that an isocentric treatment facility requires elaborate and expensive beam transport systems. Such a system has been developed for a small cyclotron, but these cyclotrons produce neutrons of an energy below what most radiotherapists want.

An easier way to generate energetic neutrons with a small accelerator is to use the D-T reaction. In this case the development of an accelerator that can be isocentrically mounted is easy, but the development of a target that withstands the large deuteron beams required for the needed intensity has been a difficult problem. Depending on target design, the power dissipation in the target is 10–100 kW.

Development of neutron sources for radiotherapy has reached the point where suitable sources are commercially available. These sources are either cyclotrons that produce a broad spectrum of neutrons or D-T sources. The D-T sources are more compact and lighter, but produce a lower neutron dose rate. A substantial increase in the intensity of compact D-T sources appears difficult. There is the possibility that linear accelerators could replace cyclotrons for broad-spectrum neutron sources. Linear accelerators would eliminate the heavy magnet of cyclotrons.

1.3   Previous Reviews

The needs of the fusion program and possible source designs for this program were the subject of an International Conference on Radiation Test Facilities for the CTR Surface and Materials Program in 1975 (8). The needs and sources for radiation therapy were discussed at an International Workshop on Particle Radiation Therapy in 1975 (9) and at an International Conference on Particles and Radiation Therapy in 1976 (10). Cross sections and yields of high energy neutron source reactions were summarized in 1977 at an International Specialists Symposium (11) and at a Symposium on Neutron Cross Sections (12). The review of "Intense Fast Neutron Source Reactions" by Lone (13) at the Neutron Cross Section Symposium contains a very complete summary and list of references that is not repeated in this review. A review of neutron sources for biomedical applications may be found in *ICRU Report 26* (14). A special issue of *Nuclear Instruments and Methods* published in August 1977 was devoted to high energy, high intensity neutron sources (15). The present review is based to a large extent on information contained in these publications; it is limited to reports published before and during 1977.

2   SOURCE REACTIONS

The requirement that the neutrons have an average energy of 10 MeV or more limits the source reactions to the interaction between light nuclei, because the interaction of even very energetic projectiles with heavy nuclei produces an energy spectrum with a median energy well below that desired for the applications in this review. Although the energy distribution of neutrons produced in the bombardment of heavy nuclei with energetic projectiles usually has a component that extends up to the bombarding energy, most of the neutrons have an energy distribution similar to a Maxwellian distribution with a temperature of 1–2 MeV. Fission neutrons likewise have too low an energy for the applications under consideration. Only the reactions of the hydrogen isotopes with targets lighter than carbon have been found to yield sufficiently high average energies.

Some of the useful reactions yield monoenergetic neutrons in a given direction for thin targets. The intensity requirements for the present applications make it necessary, however, to use thick targets, usually targets in which the bombarding particles stop.

This section summarizes information on neutron yields from various source reactions; a later section describes targets that can be used for producing high intensities of neutrons from these reactions. The monograph *Fast Neutron Physics* edited by Marion & Fowler and published in 1960 (16) gives an excellent summary of the kinematics of neutron-producing reactions, and of their neutron yields. The emphasis in this monograph is on monoenergetic sources, i.e. thin-target sources, rather than high intensity, thick-target sources. Another useful summary of the same source reactions was published in 1963 by Goldberg (17). Although more recent measurements have provided new information on thick-target yields and have extended the older measurements to higher bombarding energies, there have not been any very significant changes in our knowledge of the properties of these reactions in the last twenty years.

2.1   The D-D Reaction

The reaction $^2H + d \rightarrow {}^3He + n + 3.3$ MeV (the D-D reaction) was historically the first reaction to serve as a source of monoenergetic fast neutrons. Its usefulness as a source of monoenergetic neutrons is limited to neutron energies below about 8 MeV, since at bombarding energies above 4.5 MeV the deuterons break up and produce a continuum of much lower energy. In the forward direction the intensity of this continuum exceeds that of the monoenergetic neutrons, when the energy of the monoenergetic neutrons is above 12 MeV.

An evaluation of the measured cross sections of the reaction $^2H(d,n)^3He$ up to deuteron energies of 10 MeV was prepared by Liskien & Paulsen (18). There are additional measurements at higher bombarding energies (19, 20). For deuteron bombarding energies up to 600 keV the

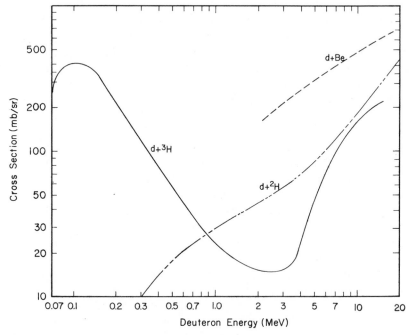

Figure 1 Differential cross sections for the production of neutrons in the forward direction as a function of deuteron bombarding energy for the reactions of deuterons with targets of ^2H, ^3H, and Be.

kinematics and cross sections for thin targets, and the spectra and yields for thick targets, have been calculated and plotted by Seagrave (21).

Several measurements of the cross section of deuteron breakup for bombarding energies up to 20 MeV have been reported. Reference 22 contains a comparison of the 0° cross sections with earlier measurements (23–25). Additional studies of the deuteron breakup have been reported for deuteron energies up to 12 MeV (26). Figure 1 shows the differential cross section for the production of neutrons in the forward direction as a function of deuteron energy for the reaction of deuterons with deuterium (27).

The neutron spectra from the bombardment of deuterium targets of various thicknesses have been measured for 10.6-MeV deuterons (28), and for 17.3-MeV deuterons (29). These measurements address particularly the use of these neutrons in radiotherapy and include measurements of dose rate and depth dose. There is also a measurement of dose rate and depth dose for 7.5-MeV deuterons on a thick target (30).

For the application of the D-D reaction to the design of high intensity, high energy neutron sources two problems must be considered. The increase in the probability of deuteron breakup with increasing bombarding energy results in a slower increase of the average neutron energy with bombarding energy between about 5 and 11 MeV, than with energy below 5 MeV, as shown in Figure 2. The second problem relates to the difficulty in designing the thick deuterium target needed to obtain

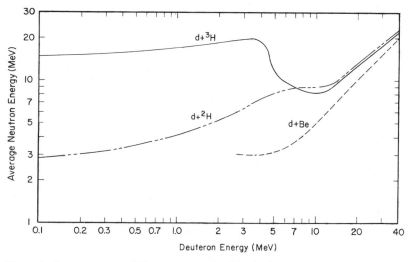

Figure 2 Average energy of the neutrons emitted in the forward direction in the bombardment of thick targets of ^{2}H, ^{3}H, and Be by deuterons as a function of deuteron energy.

sufficient intensity. Nevertheless this reaction offers the possibility of producing a fairly intense source of energetic neutrons with a small cyclotron.

2.2   The D-T Reaction

The reaction $^3H + d \rightarrow {}^4He + n + 17.6$ MeV (the D-T reaction) is of interest as an accelerator-based source for the same reason that it is the most promising reaction for a fusion reactor, i.e. it has a very large cross section at low bombarding energies, as shown in Figure 1.

For the radiation damage application its obvious advantage is that it is the same reaction as that proposed for fusion reactors. Nevertheless, in accelerator-based sources the energy of the neutrons produced may differ significantly from the energy of thermonuclear neutrons. For example, even at as low a deuteron bombarding energy as 0.4 MeV the average neutron energy from a thick tritium target in the forward direction is above 15 MeV, while the thermonuclear neutrons have an average energy of 14.1 MeV. Some important transmutation cross sections, especially those that have a threshold above 10 MeV, vary strongly between 14 and 15 MeV. For the radiotherapy application the possibility of using a low voltage accelerator is the principal reason for considering the $^3H(d,n)$ reaction as a neutron source.

The principal disadvantage of the reaction is the need for using radioactive tritium. This introduces problems in the radiation damage

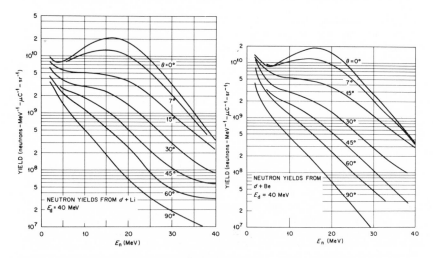

Figure 3   Energy distribution of neutrons emitted at various angles with respect to the incident deuterons when 40-MeV deuterons are incident on thick Li and Be targets (from 33).

application because of the difficulty in designing tritium targets for the large required source strengths, and in the radiotherapy application because of the difficulty of handling large amounts of tritium in a hospital.

An evaluation of the measured cross sections of the reaction ^{3}H(d,n)^{4}He up to deuteron energies of 10 MeV is contained in Reference 18. Actually for intense neutron sources the energy region of interest is below 600 keV and in this energy region Reference 21 is particularly useful, since it contains graphs of thick-target neutron yields and spectra.

Figure 2 shows the average energy of neutrons produced in the interaction of deuterons with a tritium target. The decrease in average energy above 4-MeV bombarding energy is due to neutrons from deuteron breakup (31).

Although in existing intense neutron sources based on the D-T reaction the tritium serves as target, a proposed source would use a tritium beam on a deuterium target. While the published data on the reaction are for deuteron projectiles and tritium targets, the conversion to the reverse arrangement is straightforward.

2.3   Deuterons on Lithium

Interest in the reaction of deuterons with Li for intense sources was stimulated by a proposal from the Brookhaven National Laboratory (32), which pointed out the advantages of a liquid-lithium jet target for dissipating the high power generated in the target of an intense neutron source for radiation damage studies. At the time of that proposal relatively little was known about the yield and spectra of neutrons produced in the bombardment of thick Li targets by deuterons, but several good measurements now available appear to resolve the discrepancies both in yield and spectra of earlier measurements. Such measurements are difficult because of the need to observe low energy neutrons in the presence of much more energetic neutrons.

The measurements most directly applicable to the use of the reaction as a neutron source for radiation damage studies were performed by Saltmarsh et al (33) for 40-MeV deuterons on thick Li targets. Figure 3 shows the measured neutron energy distributions for various emission angles. There are several measurements at lower deuteron energies, particularly the results obtained by Lone et al (34). These more recent data are consistent with earlier measurements by Weaver et al (22), who found a large intensity of neutrons of energy below 5 MeV, while other authors had found spectra that showed a rapid decrease of intensity at energies below 10 MeV.

Measured neutron spectra yield a value for the average energy of the neutrons emitted in the forward direction of 0.4 times the deuteron bombarding energy for deuterons above 10 MeV, but this includes only neutrons of energy above 2 MeV.

The angular distribution of the emitted neutrons is strongly forward peaked; the intensity drops to half value at about 15° at a bombarding energy of 23 MeV (34) and drops more rapidly at higher bombarding energies. The average energy of the emitted neutrons also drops off rapidly from the 0° value as the emission angle increases. This anisotropy of the reaction causes problems in its use as a neutron source for radiation damage studies because samples irradiated close to the source are exposed to neutrons that vary in both intensity and energy over the sample.

2.4   Protons on Beryllium and Lithium

Interest in the reaction of protons on Be developed on the basis of a proposal by the Fermi National Accelerator Laboratory to use the protons from the injector of the National Accelerator for producing neutrons for radiotherapy. This injector is a linear accelerator that permits extraction of protons in the energy range 37–66 MeV. Little was known about the yield and energy of neutrons from protons in this energy range on thick targets, and the information necessary to design a neutron

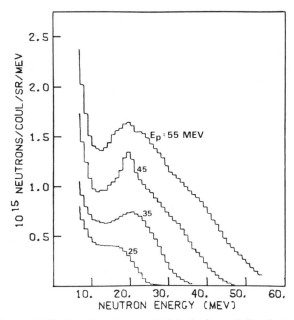

Figure 4   Energy distribution of neutrons emitted in the forward direction when protons of various energies are incident on a thick Be target (from 36).

source was obtained from measurements performed at the University of California at Davis. The results were published in three papers (35–37). Unfortunately the spectra published in References 35 and 36 differ from each other; the data reference in 36 are considered more reliable by the author (private communication). Measured neutron energy distributions at several bombarding energies are shown in Figure 4. There are also data available at lower proton energies, especially in References 34 and 38, and at higher energy (39, 40). Figure 5 shows the energy dependence of the neutron yield as a function of proton energy.

The neutron spectra from the proton bombardment of thick Li and Be targets are similar and show a large peak at energies below 5 MeV. This peak is attributed to evaporation neutrons; its intensity is fairly independent of emission angle and varies slowly with bombarding energy.

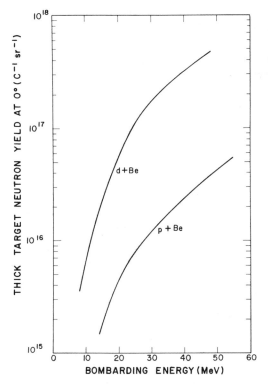

Figure 5   Thick target yields of high energy neutrons emitted in the forward direction for protons and deuterons incident on thick beryllium targets. The plotted yields are based on the observation of neutrons of energy above 5 MeV for p + Be, and above 2 MeV for d + Be. Many neutrons of lower energy are produced in both reactions, but their number is not accurately known (from 13).

At neutron energies above 5 MeV there is a broad maximum for neutrons emitted in the forward direction; this broad distribution extends up to the proton bombarding energy. The intensity of the energetic neutrons decreases rapidly with emission angle. Although the average energy of the neutrons is relatively low, at high bombarding energies the number of high energy neutrons is large enough to make the reaction useful for radiotherapy.

The average energy of the emitted neutrons for both Li and Be targets is shown as a function of proton energy in Figure 6. Because of the presence of many low energy neutrons, the observed average energy depends strongly on the lower limit of neutron energy observed in the measurements. In some applications the low energy neutrons may be filtered out. The yields plotted in Figure 5 do not include these low energy neutrons.

2.5 Deuterons on Beryllium

Traditionally, neutron sources based on cyclotrons have used deuterons on beryllium targets. Beryllium metal is a convenient target material. Since the last neutron in the Be nucleus is loosely bound, this neutron should be easily knocked out, so that bombardment of Be was expected to give a high neutron yield. In the first application of neutrons to radio-

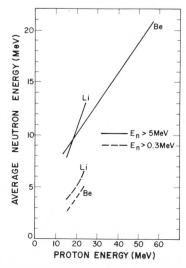

Figure 6 Average energy of the neutrons emitted in the forward direction in the bombardment of thick targets of Li and Be as a function of proton energy. The large number of low energy neutrons emitted results in a much lower average neutron energy, if neutrons below 5 MeV are included in the average (from 13).

therapy at Berkeley (3) the neutrons were produced by this reaction, and it is still used at Hammersmith. Many of the more recent clinical trials of neutron radiotherapy also use this reaction, for example at the University of Washington, the Naval Research Laboratory, and Texas A & M University.

Neutron spectra from the bombardment of thick beryllium targets by deuterons have been measured by many investigators. For neutrons emitted in the forward direction, measurements performed before 1972 agreed that the spectrum was similar to a Maxwellian distribution with a peak not far below half the deuteron bombarding energy. The more recent measurements (22, 33, 34, 41) show little or no decrease in intensity at low neutron energies. In fact, the most recent data (33, 34) agree that the energy distribution increases toward low neutron energies. For example, Lone et al (34) find a neutron distribution about four times higher at 0.5 MeV than at 2.5 MeV for a deuteron bombarding energy of 18 MeV.

Recent measurements of yields and spectra extend to bombarding energies around 80 MeV (22, 33–35, 37, 39–42). Generally the angular distributions are peaked in the forward direction; the peak becomes more pronounced at higher bombarding energies. The average energy of the neutrons emitted in the forward direction is about 0.4 times the deuteron bombarding energy, if one excludes neutrons below about 2 MeV. As the emission angle increases, the average energy of the emitted neutrons decreases. Generally spectra and yields of neutrons from the bombardment of Be targets with deuterons are similar to those from the bombardment of Li targets (Figure 3).

The thick-target yield of the neutrons emitted in the forward direction from a thick Be target is shown in Figure 5 and the average energy of the neutrons is shown in Figure 2. The neutron yield from deuteron bombardment of Be is about an order of magnitude higher than for proton bombardment at the same bombarding energy. Since cyclotrons of a given size can accelerate protons to twice the energy of deuterons, the difference in yield between the two reactions for a given cyclotron is not very large.

Because of the wide use of the d + Be reaction in biomedical applications, the dose rate of the neutrons emitted in the forward direction has been measured by several groups (43–46). The dose rate increases somewhat more slowly than the third power of the deuteron bombarding energy. For the application in radiotherapy a knowledge of the decrease of dose with penetration in tissue is of importance. The depth at which the dose drops to 50% has been measured at various neutron radiotherapy centers. In order to increase the average neutron energy and to increase the penetration of the neutrons, some of the centers use Be targets in which the deuterons do not stop. At Hammersmith, where 16-MeV deuterons bombard a 0.8-mm thick Be target, the depth at which the dose is 50% was found to be 8.8 g cm^{-2} of tissue equivalent material, while at Texas A & M University, where 50-MeV deuterons bombard a thick target, the corresponding value was 13.8 g cm^{-2} (47).

2.6   Other Reactions

A convenient target for a neutron source is carbon. Neutron yields and spectra from deuterons on thick carbon targets have been measured for deuteron energies from 12 to 50 MeV (22, 42). Although the shape of the neutron spectrum is similar for thick C and thick Be targets bombarded by deuterons, the yield from C is about 40% lower, so there is no advantage in using C over Be.

Another reaction considered as a source of energetic neutrons for radiotherapy is ^{3}He on Be. A cyclotron produces ^{3}He^{2+} ions of 2.5 times the energy of deuterons. This led to the hope that relatively high neutron energies could be obtained with a small cyclotron. Studies of the ^{3}He + Be reaction (48, 49) show, however, that the yield of the reaction is rather low, and that the reaction has no real advantage over d + Be.

Although the reaction of protons with tritons is useful as a source of monoenergetic neutrons, this reaction is not useful for producing high intensities of energetic neutrons. At a fixed deuteron energy the neutron yield generally decreases with atomic weight (41, 42), hence targets heavier than Be offer no advantages for high intensity sources.

3   ACCELERATORS

For the neutron sources in present use the development of suitable accelerators has been much easier than the development of suitable targets. Cyclotrons that produce 50 μA of external beam at 40 MeV are available for neutron sources based on the d + Be reaction. For d + ^{3}H sources an accelerator that produces 25 mA at 400 keV (50) and a single-gap accelerator that produces 200 mA at 160 keV (51) are in operation with external beams, and another system operates at 200 kV with a current of 500 mA in a closed tube (52).

For the radiotherapy application available accelerators are adequate, and the problem has been primarily in the development of targets. For the much higher source strengths desired for radiation damage studies higher current accelerators must be developed. Desired currents for two accelerators presently under consideration are 1.1 A at 0.3 MeV and 0.2 A at 35 MeV. In both cases well-focused external dc beams are needed, and at this writing there are no accelerators in operation that approach these specifications.

3.1 *Low Voltage Accelerators*

At present several high current, low voltage accelerators are being designed or constructed for neutron sources. The accelerator that has been completed (53) is designed to provide a 1-cm diameter external beam of 0.15 A, 400-keV atomic deuterium ions. An ion source (54) with 17 apertures allows extraction of up to 0.4 A dc total beam at 15 kV. A 90° double focusing magnet separates out the D^+ ions that pass through a solenoid lens into a uniform gradient acceleration tube with reentrant electrodes. The acceleration tube has three intermediate electrodes between ground and the high voltage terminal and produces a gradient of 17 kV/cm for accelerating the ions. Space charge effects are avoided by maintaining a high enough pressure for complete space charge neutralization.

3.2 *High Voltage Accelerators*

At present most neutron sources that require deuterons of energies above 5 MeV use cyclotrons to accelerate the deuterons. For the high currents needed for neutron sources planned for radiation damage studies, cyclotrons are not expected to be able to provide high enough currents, and linear accelerators are preferred (55).

Although the linear accelerator planned for the acceleration of 200 mA has not yet been designed in detail, it will undoubtedly consist of an injector similar to the low voltage accelerators described in the preceding section. The output of the injector has to be chopped and/or bunched to match the times during which the accelerator accepts ions for acceleration. If a dc beam is desired on target, the accelerated beam has to be debunched probably by allowing the ions to drift through a long distance.

The design of the accelerator is complicated by the need to minimize beam losses. Even a small fraction of the 7 MW of beam power would melt any components it might strike and/or would induce a high level of radioactivity.

4 TARGETS

4.1 *Deuterium*

4.1.1 SOLID TARGETS   Thick deuterium targets for generating neutrons with a low voltage accelerator are most easily prepared by freezing heavy water on a liquid-nitrogen-cooled metal plate (56). Because of the low thermal conductivity of ice, such targets evaporate rapidly at high beam currents. Heavy-ice targets have been used for beam currents up to 0.2 mA. Currents up to 0.5 mA can be used if the target rotates (57).

Deuterium absorbed in metals, such as Ti and Zr, has been used for producing neutrons. Such hydride targets will be discussed further for tritium. The neutron yield is lower than for a heavy-ice target by a factor of the order of three, partly because of the higher stopping power of the metals compared to oxygen, partly because the targets contain fewer than two deuterium atoms per metal atom.

Deuterium may be imbedded in the target by the deuteron beam. This method of loading the target usually produces a lower deuterium concentration than if the deuterium is loaded from gas at elevated temperature. Neutron yields have been reported to be of the order of a tenth of those from heavy-ice targets. None of the solid targets have been found to be useful for neutron sources for the applications discussed in this review.

4.1.2 LIQUID TARGETS   Although heavy water may be a useful target material for neutron sources, it does not appear to have been used. Heavy water could either be separated from the accelerator vacuum system by a foil, or it might be used in the form of a free-standing jet in the vacuum system (58). Experiments with a jet of ordinary water injected at velocities up to 300 m/sec indicate that the jet is quite stable in vacuum. The temperature rise of the water caused by 20 mA of 35-MeV deuterons is estimated to be only 1°C.

4.1.3 GAS TARGETS WITH ENTRANCE FOIL   Deuterium gas targets have been used extensively in accelerators for nuclear physics experiments (56). The gas target is separated from the accelerator by a thin window through which the beam passes. Foils of Al, Ni, Mo, and Havar, a Co-Ni-Cr alloy of high tensile strength, have served as window materials. Often the foils rupture when the beam strikes them. Such failures are most likely at high beam currents and high gas pressures, i.e. under the conditions for which the neutron output is highest.

At the Cancer Research Center in Heidelberg a deuterium gas target has been in use for biomedical research (59); 70 μA of 11 MeV deuterons from a cyclotron pass through a 10 mg/cm^2 Havar foil into a 30-cm long chamber filled with deuterium gas to a pressure of 11 atm. Typically a foil lasts 8 hours. The beam power dissipated in the target is removed by circulating the gas and cooling it with water. This arrangement yields dose rates of 0.35 Gy/min (where Gy = J/kg) 1 m from the target in the forward direction, and the dose in water decreased to 40% in 10 cm.

Kuchnir et al (60) have pointed out the advantages of cooling the target gas to liquid-nitrogen temperatures. The increased density permits the use of a shorter target, and the lower temperature increases the tensile strength of the Havar foil. A cryogenic target system has been developed for use with 200 μA of 8.3-MeV deuterons from the 30-in. cyclotron at the University of Chicago. A 10-mg/cm^2 Havar foil separates the accelerator vacuum system from the 7.5-cm long deuterium gas target, which is at a pressure of 10 atm and a temperature of 80 K.

4.1.4 DIFFERENTIALLY PUMPED GAS TARGETS   One of the earliest suggestions to use a differentially pumped high pressure gas target was made in 1957 (61). A decade later Colombant and Lidsky (62, 63) proposed a high pressure differentially pumped deuterium gas target for an intense neutron source. This was to be accomplished by bombarding the deuterium with 1 A of tritium. A neutron source based on this proposal has been designed at Los Alamos (64). A group in Canada (65) has been studying various types of deuterium gas target systems using subsonic, transonic, and hypersonic flow.

The deuterium gas target designed by the Los Alamos group (64) uses a supersonic (3300 m/sec) jet that enters the target region at a temperature of 30–40 K and experiences a 1400 K temperature rise because of the 300 kW power dissipation by the beam. Eighty grams per second of deuterium pass through the 1-cm diameter target region and produce a density of 2×10^{19} deuterium molecules/cm^3. The target system designed at Los Alamos is shown in Figure 7.

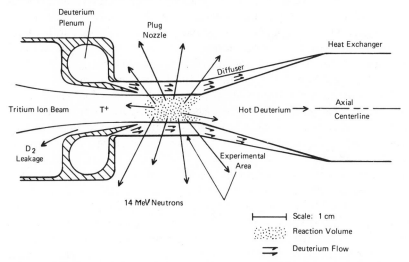

Figure 7 Schematic of the supersonic-jet deuterium target (INS). A beam of tritium ions is incident from the left (from 64).

4.2 *Tritium*

Although the same types of targets described for deuterium could be used for tritium, the radiation hazards associated with tritium and the higher cost of tritium compared with deuterium impose restrictions in its use. For example, tritiated water or ice has never been employed for targets.

Another difference is the much higher neutron yield of the D-T reaction relative to the D-D reaction at low bombarding energies. The cross section for the D-T reaction peaks at about 100 keV deuteron energy. For a thick target the neutron yield rises rapidly up to a deuteron energy of 200 keV; it increases by about 40% from 200 keV to 400 keV, but only about 10% from 400 keV to 600 keV. Hence neutron generators using this reaction operate at voltages between 150 and 400 keV, while high intensity neutron sources using the D-D reaction employ energies above 5 MeV. This difference affects the target design.

The radioactivity of tritium favors closed rather than pumped acceleration systems, especially in hospitals, so much effort has gone into the development of closed systems using mixed D-T beams.

4.2.1 SOLID TARGETS Tritiated solid targets were first used in 1949 (66). Originally tritium was absorbed in Zr, but other elements, such as Sc, Ti, and Er, form useful hydrides. Titanium is presently most widely used. The first Zr targets were backed by W, but other materials such as Cu, Ag, Ta, and Pt have also served as backings. For high current applications the cooling requirements determine the choice of the target backing, and Cu or copper alloys are most useful. In particular, a Cu-Zr alloy, Amzirc, combines good thermal conductivity and high mechanical strength. There are several suppliers of metal tritide targets, such as Oak Ridge National Laboratory in the US and Nukem in Germany.

The neutron yield from tritide targets decreases with use because tritium may be released when the target is heated by the incident beam, but also because the incident deuterons displace tritium in the target. The combination of these effects results in a decrease of the neutron yield by a factor of two when about 6 C of deuterons have bombarded a target in a typical, commercially available, 14-MeV neutron generator, as is often used for neutron activation analysis.

Tritium loss by heating can be greatly reduced by adequate cooling. The rotating target developed by Booth (67) at Livermore has provided particularly effective cooling. In this design the outside of the target is in contact with a 1-mm thick layer of water. In order to provide adequate cooling the target rotates at 1100 rpm. This rapid rotation is made possible by a specially designed vacuum seal (68, 69). The first use of the rotating targets was with 8 mA of 400-keV deuterons, and these targets had a diameter of 15 cm. When the beam current was increased to 25 mA, a target diameter of 22 cm was chosen (50). For the more intense source now under construction at Livermore, 50-cm diameter targets will be used, and the rotation speed will be increased to 5000 rpm (53). With these larger targets and the higher rotation speed the power requirement to overcome the viscous drag caused by the external cooling would be excessive. Hence the cooling is placed inside a sandwich target backing that contains convoluted channels to produce turbulent flow.

For long target life an analyzed deuteron beam is essential, because the displacement of tritium in the target occurs most rapidly where the deuterons come to rest. If both atomic and molecular ions are used, the molecular ions displace tritium at a depth where the atomic ions have the highest probability of producing neutrons (70). The fact that the loss of tritium occurs most rapidly at the depth at which the deuterons stop has been corroborated by measuring the tritium distribution in a used target (71).

For an extension of the target lifetime the rotating target can be mounted in such a way that the beam strikes different concentric rings at different times (70). This arrangement is shown in Figure 8. The lifetime

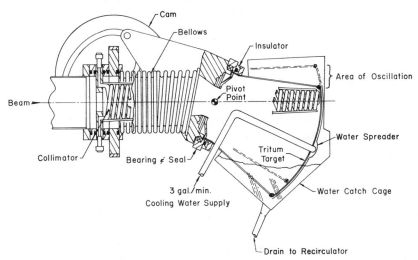

Figure 8 Schematic of the rotating target neutron source (RTNS I) at Lawrence Livermore Laboratory. The section to the right of the bearing and seal rotates at about 1100 rpm. The tritium loaded target is held with an O-ring seal at the end of the vacuum system. The section to the right of the bellows can be moved up and down so that the entire tritium loaded target area can be used. The neutron source remains fixed in space. Samples may be placed on the beam axis within a few millimeters of the neutron source (from 50).

of the target depends strongly on how sharply the beam is focused. While a sharply focused beam increases the flux density near the source, it drastically reduces target life. In the Livermore rotating target source, when a 1.6-cm diameter beam of 15 mA produces an initial source strength of 4×10^{12} sec^{-1}, the source strength decreases about 15% in 100 hr of operation (72). With a more sharply focused beam a high flux density $(1.5 \times 10^{12}$ cm^{-2} sec$^{-1})$ can be achieved on small samples placed very near the source, but the targets deteriorate more rapidly (73).

There are several other neutron sources that use rotating targets similar to Booth's design. The first of these was built by Broerse et al (74) at the Radiobiological Institute in Rijswijk, The Netherlands. It uses 6 mA of 270-keV deuterons and produces a source strength of 6×10^{11} sec^{-1} with a 15-hr half-life. Another rotating target source is in operation at the Eppendorf Hospital in Hamburg, Germany. This unit was manufactured by Radiation Dynamics (Westbury, Long Island, New York) and uses 12 mA of 500-keV deuterium ions (75). It produces 2.5×10^{12} neutrons/sec, with a half-life of less than 10 hr. A very large rotating target source is in operation at Valduc, France. A 25-cm diameter target, which rotates at 3000 rpm, is cooled with a Na-K alloy. When the target is bombarded with 110 mA of 160-keV deuterons, a source strength of 6×10^{12} sec^{-1} can be obtained, but the intensity drops to half its initial value in only 3 hours (51).

The much longer lifetimes of the rotating targets at Livermore are undoubtedly due to the use of an analyzed beam, while an unanalyzed beam that contains both atomic and molecular ions is used at the other installations.

Cranberg (76) has suggested that the target life could be extended in installations where both atomic and molecular ions are accelerated, by applying a weak magnetic field near the target so that the different species of ions impinge on separate adjoining portions of the target.

4.2.2 DRIVE-IN TARGETS   The decrease of neutron source strength from a metal hydride target may be avoided by replenishing the tritium by the charged-particle beam. For this purpose the beam should consist of approximately equal numbers of deuterium and tritium ions. Such a self-replenishing system can be sealed off. For the same concentration of hydrogen in the target a mixed deuterium-tritium beam on a mixed target gives only half the source strength of a monoisotopic beam and target, since only half the collisions are between deuterons and tritons, while collisions between deuterons and deuterons or tritons and tritons produce a relatively very small number of neutrons.

The performance and general aspects of drive-in targets have been discussed by Hillier et al (77) and more recently by Kim (78). An experimental study of solid targets for intense neutron sources is underway at the Sandia Laboratories where a special target test facility has been constructed (79).

As self-replenishing targets either metal tritide or pure metal have been used. Metal tritide targets give initially a much higher neutron yield, but they tend to blister and, in addition, to lose by diffusion the tritium originally present (77).

The neutron generators that use drive-in targets and have been reported to produce the highest neutron yields have quite different targets. One uses chromium-plated copper tubing (80, 81). These tubes are arranged in such a way that 280 mA of mixed deuterium-tritium ions from two ion sources can bombard the target from opposite directions.

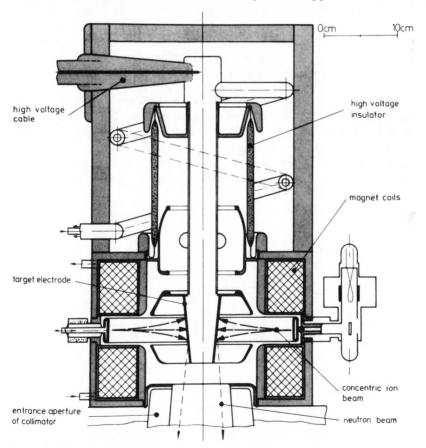

Figure 9 Sealed neutron generator tube. A mixed beam of deuterons and tritons from a toroidal ion source is focused by magnet coils onto the conical Sc target electrode at the center of the tube. Neutrons emitted downward are used (from 52).

With this arrangement a source strength of 5.6×10^{12} sec^{-1} has been achieved within a target spot 6 cm in diameter. The other system (52) shown in Figure 9 employs a conical scandium deuteride-tritide target. This cone is concentrically surrounded by a ring-shaped ion source that produces about 150 mA of mixed deuterium-tritium ions. A neutron source strength of 5×10^{12} sec^{-1} has been reported. When one looks along this axis of the cone, the neutron source appears as a ring with 3.2-cm inner diameter and 4.2-cm outer diameter.

Two types of sealed-tube neutron generators that produce 1×10^{12} neutrons/sec are commercially available; one is manufactured by Elliott in England, and two such units are in operation (at Manchester and Glasgow) (82); the other generator is manufactured by Philips in The Netherlands and is used at a hospital in Amsterdam. Both types of generators have 250-kV power supplies, the Elliott unit has an ion current of 30 mA of mixed deuterium-tritium ions incident on an Er target, while the Philips unit has 18 mA incident on a Ti target. In both generators the tube lifetimes have been more than 100 hours (83).

4.2.3 GAS TARGETS   A target of tritium gas produces three to five times more neutrons per incident deuteron than a metal tritide target, since all the collisions are between the hydrogen isotopes. Gas targets have the problem that the gas must be prevented from getting into the acceleration tube, which must be under vacuum. The currents needed for high intensity sources preclude the use of foils so that differential pumping

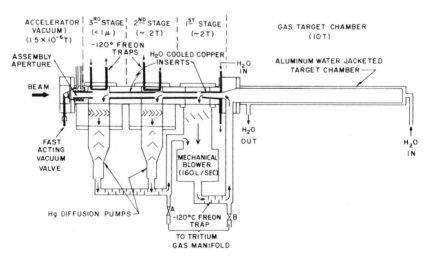

Figure 10 Differentially pumped tritium gas target. Deuterons are incident from the left. Neutrons emitted to the right are used (from 84).

must be applied between the target cell and the acceleration tube. Chenevert et al (84) have designed a system, shown in Figure 10, to accomplish this. In this design 15 mA of 250-keV deuterons pass through a 0.4-cm diameter aperture into a 50-cm long target chamber that contains tritium at a pressure of 10 Torr. So far 9.5 mA of deuterons have been used, and the highest observed source strength has been about 3×10^{12} sec^{-1}. A problem has been isotopic mixing not only between the deuterium in the beam and the target gas, but also exchange with ordinary hydrogen present in components of the target and gas handling system. This reduced the neutron yield to half its initial value in two hours, and frequent replacement of the tritium gas in the target is necessary.

4.3 Lithium

Evaporated lithium metal has been the usual target material for neutron production in low current accelerators. Because of the low melting temperature of Li, metal targets are not stable at high beam currents. A liquid Li target has therefore been selected for the intense source presently being designed for construction at Hanford. This design is expected to be based on a Brookhaven proposal (55), but the details have not yet been decided. The design aim is for a beam power dissipation of 3–10 MW in the target. According to the Brookhaven proposal liquid lithium at 200°C will flow at a rate of 18 liters per second to form a 1.5-cm thick, 12-cm wide layer.

4.4 Beryllium

Beryllium targets for cyclotron-based neutron sources usually consist of a sheet of beryllium metal thick enough to stop the incident particles. The range of 30-MeV deuterons in Be metal is about 3.5 mm. The target is usually cooled by attaching it to a water-cooled copper backing. Such targets can be used with a beam power of 1–2 kW without melting provided the beam is not too sharply focused.

In radiation damage studies a sharply focused beam is desirable for achieving high neutron flux densities near the target. In a target system designed for this application (85) the beryllium is brazed into the copper backing. The cooling water flows through convoluted channels in the backing, close to the back surface of the Be target. This target has been used with 20 μA of 30-MeV deuterons focused on a 5-mm spot, but it is designed for 100 μA.

Just as for tritium targets, much higher currents can be accommodated on a rotating Be target. Calculations (58) indicate that an internally cooled Be wheel, 32 cm in diameter, rotating at 6000 rpm could be used with a beam power of 3 MW and a beam diameter of 1 cm. This corresponds to 85 mA of 35-MeV deuterons. In such a wheel the stresses caused by centrifugal forces would be very small compared with the tensile strength of Be. The neutron yield of an internally cooled Be target could be increased by using D_2O as the coolant because the deuterium in D_2O would contribute to the neutron production.

5   STATUS OF NEUTRON SOURCES

5.1   *Radiation Damage Studies*

5.1.1   D-T SOURCES   At present most studies of radiation damage induced by D-T neutrons employ the rotating target neutron source at Lawrence Livermore Laboratory (RTNS I). The characteristics of this source and typical experiments are described in Reference 73. The experiments have included both surface and bulk radiation damage measurements, at temperatures ranging from that of liquid helium to 800°C. The highest observed neutron flux density is 1.7×10^{12} cm^{-2} sec^{-1} on a small sample; a more typical value is 1×10^{12} cm^{-2} sec^{-1} for an extended run or a fluence of 3×10^{17} cm^{-2} in an 80-hour run.

Two larger sources of similar design (RTNS II) are nearing completion at Livermore (53). These sources are expected to produce 4×10^{13} D-T neutrons per second and to provide a maximum neutron flux density of 1.2×10^{13} cm^{-2} sec^{-1}. The design should permit a future upgrading to a source strength of 10^{14} sec^{-1} and a flux density of $2–3 \times 10^{13}$ cm^{-2} sec^{-1}.

While the older (RTNS I) source was originally built for nuclear physics experiments and is in a building without adequate shielding for high intensity operation, the new sources (RTNS II) are designed for radiation damage studies and are located in a well-shielded building. The facility has provisions for remote handling for target replacements and sample positioning.

Two larger D-T neutron sources (INS) for radiation damage studies have been designed at the Los Alamos Scientific Laboratory (64, 86). These sources use a tritium beam on a deuterium jet target and are designed to produce 10^{15} D-T neutrons per second. At this source strength the neutron flux density would be more than 10^{14} cm^{-2} sec^{-1} within a 3-cm^3 volume and more than 10^{13} cm^{-2} sec^{-1} within about 100 cm^3. One of the sources will be surrounded by concentric spheres of lithium, ^{235}U, and beryllium. The uranium multiplies the number of neutrons, and the combination of shells is designed to produce, together with the primary neutrons, a neutron spectrum similar to that at the first wall of a fusion reactor. While the flux density of primary neutrons decreases with distance from the source according to the inverse square law, the flux density of secondary neutrons is fairly uniform within the spherical shell. Thus the total flux density is high over a much larger volume than for a bare source, but the ratio of the numbers of primary and secondary neutrons varies rapidly within the shell.

5.1.2 BROAD-SPECTRUM SOURCES   Many of the radiation damage studies with broad-spectrum sources have been performed with neutrons from the d + Be reaction at the University of California at Davis. In these experiments about 25 μA of 30-MeV deuterons bombard a 3.5-mm thick Be target. This source produces about 5×10^{12} neutrons sec^{-1} sr^{-1}, or a flux density of the order of 10^{12} cm^{-2} sec^{-1} at a distance of 2 cm from the target in the forward direction. For a sample that subtends an appreciable angle at the source both the flux density and the average neutron energy decrease rapidly with angle.

A large broad-spectrum neutron source is being designed for construction at the Hanford Engineering Development Laboratory (55). The plan is to bombard a liquid-lithium target with 0.1–0.2 A of 35-MeV deuterons. Such a source would produce about 3×10^{16} neutrons sec^{-1} sr^{-1} in the forward direction, and the flux density would be above 10^{14} cm^{-2} sec^{-1} within 15–20 cm from the source. This would permit the irradiation of fairly large samples at a flux density comparable to that at the first wall of a fusion reactor. The characteristics of operating and planned neutron sources for radiation damage studies for fusion reactors are listed in Table 1.

5.2 Radiotherapy

5.2.1 D-T SOURCES   While neutron sources for radiation damage studies have been designed and built at the institutions where they are used, D-T neutron sources for medical uses have been supplied by commercial manufacturers. Three of these manufacturers produce mixed-beam sealed-tube sources, one produces a pumped tube with mixed beam, and one uses a deuteron beam on a rotating target. All D-T neutron generators are isocentrically mounted, which is a great advantage in the clinical use, compared with most broad-spectrum sources.

Since these generators are commercially produced, less information about the construction of the systems has been published than for systems manufactured at research laboratories. Specifications are not necessarily indicative of actual hospital experience. Table 2 summarizes published information. Quoted neutron source strengths are usually not the number of neutrons produced at the target, but are deduced from the flux density at the exit of a collimator. This number includes scattered neutrons.

Table 1 High energy neutron sources for radiation damage studies

Location	Name	Reaction	Beam energy (MeV)	Target current (mA)	Target spot size (cm)	Source intensity (neutrons sr^{-1} sec^{-1})	Status
Livermore	RTNS I	d + ^{3}H	0.4	22	0.6 diameter	5×10^{11}	Operating
Livermore	RTNS II	d + ^{3}H	0.4	150	1 diameter	3×10^{12}	Scheduled for 1978
Valduc	Lancelot	d + ^{3}H	0.16	110	5 diameter	5×10^{11}	Operating
Los Alamos	INS	t + ^{2}H	0.3	1000	1 × 1 cylinder	8×10^{13}	Planned
University of California, Davis		d + Be	30	0.025	0.5 diameter	5×10^{12}	Operating
Hanford	FMIT	d + Li	35	100	1–10 diameter	3×10^{16}	Planned

Table 2 D-T generators for radiotherapy

Manufacturer	Voltage (kV)	Current (mA)	Target material	Source strength (10^{12} sec^{-1})	Effective lifetime (hr)	Status
Elliott	250	30	Er	1	170	Operating at Glasgow and Manchester
Philips	250	18	Ti	1	130–150	Operating at Amsterdam
Haefely	250	500	Sc	5[a]	700[a]	Operating at Heidelberg; To be installed at Zürich
Cyclotron Corporation	200	300	?	8[a]	500[a]	To be installed at Riyadh
Radiation Dynamics	500	15	Ti-T	2.5	10	Operating at Hamburg

[a] Manufacturer's specifications.

Although two of the D-T generators listed in Table 1 promise to provide much higher intensities than those listed in Table 2, these high intensity sources are not suitable for clinical use. The development of D-T sources for clinical applications with intensities much higher than those listed in Table 2 appears very difficult.

5.2.2 BROAD-SPECTRUM SOURCES   Most of the broad-spectrum sources used in radiotherapy use accelerators not constructed for this application and are located in nuclear physics research laboratories. All of them have horizontal charged-particle beams and have a fixed horizontal neutron beam. The only exceptions are two cyclotrons that were installed in Essen and in Edinburgh in 1976; these cyclotrons produce 15-MeV deuterons and have a magnetic beam transport system that permits isocentric treatment.

Table 3 summarizes information about broad-spectrum sources for radiotherapy. In all of them the accelerator is a cyclotron except at the Fermi National Accelerator Laboratory where a linear accelerator is employed. This listing does not include all the cyclotrons used for radiotherapy but gives typical parameters. For example, a 30-MeV cyclotron has been in use for neutron radiotherapy in Japan, and a cyclotron similar to that at Hammersmith has been used for neutron therapy at Dresden (GDR).

A cyclotron of a given size accelerates protons to twice the energy of deuterons. For isocentric treatment, magnetic beam transport systems of a given size can also handle protons of twice the energy of deuterons. For these reasons the p + Be reaction has advantages over d + Be for use in small cyclotrons. The last two lines of Table 3 give the expected per-

Table 3   Broad-spectrum neutron generators for radiotherapy

Location	Reaction	Beam energy (MeV)	Beam current (µA)	Dose rate in air at 125 cm (Gy/min)	Depth of 50% dose in tissue (cm)
Hammersmith	d + Be	16	80	0.35	8.8
University of Washington	d + Be	21.5	30	0.45	10.2
Naval Research Laboratory	d + Be	35	10	0.66	12.8
Texas A & M University	d + Be	50	7	0.75	13.8
Fermilab	p + Be	66	30	1.1	16
Heidelberg	d + ^{2}H	10.6	70	0.20	10
Chicago	d + ^{2}H	8.3	200	0.20	10
	p + Be	30	66	0.46	10.6
	p + Be	42	47	0.84	12.6

formance of commercially available cyclotrons if the neutrons are generated by the reaction p + Be. These cyclotrons are available with a magnetic beam transport system for isocentric treatment.

Literature Cited

1. Kulcinski, G. L., Conn, R. W. 1974. *Nucl. Fusion* Suppl., pp. 51–77
2. Kulcinski, G. L., Doran, D. G., Abdou, M. A. 1976. *Spec. Tech. Publ. 570*, pp. 329–51. Am. Soc. Test. Mater.
3. Lawrence, E. O. 1937. *Radiology* 29: 313–22
4. Stone, R. S. 1947. *Am. J. Roentgenol.* 59: 771–85
5. Sheline, G. E., Phillips, T. L., Field, S. B., Brennan, J. T., Raventos, A. 1971. *Am. J. Roentgenol.* 111: 31–41
6. Catterall, M., Sutherland, I., Bewley, D. K. 1975. *Br. Med. J.* 2: 653–56
7. Catterall, M. 1976. *Br. J. Radiol.* 49: 203–5
8. *Proc. Int. Conf. Radiat. Test Facil. CTR Surf. Mater. Progr. 1975. Rep. ANL/CTR-75-4.* 873 pp.
9. *Part. Radiat. Ther., Proc. Int. Work., Am. Coll. Radiol.* 1976. 572 pp.
10. Particles and Radiation Therapy II Conf., Berkeley, 1976; 1977. *Int. J. Radiat. Oncol. Biol.* 3
11. Lone, M. A., ed. 1977. *Rep. NBSIR 77-1279.* 34 pp.
12. Bhat, M. R., Pearlstein, S., eds. 1977. *Symp. Neutron Cross Sect. 10–40 MeV. Rep. BNL-NCS-50681.* 555 pp.
13. Lone, M. A. 1977. See Ref. 12, pp. 79–116
14. *Int. Comm. Radiat. Units Meas., Washington, DC.* 1977. *ICRU Rep.* 132 pp.
15. Ullmaier, H., ed. 1977. *Nucl. Instrum. Methods* 145: 1–218
16. Marion, J. B., Fowler, J. L., eds. 1960. *Fast Neutron Physics, Part I*, pp. 1–176. New York: Interscience
17. Goldberg, M. D. 1963. *Progress in Fast Neutron Physics*, pp. 3–21. Univ. Chicago Press, IL
18. Liskien, H., Paulsen, A. 1973. *Nucl. Data Tables* 11: 569–619
19. Van Oers, W. T. H., Brockman, K. W. 1963. *Nucl. Phys.* 48: 625–46
20. Roy, M., Bachelier, D., Bernas, M., Boyard, J. L., Brissaud, I., Detraz, C., Radvanyi, P., Sowinski, M. 1969. *Phys. Lett. B* 29: 95–96
21. Seagrave, J. D. 1958. *Los Alamos Rep. LAMS 2162.* 38 pp.
22. Weaver, K. A., Anderson, J. D., Barschall, H. H., Davis, J. C. 1973. *Nucl. Sci. Eng.* 52: 35–45
23. Cranberg, L., Armstrong, A. H., Henkel, R. L. 1956. *Phys. Rev.* 104: 1639–42
24. Lefevre, H. W., Borchers, R. R., Poppe, C. H. 1962. *Phys. Rev.* 128: 1328–35
25. Rybakov, B. V., Sidorov, V. A., Vlasov, N. A. 1961. *Nucl. Phys.* 23: 491–98
26. Coçu, F., Ambrosino, G., Guerreau, D. 1977. *Saclay Rep. CEA-R-4838.* 56 pp.
27. Barschall, H. H. 1974. *Nuclear Structure Study with Neutrons*, pp. 289–97. New York: Plenum
28. Schraube, H., Morhart, A., Grünauer, F. 1975. *Symp. Neutron Dosimetry Biol. Med.*, 2nd, ed. G. Burger, H. G. Ebert pp. 979–1003. *EUR 5273 d-e-f*
29. Edwards, F. M., Fielding, H. W., Kraushaar, J. J., Weaver, K. A. 1974. *Med. Phys.* 1: 317–22
30. McDonald, J. C., Kuo, T. Y. T., Freed, B. R., Laughlin, J. S. 1977. *Med. Phys.* 4: 319–21
31. Poppe, C. H., Holbrow, C. H., Borchers, R. R. 1963. *Phys. Rev.* 129: 733–39
32. Goland, A. N., Snead, C. L., Parkin, D. M., Theus, R. B. 1975. *IEEE Trans. Nucl. Sci.* NS-22: 1776–79
33. Saltmarsh, M. J., Ludemann, C. A., Fulmer, C. B., Styles, R. C. 1977. *Nucl. Instrum. Methods* 145: 81–90
34. Lone, M. A., Bigham, C. B., Fraser, J. S., Schneider, H. R., Alexander, T. K., Ferguson, A. J., McDonald, A. B. 1977. *Nucl. Instrum. Methods* 143: 331–44
35. Heintz, P. H., Johnsen, S. W., Peek, N. F. 1977. *Med. Phys.* 4: 250–54
36. Johnsen, S. W. 1977. *Med. Phys.* 4: 255–58
37. Amols, H. I., Dicello, J. F., Awschalom, M., Coulson, L., Johnsen, S. W., Theus, R. B. 1977. *Med. Phys.* 4: 486–93
38. Nelson, C. E., Purser, F. O., Von Behren, P., Newson, H. W. 1977. See Ref. 12, pp. 125–33
39. Harrison, G. H., Cox, C. R., Kubiczek, E. B., Robinson, J. E. 1977. See Ref. 11, pp. 19–23
40. Madey, R., Waterman, F. M., Baldwin, A. R. 1977. *Med. Phys.* 4: 322–23
41. Daruga, V. K., Matusevich, E. S., Narziev, Kh. 1972. *At. Energy* 33: 934–36
42. Meulders, J. P., Leleux, P., Macq, P. C., Pirart, C. 1975. *Phys. Med. Biol.* 20: 235–43
43. Parnell, C. J., Page, B. C., Chaudhri, M. A. 1971. *Br. J. Radiol.* 44: 63–66

44. August, L. S., Theus, R. B., Attix, F. H., Bondelid, R. O., Shapiro, P., Surratt, R. E., Rogers, C. C. 1973. *Phys. Med. Biol.* 18:641–47

45. Weaver, K. A., Anderson, J. D., Barschall, H. H., Davis, J. C. 1973. *Phys. Med. Biol.* 18:64–70

46. Howell, R. H., Barschall, H. H. 1976. *Phys. Med. Biol.* 21:643–45

47. Smith, A. R., Almond, P. R., Smathers, J. B., Otte, V. A., Attix, F. H., Theus, R. B., Wootton, P., Bichsel, H., Eenmaa, J., Williams, D., Bewley, D. K., Parnell, C. J. 1975. *Med. Phys.* 2:195–200

48. Broerse, J. J., Barendsen, G. W. 1967. *Int. J. Radiat. Biol.* 13:189–94

49. Parnell, C. J., Page, B. C., Jones, E. J. 1975. *Phys. Med. Biol.* 20:125–27

50. Booth, R., Barschall, H. H., Goldberg, E. 1973. *IEEE Trans. Nucl. Sci.* NS 20–3:472–74

51. Hourst, J. B., Roche, M., Morin, J. 1977. *Nucl. Instrum. Methods* 145:19–24

52. Schmidt, K. A., Dohrmann, H. 1976. *Atomkernenergie* 27:159–60

53. Booth, R., Davis, J. C., Hanson, C. L., Held, J. L., Logan, C. M., Osher, J. E., Nickerson, R. A., Pohl, B. A., Schumacher, B. J. 1977. *Nucl. Instrum. Methods* 145:25–39

54. Osher, J. E., Hamilton, G. W. 1974. *Rep. UCRL 76020.* 8 pp.

55. Grand, P., Goland, A. N. 1977. *Nucl. Instrum. Methods* 145:49–76

56. Coon, J. H. 1960. See Ref. 16, pp. 677–720

57. Van Dorsten, A. C. 1955. *Philips Tech. Rev.* 17:109–11

58. Logan, C. M., Anderson, J. D., Barschall, H. H., Davis, J. C. 1975. See Ref. 8, pp. 410–15

59. Scheer, K. E., Schmidt, K. A., Höver, K.-H. 1976. *Proc. Int. Conf. Interactions Neutrons Nucl.*, pp. 1162–85. *ERDA Rep. CONF 760715*

60. Kuchnir, F. T., Waterman, F. M., Forsthoff, H., Skaggs, L. S., Vander Arend, P. C., Story, S. 1976. *Conf. Use Small Accel. Res. Teach. Ind. Appl., 4th, Denton, Texas*, pp. 513–16

61. Schumacher, B. W. 1957. *Can. J. Phys.* 35:239–40

62. Colombant, D. 1969. *High intensity 14 MeV neutron source.* PhD thesis. MIT, Cambridge, MA

63. Lidsky, L. M., Colombant, D. 1967. *IEEE Trans. Nucl. Sci.* NS-14:945–49

64. Armstrong, D. D., Emigh, C. R., Meier, K. L., Meyer, E. A., Schneider, J. D.

65. Deleeuw, J. H., Haasz, A. A., Stangeby, P. C. 1977. *Nucl. Instrum. Methods* 145:119–25

66. Graves, E. R., Rodrigues, A. A., Goldblatt, M., Meyer, D. I. 1949. *Rev. Sci. Instrum* 20:579–82

67. Booth, R. 1967. *IEEE Trans. Nucl. Sci.* NS-14:938–42

68. Booth, R. 1968. *Nucl. Instrum. Methods* 49:131–35

69. Booth, R., Logan, C. M. 1977. *Nucl. Instrum. Methods* 142:471–73

70. Booth, R., Barschall, H. H. 1972. *Nucl. Instrum. Methods* 99:1–4

71. Davis, J. C., Anderson, J. D. 1975. *J. Vac. Sci. Technol.* 12:358–60

72. Booth, R., Goldberg, E., Barschall, H. H. 1974. *Br. J. Radiol.* 47:737

73. Van Konynenburg, R. A., Barschall, H. H., Booth, R., Wong, C. 1975. See Ref. 8, pp. 171–82

74. Broerse, J. J., Broers-Challiss, J. E., Mijnheer, B. J. 1975. *Strahlentherapie* 149:585–96

75. Cleland, M. R., Offermann, B. P. 1977. *Nucl. Instrum. Methods* 145:41–47

76. Cranberg, L. 1976. See Ref. 59, pp. 926–41

77. Hillier, M., Lomer, P. D., Stark, D. S., Wood, J. D. 1967. *Conf. Accel. Targets Designed Prod. Neutrons*, pp. 125–45. *Euratom Rep. EUR 3895*

78. Kim, J. 1977. *Nucl. Instrum. Methods* 145:9–17

79. Bacon, F. M., Riedel, A. A., Cowgill, D. F., Bickes, R. W., Boers, J. E. 1977. *Sandia Lab. Rep. SAND 77-1326.* 28 pp.

80. Brennan, J. T., Hendry, G. O., Block, P., Hilton, J. L., Kim, J., Quam, W. M. 1974. *Br. J. Radiol.* 47:912–13

81. Brennan, J. T., Hendry, G. O., Herring, D. F., Hilton, J. L., Kim, J., Quam, W. M. 1973. *Br. J. Radiol.* 46:233

82. Hillier, M., Lomer, P. D., Wood, J. D. L. H. 1971. *Br. J. Radiol.* 44:716–19

83. Broerse, J. J., Greene, D., Lawson, R. C., Mijnheer, B. J. 1977. *Int. J. Radiat. Oncol. Biol. Phys.* 2: Suppl. 2, pp. 361–65

84. Chenevert, G. M., De Luca, P. M., Kelsey, C. A., Torti, R. P. 1977. *Nucl. Instrum. Methods* 145:149–55

85. Logan, C. M., Booth, R. Nickerson, R. A. 1977. *Nucl. Instrum. Methods* 145:77–79

86. Dierckx, R., Emigh, C. R. 1977. *Los Alamos Rep. LA-UR-77-2187.* 18 pp.

Ann. Rev. Nucl. Part. Sci. 1978. 28:239–326

ADVANCES IN MUON SPIN ROTATION[1]

✶5596

J. H. Brewer
Department of Physics, University of British Columbia, Vancouver, British Columbia, Canada

K. M. Crowe
Department of Physics, University of California, Berkeley, California 94720

CONTENTS

1   INTRODUCTION

1.1   *Muon Spin Rotation*

The parity-violating asymmetric decay of the positive muon was first observed in 1957 by two different experimental techniques (1, 2). The first involved precession of the muon magnetic moment in an applied magnetic field, and was the percursor of the μSR (muon spin rotation) techniques used today. The original apparatus of Garwin et al (1) is pictured in Figure 1. The essential features of the technique are evident: a "μ stop" timing pulse is generated by a muon entering the target through counters 1 and 2; when the muon decays, if the positron is emitted in the direction of counters 3 and 4, a "decay e" trigger is generated. Such triggers are more likely to occur at times when the muon spin has rotated in a magnetic field in such a way that it points toward the positron telescope. In the original experiment the relative probability of a decay e trigger during a gate from 0.75 to 2.0 μsec after

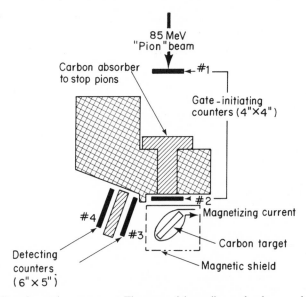

Figure 1   Experimental arrangement. The magnetizing coil was closely wound directly on the carbon to provide a uniform vertical field of 79 gauss per ampere (1).

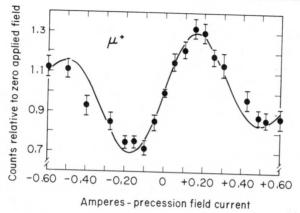

Figure 2 Variation of gated 3–4 counting rate with magnetizing current. The solid curve is computed from an assumed electron angular distribution $1 - \frac{1}{3}\cos\theta$, with counter and gate-width resolution folded in.

the μ stop was measured as a function of the magnetic field applied to the muon. The results are shown in Figure 2.

A μSR arrangement typical of those currently in use is represented in Figure 3; the main difference is that the magnetic field is fixed and the positron detection probability is measured as a function of the time the muon spends in the target. The magnetic field is produced by an electromagnet or Helmholtz coil, and timing pulses, μ stop, and decay e are

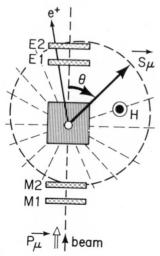

Figure 3 Schematic experimental arrangement in a transverse field. The asymmetric decay pattern is rotating past the counters.

generated as in the Garwin experiment. The time interval between these two pulses is histogrammed to produce a spectrum like that shown in Figure 4a, which represents the time distribution of the probability of detecting the positron in a fixed direction. This distribution reflects the exponential decay of the muon, but more importantly (in this context) the superimposed oscillations reflect the preference of the positrons to be emitted along the spin of the muon. The spin precesses in the local field at its Larmor frequency, $\omega_\mu = \gamma_\mu B$, where

$$\gamma_\mu = 8.514 \times 10^4 \text{ rad s}^{-1}\text{G}^{-1} \ (\gamma_\mu = 0.01355 \text{ MHz per gauss}). \qquad 1.$$

The fact that the muon precesses in the local field is dramatized by the spectrum in Figure 4b, which represents μ^+ precession in a single crystal of Co at 320°C in zero applied field. The local field is entirely provided by the microscopic magnetization of cobalt.

Such a spectrum is fitted to the functional form

$$dN(t)/dt = N_0 \{B + \exp(-t/\tau_\mu)[1 - A_0 P(t)]\}, \qquad 2.$$

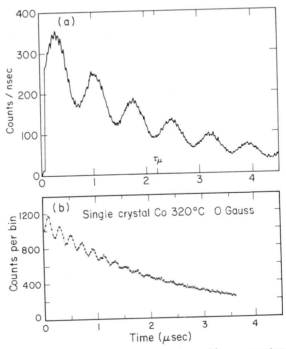

Figure 4 Typical experimental histogram. In (a), the positive muons stop in a target of CCl$_4$ in a magnetic field of 100 gauss. In (b), the stopping target is cobalt in zero external field. The mean muon decay lifetime $\tau = 2.20$ μsec is seen.

where B is a constant accidental background, τ_μ is the μ^+ lifetime (2.20 μsec), A_0 is the empirical maximum asymmetry (typically between 0.25 and 0.4), and $P(t)$ is the projection of the time-dependent polarization of the muon spin ensemble along the direction of the positron counter telescope. For simple one-frequency examples like that shown in Figure 4, $P(t)$ has the form

$$P(t) = P_0 f(t) \cos{(\omega_\mu t + \phi)}, \qquad\qquad 3.$$

where P_0 is the apparent initial muon polarization, $f(t)$ is the (time-dependent) amplitude of the oscillations, and ϕ is the apparent initial phase of the precession. For the majority of cases, $f(t)$ is an exponential decay, $f(t) = \exp{(-t/T_2)}$, where the transverse relaxation time T_2 is a measure of local field fluctuations and other dynamics of the coupling of the muon spin to the medium. Often, however, $f(t)$ may be a gaussian relaxation term or even more complicated time dependence.

In many cases the "μSR signal," $P(t)$, exhibits more than one precession frequency, in which case the simple description Equation 3 must be replaced by a sum of similar terms for each frequency. Whether constraints are imposed upon the extra parameters depends upon the theory of the multifrequency precession. In general the physics is contained in $P(t)$, both experimentally and theoretically. Usually these more complicated spectra yield their information most easily to a Fourier transform.

The above description obviously reflects a bias toward positive muon spin rotation (μ^+SR) as opposed to negative (μ^-SR). This does not mean that we consider μ^-SR unimportant. On the contrary, while the μ^-SR technique is more difficult because of small muon polarizations in the final state and the dominance of muon capture over decay for high Z, it may soon provide very important information complementary to that discussed here (3–5). However, the overwhelming majority of μSR work is still being done with positive muons, and this review reflects that fact. The treatment of μ^-SR spectra follows the above description, except that one is usually obliged to introduce several distinct components of the form of Equation 2, each characterized by a different muon lifetime for negative muons captured by different elements.

The source of the asymmetry in muon decay, the characteristics of the basic effect, formal descriptions of the parameterization of the time distribution $dN(t)/dt$, and the elements of basic μSR techniques are well documented in Reference 11; we do not duplicate the basics here, except to raise conceptual points and describe improvements to the technique.

1.2   Muon Depolarization

In the first experiments on asymmetric muon decay (e.g. 12), there was evidence of the effects of the stopping medium upon the muon polarization. Most metals left the μ^+ polarization essentially undisturbed, but other solids showed a wide variation in the "residual" μ^+ polarization, designated P_{res}. In liquids, P_{res} was a strong function of the chemical properties of the solution. From the outset, it was realized (2) that variations in P_{res} were mainly due to the temporary formation of muonium (μ^+e^-) atoms, in which the hyperfine coupling of the muon and electron spins caused a reversal of the muon spin within a fraction of a nanosecond. However, it was essential that muonium (or Mu) atoms be short-lived, since not all the polarization was lost, and since the observed precession was at the frequency of a free muon, rather than that of the triplet state of Mu (103 times faster). After some initial confusion, it was recognized that this effect probably involved chemical reactions of the hydrogen-like Mu atom, which afforded an opportunity for study of its chemistry (7). However, all of the early models were either incorrect or incomplete, and for several years progress along these lines was hampered by inability to make quantitative predictions. At the same time, however, another application of μSR techniques was developed in a quite different field of physics.

1.3   Muonium and Quantum Electrodynamics

The development of advanced μSR technology was largely the result of a program of measurements of the muon magnetic moment (8), its anomalous magnetic moment $(g-2)$ (9), and the hyperfine splitting of the muonium atom (10), which has provided the most stringent test of quantum electrodynamics (QED) to date, and which continues today (13). The hyperfine splitting of Mu due to the Fermi contact interaction coupling muon and electron spins is perfectly analogous to that of the H atom, except that in this purely leptonic atom there are no complications caused by strong-interaction-induced anomalous magnetic moments. We remark only briefly on the form and consequences of this interaction, leaving the interested reader to consult one of the many fine reviews of this topic (9, 13).

The Hamiltonian acting on the spins in the Mu atom, including an external field **B** and the contact interaction, is

$$\mathscr{H} = \tfrac{1}{2}h\,(\gamma_\mu\sigma_\mu - \gamma_e\sigma_e)\cdot\mathbf{B} + \tfrac{1}{4}h\,\omega_0(\sigma_\mu\cdot\sigma_e), \qquad\qquad 4.$$

where σ_μ and σ_e are the muon and electron spin Pauli operators, $\gamma_\mu/2\pi = 0.01355$ MHz per gauss and $\gamma_e = (m_\mu/m_e)\gamma_\mu = 206.77\,\gamma_\mu$ are the muon and electron gyromagnetic ratios, and ω_0 is the hyperfine frequency (13),

$$\omega_0 = 2.804 \times 10^{10} \text{ rad per sec } (\omega_0/2\pi = 4463 \text{ MHz}). \qquad 5.$$

This Hamiltonian has energy eigenstates and eigenvalues

$$|1\rangle = |++\rangle, \; E_1/\hbar = \frac{\omega_0}{4} + \omega_- \qquad 6a.$$

$$|2\rangle = s|+-\rangle + c|-+\rangle, \; E_2/\hbar = -\frac{\omega_0}{4} + \left(\frac{\omega_0^2}{4} + \omega_+^2\right)^{\frac{1}{2}} \qquad 6b.$$

$$|3\rangle = |--\rangle, \; E_3/\hbar = \frac{\omega_0}{4} - \omega_- \qquad 6c.$$

$$|4\rangle = s|-+\rangle - c|+-\rangle, \; E_4/\hbar = -\frac{\omega_0}{4} - \left(\frac{\omega_0^2}{4} + \omega_+^2\right)^{\frac{1}{2}}, \qquad 6d.$$

where $\quad c = \frac{1}{\sqrt{2}}\left(1 + \frac{x}{(1+x^2)^{\frac{1}{2}}}\right)^{\frac{1}{2}},$ 7a.

$$s = \frac{1}{\sqrt{2}}\left(1 - \frac{x}{(1+x^2)^{\frac{1}{2}}}\right)^{\frac{1}{2}}, \qquad 7b.$$

and $\quad \omega_\pm = \frac{1}{2}(|\omega^e| \pm |\omega^\mu|).$ 7c.

Here "$+/-$" refers to $m = +$ or $-\frac{1}{2}$ along the external field direction for the muon (first sign) and electron (second sign) spins. These energy levels are plotted as a function of external field in Figure 5, a typical Breit-Rabi diagram. In this figure, an unphysical value of the muon mass (and therefore the muon magnetic moment) is used to exhibit the curvature at low field at the same time as the asymptotic behavior at high field. The field is given in units of the dimensionless "specific field,"

$$x = 2\omega_+/\omega_0 = (g_e\mu_0^e - g_\mu\mu_0^\mu)|B|/\hbar\omega_0 = B/B_0, \quad B_0 = 1585 \text{ G}. \qquad 8.$$

Also shown in Figure 5 are the observable transition frequencies $v_{ij} = v_i - v_j$ corresponding to allowed ($\Delta m = \pm 1$) electromagnetic transitions. These are the frequencies actually measured by resonant Mu depolarization in a rf field in the QED tests. They are also observable in muonium spin rotation (MSR) experiments as precession frequencies in a transverse field (14) and are referred to again later. We give here the formula for the time dependence $\hat{P}(t)$ of the muon polarization in free muonium in a transverse magnetic field, using the complex convention in which the real part of $\hat{P}(t)$ is the muon polarization along its initial (x) direction, and the imaginary part is the muon polarization along the y direction, which makes a right-handed coordinate system with the magnetic field (z) direction (11):

$$\hat{P}(t) = \tfrac{1}{2}[c^2(e^{i\omega_{12}t} + e^{-i\omega_{34}t}) + S^2(e^{i\omega_{23}t} + e^{i\omega_{14}t})], \qquad 9.$$

where $\omega_{ij} \equiv 2\pi\nu_{ij}$. This can be reduced to the form

$$\hat{P}(t) = e^{i\omega_{-}t} \cos(\tfrac{1}{2}\omega_0 t)[\cos(\tfrac{1}{2}\omega_0 + \Omega)t - i(c^2 - s^2)\sin(\tfrac{1}{2}\omega_0 + \Omega)t] \qquad 10.$$

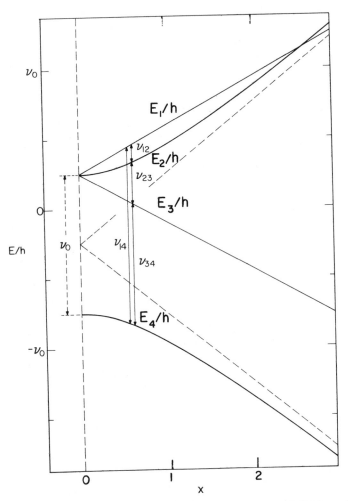

Figure 5   Energy eigenstates of $l = 0$ muonium in an external magnetic field, as a function of the dimensionless "specific field," which is given in Equation 8. A nonphysical value of the muon magnetic moment is used to display clearly the qualitative features.

where $\omega_- = \frac{1}{2}(\omega_{12}+\omega_{23})$ is the mean precession frequency of triplet muonium and Ω is the splitting frequency of triplet muonium precession,

$$\Omega = \frac{1}{2}(\omega_{23}-\omega_{12}) = \frac{1}{2}\omega_0 \left[(1+X^2)^{\frac{1}{2}}-1\right]. \qquad 11.$$

In low field, Ω is well approximated by $\Omega = \omega_-^2/\omega_0$ and $\hat{P}(t)$ has the form (averaged over the hyperfine oscillations, which are too fast to observe)

$$\langle \hat{P}(t)\rangle_{\omega_0} = \frac{1}{2}e^{i\omega_- t} \cos \Omega t. \qquad 12.$$

That is, "two-frequency" triplet muonium precession is seen.

Measurement of the hyperfine splitting in muonium required the development of methods of producing and observing stable Mu atoms in inert gases, which in turn led to the first studies of chemical reactions of Mu in gases (15). Similarly, the experimental basis of modern μSR techniques was developed to measure the muon's magnetic moment via its precession frequency in an external field (8). Because of the effects of diamagnetic shielding (16), interpretation of the results of that experiment depended upon a sure knowledge of the muon's final environment. This in turn led to an investigation of the chemical behavior of Mu in liquids (17, 18).

1.4 μSR in Chemistry and Solid State Physics

From 1957 to around 1970, the possible uses of parity-violating muon decay in materials sciences received little attention from the scientific community. However, a few groups performed important "seed" experiments (7, 12, 15, 19–22) that gave such ideas a vital experimental base; and a series of Soviet theoretical papers (23–26) pointed the way toward a quantitative understanding of the depolarization of positive muons in matter in terms of the formation, disruption, and reaction of Mu atoms. Thus in the early 1970s the field began to gel. The following section of this review describes the main lines of progress in those formative years.

Since about 1975 the volume of experimental and theoretical activity in this field has grown almost exponentially, producing a great variety of new information and clarifications of old questions. In Figure 6 we indicate schematically the current breadth of μSR research in materials science, with apologies to the omitted topics certain to have been developed during the preparation of this review. Obviously we cannot attempt a comprehensive review of such a collection of research topics, so we have selected what seem now to be centrally important subjects that represent breakthroughs in qualitative understanding and that will be of benefit to many of the subfields shown in Figure 6. These highlights of recent progress in μSR are the subject of the third section of this review.

2 THE FORMATIVE PERIOD: 1957–1975

2.1 *Depolarization and Muonium Chemistry*

2.1.1 FAST DEPOLARIZATION AND THE RESIDUAL POLARIZATION As
mentioned in the introduction, the effects of the chemical properties of
the stopping medium upon the muon's residual polarization were
noticed in the very first experiments verifying the asymmetric decay of the
muon. Those experiments, like most that followed for the next decade,
were performed in the condensed phase—most often liquids—because of
the difficulty of stopping muon beams in low density gas targets; the
experimental observable was in most cases the quasifree precession
amplitude of muons in diamagnetic environments. Thus the first
theoretical studies of the behavior of muons in matter were oriented
toward explaining the varying degrees of depolarization in terms of the

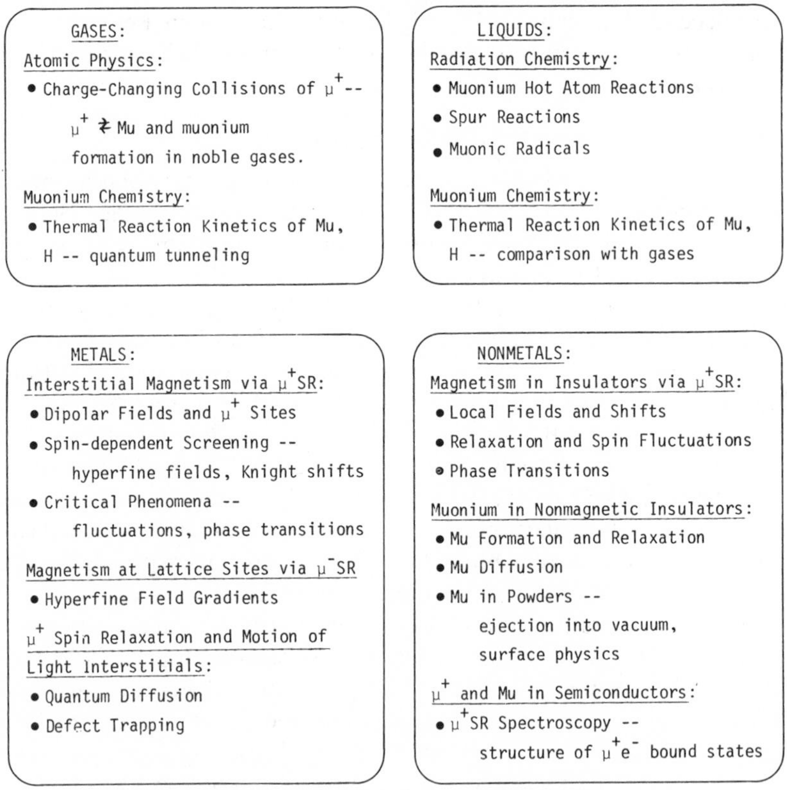

GASES:

Atomic Physics:
• Charge-Changing Collisions of μ^+ --
   $\mu^+ \not\rightleftarrows$ Mu and muonium
   formation in noble gases.

Muonium Chemistry:
• Thermal Reaction Kinetics of Mu,
   H -- quantum tunneling

LIQUIDS:

Radiation Chemistry:
• Muonium Hot Atom Reactions
• Spur Reactions
• Muonic Radicals

Muonium Chemistry:
• Thermal Reaction Kinetics of Mu,
   H -- comparison with gases

METALS:

Interstitial Magnetism via μ^+SR:
• Dipolar Fields and μ^+ Sites
• Spin-dependent Screening --
   hyperfine fields, Knight shifts
• Critical Phenomena --
   fluctuations, phase transitions

Magnetism at Lattice Sites via μ^-SR
• Hyperfine Field Gradients

μ^+ Spin Relaxation and Motion of
Light Interstitials:
• Quantum Diffusion
• Defect Trapping

NONMETALS:

Magnetism in Insulators via μ^+SR:
• Local Fields and Shifts
• Relaxation and Spin Fluctuations
• Phase Transitions

Muonium in Nonmagnetic Insulators:
• Mu Formation and Relaxation
• Mu Diffusion
• Mu in Powders --
   ejection into vacuum,
   surface physics

μ^+ and Mu in Semiconductors:
• μ^+SR Spectroscopy --
   structure of μ^+e^- bound states

Figure 6   The major active fields of μSR research as of 1977.

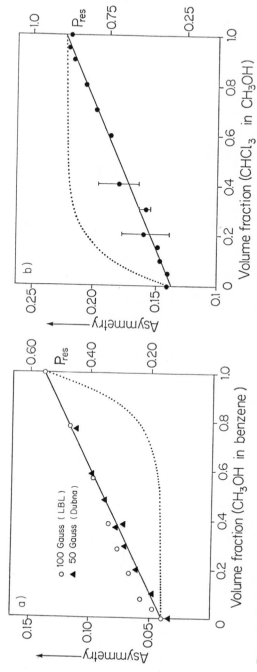

Figure 7 (*left*) Measured asymmetry of μ^+ precession and corresponding residual polarization P_{res} as a function of the volume fraction of CH_3OH in C_6H_6 in two transverse magnetic fields. Circles represent data from LBL at 100 G; triangles represent data from JINR at 50 G. The dotted line indicates how the asymmetry would be expected to change if benzene gave "sponge-like" protection to CH_3OH. (*right*) Similar data from LBL on $CH_3OH/CHCl_4$ mixtures in 100 G fields. The dotted line here indicates how the asymmetry is expected to vary if selective reaction by $CHCl_3$ were possible, or if spur scavenging by $CHCl_3$ occurred.

properties of the medium. It was immediately realized that the Mu atom was responsible, with its hyperfine interaction and rapid precession; but there was uncertainty about why this depolarizing effect was not always complete. In 1963 Nosov & Yakovleva (23) introduced a density matrix formalism describing the evolution of muon and electron spins in muonium, allowing for "spin-flipping" of the electron by outside influences; in 1969 Ivanter & Smilga (25) pointed out that fast chemical reaction of the Mu atoms placed the muons in diamagnetic molecules, which would quench the depolarizing effect of Mu formation, and offered an elegant quantitative model for the residual polarization.

The implication that depolarization studies could provide information about the chemistry of muonium atoms had already been recognized by Firsov & Byakov (7). However, their first attempts at such an application (22) were limited by an incorrect model. Data from that study are shown in Figure 7 along with more recent results exhibiting the dependence of P_{res} upon the volume fraction of mixtures of benzene, methanol, and chloroform. In 1971, it was demonstrated that the purely thermal reaction scheme of Ivanter & Smilga could not account for the observed effects, unless a high probability of epithermal reaction was included (27). That is, a significant fraction of muons destined for depolarization are saved from that fate "on the way in" by higher energy (up to a few

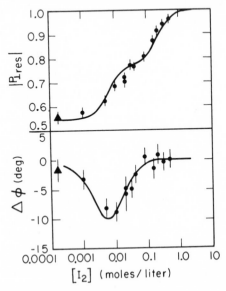

Figure 8   Magnitude $|P_{res}|$ and apparent initial phase $\Delta\phi$ of the residual μ^+ polarization in methanol solutions at 100 G, as a function of the concentration of dissolved iodine.

tens of eV) chemical reactions leaving the muons in diamagnetic molecules. Thus μSR experimenters were introduced to the field of "hot atom" chemistry. The behavior shown in Figure 7, which was attributed to thermal reactions of Mu by Babaev et al (22), is now interpreted in terms of hot atom reactions alone. At the same time, the 1971 experiment (27) verified the important role played by thermal reactions (in that case, $Mu + I_2 \rightarrow MuI + I$) in the de- or repolarization scheme. Figure 8 shows data taken later (18) on the same reaction in methanol solvent. Clearly evident is the variation of the apparent initial phase of the muon precession due to short-lived, high frequency precession of Mu atoms—a distinctive feature of the depolarization mechanism.

By 1972 the role of radicals in the general depolarization mechanism had been recognized as well (17), and it appeared that a correct qualitative description of the depolarization process had been achieved (18). Figure 9 shows data from that experiment, which demonstrates that the simpler theory excluding radicals (dashed line) was unable to explain the repolarization of muons in benzene by the addition of bromine reagent. Unfortunately, the additional parameters required in the more elaborate reaction scheme made the fitted values less unambiguous, and it became apparent that accessible parameters would have to be measured as precisely as possible before the general model could be used to measure rate constants for the various chemical reactions involved (18a).

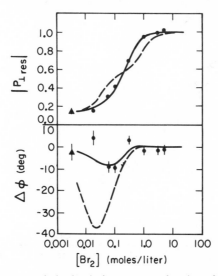

Figure 9   Residual muon polarization in benzene as a function of the concentration of dissolved bromine. Uncertainties of $|P_{res}|$ data are smaller than the dimensions of the points.

2.1.2 SLOW RELAXATION IN PARAMAGNETIC SOLUTIONS   Residual polarization studies rely upon measurements of the amplitude and apparent
initial phase of muon precession; these quantities vary because of the
fast depolarization of the muons by short-lived muonium formation.
Once the Mu atoms have reacted, placing the muons in diamagnetic
compounds with some residual polarization, the muons can still be
depolarized gradually by random local magnetic fields or spin-lattice
relaxation phenomena. The rate of this slow depolarization is characterized by the transverse relaxation time T_2 in $f(t)$ of Equation 3 (12,
18b); early studies of this parameter (28) gave μ^+SR experimenters
access to several new topics, one of which was the structure of liquids (29).

In aqueous solutions of $MnCl_2$, the residual muon polarization was
found to be independent of concentration, which indicated a lack of any
fast thermal reactions with the solute, and suggested that the observable
muon precession signal came from MuOH molecules formed epithermally
(11). Like H_2O, these "muonic water" molecules spend part of their time
in $Mu(H_2O)$ complexes, where the muons (like protons) are relaxed by
interactions with the paramagnetic ions. Figure 10 shows the observed

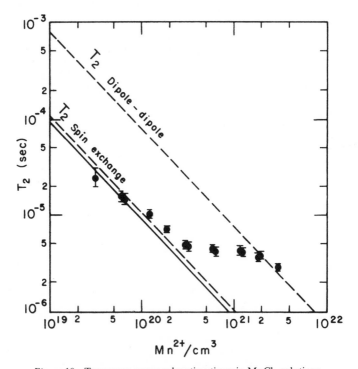

Figure 10   Transverse muon relaxation times in $MnCl_2$ solutions.

dependence of the μ^+ relaxation time on the concentration of Mn^{++}. The muon lifetime sets practical limits of about 20 nsec to about 20 μsec on conveniently measurable values of T_2, although in principle both these limits can be transcended by high statistics. In this case (29) the lower limit was set by the solubility of $MnCl_2$ in water. The relaxation at low concentrations follows the dependence predicted for spin-exchange inter- actions between the paramagnetic ion and the normally diamagnetic MuOH molecule, but at high concentrations this mechanism is broken and the slower relaxation of the muon by the dipolar field of the ion takes over. This is thought to be due to mutual interactions of neighboring Mn^{++} ions, and thus provides a measure of local structure and correlation times in the liquid. The muon offers certain advantages over proton NMR in these studies, since the region of fast relaxation where NMR has difficulties is just where μSR works best. However, this promising line of investigation has remained virtually untouched since 1972.

2.2 μ^+ Site in Solids

The sensitivity of μSR to local magnetic fields (through the muon pre- cession frequency) and their dispersion (through the relaxation rate) make the muon a different probe of the solid state, analogous to nuclear magnetic resonance (NMR) or electron paramagnetic resonance (EPR); this was recognized as soon as the asymmetric decay was discovered. Furthermore, unlike NMR, μSR is free of complications associated with rf fields and the necessity for macroscopic populations of the resonant species; μSR is a purely passive technique using no more than one muon at a time in the sample. Thus it was expected (correctly) that μSR would furnish new information about local fields in crystals, thus contributing to the microscopic theory of magnetism, among other topics in solid state physics.

However, knowing the local fields "seen" by the muon is not of great value until we have some idea where the muon is in the crystal lattice. This places great emphasis upon the determination of muon locations, a topic still of high priority today.

2.2.1 μ^+ SITE IN GYPSUM

The first μ^+ site determination was performed (21) in a single crystal of gypsum ($CaSO_4 \cdot 2H_2O$), for which the NMR of protons was well known (30). As indicated in Figure 11, the fixed location of the protons in gypsum leads to a unique contribution to the net magnetic field seen by one proton because of the dipolar field of an adjacent proton. Since the proton has spin $\frac{1}{2}$, this field either adds or subtracts from the applied external field, producing two discrete values of the net field "seen" by a given proton. The magnitude of this splitting depends on the orientation of the crystal in the external field, and in some cases a double splitting (four discrete frequencies) is observed, owing to the influence of two different proton-proton orientations (30). The analogous μ^+SR experiment led to a test of the μ^+ location in gypsum. If the μ^+ were interstitial or mobile in the crystal, the dipolar fields of the protons would simply cause relaxation of the muon spin. However, if the muons replaced protons at the appropriate lattice sites, the μSR results would show exactly the same behavior as the proton NMR had. The experiment (21) showed the split-frequency muon precession expected for muons occupying proton sites, and thus gave the first positive identification of the μ^+ site in a crystal.

2.2.2 μ^+ LOCATION IN METALS In simple metals, there is no ion site where the μ^+ will be electrically equivalent, and thus the muon must be either interstitial or trapped at vacancies, dislocations, or other defect centers. The question of which of these possibilities is preferred is crucial to interpretations of local field and relaxation measurements with muons. Recently there has been considerable progress along these lines, again relying upon the local dipolar field contributions (see Section 3.3.1), but

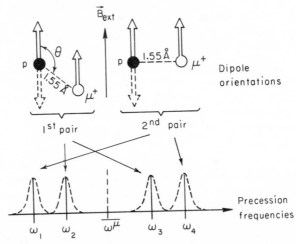

Figure 11 Muon-proton dipole-dipole interaction in a single crystal of gypsum: schematic representation of the effect of the muon-proton situation relative to the magnetic field direction on the μ^+ spin precession. $\overline{\omega}^\mu$ corresponds to the precession frequency unperturbed by dipole-dipole interactions; it is split by that interaction into two symmetrically shifted frequencies (one per proton spin orientation for each μ^+-p pair. The line broadening produced by the magnetic dipoles farther away is indicated by the dashed curves.

for many years this question was obscured by the tendency of the muon to diffuse rapidly in metals—a phenomenon of intrinsic interest independent of the site determination.

2.3 μ^+ Motion in Metals

A muon held immobile at a well-defined interstitial position in a non-magnetic metal whose nuclei have magnetic moments should experience a gaussian relaxation $f(t) = \exp(-\sigma^2 t^2)$, where σ^2 characterizes the averaged strength and orientation of dipolar fields from neighboring nuclear moments (30). Such behavior was observed by Gurevich et al (31) for μSR in copper at 77°K, indicating that the muons were not diffusing significantly at that temperature.

However, at higher temperatures the μ^+ relaxation became slower and more exponential, which indicated a "motional narrowing" effect of the μ^+ diffusion upon the relaxation rate of the muon (analogous to the linewidth in NMR). The origin of this effect is simple. As the μ^+ moves more rapidly between sites with different fields, it begins to see the average field, and the "dephasing" effects of different fields at different sites disappear (32). Figure 12 shows the temperature dependence of the relaxation rate of muons in Cu (31). At temperatures above about 120°K the data are described by a motional-narrowing model in which muon diffusion is governed by a thermally activated "hopping" between potential wells.

The results of that experiment were perplexing: The activation energy

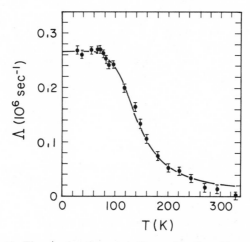

Figure 12   The μ^+ relaxation rate in Cu as a function of temperature.

extracted for the μ^+ hopping process was small compared with the values for hydrogen, and the preexponential factor (characteristic of the vibrational frequency of the muon in the potential well) was lower by a factor of a million. As it now turns out, the case of Cu studied by Gurevich et al (31) was one of the few examples of relaxation/diffusion μSR studies that exhibit the simple behavior shown in Figure 12. Other nonmagnetic metals have considerable structure in the temperature dependence of the depolarization rate (see Section 3.4).

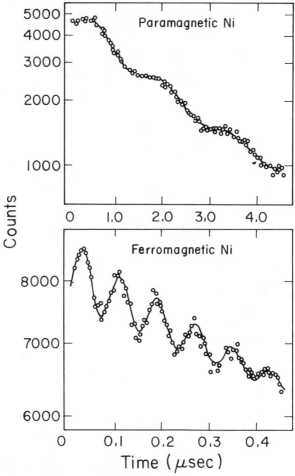

Figure 13 The μ^+ SR time spectra in paramagnetic nickel at about 670°K and ferromagnetic nickel at 551°K. The data shown are semilog plots of the number of decay positrons detected in a fixed direction as a function of time after a muon stop. Note that the time scales differ by a factor of ten.

2.4   The Muon as a Magnetic Probe

2.4.1 μ^+ IN FERROMAGNETS   Since the muon was effective as a detector of small local fields, it was suspected that it might also be used to probe the internal fields in magnetic metals; however, early survey experiments failed to reveal any μSR signals from magnetic samples. This was largely because of inadequate purity in the samples initially used; the problem was first solved by raising the temperature of the samples until the muons diffused rapidly enough to quench any depolarization. Figure 13 shows the μ^+SR time spectra from that first study by Foy et al (33). The temperature dependence of the local field in nickel follows a Brillouin function approximately the same as the temperature dependence of the saturation magnetization.

The average local field at the muon, $\mathbf{B}_\mu$, has several contributions:

$$\mathbf{B}_\mu = \mathbf{B}_{ext} + \mathbf{B}_{DM} + \mathbf{B}_L + \mathbf{B}_{dip} + \mathbf{B}_{hf},\qquad\qquad 13.$$

where $\mathbf{B}_{ext}$ is the external applied field, $\mathbf{B}_{DM}$ is the sample-geometry-dependent demagnetization field, $\mathbf{B}_L$ is the "Lorentz field" of magnetic

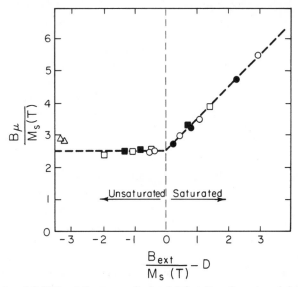

Figure 14   Local field $\mathbf{B}_\mu$ at the muon site in nickel vs $\mathbf{B}_{ext}$, the external field measured with target out. D is the sample demagnetizing factor. The points denoted by triangles are from data on an approximately spherical single crystal with [111] axis parallel to $\mathbf{B}_{ext}$ at 300 and 77°K. Other points are from data on a polycrystalline ellipsoid (4.5 × 2 × 0.5 in.). Solid and open circles refer to the 4.5-in. axis parallel to $\mathbf{B}_{ext}$ ($D = 0.69$) at 523 and 573°K respectively. Solid and open squares refer to the 2-in. axis parallel to $\mathbf{B}_{ext}$ ($D = 2.24$) at 523 and 573°K.

charge on the inner surface of an imaginary sphere centered on the muon, $\mathbf{B}_{dip}$ is the field due to dipole moments within that sphere, and $\mathbf{B}_{hf}$ is the effective field due to local interaction with polarized electrons. For high permeability material below saturation, $\mathbf{B}_{DM}$ cancels $\mathbf{B}_{ext}$, as can be seen from Figure 14, showing the dependence of $\mathbf{B}_\mu$ on $\mathbf{B}_{ext}$ in Ni (11).

All the terms in Equation 13 are easily calculable except $\mathbf{B}_{dip}$ and $\mathbf{B}_{hf}$. For Ni, the interstitial site has cubic symmetry and thus $\mathbf{B}_{dip}$ averages to zero, and for Fe there are two otherwise equivalent interstitial sites, one with a dipolar field of about 10 kG and the other (only half as numerous) with $\mathbf{B}_{dip}$ of about -20 kG. Thus in either case, as long as the μ^+ diffuses rapidly between sites, the time-averaged dipolar field is zero. (This argument applies equally well for the interstitial sites with octahedrally or tetrahedrally positioned ion neighbors.) At lower temperatures the different fields at different sites in Fe cause a dramatic relaxation of the muon, which has been the subject of several recent studies (see Section 3.3.3).

Thus the early results of Kossler et al (33) and others (34, 35) were reducible to measurements of temperature dependence of $\mathbf{B}_{hf}$ in Ni and Fe. Subsequent work has followed similar patterns. The theoretical value of a measurement of $\mathbf{B}_{hf}$ at the positive muon depends upon the eventual clarification of: (a) the site occupied by the muon most of the time; (b) the vibrational motion of the muon in that site; (c) the distortion of the lattice by the presence of the muon; (d) the screening of the muon by conduction electrons; and (e) the spin dependence of that screening, since it is largely the same polarized conduction electrons that deliver the effective contact field to the muon. Early treatments helped to explain the qualitative behavior of (d) and (e) (36), but these theoretical problems are still with us today.

2.4.2   KNIGHT SHIFTS

In nonmagnetic metals the conduction electrons are slightly polarized by the external field, and though the resulting contact field at the muon is normally small (a few gauss), the mechanism is analogous to that producing $\mathbf{B}_{hf}$ in ferromagnetic metals. Thus the theoretical problems described above apply equally to the interpretation of Knight shifts in nonmagnetic metals.

As early as 1963, muonic Knight shifts were measured with 10-ppm accuracy in various metals as part of a determination of the muon magnetic moment (20). The results of that study are shown in Table 1; the large shifts in carbon and calcium are still not understood. Recent experiments have uncovered other, even larger, anomalous Knight shifts of positive muons (see Section 3.5).

Hutchinson et al (20) also measured the magnetic moment of the

Table 1 μ^+ and μ^- Knight shifts

Target substance	μ^+ Asymmetry coefficient	μ^+ Frequency shift[a] (ppm)	μ^- Frequency shift[b] (ppm)	NMR Knight shift (ppm)
Carbon	—	+380	+ 70± 32	—
Silicon	—	—	+287±110	+ 180
Magnesium	0.12	+ 87	+334± 75	+1400
Copper	0.12	+ 81	—	+2320
Lead	0.09	+132	—	+1200
Calcium	0.09	+420	—	+3100
Lithium	0.12	+ 11	—	+ 249
Potassium	0.13	+ 88	—	+2900
Sodium	0.13	+ 79	—	+1130
Reference samples				
CH_2I_2	0.15	− 25	—	—
$CHBr_3$	0.14	− 14	—	—
H_2O	0.09	0	+540± 80[c]	—
Sulfur	—	—	+ 96±170	—

[a] The diamagnetic shielding of muons in water (~ 25.6 ppm) has not been included.
[b] Relativistic effects, and diamagnetic shielding, nuclear polarization, and coulomb radiative corrections have been applied. Reference: Ford, K. W., Hughes, V. W., Wills, J. G. 1963. *Phys. Rev.* 129:194.
[c] The μ^- is transferred to the oxygen. Reference: Bingham, G. 1963. *Nuovo Cimento* 27:1352; also Reference 20.

negative muon, and in the process studied the Knight shifts at the μ^- in various media. Their results are also included in Table 1. Here the interpretation is completely different: the μ^- is bound to a lattice nucleus and is so close that it effectively forms a $Z - 1$ substitutional nucleus, which behaves accordingly in the lattice. If the original nucleus has a magnetic moment, the $\mu^- Z$ pseudonucleus can have several spin states, and the muon motion often becomes complicated (39). When the μ^- captures on a spinless nucleus, it precesses in the local field as if free, except for relativistic corrections (high Z only) (3). Those muons that decay instead of capturing on a proton exhibit the same asymmetric decay observed in μ^+SR, except that the μ^- polarization is usually only about 15% in the final state. The field seen by the μ^-, then, is different from that seen by a $Z - 1$ substitutional impurity only if the local field has such a large spatial gradient that it is significantly different at the mean radius of the muon's orbit from the value averaged over the nucleus (3).

2.4.3 μ^- SR IN MAGNETIC CRYSTALS   Since Knight shifts are small in the first place and the difference between those seen by $\mu^- Z$ and $Z - 1$ substitutional nuclei are expected to be subtle, it may prove difficult to extract

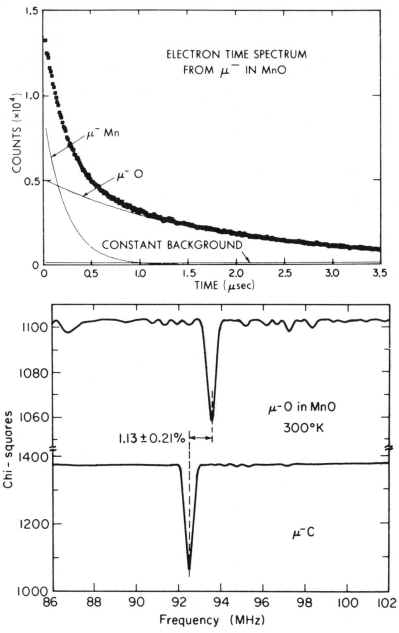

Figure 15 (*Upper*) μ^-SR time spectrum in MnO, showing the different lifetimes for μ^- bound to Mn and O. (*Lower*) μ^-SR frequency spectra in C and MnO at 6.9 kG showing a shift in the frequency of μ^-O relative to μ^-C.

new information from such data. However, in magnetic crystals the local fields at the nuclei can be huge, and the difference due to the spatial distribution of the muon may be significant.

An experiment of this sort was performed by Yamazaki et al (5) on the antiferromagnetic crystal MnO, where precession of the $\mu^- O$ component was compared with that of the $\mu^- C$ in a carbon target. Because of the weak signals in these experiments, a Fourier transform was taken of the time spectra after removal of background and correction for the exponential decay of the muon. The results are shown in Figure 15. The measured shift, $1.16 = 0.21\%$, is about a third of the O^{17} NMR shift (3.21%). Assuming a $N^{3-}(Mn^{2+})_6$ electronic configuration, the paramagnetic shift is lowered only $\sim 30\%$.

H. Kamimura and Y. Natsume (1978, private communication) have shown that if the $\mu^- O$ system forms an antisymmetrical ion configuration $N^{3-}(Mn^{2+})_5 Mn^{3+}$ as the ground state, where the Mn^{2+} has lost an electron to a defect, then the observed shift can be explained.

2.5 $\mu^+ SR$ Spectroscopy and Muonium in Semiconductors

In pure inert nonmetals, the μ^+ usually captures a single electron to form a stable neutral atom of muonium. As long as nuclear moments are small or absent and/or the Mu atom diffuses rapidly, this entity can be detected via its characteristic precession frequency, 103 times as fast as the free μ^+. The first such muonium spin rotation (MSR) measurement in solids was performed in fused quartz, which remains a reference standard (40). In moderate magnetic fields (20–150 G) the simple Mu precession described above gives way to the more general two-frequency Mu precession of Equation 12, first observed in 1971 by Gurevich et al (14). As mentioned in the introduction, the splitting of the Mu precession frequencies is inversely proportional to the hyperfine coupling strength. Thus a measurement of this splitting provides an imprecise (good to about 1% usually) but unambiguous value for the hyperfine coupling of the Mu atom in the lattice, which is in turn related to the physical size of the Mu atom.

Gurevich et al (41) were the first to study this phenomenon in semiconductor crystals, where the μ^+ also forms Mu atoms. They found a reduction in the Mu hyperfine frequency ω_0 of nearly a factor of two in germanium, as can be seen from the two-frequency precession signals in Figure 16. In fused quartz, muonium has essentially the vacuum value of ω_0, which indicates a Mu atom virtually undisturbed by the presence of the surrounding lattice. However, in Ge at 77°K the Mu atom appears to have been expanded significantly. At about the same time, Andrianov et al (44) performed a longitudinal-field "Paschen-Bach effect" experiment on Mu in Si that indicated an even more pronounced effect. This result

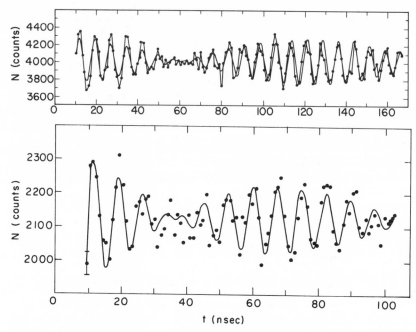

Figure 16 "Two-frequency precession" of the muon in fused quartz with a transverse field of 95 G (*upper graph*) and in cold (77°K) germanium at 98 G (*lower graph*). The smooth curves represent the best fits of the theoretical dependence to the data, which is corrected for the muon decay exponential $\exp(-t\tau)$.

was soon confirmed by the two-frequency MSR method (42). These observations stimulated theoretical interest in the electronic structure of the Mu impurity state in Si and Ge (45), where the analogous H atom impurity state, though sure to be present, has never been observed. In view of the practical importance of impurity electronic states in semi-conductors, this simplest of all interstitial impurity systems provides a valuable experimental testing ground.

2.5.1 FOURIER SPECTROSCOPY AND ANOMALOUS μ^+ PRECESSION   When more than one precession frequency is present in a single μ^+SR time spectrum, least-squares fitting becomes rather tedious and extremely model dependent. In these cases, and in general when systems with potential magnetic structure are being studied, it is wise to begin analysis with a Fourier transform of the time distribution (corrected for the back-ground term and the muon decay factor). Figure 17 shows two such frequency spectra for μ^+ in fused quartz at room temperature and Si at 77°K, both at 100 G. Both spectra show the free μ^+ signal at low frequency and the two muonium signals at about 103-times higher frequency, corresponding to the two-frequency precession as in Figure 16. Also evident in Figure 17 are at least two distinct signals at intermediate frequencies in Si, corresponding to the "anomalous μ^+ precession" reported in 1973 (42). These frequencies, representing a distinct coherent impurity state of μ^+ in Si, can be described as muonium transition frequencies ω_{12} and ω_{34} (see Figure 5 and Equation 9) as a function of applied field. This is illustrated in Figure 18a, but they will behave like this only if the hyperfine coupling is taken to be only about 2% of that in vacuum *and* the effective g factor of the electron is allowed to take a value of 13 instead of 2. Furthermore, in this model the hyperfine frequency is about 4% anisotropic with respect to the orientation of the Si crystal in

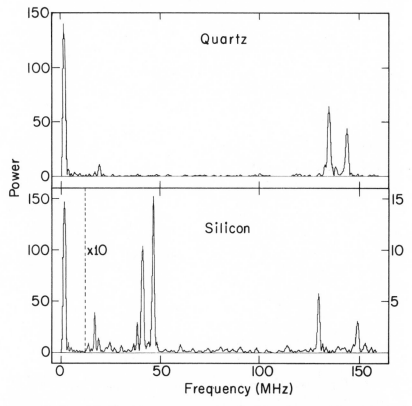

Figure 17   The μ^+SR Fourier power spectra (amplitude² vs frequency) for μ^+ in fused SiO₂ at room temperature and 100 G (*upper plot*) and in Si at 77°K and 100 G (*lower plot*).

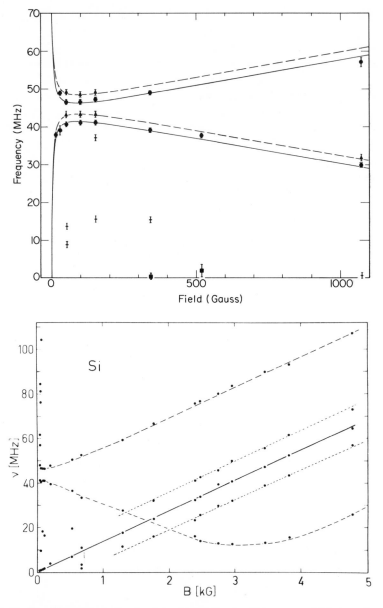

Figure 18a Dependence of anomalous frequencies in silicon upon field strength and crystal orientation. Round points and triangular points were obtained with different orientations of the crystal axes with respect to the magnetic field.

Figure 18b Field dependence of anomalous and μ^+ (*solid line*) frequencies in Si at 30°K with the [111] axis along the field (48).

the field. Several interpretations of these results were advanced, but the "shallow-donor" model was the most popular (11, 42, 46). Basically, the state was assumed to be an excited state of Mu, formed in the initial process as the μ^+ stops, and stable for the time of observation (on the order of 500 nsec) because the excited state is "shallow." That is, the electron is spread out over many lattice sites rather than being localized with the μ^+ at one interstitial site, and thus has small overlap with the ground state Mu. Several variations on this model have been put forward to explain the g factor of the electron, including an orbital angular momentum in a p-like state or a pseudo-p-like character for a "T1" state (47).

Figure 18b shows the significant precession frequencies recently measured at the Swiss Institute for Nuclear Research (SIN) (48) as a function of applied field for a mildly p-type sample at 30°K with the [111] crystal axis parallel to the field. The solid line connects frequency points attributable to the free muon component, and the sharply increasing points at low field are due to normal muonium. The dashed lines are the signals previously observed and designated Mu*. The signals connected by the dotted lines are weaker in intensity than all the others and were discovered at SIN. It is believed that they too are attributable to Mu*.

Measurements were also made at SIN on crystals with different orientations with respect to the external field. The adoption of an anisotropic Hamiltonian. The data can be fitted by the introduction of an anisotropic Hamiltonian with axial symmetry of the form:

$$\mathscr{H} = A_\perp(I_x S_x + I_y S_y) + A_\parallel I_z S_z - g_e \mu_e S \cdot H - g_\mu \mu_\mu I \cdot H$$

The z axis is a preferred direction in the crystal that coincides with the [111] directions of the diamond lattice. The SIN group obtain for the phenomenological hyperfine constants

$$|A_\perp| = 92.1 \pm 0.3 \text{ MHz}$$

$$|A_\parallel| = 17.1 \pm 0.3 \text{ MHz}$$

The electron and muon g factors are measured to be

$$g_e = -2.2 \pm 0.2$$

$$g_\mu = 2.01 \pm 0.01.$$

The μ^+SR spectroscopy of Si is thus a source of information about the structure of $\mu^+ e^-$ bound states in Si, of which there are apparently an unexpected variety. This represents a very interesting body of information, which has certainly not been adequately studied yet from either an experimental or a theoretical point of view. The last few years, far from clearing up the questions raised in early experiments, have mainly exposed more structure and raised new questions. Soviet work on Ge crystals has followed different lines (49, 50) but seems no less perplexing.

3   RECENT ADVANCES IN μSR

3.1   *Technology: The Surface Muon Beam*

Perhaps the most important new aid to μ^+SR in the past few years has been the "surface muon beam," also called the "Arizona beam" after the group that designed and built the beamline at Berkeley (51). This 4.1-MeV μ^+ beam, collected from the decay of positive pions at rest in the surface of the production target, is nearly 100% polarized, has a range of only about 150 mg cm^{-2} in air, and can be stopped completely in about 30 mg cm^{-2} of air. This can be seen in Figure 19, which shows the residual beam intensity following various thicknesses of Mylar degrader for surface muons produced at the Tri-University Meson Facility (TRIUMF).

While the very short range of surface muons creates problems in extraction of the beam from a vacuum pipe, passing it through a scintillator to form a μ-stop pulse, and injecting it into a target, the same effect permits a mass-stopping density orders of magnitude higher than

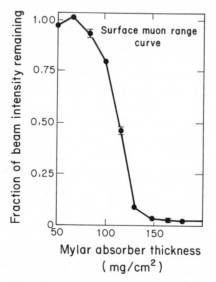

Figure 19   Range curve for surface muons at TRIUMF: Intensity of μ^+ beam as a function of Mylar degrader traversed.

previously possible. These beams have thus opened up the field of muonium chemistry by allowing convenient gas-phase studies at atmospheric pressure. They also show promise for solid state applications, providing potentially superior luminosities and offering incomparable stop rates per total mass of target. Several new facilities are being developed to exploit these possibilities, and these will give μ^+SR experimenters a chance to efficiently study targets a few millimeters in size for the first time.

3.2 Muonium Chemistry

3.2.1 THE MUONIUM SPIN ROTATION METHOD   While the first studies of muonium chemistry were primarily dependent upon the indirect, residual polarization method described in the introduction (and discussed again below), the recent progress in this branch of μ^+SR research has developed around the more direct and less ambiguous technique of precessing the triplet state of muonium in a weak magnetic field and observing the amplitude and relaxation of that MSR signal. We begin by discussing some applications of this technique.

3.2.1.1 *Muonium formation in gases*   It is believed that muonium atoms are formed as positive muons come to rest in almost all materials (11). There is adequate justification for this expectation in that Mu has a higher ionization potential than most other atoms or molecules, and can thus capture an electron even after it comes to rest. However, there are exceptions to this rule. In metals, the μ^+ is screened by conduction electrons collectively. In the lighter noble gases, the ionization potential is too high for the muon to capture an electron once it thermalizes, so any muonium must be formed as it slows down. What is the probability that this will occur? In what energy region is it most likely? These questions are not only central to the related topic of muonium hot atom chemistry, but also very relevant to the atomic physics of charge-changing collisions. Only recently has there been any experimental evidence brought to bear on this problem.

In order to study muonium formation, one has to be able to "see" muonium; that is, there must be an observable that relates directly to the fraction of Mu formed. The triplet-Mu precession signal described above is an obvious candidate. However, in solids and liquids, where most early MSR studies were made (because of stopping densities), Mu precession is generally difficult to observe. Most solids have depolarizing fields or conduction electrons that disrupt Mu precession, and all but the purest liquids have enough dissolved impurities to scavenge or depolarize muonium before it can be observed directly (11, 52). Gases, on the other hand, are easy to purify, and their lower density reduces the deleterious effect of what contaminations remain. For these reasons the first unambiguous information about the dynamics of Mu formation has come from MSR studies in inert gases.

A typical gas-phase MSR apparatus is pictured in Figure 20 (53). The muon enters the target with a very large energy on the scale relevant to atomic collision processes. It then slows down by ionization, and later by elastic scattering, passing through every energy region down to thermal velocities. Only the net Mu formation can be measured directly, which makes it difficult to untangle the competing mechanisms at different energies. This is accomplished by comparing the precession signals of muonium atoms and quasifree muons in the same sample, for a series of similar samples in which variation of an experimental parameter causes muonium formation to appear and disappear systematically. Information about the formation mechanism(s) is obtained indirectly from analysis of these phenomena. Early studies of muonium formation used conventional muon beams (from pion decay-in-flight) and high pressure gas targets. This technique is still used with no ill effects on the physics (epithermal processes are not expected to be strongly pressure dependent). However, the advent of the surface muon beam has made these studies feasible in low pressure gases (around 1 atm for all but the lightest gases), with a marked increase in convenience.

In either case, a transverse-field time spectrum is accumulated for muons stopping in the gas target in a field of a few gauss, where muon

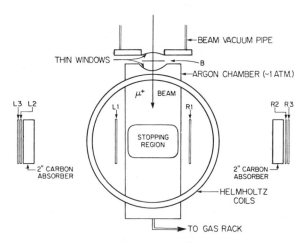

Figure 20 Diagram of the gas-phase MSR apparatus used at LBL and TRIUMF. The μ-stop pulse is generated by one or more thin (0.015 in.) scintillator Bi, and decay-e pulses are formed from either $e_L = L1 \cdot L2 \cdot L3$ or $e_L = R1 \cdot R2 \cdot R3$ coincidences.

Table 2   Fraction of free μ^+ and of Mu formed in rare gases[a]

Target gas	Pressure (atm)	f_{μ^+} (%)	f_{Mu} (%)
He	50	99 ($\pm$ 5)	1 ($\pm$5)
He +0.015% Xe	50	83 ($\pm$15)	—
He +0.09% Xe	50	25 ($\pm$ 9)	75 ($\pm$9)
Ne	26	100 ($\pm$ 2)	0 ($\pm$2)
Ne +0.15% Xe	26	19 ($\pm$ 3)	81 ($\pm$3)
Ar	30	35 ($\pm$ 5)	65 ($\pm$5)
Xe	44	10 ($\pm$ 5)	100[b]

[a] From Reference 54.
[b] No error estimate given.

precession is too slow to observe but Mu precession has a convenient period (usually about 100 nsec). The amplitude of this MSR signal establishes the fraction of muons thermalizing as muonium. The field is then raised to a few hundred gauss to produce a μ^+SR signal whose amplitude is proportional to the fraction of muons thermalizing in diamagnetic environments (the simplest of which would be a free μ^+). These two signals are expected to obey the normalization

$$P_D + P_{Mu} = 1 \qquad\qquad 14.$$

where $P_D = A(\mu^+)/A_0$ and, from Equation 12, we find $P_{Mu} = 2A(Mu)/A_0$ (A_0 is the experimental maximum asymmetry). This normalization may not always be the case, however (52).

A summary of the results of the only published study of this sort (54) is presented in Table 2. Certain qualitative features can be seen at a glance. In xenon, muonium is formed by virtually every muon entering the target. In argon, Mu formation seems somewhat less efficient. In helium and neon, Mu is not formed at all unless impurities with less tightly bound electrons are added.

From this limited information a surprising number of conclusions can be drawn. Beginning with the simplest, it is clear that the naive expectation that Mu will be less likely to be formed in gases with ionization potentials higher than 13.5 eV is qualitatively justified. However, there is no sharp demarcation on this basis; in Ar, with ionization potential 2.3 eV higher than Mu, most muons still form muonium; and in nitrogen, Mu formation is essentially complete, even though it is endothermic by 2.1 eV. For positronium most of the formation occurs when the positron's energy falls between the threshold for removal of the outer electron and the first excitation level, which region is known as the "Ore gap" after its proponant Aadne Ore (54a). Thus any analog of the "Ore gap" phenomenon in positronium formation is less important for Mu formation than the details of the charge-changing collision cross sections in the many-eV region. How can more be learned about those cross sections?

One approach has been to investigate the effect of impurity gases on Mu formation in helium and neon. To avoid "chemical" effects, this was done first with xenon, the only inert gas in Table 2 for which Mu formation is exothermic. The results of a recent study at TRIUMF are shown in Figure 21, in which the muon and muonium asymmetries are plotted as a function of Xe concentration (55). As can be seen from Figure 21, 100 ppm of Xe were sufficient to cause significant Mu formation in Ne. It was at first difficult to explain how so little xenon could affect the Mu formation probability. At 100-ppm Xe in Ne at 1 atm, assuming a cross section of $\sim 10^{-15}$ cm^2, thermal collisions between free muons and Xe atoms should occur only about every 0.5 μsec at room temperature. If Mu is formed in a thermal collision with Xe (as was first proposed), then the μ^+ precession signal should be observed to decay on this time scale as free muons are "lost" to the fast-precessing Mu state over a period of time. Figure 22 shows the relaxation rate of the muon signal as a function of Xe concentration, and in fact shows some evidence for such a thermal reaction

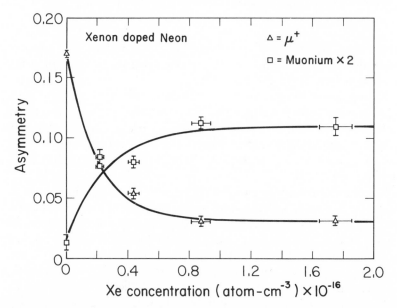

Figure 21   Muon and muonium ×2 asymmetries in neon doped with xenon as a function of xenon concentration. The nonzero asymptotic muon asymmetry at high xenon concentration may be due to muons that have scattered into the walls of the aluminum gas target vessel.

(55). However, the rate is too slow to account for the muonium formation: Because of the dephasing of Mu atoms starting to precess at different times, no observable muonium signal could result from such a process.

Contrary to these expectations, both μ^+ and Mu precession signals were observed simultaneously in Ne with minute Xe impurities, as can be seen from Figure 21. This indicates that the process

$$\mu^+ + Xe \rightarrow Mu + Xe^+$$

15.

responsible for Mu formation in this mixture must be an epithermal charge-changing collision, and that the rate for any thermal analog of reaction (15) must be comparatively slow. This is not surprising until one considers the "collision budget" for a μ^+ slowing down in Ne. In a simple "hard sphere" elastic scattering model, it should take the muon only about 700 elastic collisions with Ne atoms to slow down through the energy region where Reaction 15 is apt to be important. With a 100-ppm Xe impurity, the number of collisions with Xe atoms over the same range

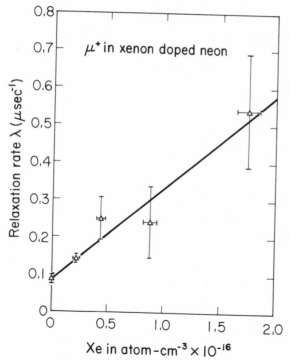

Figure 22   The rate of free muon relaxation in xenon doped neon as a function of xenon concentration at 295°K. This corresponds to the thermal production of muonium with a bimolecular rate constant $k = (2.4 \pm 0.3) \times 10^{-11}$ cm^3 atom^{-1} sec^{-1}.

should be only 0.08 times the ratio of cross sections for Mu on Xe and Ne. The muon indeed captures an electron from Xe a large fraction of the time in this situation. Thus we must conclude that the cross section for Reaction 15 is at least 10 times that for elastic hard-sphere scattering of muons with Ne, even assuming a constant cross section over the entire energy range of interest. This is plausible, since the electron is most likely to be captured from the attractive tails of the Lennard-Jones potential, which extend far beyond the repulsive core responsible for the hard-sphere scattering.

Further support for the conclusion that Mu is formed in epithermal collisions with Xe comes from the data for He with Xe impurities. While 100-ppm Xe causes significant Mu formation in Ne, it has no measurable effect in He. This would be unlikely for a thermal process; but since the μ^+ is moderated about 5 times faster by scattering from He than from Ne atoms, such a result is expected for an epithermal formation process.

Since charge-changing cross sections are generally assumed to be functions only of the velocity of the impinging light species (56), a μ^+ or

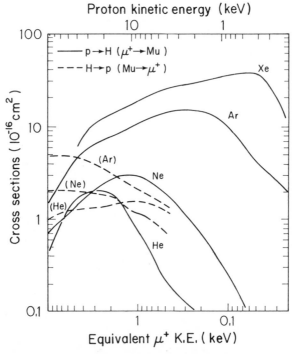

Figure 23   Charge-exchange total cross sections for H (and Mu) in various gases. Energy decreases from left to right; proton kinetic energy is plotted at the top, with the equivalent muon energy at the bottom. Solid lines are electron-capture cross sections, and dashed lines are electron-loss cross sections. From Reference 56.

Mu atom with kinetic energy T_μ may be expected to experience the same such interactions as a proton or H atom with $T_H = (M/m)\ T_\mu$, where $(M/m) = 8.85$, the ratio of proton to muon masses, assuming that positive muons behave exactly like light protons in the realm of atomic physics. Thus the extensive literature on H atom collision processes can be consulted for predictions of the detailed behavior of positive muons stopping in gases. In Figure 23, for instance, the charge-changing cross sections of interest here are plotted vs proton (upper ordinate) and/or muon (lower ordinate) kinetic energy. (The data are taken from H atom measurements reviewed in Reference 56.)

The curves in Figure 23 suggest that there are probably very few neutral atoms left in a μ^+ beam slowing down in He once the energy drops below about 200 eV; in Ne the same appears likely to be the case for $T_\mu < 50$ eV. This prediction corresponds nicely to the observed behavior. The Mu formation data in Ar and N_2 further suggest that the neutral fraction continues to rise with decreasing T_μ in N_2 but declines again below about 200 eV in Ar.

The conclusion that the electron capture cross section for μ^+ on Xe must be large compared to that on Ne or He in the many-eV energy range also agrees with the measured H atom cross sections shown in Figure 23.

The apparent slowness of the thermal Reaction 15, even though it would be exothermic by 1.4 eV, is surprising. One possible explanation is that μ^+Ne molecular ions are formed as the μ^+ comes to rest (54). Such diamagnetic ions would give a "free" μ^+ signal, as observed, but would be slow to form Mu in collisions with Xe atoms. From the measured and calculated properties of the H^+Ne molecular ion (57, 58), the binding energy of μ^+Ne can be estimated to be about 1.8 eV; the first excited state should also be bound by about 1 eV. This is just enough to reverse the energy budget for Mu formation, so that the reaction

$$\mu^+Ne + Xe \rightarrow Ne + Mu + Xe^+ \qquad\qquad 16.$$

from the ground state is endothermic by 0.4 eV, but the same reaction from the excited state will be exothermic by about the same amount. If this is indeed the explanation for the comparative stability of the "free" μ^+ signal in the presence of Xe, then some very interesting studies of the μ^+Ne ion, its ionization potential, reactivity, excited states, etc, may prove feasible.

As a tool for probing the details of atomic physics, the muon offers little competition for existing atomic beam techniques, as a glance at the literature shows (59). However, the μ^+SR technique is particularly sensitive to one parameter of general practical interest: When high energy hydrogen isotopes thermalize in various gases, what fraction comes to equilibrium as neutral atoms? By answering this question, the μ^+SR technique may prove useful to atomic physicists.

3.2.2   MUONIUM CHEMISTRY IN GASES   A very favorable consequence of the propensity of positive muons to form Mu atoms in most inert gases is the convenience with which thermal chemical reactions of Mu can be quantitatively studied. This fact was recognized early in the history of μ^+SR studies (15), but only recently did gas-phase muonium chemistry begin in earnest. New techniques developed in the last three years make such studies straightforward. The resultant data on reaction rates of Mu are of interest in the theory of isotope effects in elementary chemical reactions by virtue of the direct analogy with H atom reactions (60). This is primarily because of the unambiguous interpretation of the experimental observables and the tractability of theoretical calculations in the gas phase. Earlier results in liquids demonstrated the potential of μ^+SR techniques for measuring rate constants of Mu, but relied upon the indirect, residual polarization technique and produced rate constants in the liquid phase, for which no absolute calculations have been attempted.

These first experiments, then, demonstrated the applicability of MSR methods to physical chemistry, not as a qualitative or esoteric test, but as a precise and widely useful tool for studying isotope effects in absolute reaction rate theory.

3.2.2.1   *Experimental techniques*   The apparatus for the study of muonium chemistry in gases is the same as for Mu formation studies (see Figure 20): 4.1-MeV surface muons are stopped in argon or nitrogen gas at approximately 1 atm, where they form muonium almost without exception. The Mu atoms precess in a weak applied magnetic field (typically a few gauss) at an amplitude half the maximum for free muon precession. In the absence of field inhomogeneity (less than about 0.1 gauss) or impurities (less than about 5 ppm) this precession is long lived. However, as measured contaminants are added, the Mu precession signal decays exponentially as Mu atoms react to place the muon in diamagnetic molecules (or are depolarized by spin exchange as in the case of Mu + O_2). This relaxation rate λ can be fitted from the experimental time spectrum, and the dependence of the relaxation rate on the impurity concentration [X] can be fitted to a linear dependence whose constant of proportionality is the rate constant k for the reaction: $\lambda = k[X]$.

3.2.2.2   *Measured reaction rates of muonium in gases*   A typical MSR spectrum is shown in Figure 24a for Mu precession in pure Ar, with the constant background and exponential muon decay removed. Figure 24b shows the MSR signal in Ar to which 19×10^{-4} moles liter^{-1} Cl_2 was added; the resultant relaxation of the Mu signal is evident (52). Figure 25 shows the dependence of the relaxation rate so observed upon the HBr concentration in Ar (61). The rate constant extracted from this fit is listed in Table 3 along with several others measured in the same manner (61). The corresponding rate constants for H atoms are listed for comparison.

The recent results of Garner et al (62) on the temperature dependence of $k(\text{Mu} + \text{Cl}_2)$ and $k(\text{Mu} + \text{F}_2)$ are shown in Figure 26. These data can be tentatively fitted to the form $k = A \exp(E_a/kT)$, with activation energies E_a shown in Table 3. At room temperature, both $k(\text{Mu})/k(\text{H})$ and the apparent Arrhenius activation energy E_a for the F_2 reaction are in agreement with the recent theoretical calculation of Connor et al (60). The

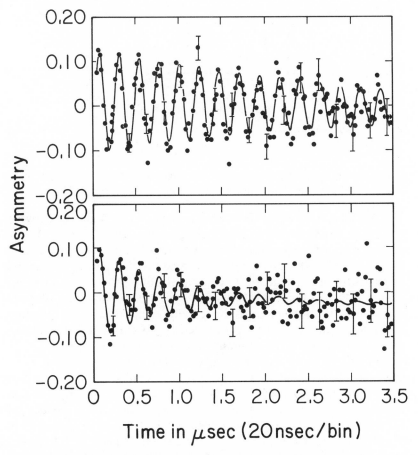

Figure 24   MSR signals in argon at 3.0 gauss, 1 atm, room temperature: (*top*) pure Ar; (*bottom*) Ar + 19 × 10^{-4} M Cl_2.

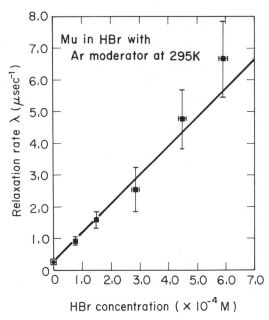

Figure 25   Muonium relaxation due to chemical reaction with HBr in Ar at room temperature corresponding to a bimolecular rate constant of $(9.1 \pm 1.0) \times 10^9$ (mole-sec)$^{-1}$.

lower activation energy for Mu $+ \text{F}_2$ compared with H $+ \text{F}_2$ is striking evidence for quantum tunneling in the chemical reaction of H isotopes.

3.2.2.3   *Comparison with H atom reaction rates*   The similarity between Mu and H atomic physics in the energy region below a few keV (see previous section) supports the expectation that Mu can be properly and precisely treated as a light isotope of hydrogen. In this light, a comparison between Mu and H reaction rates in analogous processes is possible, because in the gas phase it is feasible to calculate absolute reaction rates almost *ab initio*, predicting dynamic isotope effects such as quantum mechanical tunneling, which can be tested by experiment (60). Previously, these tests relied upon the naturally occurring isotopic differences (H, D, and sometimes T) for experimental data. The Mu atom offers a mass difference nine times larger, without loss of generality.

In a naive hard-sphere kinetic theory, one can treat the cross section σ for reaction of Mu (or H) with reagent X as a constant independent of thermal velocity v, which allows the conceptual separation of kinetics from dynamics in writing $k = \sigma \langle v \rangle$ for the rate constant, where $\langle v \rangle$ is the mean thermal velocity of the light atom (Mu or H). (For Mu, $\langle v \rangle = 0.75 \times 10^6$ cm sec^{-1} at room temperature.) While such a model wrongly ignores the fact that reaction cross sections are generally very strong functions of energy, it serves to emphasize the qualitative expectation that Mu reaction rates will be faster by about a factor of three faster than those of H, just because the mean thermal velocity, which varies as $m^{-\frac{1}{2}}$, is faster by a factor of three. This approximation is not as arbitrary as it sounds, since the energy distribution of Mu and H are the same at the same temperature. The muonium atom just gets there faster. Thus it is even rigorously true that the ratio between Mu and H rate constants will be 3 unless there is a genuine dynamic isotope effect — that is, unless the detailed cross sections are different for Mu and H. In general, a proper theoretical treatment requires rather difficult trajectory calculations, in effect the calculation of the orientation-averaged energy dependence of the reaction cross section for Mu and H with X; but in perusing Table 3 one should first look for $k(\mathrm{Mu})/k(\mathrm{H})$ ratios significantly different from 3.

The first muonium chemistry measurements in the gas phase included a measurement of the rate of Mu depolarization in the presence of minute

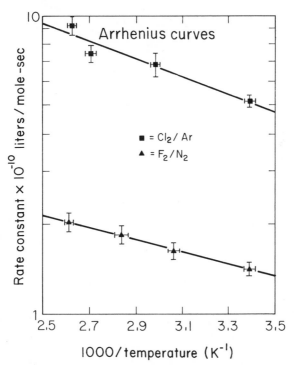

Figure 26   Rate constants for muonium reacting with halogens as a function of inverse temperature, showing Arrhenius Law dependence. (*Solid squares*) $\mathrm{Mu} + \mathrm{Cl}_2 \rightarrow \mathrm{MuCl} + \mathrm{Cl}$. (*Solid triangles*) $\mathrm{Mu} + \mathrm{F}_2 \rightarrow \mathrm{MuF} + \mathrm{F}$.

Table 3　Reaction rate parameters for Mu and H in the gas phase

Reaction	Muonium		Hydrogen			H atom ref.
	k (295°K)[a]	E_a (kcal mole^{-1})	k (295°K)[a]	E_a (kcal mole^{-1})	k_{Mu}/k_H (295°K)	
F_2	1.4 ± 0.1	0.92 ± 0.23	0.20 ± 0.05	$2.4\ \pm0.2$	5.2–10	124, 125
			0.09 ± 0.01	$2.2\ \pm0.1$	13–19	126
Cl_2	5.1 ± 0.2	1.36 ± 0.21	$1.7\ \pm0.6$	$1.8\ \pm0.3$	2.1–4.8	124
			0.41 ± 0.04	$1.4\ \pm0.2$	11–14	127
			1.15 ± 0.15	1.15 ± 0.1	3.8–5.3	128
Br_2	$24\ \ \pm3$	—	$2.2\ \pm1.5$	$1.0\ \pm0.5$	5.7–38	129
HCl	$\leq0.000034^{b}$ ±0.000005	f	$2(0.0021^{c}$ $\pm0.0002)$	$3.1\ \pm0.3$	$\frac{1}{2}(0.013\text{–}0.020)^{c}$	130, 131
		g	0.000018^{d} ±0.000018	$\geqq4.0$	~1.9	132
HBr	$0.9\ \pm0.10$	f	0.21 ± 0.02	$2.6\ \pm0.1$	3.5–5.3	133
		g	$\leqq0.0023$	5–6	350–440	133
HI	2.53 ± 0.13	f	0.86 ± 0.39	$1.2\ \pm0.4$	1.9–5.7	134
O_2	16.0 ± 0.7^{e}					

[a] K ($\times10^{10}$ l/mole-sec).
[b] Upper limit only.
[c] There exists a stoichiometric ambiguity of a factor of 2 (see Reference 53).
[d] Upper limit (i.e. $<3.6\times10^{5}$ l/mole-sec).
[e] Probably spin exchange, not chemical reaction.
[f] H atom reaction type is abstraction.
[g] H atom reaction type is exchange.

O_2 impurities in Ar (15). In that study it was not clear whether a significant role was played by the actual chemical combination of Mu and O_2, or if the spin-exchange interaction was completely dominant (11). It is now believed that the chemical reaction is relatively slow, so that the rate constant in Table 3 can be compared directly with the spin-exchange rate for H. Following References 15 and 11, the rate constant can be reduced to a spin-exchange cross section σ_{SE} by the following formula: $k = (\frac{32}{54})\sigma_{SE}\langle v \rangle$, which gives $\sigma_{SE} = (5.72 \pm 0.3) \times 10^{-16}\,cm^2$ for Mu + O_2, as opposed to $(21 \pm 2.1) \times 10^{-16}\,cm^2$ for H + O_2 (15). The smaller cross section for Mu is surprising; since only a spin-exchange interaction is expected, there should be little "dynamics" involved, and the hard-sphere approximation should hold reasonably well.

3.2.2.4 *Muonium in vacuum and Mu chemistry in powders*   In the last several years, many experimenters have demonstrated that positronium

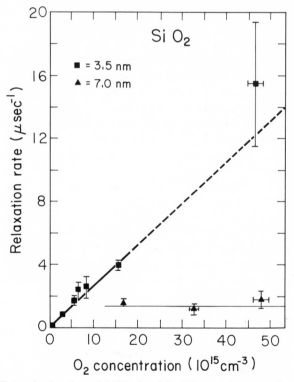

Figure 27 Rate of relaxation of MSR signal in SiO_2 powders as a function of the concentration of O_2 added to previously evacuated sample container. Square points: SiO_2 powder grain, nominal radius $r = 3.5$ nm; triangular points: nominal grain radius $r = 7.0$ nm.

atoms (Ps $= e^+e^-$) formed in very small grains of powdered SiO_2, MgO, and Al_2O_3 will very quickly diffuse out of the grains into vacuum, where they remain, affected only by gas molecules and collisions with the powder grains (63, 64). A recent study at TRIUMF has shown that the same thing happens to muonium atoms (65).

This was verified by stopping a beam of surface muons in a fine silica powder, where a long-lived Mu precession signal could be seen at low field, but only if the powder sample had been pumped to a hard vacuum. The addition of as little as 10 Torr of O_2 gas pressure resulted in the destruction of the Mu signal, as shown in Figure 27. In fact, the disappearance rate of the Mu signal had exactly the same dependence upon O_2 partial pressure in the quartz powder as it did in Ar moderator gas. It was concluded that the powder grains played essentially the same role as the molecules or atoms of the inert moderator gas—that the powder was more or less a special moderator gas with very large molecules (65).

The Mu atoms diffuse rapidly to the surface of the SiO_2 grains. In order to add some quantitative precision to this observation, a coarse-grained powder (140Å diameter) was run with high O_2 pressure (12 Torr). Whereas the 70-Å grains gave up their Mu to the depolarizing effects of the O_2 within a few nanoseconds, the coarse grains held on to it long enough for a Mu precession signal to be observed experimentally, as indicated in Figure 27. A calculation of the probability that a Mu atom formed at a random volume element inside a spherical grain will still be within that grain at time t gives the following formula which should match the precession envelope:

$$2P(t) = \text{erf}(\beta t)^{\frac{1}{2}} - \frac{3(\beta t)^{\frac{1}{2}}}{\pi^{\frac{1}{2}}} \left\{ 1 - \frac{\exp(-1/\beta t)}{3} + \frac{2\beta t}{3} [\exp(-1/\beta t)] \right\} \qquad 17.$$

where $\beta = D/R^2$, D is the diffusion constant, and R is the grain radius.

From this the diffusion constant can be obtained by fitting the experimental data. A preliminary measurement gives $D = (2.9 \pm 0.7) \times 10^{-7}$ cm^2 sec; further experiments with more uniform grain size should allow this number to be determined with better accuracy.

3.2.3   MUONIUM CHEMISTRY IN LIQUIDS   The gas-phase work described above has shown the advantage of measuring reaction rates via the relaxation of the muonium precession signal itself. However, the liquid phase presents some obstacles absent in gases.

3.2.3.1   *The direct MSR technique*   The prime impediment to direct observation of Mu in liquids is the presence of dissolved O_2. At STP, the partial pressure of O_2 over a liquid is about 0.2 atm; in water, this leads to a dissolved oxygen concentration of about $2.5 \times 10^{-4} M$ at 25°C. Assuming the same bimolecular rate constant for Mu + O_2 spin exchange as measured in gases $[1.5 \times 10^{11}$ liter/(mole sec)$^{-1}]$ we thus expect Mu to relax within about 20 nsec in water exposed to air—too fast to observe any MSR signal directly. However, techniques for removal of this dissolved O_2 are known, and it is possible to obtain pure enough water samples to permit the study of long-lived MSR signals.

Percival et al (66) were the first to observe Mu precession in a liquid, ultrapure water in a sealed glass bulb. They found a long-lived (more than a muon lifetime) MSR signal, which was positively identified as muonium by a measurement of the splitting of the precession into two frequencies at high field (recall Equation 12). The extracted hyperfine frequency was the same as that of Mu in vacuum. Muonium precession in water has since been observed at TRIUMF as well.

Even with the purification challenge met, there is a serious obstacle to direct MSR studies in most liquids, in the form of the "hot fraction"— those muons (usually a substantial percentage) that react epithermally at essentially $t = 0$. In water, about 60% of the muons are thus prevented from ever contributing to the MSR signal. Since the amplitude of a fully polarized Mu signal is only half that of an equivalent μ^+ signal anyway, this leaves only 30% of A_0 for a muonium amplitude, requiring at least 10 times as many events as a gas-phase measurement for an MSR signal of the same statistical significance.

Worse luck, Percival et al (66) found a muonium signal in water that was reduced by yet another factor of two. Taking CCl_4 and Al as asymmetry standards and correcting for the factor of two lost in the original differentiation of Mu into singlet and triplet states, they extracted the fractional populations of (magnetically) free μ^+ and free Mu in several liquids, listed in Table 4, and in water as a function of temperature, shown

Table 4 Muon polarizations in pure substances

Sample	P_D[a]	P_M[a]
H_2O, liquid	0.622 ± 0.006	0.196 ± 0.003
H_2O, ice ($T > 160°K$)	0.480 ± 0.004	0.52 ± 0.02
D_2O, liquid	0.57 ± 0.03	0.18 ± 0.01
D_2O, ice ($T > 160°K$)	0.393 ± 0.005	0.63 ± 0.01
CH_3OH	0.61 ± 0.01	0.19 ± 0.02
CD_3OD	0.51 ± 0.02	0.31 ± 0.05
C_2H_5OH	0.59 ± 0.03	0.20 ± 0.04
Al, granular	0.990 ± 0.007	—

[a] Relative to $P_D = 1.0$ for CCl_4.

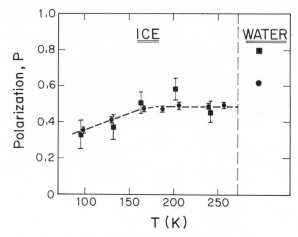

Figure 28   Fraction P_D of μ^+ in diamagnetic environments (*circles*) and fraction $(1 - P_{Mu})$ not accounted for in Mu signals (*squares*) as a function of temperature for μ^+ in H_2O. Unless $P_D = (1 - P_{Mu})$, part of the original μ^+ ensemble is unaccounted for (67).

in Figure 28. In each case a substantial fraction (18% for H_2O at room temperature) of the polarization is unaccounted for (except in ice, where all the polarization is accounted for). These unexpected "missing muons" may be in the form of epithermally formed radicals, which relax too rapidly to observe directly. In any case their discovery has important consequences for the interpretation of earlier residual polarization studies.

Table 5A   Comparison of $k_{(Mu)}$ with $k_{(H)}$, $k_{(e_{aq}^-)}$, and $k_{(Ps)}$ (in m^{-1} sec^{-1})

Solute	$k_{(Mu)}$[a]	$k_{(H)}$[b]	k_{Mu}/k_H	$k_{(e_{aq}^-)}$[c]	$k_{(Ps)}$[d]
Phenol	7×10^9	2×10^9	3.5	1.8×10^7	$<10^8$
p-Nitrophenol	8×10^9	$(3 \pm 1) \times 10^{9e}$	2.7	3.5×10^{10}	9×10^{9h}
Tl^+	8×10^8	$[1.2 \times 10^8]^f$	~ 7	3×10^{10}	$<10^8$
CNS^-	6×10^7	$[2 \times 10^8]^f$	~ 0.3	$<10^6$	$<10^8$
Zn^{2+}	$<10^7$	$<10^5$	—	1.5×10^9	$<10^7$
$Na^+/SO_4^=$	$<10^7$	very small[g]	—	$<10^6$	$<10^7$

[a] Jean, Y. C., Brewer, J., Fleming, D. G., Garner, D. M., Mikula, R. J., Vaz, L. C., Walker, D. C. Private communication (1978).
[b] Data from NSRDS-NBS 51 by Anbar, M., Farhatziz, and Ross, A. B. (Pulse data used when available).
[c] Data from NSRDS-NBS 43 by Anbar, M., Bambenek, M., Ross, A. B.
[d] Data estimated from *At. Energy Rev.* 6:(1968) by Goldanskii, V. I.
[e] Estimated from: K(H + nitrobenzene) = 3×10^9 and k(H + phenol) = 2×10^9.
[f] Obtained from only one source, and that (unpublished) is based on the rate relative to 2-propanol, which has been shown to be a poor basis for comparison in the case of some other solutes.
[g] Not recorded but most probably $<10^7$.
[h] Data from Jean, Y. C., Ache, H., *J. Phys. Chem.* 81:2093 (1977).

Table 5B   Muonium and hydrogen atom rate constants in aqueous solutions[a]

Substrate	k_{Mu} (m^{-1} sec^{-1})	k_H[b] (m^{-1} sec^{-1})	k_{Mu}/k_H
Methanol	$<3 \times 10^4$	2.5×10^6	0.01
Ethanol	$<3 \times 10^5$	2.1×10^7	<0.015
2-Propanol[c]	$\approx 3 \times 10^6$	6.8×10^7	~ 0.04
2-Butanol[c]	$\approx 2 \times 10^6$	1.3×10^8	~ 0.02
Formate ion (pH >7)	7.8×10^6	1.2×10^8	0.07
Maleic acid (pH 1)	1.1×10^{10}	8×10^9	1.4
Fumaric acid (pH 1)	1.4×10^{10}	7×10^9	2
MnO_4^-	2.5×10^{10}	2.4×10^{10}	1
Ag^+	1.6×10^{10}	$(1-3) \times 10^{10}$	~ 1
NO_3^-	1.5×10^9 $[1.2 \times 10^{11}]$[d] $[1.7 \times 10^{10}]$[e]	9×10^6	~ 170
Acetone	8.7×10^7	2.8×10^6	30
Ascorbic acid (pH 1)	1.8×10^9	1.7×10^8	10
OH^-	1.7×10^7 $[1.8 \times 10^9]$[e]	1.8×10^7	1
ClO_4^-	$<10^7$ $[3.8 \times 10^{10}]$[d] $[1.2 \times 10^9]$[e]	—	—
H^+, Na^+, Cl^-	$<2 \times 10^5$	$<10^5$	—

[a] Ref. 56; Percival, P., Private communication; and Percival, P. W., Roduner, E., Fischer, H., Camani, M., Gygax, F. N., Schenck, A. 1977. *Chem. Phys. Lett.* 47:11.
[b] Anbar, M., Farhataziz, Ross, A. B. 1975. NSRDS-NBS 50.
[c] Muonium values are preliminary.
[d] Rate constants were obtained from Reference 18 and P_{res}, which may involve radical reactions etc.
[e] Minaichev, E. V., Myasischeva, G. G., Obukhov, Yu. V., Rogonov, V. S., Savel'ev, G. I., Smilga, V. P., Firsov, V. G. 1974. *Sov. Phys. JETP* 66:1926.

These effects reduce the maximum available MSR signal in water to one fifth that in N_2 gas. Although this has no effect on the principles of measurement of reaction rates as used in the gas phase, the practical effect is serious: One must measure 25 times as many events to obtain the same statistical precision. Without the high muon fluxes available at new facilities such as SIN and TRIUMF, this disadvantage would be insurmountable. However, Percival et al succeeded in measuring several rate constants in this way, clarifying the reaction mechanisms at work in aqueous results. Table 5B lists some of their recent results. Figure 29 shows the dependence of the rate of disappearance of the MSR signal upon the concentration of fumaric acid in water (from Reference 66). The rate constant k_{Mu} corresponding to the straight line fit to λ as a function of reagent concentration is listed in Table 5A along with rate constants for

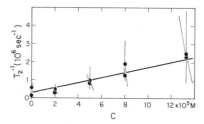

Figure 29   Muonium relaxation rate as a function of fumaric acid concentration in water (67).

analogous reactions of H, e_{aq}^-, and Ps. These results indicate that Mu is indeed best characterized as a light isotope of the H atom, even in aqueous solutions. In this context, a comparison of Mu and H rate constants in aqueous solution, shown in Table 5B, illuminates several qualitative features of the reaction mechanisms in water: H reacts with the first five substrates by abstracting another H atom from a C–H bond; apparently Mu is much less efficient for such abstraction reactions; for the next four substrates, both H and Mu apparently react at the expected diffusion-controlled rate limit; reactions with nitrate, acetone, and ascorbic acid are more efficient for Mu than for H; and the similarity of the Mu and H rates with OH$^-$ suggest that the reaction is an interconversion of Mu (H) into its conjugate base, e_{aq}^-:

$$Mu + OH^- \rightarrow e_{aq}^- + MuOH.$$

3.2.3.2   *The residual polarization method*   It is preferable to measure reaction rates of Mu by the direct MSR method described above, when feasible; for one thing, only by this method can slow ($< 10^8 \, sec^{-1}$) reaction rates be studied. However, even under optimal conditions the MSR method is tedious in liquids, and in many situations of interest no Mu signal may be seen. Therefore this technique will be most useful in cooperation, rather than competition, with the residual polarization μ^+SR method developed earlier. Indeed, there are phenomena, such as hot atom reactions of Mu*, that can only be studied by this method.

Although longitudinal-field measurements (11) can provide valuable information to supplement that obtained in the transverse-field technique described here (and a thorough experimental approach should utilize both methods), we restrict our discussion to the latter because it contains the most information. In this arrangement the positron counters and the muon spin are in a plane perpendicular to the applied field, and the amplitude of the sinusoidal precession signal is proportional to the magnitude of the residual muon polarization. The initial phase of the precession depends upon the geometry and the initial polarization direction of the beam, of course, but this part of the phase is usually removed by subtraction of the value measured when there is no depolarization. It is the deviation $\Delta\phi$ from this reference phase that provides information about the depolarization mechanism. Similarly the amplitude of the μ^+SR signal is the product of P_{res} and a maximum amplitude A_0 corresponding to no depolarization. However, an unambiguous establishment of A_0 is a more delicate matter.

There are problems associated with thick-target measurements. Some materials such as Cu metal are not thought to have any depolarizing effect on the muons at early times, and can therefore be used to establish an asymmetry standard. In liquid-phase studies, CCl_4 is often used for the same purpose, although with less confidence that P_{res} is exactly 1.0. In any case, great care must be taken in such normalizations because of the effect of density upon the experimental asymmetry for a fixed polarization.

There are two sources of this density effect. The first and lesser part of the effect is the consequence of the correlation of the polarization with the momentum of muons in the beam. Whether conventional muons are derived from "backward" or "forward" decay-in-flight of pion beams, the angle of the decay in the c.m. frame is translated into a momentum increment in the lab frame. Thus (for instance), higher momentum "forward" muons have higher polarization. Since a denser target of fixed thickness will stop more high momentum muons, it will produce a slightly larger asymmetry, even though its depolarizing properties are identical.

There is an even larger density effect due to the absorption of decay positrons. The decay asymmetry is a function of the positron energy, because of the dynamics of the weak interaction (11). To contribute to the μ^+SR signal, the positrons must escape from the target and penetrate several counters. Thus many low energy positrons are lost by absorption, raising the average experimental asymmetry. This is advantageous in the sense that the higher energy positrons have the larger asymmetries, but if the target density increases, then still more positrons are lost, which influences the experimental asymmetry.

These effects can be calculated and a correction applied to the residual polarization results; more commonly an empirical survey is made of the dependence of the asymmetry upon target thickness (40). Such corrections are less important with smaller targets. However, much older data (see for instance the asymmetry table in Reference 11) are subject to systematic fluctuations of up to 25% as a result of density effects and variations in beam polarization. Thus it would be advantageous to use a technique that makes density an unimportant parameter in residual polarization measurements. This is now available.

As a result of the low energy (4.1 MeV) and concomitant short range (less than 1 mm of water) of "surface muons," liquid targets can be built that are of negligible thickness in the direction of positron detection and yet they will stop the entire surface muon beam in the solution of interest. The apparatus used at TRIUMF is shown in Figure 30. Preliminary studies with this technique indicate its reliability for residual polarization studies. A few such measurements are shown in Table 6.

Residual μ^+ polarizations measured in various liquids by several techniques are summarized in Table 7. Care must be taken in the interpretation of these values of P_{res}, since it is generally possible for both thermal and hot atom reactions of Mu to help prevent the complete destruction of the μSR signal (17). In addition, there is some indication

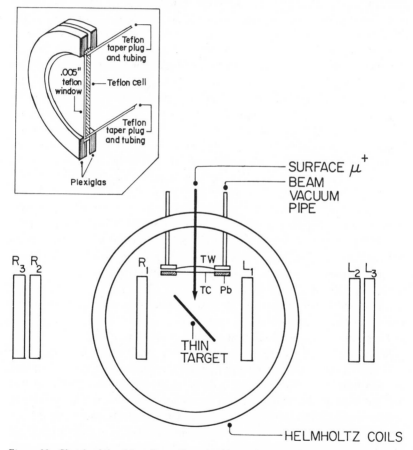

Figure 30   Sketch of the thin teflon cell used with "surface" muons and its arrangement with respect to the counters. TC is the thin counter that defined the μ^+ beam; L1, L2, L3 are the left-hand-side counters; and R1, R2, and R3 the right-hand-side counters.

Table 6 Comparison of data obtained on P_{res} from TRIUMF using surface muons with density-corrected data from LBL, JINR,[a] and SIN, using conventional muons[b]

Target liquid	TRIUMF	LBL	JINR	SIN
CCl_4	1	1	1	1
$CHCl_3$	0.86 ± 0.04	0.85 ± 0.04	0.80 ± 0.06	—
H_2O	0.61 ± 0.02	0.59 ± 0.02^c	0.62 ± 0.04	0.62 ± 0.01
D_2O	0.57 ± 0.04	0.59 ± 0.02	—	0.58 ± 0.03
CH_3OH	0.56 ± 0.04	0.54 ± 0.02^c	0.58 ± 0.05	0.61 ± 0.01
$(CH_3)_2CHOH$	0.61 ± 0.04	0.64 ± 0.01	—	—
$C\text{-}C_6H_{12}$	0.68 ± 0.04	0.67 ± 0.02	0.68 ± 0.05	—
$C\text{-}C_6H_{10}$	0.47 ± 0.03	—	0.48 ± 0.05	—
C_6H_6	0.12 ± 0.02	0.13 ± 0.01^c	0.15 ± 0.03	—
CS_2	0.16 ± 0.03	0.11 ± 0.01	—	—
$(CH_3)_2CO$	0.54 ± 0.05	—	—	—
$Si(CH_3)_4$	0.54 ± 0.03	—	—	—
C_5H_{12}	0.64 ± 0.04	—	—	—
C_7H_{16}	0.65 ± 0.04	—	—	—
$C_{10}H_{22}$	0.67 ± 0.05	—	—	—
2,2,4 Trimethyl-pentane	0.61 ± 0.05	—	—	—

[a] JINR is the Joint Institute for Nuclear Research, Dubna, USSR; LBL is the Lawrence Berkeley Laboratory.

[b] The errors quoted arise from the uncertainty due to the statistics of the experiment and do not take into account other possible sources of error.

[c] These data were obtained as limiting values in titration curves and therefore required no density corrections.

that "spur" reactions in the region ionized by the passage of the incoming μ^+ may play an important role (67). However, in most inert solvents the thermal reaction rate of Mu can be expected to be too small ($\lesssim 10^8 \ sec^{-1}$) to save the μ^+SR signal, and the residual polarization can be tentatively equated to the hot fraction h; that is, the fraction of Mu* atoms reacting epithermally to place the μ^+ in a diamagnetic molecule before Mu* thermalizes in $\sim 10^{-12}$ sec (11). In some cases, magnetic field independence of P_{res} provides direct experimental support for the assumption that $P_{res} = h$, since destruction of more than 50% of the μ^+SR signal is the result of Mu precession, the period of which decreases with field (11).

Treating the P_{res} values listed in Table 7 as pure hot atom fractions, we notice several trends. First, even the qualitative expectation that homologous compounds such as the group IV tetrachlorides will show a systematic trend in P_{res} as a function of mass or bond energy is not verified. Evidently the behavior of Mu* in these media cannot be characterized even parametrically in these terms. However, there is a consistent tendency for compounds with a higher degree of π bonding to have smaller values

Table 7   Collection of P_{res} data from the various sources referred to in the text

Target substance	P_{res}	Reference and comments
CCl$_4$	1.0	By definition, Bond Energy (BE) = 78 kcal/mole
SiCl$_4$	0.48	TRIUMF (conventional muons) BE = 91 kcal/mole
SnCl$_4$	0.99	TRIUMF (conventional muons) BE = 76 kcal/mole
TiCl$_4$	1.00	TRIUMF (conventional muons) BE = 102 kcal/mole
Cyclohexane	0.68	See table 6
Cyclohexene	0.55	JINR normalized with respect to CHBr$_3$
1,4 Cyclohexadiene	0.47	JINR normalized with respect to CHBr$_3$
1,3 Cyclohexadiene	0.38	JINR normalized with respect to CHBr$_3$
Benzene	0.18	JINR normalized with respect to CHBr$_3$
Hexane	0.62	LBL
Hexene	0.50	LBL
Hexyne	0.43	LBL
2-Propanol	0.62	Table 6
Acetone	0.54	Table 6
C$_6$H$_6$	0.18	JINR
C$_6$H$_5$Cl	0.27	JINR
C$_6$H$_5$Br	0.45	JINR
C$_6$H$_5$I	0.59	JINR
C$_6$H$_5$CH$_2$Cl	0.42	JINR
C$_6$H$_5$CHCl$_2$	0.55	JINR
C$_6$H$_5$CCl$_3$	0.68	JINR
CHCl$_3$	0.85	See table 6
CH$_2$Cl$_2$	0.70	LBL
C$_6$H$_5$OH	0.38	LBL
Glycerol	0.75	LBL

of P_{res} (42). This may be related to the higher efficiency of these compounds for slowing down the Mu* by excitation of low lying electronic levels, or it may support the view that spur reactions are important, since the thermal Mu atom may be more likely to combine with such molecules to form radicals rather than diamagnetic species.

The state of theoretical interpretation of these data can only be described as preliminary and confusing; perhaps new experimental tests will lead to further progress in the next few years. Certainly gas-phase experiments should help to determine if spurs (virtually absent in gases) really play an important role in the liquid-phase hot atom effects.

3.2.4 μ^+ SR FOURIER SPECTROSCOPY OF MUONIC RADICALS   As early as 1972, residual polarization studies (18, 42) indicated the role of muonic radicals (paramagnetic molecules incorporating the μ^+) in the μ^+ depolarization mechanism (see Section 2.1.1). In the following five years, efforts were made to observe these radicals directly via their expected muonium-like precession in low fields; the beat frequency Ω (see Equations 11 and 12) would, if observed, give a direct measurement of the isotropic average hyperfine coupling of the μ^+ to the unpaired electron, $h\omega_r$. These efforts were frustrated by the fact that the large unsaturated organic molecules most convenient for radical formation by addition of Mu (e.g. benzene) contain proton spins with which the unpaired electron is $\sim \frac{1}{3}$ as strongly coupled as with the μ^+, which results in a complicated spin Hamiltonian and a spread over many precession frequencies in low field— effectively a fast relaxation of the unpaired electron.

Attempts to avoid this complication by using unsaturated molecules with no nuclear moments (CS_2, CO_2, SO_2) were unsuccessful in the liquid phase, perhaps because of the effective magnetic fields produced by the more rapid rotations of these light molecules, again splitting the precession frequencies.

By a new approach the recent experiment at SIN (67a) has demonstrated the existence of muon radicals in the Paschen-Bach limit of high magnetic fields.

Recalling the general equation (9) for muonium precession, we can write the high field ($x \gg 1$) form:

$$P(t) = \tfrac{1}{2}(e^{i\omega_{12}t} + e^{i\omega_{34}t})$$

with

$$\left|\omega_{\substack{12\\34}}\right| = \omega_\mu \mp \frac{\omega_0}{2};$$

that is, precession at two frequencies split by ω_0 about the normal μ^+ Larmor frequency. For most radicals, ω_r can be expected to be a small fraction of ω_0 for free muonium, so that the critical field becomes much less than $B_0 = 1592$ G, e.g. $B_r \lesssim 100$ G, and the splitting is reduced from $\omega_0/2\pi = 4463$ MHz to $\omega_r/2\pi \sim 200$ MHz. More important, for fields $B \gg B_r$, the electron is essentially decoupled from the various nuclear spins, in the sense that there is no longer any appreciable mixing of hyperfine states; thus only the frequencies appear in the time dependence to first order, and the signals are long lived. This behavior is familiar in the ENDOR spectra of free radicals with single spin $\frac{1}{2}$ nuclei at high fields (67b), and is qualitatively similar to the high field precession of the "anomalous muonium" state in silicon (see Section 2.5.1).

Using this high field technique, Roduner et al (67a) were able to identify and measure couplings of several radicals listed in Table 7A. The frequency spectrum which they observe for muons in tetramethylethylene is shown

Table 7A Hyperfine coupling constants of muonium-substituted radicals and comparison with hydrogen analogues

Compound	Radical	A_μ [MHz][a]	$A_\mu \dfrac{\mu_p}{\mu_\mu}$ [MHz]	A_p [MHz][b]	Reference
2,3-Dimethyl-2-butene	$(CH_3)_2CMu\dot{C}(CH_3)_2$	160.9	50.5	30.18 (298)	135
2-Methyl-butadiene	$CH_2 = C(CH_3)\dot{C}HCH_2Mu$	180.7 or 199.5	56.8 or 62.7	37.8[c] (300)	136
	$CH_2Mu\dot{C}(CH_3)CH = CH_2$	199.5 or 180.7	62.7 or 56.8	42.91[d] (363)	137
1,3-Pentadiene	$CH_2 = CH\dot{C}HCHMuCH_3$	182.7 or 169.0	57.4 or 53.1	39.74 (140)	138
	$CH_2Mu\dot{C}HCH = CHCH_3$	169.0 or 182.7	53.1 or 57.4	[e]	
Benzene	Cyclohexadienyl	514.6[f]	161.6	133.71 (288)	139
Acetone	$(CH_3)_2\dot{C}OMu$	26.0	8.2	0.90 (300)	140

[a] Values at ambient temperature, ±0.2 MHz if not specified otherwise.
[b] Temperature (°K) indicated in parentheses.
[c] Tentative.   [d] exo-CH$_3$ group.   [e] Not determined.   [f] ±0.6 MHz.

in Figure 30a. There is no reason to doubt that the technique will be applicable to a wide variety of radicals, and that impacts on the physics and chemistry of these species should be far reaching.

3.3   The Muon as a Magnetic Probe

3.3.1   FERROMAGNETIC MATERIALS   The major advance in the technique for μSR in ferromagnetic samples is the observation of μSR in zero external fields. The internal field of an unmagnetized sample will point in a random

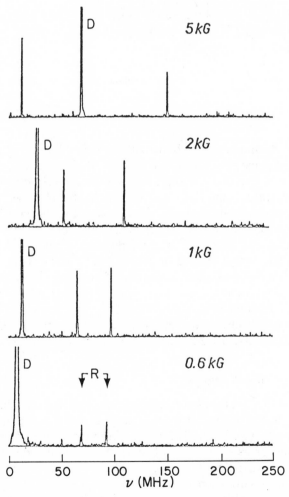

Figure 30a   Muon precession frequencies for tetramethylethylene at various magnetic fields. D = muons in diamagnetic environments. R = muonium substituted radical.

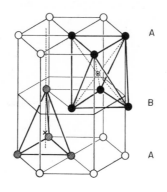

⊛   OCTAHEDRAL SITE

×   TETRAHEDRAL SITE

Figure 31   The two muon interstitial sites in the hcp crystal structure of cobalt.

direction. If the muon is polarized along the beam direction, the spin will precess in the plane perpendicular to the local field and the spin component along the beam direction will appear to have a coherent modulation at the frequency corresponding to the average internal field it experiences.

Recent results have allowed an assignment for the location of the muon in the crystal lattice of ferromagnetic cobalt (70, 71). If the external magnetic field $\mathbf{B}_{ext}$ is zero, the muon precesses in a field $\mathbf{B}_\mu = \mathbf{B}_L + \mathbf{B}_{dip} + \mathbf{B}_{hf}$, where $\mathbf{B}_L$ is the Lorentz field. The latter two terms will in general depend on the orientation of the local magnetization field in the crystal lattice at the site of the muon. Figure 31 shows the possible interstitial sites in the hexagonal crystal (68).

If θ is the angle between the local magnetization in the domain and the $\hat{c}$ axis of the crystal (perpendicular to the basal plane), the magnetic energy density is

$$E = K_1 \sin^2 \theta + K_2 \sin^4 \theta, \qquad\qquad 18.$$

where K_1 and K_2 are the anisotropy constants (69). The equilibrium value of θ is determined by minimizing the energy expression, Equation 18.

Figure 32 shows the field $\mathbf{B}_\mu$ extracted from the measured frequency of muon precession as a function of temperature in a pure single crystal of cobalt in zero external field (70, 71). At 690°K, it is known that the crystal structure changes from hcp to fcc. At this phase change a sharp change in local field is seen by the muon. The Lorentz field varies from a maximum of 6000 gauss at 0°K to zero at the Curie temperature following an approximately Brillouin function with a small discontinuity at the phase transition. Neutron diffraction measurements have shown that the direction of the magnetization relative to the crystal axis undergoes a gradual rotation between 450°K and 500°K (72). The anisotropy energy minimum changes the direction of the magnetization from a position in the basal plane above the transition to along the $\hat{c}$ axis at low temperature. This rotation of the easy axis of magnetization occurs between 450 °K and 500 °K, as shown in Figure 33.

Depending on whether the muon is in an octahedral site or a tetrahedral site, the dipolar field calculated assuming classical point dipoles follows the solid or dotted curve in Figure 32. In the calculation the dipole moment for cobalt

$$\mu_{Co} = 1.72\mu_B \frac{M(T)}{M(O)}.$$

19.

This calculation shows an increase with increasing temperature in the

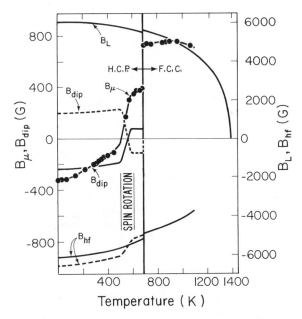

Figure 32 Temperature dependence of $\mathbf{B}_\mu$ in Co. Shown also are the calculated dipolar fields (projected on the local magnetization direction for different site assignments of the μ^+ (solid: octahedral; dashed: tetrahedral), the Lorentz field $\mathbf{B}_L$ and the resulting hyperfine field $\mathbf{B}_{hf}$. Only the octahedral interstitial site assignment leads to a smooth temperature dependence of $\mathbf{B}_{hf}$. Note that $\mathbf{B}_\mu$ and $\mathbf{B}_{dip}$ are plotted on a different scale from $\mathbf{B}_L$ and $\mathbf{B}_{hf}$.

dipolar contribution for the octahedral assignment vs a reduction in the tetrahedral assignment. If the hyperfine field is obtained by subtraction, the result is given in the lower curve. One would expect that the hyperfine field would be independent of the magnetization direction, since it involves only the conduction electron-spin density at the muon site in this simplified view. Therefore, there is a clear preference for the octahedral interstitial site assignment. There is a small step in B_{hf} at the crystal phase transition, as shown.

This qualitative analysis has been applied to another example of hcp structure, ferromagnetic gadolinium (71, 73, 74) and a comparison with the measurements gives again a good fit with the octahedral site assignment. In these analyses it is not determined whether the muon is localized at a particular octahedral site or diffuses among various 0-sites.

Iron has a bcc structure, but apparently the motion of the muon between sites averages out the differences in dipolar fields in the temperature ranges that have been studied. We shall return to iron in connection with the muon diffusion in the following section.

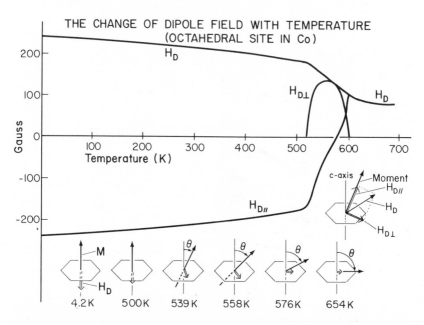

Figure 33 The variation with temperature of the dipolar field at the octahedral site in cobalt. The easy axis of magnetization is rotated by an angle θ with respect to the $\hat{c}$ axis. At $0°K$, **M** is parallel to the $\hat{c}$ axis and the dipole field is antiparallel. As the temperature is elevated above $520°K$, **M** tips into the basal plane, and H_{dip} lies along **M**, both lying in the basal plane.

3.3.1.1 *Hyperfine fields* The hyperfine field that arises from the polarization of the conduction electrons has been the subject of a number of experimental and theoretical studies, which are in progress at this writing. It will suffice to summarize the status experimentally and indicate briefly the difficulties various quantitative interpretations have experienced. The values of the hyperfine field at $T = 0$ have been collected (76) in Table 8 for nickel, iron, cobalt, and gadolinium, together with the results of the neutron diffraction measurements for the magnetic field present at the octahedral site due to the polarized electron-spin density. The ratios of the observed muonic hyperfine field to the unperturbed interstitial magnetization vary from 1.04 in nickel to 8.41 in iron. The models have attempted to explain this enhancement factor in terms of a distortion of the electron density distribution due to the muon's positive charge. The results can be summarized as follows: (*a*) The hyperfine field is negative in agreement with the reversal of magnetization at the

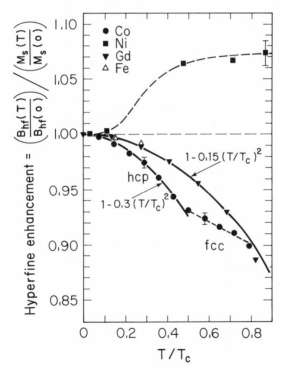

Figure 34 The variation with temperature of the normalized hyperfine enhancement for μ^+ in ferromagnetic metals. $\mathbf{B}_c$ is the hyperfine field at the μ^+ while M_s is the saturation magnetization. Nickel shows an increase of $\mathbf{B}_c$ compared to M_s whereas for cobalt, iron, and gadolinium there is a reduction.

Table 8 Values of the hyperfine field at $T = 0$

| | Curie temperature T_c (°K) | Crystal structure | Lorentz field at $T=0$ $\frac{4}{3}\pi M_S(0°K)$ (kG) | $B_a(0°K)$ (kG) | B_{hf} (kG) | B_{dip} (kG) | $\mu B/Å^3$ (kG) | Neutron diffraction results | | | Calculated hyperfine fields (kG) | | | |
								M_{local} (kG)	$B^0 = \frac{8}{3}M_{local}$ (kG)	Enhancement factor B_{hf}/B^0	Patterson & Falicov (36)	Jena I II (81,82)	Meier (83)	Petzinger & Munjal (84)
Ni	630	fcc	2.1826	1.4926	−0.6883	0	−0.0085	−0.079(77)	−0.66	1.04	−0.924	−0.6 −8.2	—	−8.25
Fe	1045	bcc	7.33	−3.77	−11.10	0	−0.0168	−0.159(78)	−1.32	8.41	—	−5.6 +82/−164	—	—
Co	1390	hcp→fcc	6.0569	−0.31	−6.1	≠0	−0.021	−0.194(79)	−1.62	3.76	—	−2.5 −84	—	—
Gd	289	hcp	8.4	+1.1	−7.3(±2)	≠0	−0.037	−0.342(80)	−2.87	2.5	—	−26	−28.7	—

interstitials. (b) The observed enhancements are all greater than one, which indicates a buildup of negative charge. (c) The size of enhancement calculated varies by the same amount as the effect, depending on the assumptions. (d) None of the theories claim sufficient accuracy to indicate any significant discrepancy, but more extensive and careful calculations are being pursued. (e) The experimental results for the normalized hyperfine enhancement as a function of temperature are shown in Figure 34 (76). Possible interpretations of these data include effects due to increased vibrational motion of the μ^+, volume expansion, and formation of a localized moment (i.e. an induced polarization of electrons near the Fermi surface). Interest in these measurements and calculations is based on the premise that a thorough understanding of these phenomena will increase our understanding of the magnetic properties on a microscopic level.

3.3.2 μ^+ KNIGHT SHIFTS   There is a Fermi contact interaction of the μ^+ with the conduction electrons in any metal. The Hamiltonian is

$$\mathcal{H}_c = \tfrac{8}{3}\pi\gamma_\mu\gamma_e \hbar^2\, \mathbf{S}_\mu \cdot \sum_i \mathbf{S}_e^i\, \delta(\mathbf{r}_i).\qquad 20.$$

The subscripts μ and e refer to the muon and electron and the sum is taken over all the conduction electrons; $\delta(r)$ is the Dirac δ-function. This term produces an effective field $\Delta\mathbf{B}$, which is the familiar Knight shift (76a):

$$\Delta\mathbf{B} = \frac{8\pi}{3}\gamma_e\, \hbar \sum_i \mathbf{S}_e^i\, \delta(\mathbf{r}_i)$$

$$= \frac{8\pi}{3}\rho_F\chi\mathbf{B} \qquad 21.$$

Here the density $\rho_F = \langle |U_k(0)|^2 \rangle_F$ is the average over the periodic part of the conduction electron Bloch functions taken at the Fermi level, and χ is the Pauli electron-spin susceptibility.

The Knight shift constant K is

$$K = \left(\frac{\Delta\mathbf{B}}{\mathbf{B}}\right) = \left(\frac{8\pi}{3}\chi\rho_F\right). \qquad 22.$$

For a free Fermi gas one obtains for the susceptibility

$$\chi^F = \mu_B^2\, \frac{mk_F}{\pi^2\hbar^2}\, V. \qquad 23.$$

Here k_F is the Fermi momentum, and V is the sample volume. Since the μ^+ produces a change in the electron density because of its charge, there

Table 9 Knight shift enhancement data

Element	Valence state	Crystal structure	Electron density (r_s)	Susceptibility		Knight shifts		Enhancement factor E[b]
				$\chi_p^F \times 10^6$ Calculated	$\chi_p \times 10^6$ Observed	K_{μ^+} (ppm)[a]	K_p (ppm)	
Li	+	bcc	3.25	—	2.0	-9.5 ± 19	—	~ 1
Na	—	bcc	3.93	0.66	1.1	55 ± 11	—	~ 6
K	—	bcc	4.86	0.53	0.84	64 ± 11	—	~ 9.1
Cu	—	fcc	2.67	0.97	0.96	55 ± 11	—	~ 6
						58 ± 6		
						48 ± 5		
Mg	2+	hcp	2.65	0.97	—	63 ± 11	—	~ 7.8[a]
Ca	—	fcc	3.27	0.79	—	400 ± 15	—	~ 60.4[a]
Al	3+	fcc	2.07	1.25	1.8	15 ± 15	—	~ 1
Pb	4+	fcc	2.30	1.13	1.4	110 ± 13	—	~ 9.4
Pd	2+	fcc	—	—	—	-400 ± 20	-350 ± 10	—

[a] The diamagnetic shielding correction for the μ^+ in the standard, H_2O, was applied and a demagnetizing field $= -4\pi M_s (0.5 \pm 0.5)$ was assumed.
[b] The enhancement of the observed Knight shift due to the response of the metal. E is given in Equation 28.

is an enhancement

$$E(k) = \frac{|\Psi_k(0)|^2}{|\Psi_k^0(0)|^2},$$ 24.

where the super zero indicates the undisturbed electron density. Table 9, taken from Schenck's review (85), gives a survey of the metals for which Knight shifts have been measured together with the corresponding enhancement observed for the μ^+.

To calculate the enhancement due to the presence of the μ^+ impurity, one must make numerous assumptions about the response of the crystal, the shielding due to screening of the conduction electrons, which depends on their density ρ_s.

Figure 35 gives the dependence of the enhancement on the electron spacing $r_s \propto (\frac{4}{3}\pi\rho_s)^{-\frac{1}{3}}$ as calculated by several authors (86, 87, 89–92) using a nonlinear approach to include the possible bound states. The more detailed calculations use the Hohenberg-Kohn-Sham (88) or HKS density

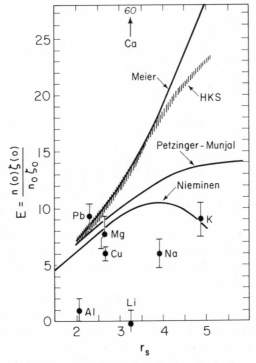

Figure 35 Comparison of theoretical predictions for the spin-density enhancement factor with experimental results from Knight shift measurements. The parameter r_s is a measure of the electron density $n : r_s = (3/4\pi n)^{\frac{1}{3}} a_B$.

functional formalism and attempt to include exchange and correlation effects. The situation can be described as confusing at this time. Clearly there is a need for improving many of the measurements to indicate whether there are real discrepancies with these screening estimates. One can also expect rapid progress in the theoretical predictions as more complete calculations are made.

3.3.2.1 *Knight shifts in ferromagnetic metals*   The behavior of the local field, $\mathbf{B}_\mu$, above saturation is shown in Figure 14. The slope can be used to find the Knight shift as follows: The magnetization within a domain with an internal field $\mathbf{B}_{int} = \mathbf{B}_{ext} - \mathbf{B}_{DM}$ is increased proportional to the total magnetic susceptibility χ_T

$$M = M(0) + \chi_t \mathbf{B}_{int}.$$ 25.

χ_t can be measured independently (93). The hyperfine field is increased by the Knight shift K,

$$\mathbf{B}_{hf} = \mathbf{B}_{hf}(0) + K\mathbf{B}_{int}.$$ 26.

Using Equation 13 and taking $\mathbf{B}_{dip} = 0$, the field at the muon will become

$$\mathbf{B}_\mu(\mathbf{B}_{ext}) = \mathbf{B}_{hf}(0) + \frac{M_d(0)\left[\frac{4}{3}\pi - N(1+K)\right]}{1 + \chi_t N} + \frac{1 + \frac{4}{3}\pi\chi_t + K}{1 + \chi_t N}\mathbf{B}_{ext}.$$ 27.

The change in field $\Delta \mathbf{B}_\mu$ with applied field $\mathbf{B}_{ext}$ will be

$$\frac{\Delta \mathbf{B}_\mu}{\mathbf{B}_{ext}} = \left(1 + \frac{4\pi}{3}\chi_t + K\right)(1 + \chi_t N)^{-1}.$$ 28.

For a spherical nickel single crystal the results obtained by Camani et al (94) give a positive shift $K = +0.0025$ (3). Preliminary results reported by Schenck (85) indicate that this Knight shift is temperature dependent going to zero at $\sim 400°$K and even becoming negative above T_c.

3.3.3 μ^+ RELAXATION IN MAGNETIC METALS   The μ^+ depolarization rate has been measured in a number of samples of iron. Figure 36 shows a compilation of the results obtained by different groups (33, 35, 95–97).

Given a local dipolar field of about ± 10 kG, these data imply a μ^+ hopping rate of $\gtrsim 10^{11}$ sec^{-1} in the purest Fe crystals at room temperature. The temperature dependence of Λ is approximately exponential in the regions $T > 100°$K and $50°$K $< T < 100°$K but depends upon the sample quality. The observation of a short-lived signal at $23°$K is taken to be strong evidence for an onset of quantum diffusion at low temperature. This result is discussed in the context of the motion of muons in pure crystals in the following section.

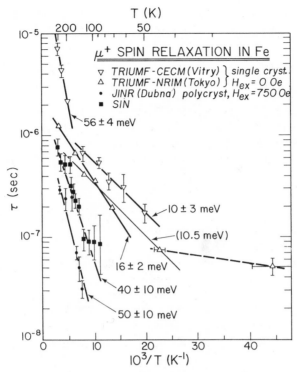

Figure 36 Positive muon spin depolarization time constant vs inverse temperature for different purity iron samples. CECM: Vitry iron ~99.995%; NRIM: Tokyo iron ~99.98%; JINR: Dubna iron ~99%.

3.4 μ^+ *Relaxation and Motion in Metals*

3.4.1 COPPER   The interpretation of the anisotropic behavior of μ^+ relaxation in a single crystal of metallic copper has provided evidence for a unique assignment of the site for the muon in that crystal lattice.

The temperature dependence of the depolarization rate of muons in a Cu single crystal was measured by Gurevich (31, 98). Copper forms a fcc structure. The lattice constant is 3.6 Å. (By chance, the orientation chosen reproduced the result obtained in the polycrystal material.) The two isotopes present in copper have the same spin and approximately the same magnetic moment and electric quadrupole moments.

The motion of the μ^+ together with the action of random magnetic field fluctuations on the μ^+ spin discussed previously, leads to a time dependence for the μ^+ asymmetry

$$A(t) = \exp\left[2\sigma^2\tau^2(e^{-t/\tau} - 1 + t/\tau)\right], \qquad 29.$$

where σ^2 is proportional to the second moment of the host nuclear dipole field $\langle M_2 \rangle$ and τ is the correlation time associated with the muon diffusion:

$$\frac{1}{\tau} = \frac{\langle v \rangle}{a} = \frac{D}{a^2} \quad \text{and} \quad \sigma^2 = \tfrac{1}{2}\gamma_\mu^2 \left[\langle M^2 \rangle = \langle \Delta B_{\text{dip}}^2 \rangle \right]. \qquad 30.$$

Figure 12 shows that at temperatures below $100°K$ the depolarization of the μ^+ appears to be consistent with that due to the fields produced by the nuclear magnetic moments of the host copper nuclei.

The line broadening is proportional to the square of the dipolar field:

$$\Delta B_{\text{dip}}^2 = \tfrac{1}{3}\gamma_s^2\, h^2(S)(S+1) \sum_i (3\cos^2\theta_i - 1)^2\, r_i^{-6}, \qquad 31.$$

where the sum is over all the host nuclei at distances $|\mathbf{r}_i|$ from the muon with moments inclined at angles θ_i with respect to $\mathbf{r}_i$. This was seen in the case of gypsum (30), where the sum of nearest neighbors produced several fields, depending on the orientation. As the temperature is raised, the muon begins to diffuse and the motional narrowing (32) averages out the field inhomogeneity and the depolarization disappears. We discuss this diffusion in detail subsequently. Camani et al (99), recognizing that the contributions of the dipolar fields would vary with the crystal orientation when the muon was fixed in a given site, lowered the temperature and

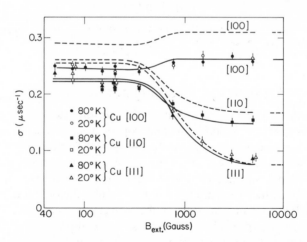

Figure 37   The field variation of the measured damping constant σ for μ^+ in a single crystal of copper oriented with the external field $\mathbf{B}_{\text{ext}}$ along different crystal axes. Assuming an interstitial octahedral site for the μ^+ the dashed curve was calculated including the quadrupole electric field gradient energy shift with no lattice distortion. The solid curve assumes a 5% dilation of the nearest neighbor separation. The pure magnetic splitting (Van Vleck values) are shown on the right.

Table 10   Damping rate ($\mu\sec^{-1}$) due to nearest neighbor Cu nuclei

Site	External field parallel to		
	(100) axis	(110) axis	(111) axis
Octahedral	0.308	0.165	0.0674
Tetrahedral	0.0771	0.279	0.319

varied both the orientation of the crystal and the applied field. Figure 37 shows the resulting variation of the relaxation rate. At high fields the calculation of the Van Vleck terms (100) for the local field can be made for each assignment of muon site. The results are given in Table 10.

As the magnetic field is lowered to zero the electric field gradient (EFG) produced by the muon interacts with the quadrupole moment of the nuclei and alters the precession of the magnetic moment. Hartmann (101), recognizing that the same phenomenon occurred in Mössbauer NMR studies (102) where EFG were produced by substitutional impurities, showed that the quadrupole effect tends to alter the field broadening at low fields in agreement with the observations. The quadrupole splitting is given by

$$\frac{\omega E}{2} = \frac{eQ}{h} \cdot \frac{V_{zz}}{4S(2S-1)} = 0.16 \pm 0.02 \text{ MHz};$$

from this one obtains for the electric field gradient, V_{zz},

$$\frac{V_{zz}}{e} = 0.27 \pm 0.15 \text{ Å} \qquad\qquad 32.$$

Quantitatively the magnitude of the depolarizing magnetic field σ^2 was found to be too large if one assumed that the host lattice was undistorted. Good agreement was obtained if the lattice was allowed to expand by $\sim 5\%$. Finally it should be noted that the extraction of the EFG from the data is limited by the uncertainty of the electric quadrupole moment. Jena et al (103) calculated the EFG from first principles, using the self-consistent density formalism. The Bloch enhancement factor computed with a band structure model agrees with the measured results.

Even with all these minor uncertainties, Table 10 shows that depolarization rates are clearly interchanged for the [111] and the [100] axes for the tetrahedral interstitial site assignment as compared with the octahedral site assignment. The experimental result leads to the unambiguous conclusion that the μ^+ is trapped in the octahedral interstitial site.

3.4.2   COMPARISON OF COPPER AND OTHER PURE METALS   The application of this technique to other metals with large nuclear spins and magnetic moments is clearly to be desired to determine the muon site, and at this time only a few have been tried. The time dependence of the depolarization can be used to study the diffusion of μ^+. Figure 38 shows a summary of data on the temperature dependence of the depolarization rate (98, 104). One can easily show that the expression for the motional narrowing (Equation 29) simplifies for the case where (t/τ) is large or small.

For short correlation times (fast hopping), $A(t) = \exp(-2\sigma^2\tau t)$, one has exponential decay with the damping rate $\Lambda = 2\sigma^2\tau$. For long correlation times, $A(t) = \exp(-\sigma^2t^2)$, one has a Gaussian time dependence. In the actual situation the fits to either Gaussian or exponential are usually not sufficiently sensitive to differentiate the two different forms. If one takes the characteristic time for $1/e$ reduction, then one has either $t_{\text{relax}} = \Lambda^{-1} = (2\sigma^2\tau)^{-1}$ or σ^{-1} in the two extremes.

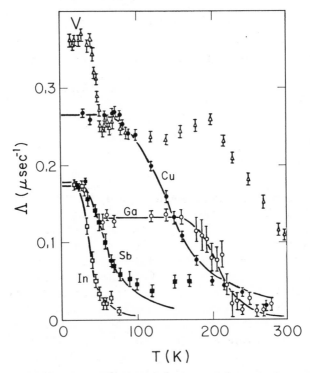

Figure 38   The temperature variation of the μ^+ depolarization rate for vanadium, copper, gallium, antimony, and indium. $\Lambda(T)$ was obtained using a Gaussian fit for the asymmetry time variation $A(t) = \exp(-\Lambda^2t^2)$.

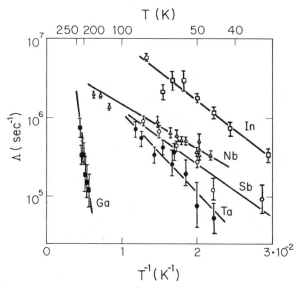

Figure 39   A summary of the Arrhenius plots for the hopping rate, inversely proportional to the correlation time τ vs reciprocal temperature.

If we characterize the data with classical Arrhenius description for the transition rate, we find $W = \tau^{-1} = Da^{-2} = v_0 \exp(-QkT)$, where τ is the time between collisions, D is the diffusion coefficient, a is the lattice constant, v_0 is the "preexponential" attempt frequency, and Q is the height of the barrier or activation energy.

The frequency v for hydrogen in copper is known (105) to be $\sim 10^{14}$ sec^{-1} at $T \sim 1000°C$. Figure 39 (104) shows the temperature dependence of Λ as an Arrhenius plot. Table 11 lists the extracted parameters. For

Table 11   The preexponential attempt frequency v_0 and the activation energy Q obtained from the muon relaxation rates

Element	$\log_{10} v$	Q/k (°K)	Q (10^{-3} ev)
Cu	7.61 ± 0.04	560	48.2
Sb	6.97 ± 0.14	173	14.9
Be	9.5 ± 0.4	1200	103.4
In	6.7 ± 0.2	155	13.3
Ta	6.9 ± 0.2	200	17.2
Ga	10.1 ± 0.8	2130	183.5
Nb	6.5	100	8.61
Bi	11.2	1400	120.6

μ^+ the preexponential factor in copper is $\sim 10^{-7}$ of the hydrogen value, and the variation of v_0'' between different metals is larger than 10^5. The (highly variable) activation energies are also smaller for muons than for hydrogen.

The theoretical understanding of the diffusion of μ^+ in metals has recently become the focus of considerable activity. Schenck (106) reviewed the principal mechanisms of diffusion, which we will summarize briefly. At high temperature, the classical "hopping" behavior can be pictured as the tail of the Boltzmann μ^+ energy distribution leaking over the barrier, which corresponds in height to the saddle point in the μ^+ potential energy surface between adjacent interstitial sites.

As the temperature is lowered, the classical hopping will cease, but quantum tunneling is expected to take over. For a naive model of uncomplicated tunneling through a single barrier, one would expect the "hop" rate to be given directly by a barrier penetration factor

$$v \to v_0 \exp\left(-\frac{2l}{\hbar}\sqrt{2m\,U}\right),$$

where v_0 is taken as the zero point frequency, typically $\sim 10^{12}$ sec^{-1}. Using this value of v_0 and assuming a tunneling barrier-height $U \sim 0.5$ ev and a width $l \sim 1$ Å, one obtains $v \sim 10^6$ sec^{-1}. Even in this simple picture, complications enter as soon as one asks whether the μ^+ tunnels from ground state to ground state of its approximate rigid-lattice harmonic potential or rather between excited states. In the latter case, the μ^+ must first be thermally excited to the state that acts as a "tunneling channel" (Orbach-type process), and an Arrhenius behavior with classically unreasonable parameters can easily result (106). No doubt the results on μ^+ diffusion in Cu can be consistently treated with this picture. The assumption of "weak coupling" between μ^+ and phonons necessary for such a model is probably too naive, but the notion of a quantum tunneling process which requires assistance from lattice vibrations is widespread in current theory.

It is also naive to picture all tunneling processes as penetrations of single barriers, since the lattice consists of a periodic array of approximately identical barriers and wells; if they were perfectly identical, the tunneling process would eventually leave the μ^+ wave function spread out over many lattice sites, giving a true "muon band" in the "coherent quantum tunneling" picture of Kagan & Klinger (109). However, dislocations, impurities, single-phonon scattering, and multiphonon processes (e.g. "self-trapping") will all disrupt the coherent tunneling ("broaden the band") and slow down the diffusion. The μ^+ then relies again on "phonon-assisted incoherent tunneling" for its motion. We now examine some of the specific predictions of such models.

If the temperature is large compared to the Debye temperature, Flynn & Stoneham (107) have shown that incoherent phonon-assisted tunneling exhibits the classical Arrhenius temperature dependence if one includes in the exponential the energy E_a associated with the distortion of the lattice, due to the presence of the μ^+ (proton) impurity, and the energy E_s due to the phonon excitations involved in the transition from site to site.

As the temperature is lowered there is a region where the pre-exponential factor has a slow temperature dependence $T^{-\frac{1}{2}}$:

$$W = \left(\frac{\pi}{4\hbar^2 E_a kT}\right)^{\frac{1}{2}} |J|^2 \exp(-E_a/kT).$$

Here J is the muon transfer matrix element.

At still lower temperature this incoherent process decreases as T^7

$$W^P = 5.76 \times 10^4 \pi \omega_D (\hbar \omega_D)^{-4} |J|^2 E_a^2 \left(\frac{T}{\theta_D}\right)^7 \exp(-5E_{a/\hbar\omega_D}),$$

where $\hbar\omega_D = k\theta_D$, and the typical energy characteristic of the phonon θ_D is the Debye temperature.

These calculations are in many ways similar to the polaron description for an electron and its strain field moving in the crystal lattice. Teichler (108) has applied an analysis of this type to the μ^+ quantum diffusion in copper. The analysis assumes phonon-assisted μ^+ tunneling between

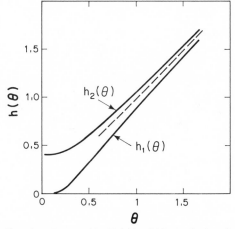

Figure 40 The auxillary functions used by Teichler (108) to calculate the quantum diffusion of muons in fcc metals when σ becomes small. For large σ, the classical (over the barrier hopping) result is the limit when $h_1 = h_2 = \theta = T/T_0 \approx 2T/\theta_D$. θ_D is the Debye temperature.

adjacent octahedral sites in the fcc lattice with suitable forces introduced to describe the lattice interaction and the muon-lattice interaction.

The universal functions $h_1(T/T_0)$ and $h_2(T/T_0)$ shown in Figure 40 are introduced to include the phonon excitations

$$h_1(T/T_0) = \frac{1}{8E_a k T_0} \frac{1}{N} \sum_q |\gamma_q|^2 (\hbar\omega_q)^2 \operatorname{csch}\left(\frac{\hbar\omega_q}{2kT_0}\frac{T_0}{T}\right),$$

and

$$\frac{1}{h_2(T/T_0)} = \frac{kT_0}{E_a} \frac{1}{N} \sum_q |\gamma_q|^2 \tanh\left(\frac{\hbar\omega_q}{4kT_0}\frac{T_0}{T}\right).$$

Here $kT_0 \simeq \frac{1}{2}k\theta_D$, and γ_q is the μ^+-phonon coupling energy. The resultant diffusion constant depends on temperature as

$$D = \frac{a^2}{\hbar}|J^2|\sqrt{\pi/4E_a k T_0}\,[h_1(T/T_0)]^{-\frac{1}{2}}\exp\left[-\frac{E_a}{kT_0 h_2(T/T_0)}\right].$$

In the calculations Teichler has included 768 points in the Brillouin zone for his q-space summation.

The results of this calculation are shown in Figure 41, together with the data of Gurevich (98). The parameters of the theory for this fit are $E_a = 75.2 \times 10^{-3}$ eV and $J = 18.4 \times 10^{-6}$ eV. The low temperature region is especially interesting as it has been predicted that the coherent quantum tunneling may be observable. Kagan & Klinger (109) investigated this problem and showed the coherent diffusion constant for protons to be

$$D_c = \frac{Za^2 J^2}{3\mathbf{B}\hbar^2\omega_D}\left(\frac{\hbar\omega_D}{T}\right)^9 e^{-2\phi s(T)}.$$

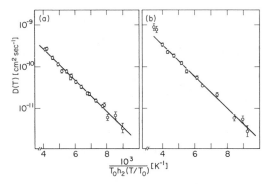

Figure 41 Teichler's calculation (108) for the diffusion of muons in copper by phonon-assisted incoherent quantum tunneling compared with the data (*a*) polycrystal and (*b*) monocrystal from Grebinnik (98).

The essential difference between the coherent and noncoherent diffusion is that for coherent diffusion the occupation numbers of the phonon states do not change with the transition. The rapid coherent process is analogous to the tunneling transition for the ammonia molecule. It will occur when the energy levels in adjacent wells coincide, i.e. the coherent process will be destroyed if (a) the energy levels are shifted by impurities or other

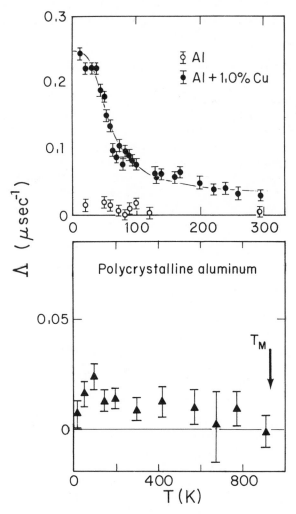

Figure 42 Temperature dependence of the depolarization for polycrystalline aluminum and aluminum alloyed with copper. The muon in pure aluminum diffuses rapidly at the lowest temperature measured. In the alloy, the muon presumably depolarizes as it is trapped by the impurity.

defects, destroying the symmetry of the crystal site; or if (*b*) the phonons are so numerous as to broaden the band until the density of states at exact resonance required by energy conservation is suppressed. In the context of (*a*) it is noteworthy that very fast diffusion has so far been observed only in metals (Al, Au, Fe, Bi,...) with monoisotopic nuclei. Self-trapping, it is asserted (109), does not enter the calculations, since the states on each side of the barrier have the same E_a. This argument would seem to depend upon the responsiveness of the lattice—i.e. the frequency spectrum of the virtual phonons responsible for the self-trapping.

The prediction of a T^{-9} dependent coherent tunneling at low temperatures with near perfect crystals is clearly an exciting prospect, and although there are possible anomalies in the hydrogen diffusion, to our knowledge no evidence has appeared in support of this prediction from hydrogen diffusion measurements. However, this phenomenon may be out of reach experimentally for muons as well.

As is clear from the above discussion, the application of the μ^+, which has a longer wavelength than the proton, to the study of quantum diffusion is an obvious challenge to μSR experimentalists; at this writing much effort is being concentrated along these lines.

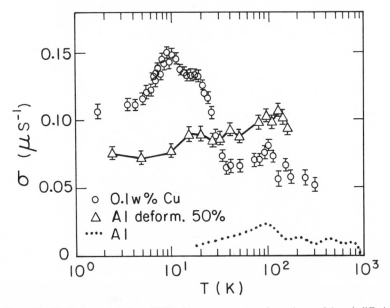

Figure 43   Preliminary data from SREL shows temperature dependence of the μ^+ diffusion in aluminum, deformed aluminum, and alloyed with copper.

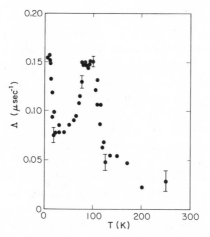

Figure 44 The depolarization rate for muons in polycrystalline bismuth has a minimum near 50°K, which implies an increased diffusion rate. The authors suggest that these data indicate a coherent quantum diffusion process. See Reference 110.

3.4.3 IMPURITY EFECTS In the studies of μ^+ diffusion in metals, aluminum seems to give anomalous results. In the Flynn-Stoneham model (107), the activation energy is dominated by the self-trapping energy if there is no neighbor host to block the transition. For example, octahedral sites

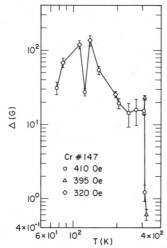

Figure 45 The muon depolarization line width from fits to an exponentially damped sine wave are plotted against temperature for polycrystalline chromium. Between the spin flip temperature 122°K and the Neel temperature 308°K, the antiferromagnetic crystal has a high local field at most interstitial sites. To fit the data in this region, a hopping frequency $\sim 0.5 \times 10^{11}$ sec^{-1} is required (110).

in bcc lattices have small hindrances compared with the octahedral site in fcc crystals. The tunneling between octahedral site in copper (fcc) will therefore (in this model) involve the lattice phonon energy E_s.

Another fcc crystal is aluminum; the magnetic moment of 3.64 μ_B and spin $= \frac{5}{2}$ assure that the magnetic interaction is ~ 3 times that of copper. The Debye temperature of aluminum is higher than copper (428°K vs 343°K). However, the data of Figures 42 and 43 show that Λ is less than 0.02 μsec^{-1} in pure Al at temperatures as low as 20°K (110–112). Apparently the muon continues to diffuse rapidly. Very slight amounts of impurities have a profound effect on this diffusion. With the addition of only 0.1% copper, the μ^+ becomes depolarized with a $Q \sim 100$°K (8 meV). Figure 44 shows the first clear evidence for trapping of μ^+ at dislocations in an otherwise pure sample of Al (111).

There are a number of other irregularities, which are illustrated in Figures 44–47. The data indicate that there are several cases where structure in the relaxation rate is observed. The interpretation of these data is uncertain at this writing. However, it is instructive to look at the different postulated mechanisms to understand what may be happening.

The high temperature line narrowing is presumably phonon activated diffusion. As the temperature is lowered, the μ^+ becomes trapped in a site connected with either defects, impurities, or vacancies; or it becomes self-trapped because of the distortion of the lattice connected with the interaction of the μ^+ with the core electrons of the host.

As the temperature is raised the last effective trapping site will be the

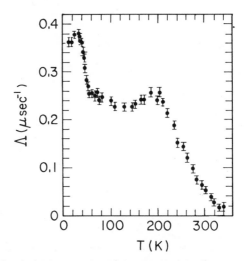

Figure 46 The depolarization rate for μ^+ in vanadium is plotted against temperature. The plateau at 200°K has a lower effective depolarizing field than at the muon site at ~ 40°K. Either a different type of site or higher μ^+ mobility can explain this behavior (113).

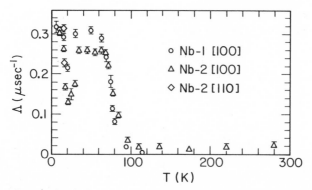

Figure 47 The μ^+ depolarization in a single crystal of niobium is plotted against temperature for different orientations relative to the external field of 102 Oe. The sharp dip at 20°K indicates a rapid increase in diffusion rate in a limited temperature region (114).

lowest or deepest trap. If the coherent diffusion becomes important the μ^+ may begin to tunnel as the temperature is lowered, and this may be responsible for the dips in the relaxation. For the case of bismuth it is speculated (110) that as the rate of coherent diffusion increases the μ^+ is captured on defects, and the μ^+ remains caught as soon as the diffusion rate is high enough for the μ^+ to find the defect within the time scale of the experiment. Note that either σ^2 or τ_c (cf Equation 29 and following) may independently undergo temperature variation. If the muon wave function is spread over a small number of sites, there will be a reduction of the average field.

Figure 47 shows the diffusion rate vs temperature in niobium. Birnbaum et al (116) rotated the crystal axes to show that the tetrahedral site is preferred. This tentative assignment is less definite than the octahedral assignment for copper. The magnitude of the σ^2 damping at low temperatures is lower than predicted by the dipolar sum (117). If quantum tunneling occurs between four interstitial sites (115), there is approximate agreement with the line broadening—only a 3% expansion of the lattice is necessary rather than the 15% expansion required if the μ^+ is trapped at a single site. The correlation time can be extracted assuming that σ^2 is fixed with the low temperature points. Figure 48 shows the diffusion rate, $1/\tau_c$ vs temperature.

The authors speculate that at low temperature the μ^+ is self-trapped in a cyclic tunneling configuration between four sites, the T_4 state, and as the temperature is elevated, jumping between these rings increases until the diffusion is sufficient to find impurity states. In corroboration of this description the addition of N_2 in small quantities (~ 100 ppm) was seen to gradually eliminate the notch at 20°K (118, 119). This is seen in Figure 49. These phenomena may have been clarified since this writing, and perhaps other mechanisms may be involved to explain these interesting results. There is a hint that the μ^+ may diffuse coherently either within a ring or in a translational mode at low temperatures in the highest purity crystal. The confirmation and elucidation of these hints will be a challenging task.

3.5   Knight Shifts

3.5.1 μ^+ IN ANTIMONY   Recently the study of μ^+ frequency shifts in semimetal crystals (120) has resulted in an observation of a very large Knight shift that is anisotropic. Figure 50 shows the shift in a

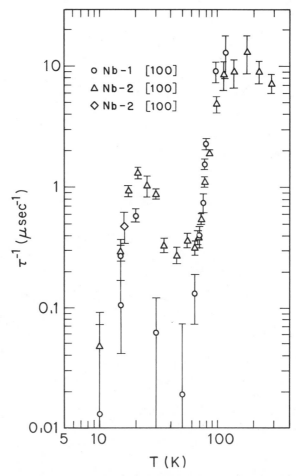

Figure 48   The correlation time for μ^+ in niobium can be extracted if one fixes the mean square field parameter $\sigma_s = 0.321$ sec^{-1} from the low temperature plateau in the previous figure (115).

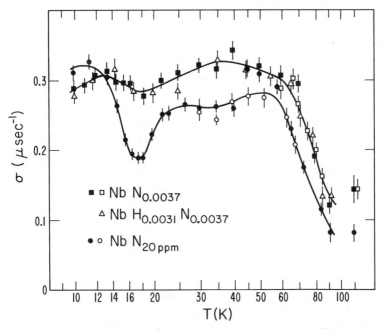

Figure 49 The relaxation rate for μ^+ in niobium with varying impurity concentrations is plotted against temperature. The presence of nitrogen of 0.37% seems to quench the muons diffusion at $\sim 18°$K. Preliminary data obtained at CERN (119).

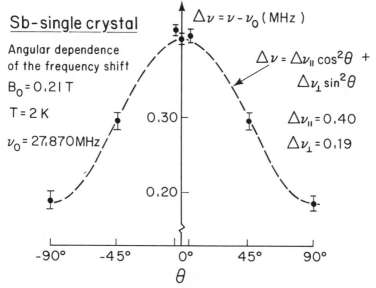

Figure 50 Dependence of the frequency shift of μ^+ in antimony single crystal on the angle between $\mathbf{B}_0$ and the $\hat{c}$-axis.

rhombohedral single crystal of antimony, which is positive amounting to 0.40 MHz and varies with angle as shown. The variation with angle fits an equation $\Delta v = \Delta v_{\parallel} \cos^2 \theta + \Delta v_{\perp} \sin^2 \theta$, with $\Delta v = 0.4$ MHz, and $\Delta v_{\perp} = 0.19$ MHz at $\mathbf{B}_{ext} = 2100$ G and $T = 2°$K. The frequency of normal μ^+ rotation is ~ 28 MHz. The shift decreases approximately linearly with increasing temperature and vanishes at $\sim 150°$K. For a polycrystalline sample a linear decrease of the average shift with applied field as the field is dropped to 400 G was reported. It has been suggested (120) that the variations of the shifts are related to the anisotropies in the Fermi surfaces with variations of the electronic g factor (i.e., $g \simeq 15$ for the $\hat{Z}$ axis).

The existence of anisotropies in antimony like silicon suggests a p-like muonic system, which arises from the symmetry of the crystal. This implies a system with long-lived collective states extending over several lattice dimensions. The semimetal differs from the semiconductor by having a smaller gap between valence and conduction bands. A quantitative inter-pretation of these results has not been presented at this time.

3.5.2 μ^+ IN MANGANESE SILICIDE   Another anomalously large Knight shift has been observed for a weakly helimagnet intermetallic crystal MnSi (121). Below 29°K the long period helical spin structure results from the weak polarization of itinerant d-electrons, which may produce both large static and fluctuating local fields at interstitial sites. Figure 51 shows the comparison with the μSR and NMR Knight shifts temperature dependence expressed parametrically in terms of the susceptibility.

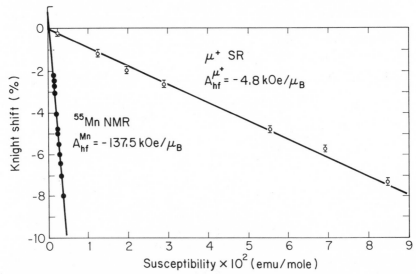

Figure 51 The Knight shift of both ^{55}Mn and μ^+ in a single crystal MnSi vs the host susceptibility with temperature as the implicit parameter.

The Knight shift is given by

$$K_\mu(T) = \frac{\mathbf{B}_{hf}(T)}{\mathbf{B}_{ext}} = \frac{\mathbf{B}_\mu - \mathbf{B}_{ext}}{\mathbf{B}_{ext}} - \left(\frac{4\pi}{3} - N_z\right)\frac{M(T)}{\mathbf{B}_{ext}} = A_{hf}^\mu \chi(T).$$

The slope A_{hf}^μ is -4.8 kG per μ_B. For the NMR at $\chi = 0$, the Knight shift is slightly positive because of the electron orbital susceptibility. For μSR the shift is close to zero and the hyperfine field is proportional to the d-spin magnetization and is larger than the other ferromagnets. Quantitative interpretation of this system is desired.

3.6 μ^+ in Magnetic Insulators

There have been two observations of μSR in antiferromagnetically aligned crystals. The crystal αFe_2O_3 (hematite/rust) is an insulator that has a corundum D_{3d}^6 structure. The Fe ions lie on trigonal axes that form the hexagonal two-dimensional array shown in Figure 52. Neutron scattering analysis shows that the magnetic moment lies in the (111) plane below the Neel temperature (945°K) and rotates at the Morin temperature (263°K) so that the moments lie close to the [111] (trigonal) direction.

Figure 53 (122) shows the local field variation with temperature in the insulator αFe_2O_3. Below the Morin temperature T_M the precession frequency is shown in Figure 54, with and without external field. The maximum asymmetry is observed when the trigonal axis of the crystal is perpendicular to the initial muon polarization, and the asymmetry vanishes with this axis parallel to the initial spin. This indicates that the local field must be parallel to this axis. Its magnitude is ± 15.7 kG. If one applies a field of 1.9 kG the line splits. Above the Morin transition $\mathbf{B}_\mu$ drops

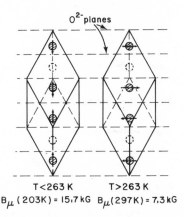

O^{2-} planes

T < 263 K   T > 263 K
B_μ (203K) = 15.7 kG B_μ(297K) = 7.3 kG

Figure 52   Spin structure of αFe_2O_3 below and above the Morin temperature. The likely muon stopping sites are indicated by the dashed circles.

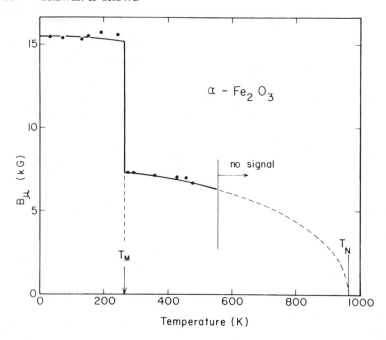

Figure 53   The measured local field **B** in single crystal αFe$_2$O$_3$ as a function of temperature. At the Morin temperature T_M, **B**$_\mu$ changes by a factor of two. Above 500°K, diffusion causes the local field to average to zero.

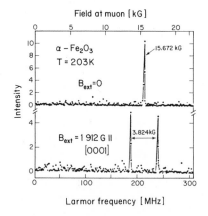

Figure 54   Below the Morin temperature the μ^+ precesses in the local fields of the antiferromagnet. At the top, frequency spectra of the muon precession are shown in zero external field. At the bottom 1.912 kG applied along the trigonal axis causes a symmetrical splitting depending on the μ^+ site.

to ~7 kG and a lower asymmetry is present at all angles. Above 430°K the signal begins to be depolarized and disappears at ~500°K. No muonium is observed in this insulator.

The interpretation of these results in terms of the muon location is as follows: Below 430°K the μ^+ is localized at distinct interstitial sites. The μ^+ diffusion above 430°K relaxes the asymmetry, and above 500°K the μ^+ asymmetry disappears. The averaging of the two sublattice fields destroys the symmetry. Below the Morin transition there is a strong indication that the muon is localized on or near the trigonal axis (Figure 53). The local field, which has dipolar symmetry, changes by a factor of two as the spins rotate by 90°. The preliminary model calculations of $\mathbf{B}_\mu$ require rapid diffusion in the cyclic orbit shown in Figure 52 to average out the perpendicular field components. The longitudinal field estimated from the dipole sum is ~23 kG. Refined calculations are in progress at the time of this review.

3.7 μ^+SR in Antiferromagnets

μSR has been seen in annealed polycrystalline dysprosium (123). Dysprosium is ferromagnetic below 85°K with a saturation magnetization

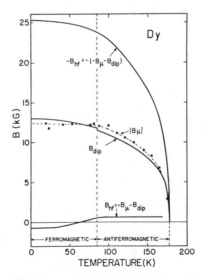

Figure 55   The measured field $\mathbf{B}_\mu$ as a function of temperature (*dots with dashed line*) for polycrystalline dysprosium. No discontinuity of $\mathbf{B}_\mu$ was observed at the Curie temperature. $\mathbf{B}_{dip}(T)$ is the calculated dipolar field for the octahedral interstitial site. (The curve for the tetrahedral site differs by less than 100 G from the one given.) From $\mathbf{B}_\mu$ and $\mathbf{B}_{dip}$ the hyperfine field $\mathbf{B}_{hf}$ is evaluated for the two possible signs of $\mathbf{B}_\mu$. Since $\mathbf{B}_{hf}$ shows no abrupt change at T_C, there is an indication that $\mathbf{B}_\mu$ is positive and $\mathbf{B}_{hf}$ is small.

of 10.2 μ_B per atom. Between 85°K and 179°K, it has a helical anti-ferromagnetic ordering with the helix axis parallel to the hexagonal axis $\hat{c}$. In the antiferromagnetic case, the 4f moments lie in the hexagonal basal plane. The crystal structure is hcp. The moment is rotated from plane to plane by an angle θ, which depends on the temperature.

In Figure 55 the local field is followed through the ordering phase transition and above. No discontinuity in $\mathbf{B}_\mu$ is observed. The amplitude of precession is small, ~ 0.03. The relaxation time, $T_2 \simeq 150$ nsec is approximately constant in the range investigated. $\mathbf{B}_{dip}$ is calculated using the known magnetization density from other data. In the interpretation of this data the authors conclude: (a) The μ^+ is probably located in the octahedral site. (b) The $\mathbf{B}_\mu$ is probably parallel to the magnetization M_{local} and the hyperfine field is small. The antiparallel choice implies a large $\mathbf{B}_{hf}$ that would change rapidly at T_c when the interlayer spin turn angle is known to shift abruptly $\sim 26°$. And (c) Single crystal studies would presumably resolve these ambiguities.

4 SUMMARY

The most significant recent progress in μSR has been conceptual. In each field of application, new data and new interpretations have combined to clarify our understanding of the behavior of the muon in matter, leaving a better picture of its potentials as a probe. In gases, it is clear that Mu behaves just as expected of a light isotope of the H atom, both in its formation and in its subsequent reactions. However, its mass is light enough to make quantum tunneling through reaction barriers an important process. Such an unprecedented mass difference in true chemical isotopes provides a testing ground for elementary rate theories in physical chemistry. Chemical reactions of Mu in liquids, after years of mainly qualitative study via the residual polarization technique, have now also been opened up for precise quantitative investigation by the successful observation of Mu precession in liquids. So far the results support the same characterization of Mu in solution as in the gas phase: a light isotope of H.

Our understanding of the motion of the μ^+ in metals has improved rapidly in the last three years, thanks to numerous experiments on μ^+ spin relaxation and extensive theoretical work relating the temperature dependence of the relaxation to diffusional motion of the μ^+. The familiar "motional narrowing" picture has now been refined to include detailed models of quantum tunneling, self- and defect-trapping, and other phenomena of general importance in the motion of light interstitials. One lesson that emerges from these efforts is that the μ^+-lattice interaction, often an annoyance when one hopes to probe the intrinsic properties of the crystal, is actually a rich source of information about the coupling of impurities to their environment. This has been evident in the theoretical work on hyperfine fields at the μ^+ in magnetic metals, where attempts to understand the experimental data in terms of spin-dependent screening, etc, have led to fresh original work in the area of interstitial magnetism.

Certain experimental milestones can be expected to mark new avenues of μSR research. One example is the successful observation of muonic radicals and their hyperfine structure via Paschen-Bach (high transverse field) Fourier spectroscopy, which should permit extensive comparisons between muonic radicals and the analogous hydrogenic radicals. Another is the empirical characterization of the anisotropic spin Hamiltonian for anomalous muonium in silicon, which must still be combined with a physical model of that entity, but should then make a real contribution to our understanding of the electronic structure of impurity states in semiconductors. A third discovery whose potentials are still largely untapped is the diffusion of Mu out of fine grains of powdered insulators and into vacuum, which may make possible a study of the physiochemistry of surfaces with muons. In solids, the observation of large anisotropic Knight shifts in semimetals, large local fields in antiferromagnetic insulators, magnetic phase transitions in magnetic media and spin glasses, all lead to potential programs of investigation with unpredictable results.

We would like to conclude by reiterating what we consider to be the most urgent unanswered questions in μSR. In both gas and liquid phases, the fraction of muons that thermalize in diamagnetic states is still mysterious. In Ne and He gases, does the μ^+ precession signal come from actual "bare" muons or from molecular ions $Ne\mu^+$ and $He\mu^+$? If the slow relaxation of the μ^+ signal with added Xe is really due to thermal electron capture, temperature-dependent studies may expose the identity of this component. In liquids, does the μ^+ reach a diamagnetic state at early times primarily via hot atom reactions or through short-lived processes, in the radiation "spur" caused by its ionization? A major controversy has arisen over this issue; it should be resolved, at least in part, by studies of hot atom reactions in the gas phase, where there is no "spur." In semiconductors the physical identity of the "anomalous muonium" state must be determined. Straining the crystals may help. Will such states be observed in other crystals besides Si? These questions may be difficult to answer, but the potential rewards make the effort worthwhile. In insulators as well as semiconductors, Mu forms in some crystals but not in others. Why is this? So far there is not even a consistent empirical rule governing this differentiation. A broader survey may be helpful. In all crystals, but especially in metals, the relative importance of quantum tunneling, polaron-type effects, and impurity trapping must be better understood before μ^+SR can have its full impact on studies of the motion of light interstitials. Various experiments determining the μ^+ site in different crystal lattices have been performed, but more are needed. When the zero-point motion of the μ^+ and the distortion of the lattice by its presence are properly combined with a correct description of spin-dependent screening, the contribution of μ^+SR hyperfine studies to magnetism theory will be more evident. These theoretical challenges have been accepted by the solid state physics community, but much remains to be explained.

ACKNOWLEDGMENTS   A complete list of the people who have actively contributed to this review would be longer than the reference list. We offer our greatest appreciation to the many members of the μSR community who have lent us their ideas, data, and friendly criticisms for this work. However, certain of these people have made such important contributions that they must at least be mentioned by name.

Don Fleming (Chemistry Department, University of British Columbia) and Ken Nagamine (Physics Department, University of Tokyo) were so intimately involved with the initial stages of this review that they must be considered virtual co-authors, and would have been formal co-authors had circumstances not removed them from North America during the final writing. Their respective contributions to the discussions of muonium chemistry and μ^+ in metals cannot be over emphasized.

Numerous experimenters have allowed us to "steal" their unpublished data; the current rapid growth of μSR makes this essential for a review that will not be hopelessly outdated before it can be published. We offer special thanks to Dave Garner, Jerry Jean, Randy Mikula and David Walker of UBC (Chemistry); to Glen Marshall, Rob Kiefl, and John Warren of UBC (Physics); to Nobu Nishida, Ryu Hayano, Toshi Yamazaki, and other members of the University of Tokyo group at TRIUMF; to Paul Percival, Emil Roduner, and Hanns Fischer of the Institute of Physical Chemistry, University of Zurich; to Bruce Patterson (now at TRIUMF) and W. Kundig of the Physical Institute, University of Zurich; to Alex Schenck and Fred Gygax of ETH; to Tony Fiory of Bell Labs; to Jack Kossler of the College of William and Mary; and to Bob Heffner, Will Gauster, and Chao-Yuan Huang of LAMPF.

We are particularly indebted to the many theorists who have patiently helped us understand the problems treated herein. Special thanks to Birger Bergerson of UBC (Physics), Tom McMullen of Queens University,

Alan Portis and Leo Falicov of Berkeley, Peter Meier of Zurich, Puru Jena of Argonne National Laboratory, and Yu. V. Kagan and V. G. Smilga of Moscow.

Finally, we would like to thank Pat Schoenfield of LBL and the editorial staff of ARNPS for patience and cooperation in the assembly of this text.

Literature Cited

1. Garwin, R. L., Lederman, L. M., Weinrich, M. 1957. *Phys. Rev.* 105: 1415-17
2. Friedman, J. I., Telegdi, V. L. 1957. *Phys. Rev.* 105: 1681-82
3. Yamazaki, T., Nagamiya, S., Hashimoto, O., Nagamine, K., Nakai, K., Sugimoto, K., Crowe, K. M. 1974. *Phys. Lett B* 53: 117
4. Yamazaki, T. 1977. *Phys. Scr.* 11: 133; *Phys. Scr.* 86-88B: 1053
5. Nagamiya, S., Nagamine, K., Hashimoto, O., Yamazaki, T. 1975. *Phys. Rev. Lett.* 35: 308
6. Deleted in proof
7. Firsov, V. G., Byakov, V. M. 1965. *Sov. Phys. JETP* 20: 719
8. Crowe, K. M., Hague, J. F., Rothberg, J. E., Schenck, A., Williams, D. L., Williams, R. W., Young, K. K. 1972. *Phys. Rev. D* 5: 2145
9. Combley, F., Picasso, E. 1976. *Int. Sch. Phys. Meteorol. Fundam. Constants, Varenna, Italy, July 1976*; Bailey, J., Borer, K., Combley, F., Drumm, H., Farley, F., Field, J., Flegel, W., Hattersley, P., Krienen, F., Lange, F., Picasso, E., Ruden, W. von. 1977. *Phys. Lett. B* 68: 191
10. Casperson, D. E., Crane, T. W., Denison, A. B., Egan, P. O., Hughes, V. W., Mariam, F. G., Orth, H., Reist, H. W., Souder, P. A., Stambaugh, R. D., Thompson, P. A., zu Putlitz, G. 1977. *Phys. Rev. Lett.* 38: 956; Hughes, V. W. 1966. *Ann. Rev. Nucl. Sci.* 16: 445
11. Brewer, J. H., Crowe, K. M., Gygax, F. N., Schenck, A. 1975. *Muon Physics*, Vol. III, ed. V. W. Hughes, C. S. Wu, pp. 3-139. New York: Academic
12. Swanson, R. A. 1958. *Phys. Rev.* 112: 580
13. Hughes, V. W. 1975. *High Energy Phys. Nucl. Struct., AIP Conf. Proc.* 26: 515
14. Gurevich, I. I., Ivanter, I. G., Makariyna, L. A., Mel'eshko, E. A., Nikol'skii, B. A., Roganov, V. S., Selivanov, V. I., Smilga, V. P., Sokolov, B. V., Shestakov, V. D., Yakovleva, I. V. 1969. *Phys. Lett. B* 29: 387
15. Mobley, R. M., Bailey, J. M., Cleland,

W. E., Hughes, V. W., Rothberg, J. E. 1966. *J. Chem. Phys.* 44: 4354; Mobley, R. M. 1967. PhD thesis, Yale Univ.
16. Ruderman, M. A. 1966. *Phys. Rev. Lett.* 17: 794
17. Brewer, J. H., Gygax, F. N., Fleming, D. G. 1973. *Phys. Rev. A* 8: 77
18. Brewer, J. H., Crowe, K. M., Gygax, F. N. Johnson, R. F., Fleming, D. G., Schenck, A. 1974. *Phys. Rev. A* 9: 495
18a. Percival, P. W., Fischer, H. 1976. *Chem. Phys.* 16: 89
18b. Minaichev, E. V., Myasishcheva, G. G., Obukhov, Yu. V., Roganov, V. S., Savel'ev, G. I., Firsov, V. G. 1970. *Sov. Phys. JETP* 30: 230
19. Eisenstein, B., Prepost, R., Sachs, A. M. 1966. *Phys. Rev.* 142: 217
20. Hutchinson, D. P., Menes, J., Shapiro, G., Patlach, A. M. 1963. *Phys. Rev.* 131: 1351, 1362
21. Schenck, A., Crowe, K. M. 1971. *Phys. Rev. Lett.* 26: 57
22. Babaev, A. I., Balats, M. Ya., Myasishcheva, G. G., Obukhov, Yu. V., Roganov, V. S., Firsov, V. G. 1966. *Sov. Phys. JETP* 23: 583
23. Nosov, V. G., Yakovleva, I. V. 1965. *Nucl. Phys.* 68: 609
24. Ivanter, I. G., Smilga, V. P. 1968. *Sov. Phys. JETP* 27: 301
25. Ivanter, I. G., Smilga, V. P. 1969. *Sov. Phys. JETP* 28: 796
26. Ivanter, I. G., Smilga, V. P. 1971. *Sov. Phys. JETP* 33: 1070
27. Brewer, J. H., Crowe, K. M., Johnson, R. F., Schenck, A., Williams, R. W. 1971. *Phys. Rev. Lett.* 27: 297
28. Schenck, A. 1970. *Phys. Lett. A* 32: 19
29. Schenck, A., Williams, D. L., Brewer, J. H., Crowe, K. M., Johnson, R. F. 1972. *Chem. Phys. Lett.* 12: 544
30. Pake, G. E. 1948. *J. Chem. Phys.* 16: 327
31. Gurevich, I. I., Meleshko, E. A., Muratova, I. A., Nikol'skii, B. A., Roganov, V. S., Selivanov, V. I., Sokolov, B. V. 1972. *Phys. Lett. A* 40: 143
32. Abragam, A. 1961. *The Principles of Nuclear Magnetism* Oxford: Clarendon. 439 pp.
33. Foy, M. L. G., Heiman, N., Kossler,

W. J., Stronach, C. E. 1973. *Phys. Rev. Lett.* 30:1064
34. Patterson, B. D., Crowe, K. M., Gygax, F. N., Johnson, R. F., Portis, A. M., Brewer, J. H. 1974. *Phys. Lett. A* 46: 453
35. Gurevich, I. I., Klimov, A. N., Maiorov, V. N., Meleshko, E. A., Nikol'skii, B. A., Roganov, V. S., Selivanov, V. I., Suetin, V. A. 1973. *JETP Lett.* 18:332; Gurevich, I. I., Klimov, A. I., Maiorov, V. N., Meleshko, E. A., Muratova, I. A., Nikol'skii, B. A., Roganov, V. S., Selivanov, V. I., Suetin, V. A. 1974. *Zh. Eksp. Teor. Fiz.* 66:374; *Sov. Phys. JETP* 39:178
36. Patterson, B. D., Falicov, L. M., 1974. *Solid State Commun.* 15:1509
37. Deleted in proof
38. Deleted in proof
39. Favart, D., Brouillard, F., Grenacs, L., Igo Kemenes, P., Lipnik, P., Macq, P. C. 1970. *Phys. Rev. Lett.* 25:1348
40. Myasishcheva, G. G., Obukhov, Yu. V., Roganov, V. S., Firsov, V. G. 1968. *Sov. Phys. JETP* 26:298
41. Gurevich, I. I., Ivanter, I. G., Meleshko, E. A., Nikol'skii, B. A., Roganov, V. S., Selivanov, V. I., Smilga, V. P., Sokolov, B. V., Shestakov, V. D. 1971. *Sov. Phys. JETP* 33:253
42. Brewer, J. H., Crowe, K. M., Gygax, F. N., Johnson, R. F., Patterson, B. D., Fleming, D. G., Schenck, A. 1973. *Phys. Rev. Lett.* 31:143
43. Deleted in proof.
44. Andrianov, D. G., Minaichev, E. V., Myasishcheva, G. G., Obukhov, Yu. V., Roganov, V. S., Salel'ev, G. I., Firsov, V. G., Fistul', V. I. 1970. *Sov. Phys. JETP* 31:1019
45. Wang, J. Shy-Yih, Kittel, C. 1973. *Phys. Rev. B* 7:713
46. Johnson, R. F. Unpublished PhD thesis, Univ. Calif., Nov. 1976; LBL Rep. 5526
47. Kohn, W., Luttinger, J. M. 1955. *Phys. Rev.* 98:915
48. Patterson, B. D., Hintermann, A., Kundig, W., Meier, P. F., Waldner, F., Graf, H., Recknagel, E., Weidinger, A., Wichert, Th. 1978. *Phys. Rev. Lett.* 40: 1347
49. Gurevich, I. I., Nikol'skii, B. A., Selivanov, E. I., Sokolov, B. V. 1976. *Sov. Phys. JETP* 41:401
50. Kudinov, V. I., Minaichev, E. V., Myasishcheva, G. G., Obukhov, Yu. V., Roganov, V. S., Savel'ev, G. I., Samoilov, V. M., Firsov, V. G. 1975. *Sov. Phys. JETP* 21:22; 1976. *Sov. Phys. JETP* 43:1065
51. Pifer, A. E., Bower, T., Kendall, K. R. 1976. *Nucl. Instrum. Methods* 135:39

52. Percival, P. W., Fischer, H. Camani, M., Gygax, F. N., Ruegg, W., Schenck, A., Schilling, H., Graf, H. 1976. *Chem. Phys. Lett.* 39:333
53. Fleming, D. G., Brewer, J. H., Garner, D. M., Pifer, A. E., Bowen, T., DeLise, D. A., Crowe, K. M. 1976. *J. Chem. Phys.* 64:1281
54. Stambaugh, R. D., Casperson, D. E., Crane, T. W., Hughes, V. W., Kaspar, H. F., Souder, P., Thompson, P. A., Orth, H., zu Pulitz, G., Denison, A. B. 1974. *Phys. Rev. Lett.* 33:568
54a. Green, J., Lee, J. 1964. New York: Academic (and references therein); Goldanskii, V. I., Firsov, V. G. 1971. *Ann. Rev. Phys. Chem.* 22:209
55. Jean, Y. C., Brewer, J. H., Fleming, D. G., Garner, D. M., Mikula, R. A., Vaz, L. C., Walker, D. C. 1978. Private communication; Mikula, R. A. 1978. Private communication
56. Allison, S. K. 1958. *Rev. Mod. Phys.* 30: 1137; Massey, H. S. W., Burhop, E. H. S. 1969. *Electronic and Ionic Impact Phenomena.* London: Oxford Univ. Press; Tawara, H., Russek, A. 1973. *Rev. Mod. Phys.* 45:178
57. Chupka, W. A., Russel, M. E. 1968. *J. Chem. Phys.* 49:5426
58. Bondybey, V., Pearson, P. K., Schaefer, H. F. III. 1972. *J. Chem. Phys.* 57:1123
59. Tawara, H., Russek, A. 1973. *Rev. Mod. Phys.* 45:178
60. Connor, J. N. L., Jakubetz, W., Manz, J. 1977. *Chem. Phys. Lett.* 45:265
61. Garner, D. M. 1978. Private communication; Fleming, D. G., Garner, D. M., Vaz, L. C., Walker, D. C., Brewer, J. H., Crowe, K. M. 1978. In press; In *Positronium and Muonium Chemistry*, ed. H. J. Ache. ACS
62. Garner, D. M., Fleming, D. G., Brewer, J. H., Warren, J. B., Marshall, G. M., Clark, G., Pifer, A. E., Bowen, T. 1977. *Chem. Phys. Lett.* 48:393; Garner, D. M., Fleming, D. G., Brewer, J. H. 1978. *Chem. Phys. Lett.* 55:163
63. Brandt, W., Paulin, R. 1968. *Phys. Rev. Lett.* 21:193
64. Gidley, D. W., Marko, K. A., Rich, A. 1976. *Phys. Rev. Lett.* 36:395
65. Marshall, G. M., Warren, J. B., Garner, D. M., Clark, G. S., Brewer, J. H., Fleming, D. G. 1978. *Phys. Lett.* 65A:351
66. Percival, P. W., Fischer, H., Camani, M., Gygax, F. N., Ruegg, W., Schenck, A., Schilling, H., Graf, H. 1976. *Chem. Phys. Lett.* 39:333
67. Percival, P. W. Roduner, E., Fischer, H. 1978. See Ref. 61; Percival, P. W. 1977. Private communication
67a. Roduner, E., Percival, P. W., Fleming,

D. G., Hochmann, T., Fischer, H. 1978. Private communication

68. Birss, R. R., Martin, D. J. 1975. *J. Phys. C* 8:189

69. Sucksmith, W., Thompson, J. E. 1954. *Proc. R. Soc. London Ser. A* 225:362

70. Graf, H., Kundig, W., Patterson, B. D., Reichart, W., Roggwiller, P., Camani, M., Gygax, F. N., Ruegg, W., Schenck, A., Schilling, H., Meier, P. F. 1976. *Phys. Rev. Lett.* 37:1644

71. Nishida, N., Nagamine, D., Hayano, R. S., Yamazaki, T., Fleming, D. G., Duncan, R. A., Brewer, J. H., Ahktar, A., Yasuoka, H. 1978. *Jpn. J. Phys. Soc.* In press

72. Bertaut, E. F., Delapalme, A., Pauthenet, R. 1963. *Solid State Commun.* 1:31

73. Gurevich, I. I., Klimov, A. I., Maiorov, V. N., Meleshko, E. A., Nikol'skii, B. A., Purogov, A. V., Roganov, V. S., Salivanov, V. I., Suetin, V. A. 1975. *Zh. Eksp. Teor. Fiz.* 9:1453; *Sov. Phys. JETP* 42:741

74. Graf, H., Hofmann, W., Kundig, W., Meier, P. F., Patterson, B. D., Reichart, W. 1977. *Solid State Commun.* 23:653

75. Deleted in proof

76. Schenck, A. 1977. *Int. Sch. Phys. Exotic Atoms*, Erice, Italy

76a. Slichter, Charles P. 1963. *Principles of Magnetic Resonance*. New York: Harper & Row

77. Mook, H. A., Shull, G. G. 1966. *J. Appl. Phys.* 37:1034

78. Shull, G. G., Mook, H. A. 1966. *Phys. Rev. Lett.* 16:184

79. Moon, R. M. 1964. *Phys. Rev. A* 136:195

80. Moon, R. M. et al. 1972. *Phys. Rev. B* 5:997

81. Jena, P. 1976. *Solid State Commun.* 19:45

82. Jena, P., Singwi, K. S., Nieminen, R. M. 1978. *Phys. Rev. B* 17:301

83. Meier, P. F. 1975. *Solid State Commun.* 17:987

84. Petzinger, K. G., Munjal, R. 1977. *Phys. Rev. B* 15:1560

85. Schenck, A. 1977. *Hyperfine Interactions, 4th*, Madison, NJ, pp. 282–99

86. Meier, P. F. 1975. *Helv. Phys. Acta* 48:227

87. Nieminen, R. M. 1977. Preprint NORDITA-77/16

88. Kohn, W., Sham, L. J. 1965. *Phys. Rev. A* 140:1133

89. Petzinger, K. G., Munjal, R. 1977. *Phys. Rev. B* 15:1560

90. Popovic, Z. D., Stott, M. J. 1974. *Phys. Rev. Lett.* 33:1164

91. Almbladh, C. O., et al. 1976. *Phys. Rev. B* 14:2250

92. Jena, P., Singwi, K. S. 1978. Private communication

93. Foner, S., et al. 1969. *Phys. Rev.* 181:863

94. Camani, M., Gygax, F. N., Ruegg, W., Schenck, A., Schilling, H., 1977. *Phys. Lett. A* 60:439

95. Nishida, N., Hayano, R. S., Nagamine, K., Yamazaki, T., Brewer, J. H., Garner, D. M., Fleming, D. G., Takeuchi, T., Ishikawa, Y. 1977. *Solid State Commun.* 22:235

96. Gurevich, I. I., Klimov, A. I., Maiorov, V. N., Meleshko, E. A., Nikol'skii, B. A., Pirogov, A. V., Selivanov, V. I., Suetin, A. V. 1974. *JETP Lett.* 20:254

97. Graf, H., Kundig, W., Patterson, B. D., Reichart, W., Roggwiller, P., Camani, M., Gygax, F. N., Ruegg, W., Schenck, A., Schilling, H. 1976. *Helv. Phys. Acta* 49:730

98. Grebinnik, V. G., Gurevich, I. I., Zhukov, V. A., Manych, A. P., Meleshko, E. A., Nikol'skii, B. A., Selivanov, V. I., Suetin, V. A. 1975. *Zh. Eksp. Teor. Fiz.* 68:1548 *Sov. Phys. JETP* 41:777

99. Camani, M., Gygax, F. N., Ruegg, W., Schenck, A., Schilling, H. 1977. *Phys. Rev. Lett.* 39:836

100. Van Vleck, J. H. 1948. *Phys. Rev.* 74:1168

101. Hartmann, O. 1977. *Phys. Rev. Lett.* 39:832

102. Matthias, E., Schneider, W., Steffen, R. M. 1962. *Phys. Rev.* 125:261

103. Jena, P., Das, S. G., Singwi, K. S. 1978. *Phys. Rev. Lett.* 40:264

104. Grebinnik, V. G., et al. 1977. Proc. Int. Symp. Meson Chem. Mesomolec. Processes Matter, Dubna, USSR, June 1977, p. 266

105. Sicking, G. 1972. Int. Meet. Hydrogen Metal, Julich, Germany, 1972. 2:408

106. Schenck, A. 1976. *On the Applications of Polarized Positive Muons in Solid State Physics. Nuclear and Particle Physics at Intermediate Energy*, ed. J. B. Warren. New York: Plenum

107. Flynn, C. P., Stoneham, A. M. 1970. *Phys. Rev. B* 1:3966; Stoneham, A. M. 1972. *Ber. Bunsenges. Phys. Chem.* 76:816

108. Teichler, H. 1977. *Phys. Lett. A* 64:78

109. Kagan, Yu., Klinger, M. I. 1974. *J. Phys. C Solid State Phys.* 7:2791

110. Grebinnik, V. G., Gurevich, I. I., Zhukov, V. A., Klimov, A. I., Manych, A. P., Maiorov, V. N., Meleshko, E. A., Nikol'skii, B. A., Pirogov, A. V., Ponomarev, A. I., Selivanov, V. I., Suetin, V. A. 1977. Proc. Int. Symp. Meson Chem. Mesomolec. Processes Matter, Dubna, USSR, June 1977. p. 272

111. Lankford, W. F., Kossler, W. J.,

Lindemuth, J., Stronach, C. E., Fiory, A. T., Minnich, R. P., Lynn, K. G. 1978. *Bull. Am. Phys. Soc.* 23:361

112. Gauster, W. B., Heffner, R. H., Huang, C. Y., Hutson, R. L., Leon, M., Parkin, D. M., Schillaci, M. E., Triftshauser, W., Wampler, W. R. 1977. *Solid State Commun.* 24:619

113. Kossler, W. J., Fiory, A. T., Murnick, D. E., Stronach, C. E., Lankford, W. F. 1977. *Hyperfine Interactions* 3:287

114. Nikol'skii, B. A. 1977. See Ref. 104, p. 246

115. Lankford, W. F., Birnbaum, H. K., Fiory, A. T., Minnich, R. P., Lynn, K. G., Stronach, C. E., Bieman, L. H., Kossler, W. J., Lindemuth, J. 1977. See Ref. 85, pp. 833–37

116. Birnbaum, H. K., Camani, M., Fiory, A. T., Gygax, F. N., Kossler, W. J., Ruegg, W. Schenck, A., Schilling, H. 1978. *Phys. Lett. A* 65:435

117. Birnbaum, H. K., Flynn, C. P. 1976. *Phys. Rev. Lett.* 37:25

118. Hartmann, O., Karlsson, E., Norlin, L.-O., Pernestål, K., Borghini, M., Niinikoski, T. O. 1977. *Phys. Lett. A* 61:141

119. Hartmann, O., Karlsson, E., Norlin, L. O., Pernestål, K., Borghini, M., Niinikoski, T. 1977. See Ref. 85, pp. 824–27; Borghini, M., Hartmann, O., Karlsson, E., Kehr, K. W., Niinikoski, T., Norlin, L. O., Pernestål, K., Richter, D., Soulie, J. C., Walker, E. 1978. Private communication

120. Hartmann, O., Karlsson, E., Norlin, L. O., Pernestål, K., Borghini, M., Niinikoski, T. 1977. See Ref. 85, pp. 828–31

121. Yasuoka, H., Hayano, R. S., Nishida, N., Nagamine, K., Yamazaki, T., Ishikawa, Y. 1978. Pricate communication

122. Graf, H., Hofmann, W., Kundig, W., Meier, P. F., Patterson, B. D., Reichart, W. 1977. See Ref. 85, pp. 452–56

123. Hofmann, W., Kundig, W., Meier, P. F., Patterson, B. D., Ruegg, K., Echt, O., Graf, H., Recknagel, E., Weidinger, A.,

Wichert, T. 1978. *Phys. Lett. A* 65:343

124. Albright, R. G., Dodonov, A. F., Lavrovskaya, G. K., Morosov, I. I., Tal'roze, V. L. 1969. *J. Chem. Phys.* 50:3632

125. Rabideau, S. W., Hecht, H. G., Lewis, W. B. 1972. *J. Magn. Reson.* 6:384; Levy, J. B., Copeland, B. K. W. 1968. *J. Phys. Chem.* 72:3168

126. Homann, K. H., Schweinfurth, H., Warnatz, J. 1977. *Ber. Bunsenges. Phys. Chem.* 31:724

127. Ambidge, P. F., Bradley, J. N., Whytock, D. A. 1976. *J. Chem. Soc. Faraday Trans. 1* 72:1157

128. Bemand, P. P., Clyne, M. A. A. 1977. *J. Chem. Soc. Faraday Trans. 2* 73:394; Wagner, H. Gg., Welzbacher, U., Zellner, R. 1976. *Ber. Bunsenges. Phys. Chem.* 80:902

129. Fleming, D. G., Brewer, J. H., Garner, D. M., Pifer, A. E., Bowen, T., DeLise, D. A., Crowe, K. M. 1976. *J. Chem. Phys.* 64:1281

130. Ambidge, P. F., Bradley, J. N., Whytock, D. A. 1976. *J. Chem. Soc. Faraday Trans 1* 72:2143

131. Westenberg, A. A., deHaas, N. 1968. *J. Chem. Phys.* 48:4405; Clyne, M, A. A., Stedman, D. H. 1966. *Trans. Faraday Soc.* 62:2164

132. Endo, H., Glass, G. P. 1976. *Chem. Phys. Lett.* 44:180

133. Endo, H., Glass, G. P. 1976. *J. Chem. Phys.* 80:1519

134. Jones, W. E., MacKnight, S. D., Teng, L. 1973. *Chem. Rev.* 73:407; Sullivan, J. H. 1962. *J. Chem. Phys.* 36:1925

135. Griller, D., Ingold, K. U. 1974: *J. Am. Chem. Soc.* 96:6203

136. Linder, R. E., Winters, D. L., Ling, A. C. 1976. *Can. J. Chem.* 54:1405

137. Krusic, P. J., Meakin, P., Smart, B. E. 1974. *J. Am. Chem. Soc.* 96:6211

138. Krusic, P. J., Kochi, J. K. 1971. *J. Am. Chem. Soc.* 93:846

139. Eiben, K., Fessenden, R. W. 1971. *J. Phys. Chem.* 75:1186

140. Zeldes, H., Livingston, R. 1966. *J. Phys. Chem.* 45:1946

Ann. Rev. Nucl. Part. Sci. 1978. 28:327–86

QUARKS   ×5597

O. W. Greenberg[1]

Laboratory for High Energy Astrophysics, NASA/Goddard Space Flight Center, Greenbelt, Maryland 20771; and Center for Theoretical Physics, Department of Physics and Astronomy, University of Maryland, College Park, Maryland 20742

CONTENTS

[1] On sabbatical leave from the University of Maryland. Supported in part by the National Aeronautics and Space Administration and by the National Science Foundation.

0066-4243/78/1201-0327$01.00

Dedicated to the memory of my mother, Betty (Sklower) Greenberg, April 15, 1891–May 8, 1978.

1   INTRODUCTION AND SUMMARY

Quarks, introduced by Gell-Mann (1) and Zweig (2) in 1964, now play a central role in elementary particle and high energy physics. Quarks as constituents provide a basis for hadron spectroscopy, including static properties and single-particle transition matrix elements of hadrons. Quark-partons provide a description of electromagnetic, weak, and hadronic inclusive scattering cross sections, such as e^+e^- annihilation to hadrons, electroproduction, inelastic neutrino scattering, and large momentum transfer hadronic scattering. Quark fields serve to construct the currents in current algebra and light-cone algebra, providing derivations of electromagnetic and weak properties of hadrons that are less model dependent than those of the constituent quark model. Matrix elements of quark fields lead to relativistic bound-state wave functions (or amplitudes) and equations that allow the hadrons to be studied without using the assumptions of the naive nonrelativistic constituent quark model. The local gauge invariant [Yang-Mills (3)] quantum field theory of colored quark and gluon fields, quantum chromodynamics (QCD) (4), serves, at least provisionally, as the fundamental theory of hadron dynamics, and may provide a mechanism to confine quarks and other color-carrying particles in hadrons permanently. Even without solving the problem of confinement, there is progress in the program to derive results of the quark-parton model from QCD. We review[2] quarks, focusing on applications to hadron spectroscopy, the quark-parton model, and the relations between them, as well as briefly surveying work on bound states, quark confinement, astrophysical and cosmological implications of quarks, and the status of quark searches, among other topics.

At the most naive level, quarks (q) and antiquarks ($\bar{q}$) provide a mnemonic that accounts for the spin, parity, charge conjugation, isospin, hypercharge (or strangeness), charm, SU (3) representation, electric charge, and baryon number of the observed hadrons as being composed of qqq for baryons and $q\bar{q}$ for mesons, where the quarks have the properties given in Table 1. In the naive model, quarks are finite mass, spin $\frac{1}{2}$ particles with two types of internal degrees of freedom: flavor and color. Flavor concerns quantum numbers, such as electric charge, isospin and hypercharge, and weak isospin, which are directly observed (i.e. "tasted") in

---

[2] For earlier reviews of quarks see Reference 5.

Table 1 Properties of quarks and antiquarks

Flavor	J^P	I	I_3	Y	S	Z^a	C	SU(3)$_f$	SU(4)$_f$	Q	B	SU(3)$_c$
u	$\frac{1}{2}^+$	$\frac{1}{2}$	$\frac{1}{2}$	$\frac{1}{3}$	0	$\frac{1}{4}$	0	3	4	$\frac{2}{3}$	$\frac{1}{3}$	3^b
d	$\frac{1}{2}^+$	$\frac{1}{2}$	$-\frac{1}{2}$	$\frac{1}{3}$	0	$\frac{1}{4}$	0	3	4	$-\frac{1}{3}$	$\frac{1}{3}$	3
s	$\frac{1}{2}^+$	0	0	$-\frac{2}{3}$	-1	$\frac{1}{4}$	0	3	4	$-\frac{1}{3}$	$\frac{1}{3}$	3
c	$\frac{1}{2}^+$	0	0	0	0	$-\frac{3}{4}$	1	1	4	$\frac{2}{3}$	$\frac{1}{3}$	3
ū	$\frac{1}{2}^-$	$\frac{1}{2}$	$-\frac{1}{2}$	$-\frac{1}{3}$	0	$-\frac{1}{4}$	0	3*	4*	$-\frac{2}{3}$	$-\frac{1}{3}$	3*
d̄	$\frac{1}{2}^-$	$\frac{1}{2}$	$\frac{1}{2}$	$-\frac{1}{3}$	0	$-\frac{1}{4}$	0	3*	4*	$\frac{1}{3}$	$-\frac{1}{3}$	3*
s̄	$\frac{1}{2}^-$	0	0	$\frac{2}{3}$	1	$-\frac{1}{4}$	0	3*	4*	$\frac{1}{3}$	$-\frac{1}{3}$	3*
c̄	$\frac{1}{2}^-$	0	0	0	0	$\frac{3}{4}$	-1	1	4*	$-\frac{2}{3}$	$-\frac{1}{3}$	3*

^{a} Supercharge Z is to SU(4)$_f$ as hypercharge Y is to SU(3)$_f$: it is traceless in an SU(4)$_f$ irreducible representation and distinguishes the charm quark from the others.
^{b} We choose the quark to be a 3 rather than a 3* by convention.

elementary particle interactions. There are four known flavors, f, associated with the up (u), down (d), strange (s), and charm (c) quarks; however, theory allows and experiment may demand additional flavors.[3] Isospin (flavor) symmetry, SU(2)$_I$, is associated with the approximate equivalence of the u and d quarks. The more strongly broken symmetry, SU(3)$_f$ (7), is useful when considering the u, d, and s quarks; with the c quark this symmetry is enlarged to the even more strongly broken SU(4)$_f$ (8). Color, c, was introduced to resolve a conflict with the generalized Pauli exclusion principle for the quarks in the lowest-lying baryons (9–11). In contrast to flavor it must be precisely three valued and is never observed directly in the type of quark model (with permanent color and quark confinement) (4) we are reviewing here. This three-valued color degree of freedom that quarks carry transforms as the fundamental representation 3 under SU(3)$_c$. The color confinement restricts the observed hadrons to the following color-singlet states: the totally antisymmetric state $\varepsilon_{ijk}q_iq_jq_k$ for baryons, and the state $q_i\bar{q}^i$ for mesons. Models in which color can be excited are discussed elsewhere (11–13). The word color and the names red, white, and blue often chosen for the colors have no logical connection to the concept.

The hadrons have values of spin, parity, etc that are composed additively [for I_3, Y (S), Q, and B], via appropriate vector addition [for J, I, SU(3)$_f$, SU(3)$_c$], or, roughly speaking, multiplicatively (for P and C) from the properties of their constituent quarks and antiquarks, together with their orbital angular momentum and permutation symmetry. All well-established hadronic states and their properties are accounted for by the quark mnemonic, a remarkable regularity that is a great success of the

[3] At least a fifth flavor, associated with a b quark of electric charge $-e/3$ seems likely (6).

quark model. For example, vector addition of the spin $\frac{1}{2}$ angular momenta of the quarks and antiquarks, together with integral values of orbital angular momenta, leads to odd half-integral angular momenta for baryons and antibaryons and integral angular momenta for mesons. The $SU(3)_f$ reductions

$$B \sim qqq \sim 3 \times 3 \times 3 = 1 + 8 + 8 + 10 \qquad \qquad 1.1$$

and

$$M \sim q\bar{q} \sim 3 \times 3^* = 1 + 8 \qquad \qquad 1.2$$

imply that baryons are in 1-, 8- and 10-dimensional representations, and mesons are in 1- and 8-dimensional representations of $SU(3)_f$.[4] [We discuss calculations with representations of $SU(n)$ groups in Section 3.5 below.]

The SU(6) theory (14, 15), considered in the context of nonrelativistic constituent quarks, combines the $SU(3)_f$ or unitary spin and $SU(2)_S$ or spin degrees of freedom to form a 6 under $SU(6)_f$:

$$q = (u^\uparrow, u^\downarrow, d^\uparrow, d^\downarrow, s^\uparrow, s^\downarrow) \qquad \qquad 1.3$$

with the reduction dim $SU(6)_f \to [\dim SU(3)_f, \dim SU(2)_S]$,

$$6 \to (3,2) \text{ under } SU(6)_f \to SU(3)_f \times SU(2)_S. \qquad \qquad 1.4$$

The possible $SU(6)_f$ content of baryons and mesons is given by the reductions

$$B \sim qqq \sim 6 \times 6 \times 6 = 56 + 70 + 70 + 20 \qquad \qquad 1.5$$

$$\square \times \square \times \square = \boxed{\square\square\square} + \boxplus + \boxplus + \boxvline$$

and

$$M \sim q\bar{q} \sim 6 \times 6^* = 1 + 35, \qquad \qquad 1.6$$

$$\square \times \boxvline = \,\cdot\, + \boxplus\,,$$

where the Young tableaux, with rows indicating symmetry and columns indicating antisymmetry under permutations, are shown under the reductions.

With charm, $SU(6)_f$ is replaced by $SU(8)_f$, which has the subgroup $SU(4)_f \times SU(2)_S$, and the quark transforms as an 8 with the reduction

$$8 \to (4,2) \text{ under } SU(8)_f \to SU(4)_f \times SU(2)_S. \qquad \qquad 1.7$$

[4] See References 5 and 7 for further discussion of SU(3), including diagrams illustrating the mathematical and physical content of the $SU(3)_f$ representations.

Baryons then have the content

$$B \sim qqq \sim 8 \times 8 \times 8 = 120 + 168 + 168 + 56 \qquad 1.8$$

and

$$M \sim q\bar{q} = 8 \times 8^* = 1 + 63. \qquad 1.9$$

To understand the necessity for the color degree of freedom within the framework of the generalized Pauli principle, consider the lowest-lying baryonic states built up from these quarks. We expect the ground-state baryon $SU(6)_f$ supermultiplet to have the three quarks in a symmetric s state; then, without other degrees of freedom, the Pauli principle for the spin $\frac{1}{2}$ quarks implies that the $SU(6)_f$ representation is the antisymmetric 20 (15). The flavor unitary spin–ordinary spin reduction of the 20 is

$$20 \rightarrow (8,2) + (1,4), \text{ under } SU(6)_f \rightarrow SU(3)_f \times SU(2)_s; \qquad 1.10$$

but there is no even-parity, unitary singlet, spin $\frac{3}{2}$ resonance at low mass, and $SU(6)_f$ predicts magnetic moments for the spin $\frac{1}{2}$ octet in the 20 that disagree in both magnitude and sign (they point the wrong way!) with the observed magnetic moments in the nucleon octet. The *symmetric* 56 has the reduction

$$56 \rightarrow (8,2) + (10,4), \text{ under } SU(6)_f \rightarrow SU(3)_f \times SU(2)_s, \qquad 1.11$$

which includes the observed, positive-parity, spin $\frac{1}{2}$ nucleon octet containing N, Λ, Σ, and Ξ and the spin $\frac{3}{2}$ decuplet containing Δ, Σ^*, Ξ^*, and Ω. Both $SU(6)_f$ and the nonrelativistic quark model predict $\mu(p)/\mu(n) = -\frac{3}{2}$ in striking agreement with experiment, and the other magnetic moment ratios of baryons in the 56 are in reasonable agreement with experiment, where measurements exist. Further, the magnetic moments themselves agree with experiment if the quarks are assumed to have effective magnetic moments

$$\mu_i = 2.79 \; q_i e \hbar/(2m_p c), \qquad 1.12$$

where $q_i e$ $(e > 0)$ is the electric charge of the ith quark and m_p is the proton mass. A quark would have such an effective magnetic moment if the quark were a Dirac particle with an effective mass

$$m_{\text{quark}}^{\text{eff}} = m_p/2.79; \qquad 1.13$$

thus, heuristically, the three quarks in the proton act as though each quark has about a third of the proton mass.

Since the $SU(6)_f$ and space symmetries of the s-wave 56 are both symmetric, there is an apparent conflict with the generalized Pauli principle for quarks. The three-valued internal color degree of freedom for quarks resolves this conflict. The three quarks in the space s state and $SU(6)_f$ 56 will obey the Pauli principle if, as mentioned above, the quark wave function in the new color degree of freedom is antisymmetric. This three-valued degree of freedom was first introduced using para-Fermi statistics of order three for quarks (9); from the standpoint of quantum mechanics the para-Fermi statistics is equivalent[5] to color using three identical quark multiplets which transform as a 3 under $SU(3)_c$. We discuss here only the theory of the color degree of freedom in which an exact $SU(3)$ symmetry, $SU(3)_c$, relates the three colors, and in which all observed hadrons belong to the singlet of $SU(3)_c$. [Other theories of color (10–13) give the same results for hadron spectroscopy. Indeed, except for the possible detection of particles not in a singlet of $SU(3)_c$, it is very difficult experimentally (11–13) to distinguish the QCD theory in which (nonsinglet) color and all color-carrying particles, such as fractionally charged quarks and antiquarks and neutral gluons are permanently confined (4) and the color degree of freedom is hidden, from the Pati-Salam gauge theory (12) in which color-carrying particles such as integrally charged quarks, antiquarks, and gluons can occur as free particles and the color degree of freedom is explicit.] Since the baryons then have a symmetric wave function in the visible degrees of freedom [space and $SU(6)_f$ or $SU(3)_f$ and spin], this model applied to baryons is called the symmetric quark model (16). This model assigns higher mass baryons to states with orbital excitations of the quarks and with the permutation symmetry of the $SU(6)_f$ and space wave functions of the quarks correlated so that total wave function is symmetric. Most known baryons have been assigned to states of this model; phenomenological $SU(6)_f$ mass formulas have been derived, first for the $(56, L^P = 0^+)$ (14), for the $(70, L^P = 1^-)$ (16), and for higher multiplets (17), in good agreement with experiment; mixing angles and decay rates have been predicted, with less good agreement with data (17, 18). Recently, the specific spin-spin interactions that follow from perturbative QCD at short distances have been used to analyze the lowest positive- and negative-parity baryon supermultiplets with striking success (19, 20). The weight of the spectroscopic evidence concerning both the low-lying and higher baryonic states overwhelmingly supports the introduction of the color degree of freedom for quarks and the symmetric quark model for baryons.

Note that the accumulation of good spectroscopic data depends on the existence of formation experiments (in which the total quantum

---

[5] See Reference 11 for a review and citations to the work of Haag and collaborators.

numbers of the initial state are the same as those of the particle or resonance formed) for the N, Δ, Λ, Σ, and their excited states. There are no formation experiments for Ξ, Ω, and their excited states, nor for most mesons (except for mesons formed in electron-positron annihilation), and the data for these resonances are not nearly as good as for the $Y = 1$ and 0 baryons. Nonetheless, the $SU(6)_f$ theory for the ground-state mesons is well verified. The 35 decomposes to

$$35 \rightarrow (8,1) + (1+8,3), \text{ under } SU(6)_f \rightarrow SU(3)_f \times SU(2)_S, \qquad 1.14$$

which includes the octet of pseudoscalar mesons containing π, K, $\bar{\text{K}}$, and η, and the nonet of vector mesons containing ρ, ω, K*, $\bar{\text{K}}$*, and ϕ with "ideal" mixing, and, for example, predicts the vector meson leptonic width ratios $9:1:2$ for $\rho:\omega:\phi$ in agreement with experiment.

Quarks as relativistic "parts" of hadrons called "partons" (21) play the central role in the quark-parton model of electromagnetic and weak inclusive scattering cross sections at high energies as well as in the developing applications of the model to hadron scattering. For an exclusive cross section, each particle in the final state is detected; for an inclusive cross section only some of the particles in the final state are detected, so that the cross sections to produce the detected particles plus varying numbers of undetected particles are summed by an inclusive cross section. The simplest inclusive cross section is the total cross section, $\sigma(ab \rightarrow X)$, in which none of the particles in the final state X are detected. A more typical inclusive cross section is the inclusive electro-production cross section $d\sigma(eN \rightarrow eX)/dvdQ^2$ in which only the electron is detected in the final state and v, the electron energy loss in the laboratory frame, and $-Q^2$, the electron invariant momentum transfer, are kinematic variables that depend on the incident electron's energy, E, the outgoing electron's energy, E', and θ, the electron scattering angle, all in the laboratory frame.

In its naive form, the parton model describes a high energy hadron as a collection of parallel-moving free, point-like partons (i.e. transverse momentum and binding forces are neglected). In the quark-parton model, the partons are quarks, antiquarks, and flavor-singlet color-octet gluons. In the parton model of a high energy electromagnetic or weak inclusive collision, an electromagnetic gauge boson (photon) or a weak gauge boson [a $W^{\pm}$ or Z^0 in the Weinberg-Salam model (22)] emitted by a lepton scatters incoherently from the quarks and antiquarks in a hadron with a cross section given by lowest-order perturbation theory for point-like, spin $\frac{1}{2}$ particles in a frame in which the hadron has a large momentum. This cross section, weighted by the quark-parton distributions for the number of quarks and antiquarks of each flavor with a fraction x of the total hadron momentum, gives an impulse approximation to the structure functions, $W_i(\nu, Q^2)$, which up to kinematic factors are the inclusive cross sections. In the naive parton model, this fraction x is identified with the Bjorken scaling variable x defined below. Here, as in most high energy experiments, the only available hadronic targets are the proton and neutron; the proton occurring by itself in hydrogen, and both proton and neutron occurring in nuclei. Isospin symmetry relates the quark-parton distribution functions in the neutron to those in the proton. The gluons carry momentum, but are inert with respect to weak and electromagnetic processes, and do not directly contribute to the structure functions. Thus the quark-parton model uses the parton distribution functions u(x), d(x), s(x), c(x), $\bar{u}$(x), $\bar{d}$(x), $\bar{s}$(x), $\bar{c}$(x) in the proton to describe electromagnetic and weak inclusive processes such as $e^{\pm}N \rightarrow e^{\pm}X$, $\mu^{\pm}N \rightarrow \mu^{\pm}X$, $\nu_e N \rightarrow e^- X$, $\nu_\mu N \rightarrow \mu^- X$, $\bar{\nu}_e N \rightarrow e^+ X$, $\bar{\nu}_\mu N \rightarrow \mu^+ X$, $\nu_e N \rightarrow \nu_e X$, $\nu_\mu N \rightarrow \nu_\mu X$, $\bar{\nu}_e N \rightarrow \bar{\nu}_e X$, $\bar{\nu}_\mu N \rightarrow \bar{\nu}_\mu X$ at high energies. This description leads to Bjorken scaling (23), which is the statement that the structure functions depend only on the ratio $x = Q^2/(2m\nu)$, where m is the nucleon mass, instead of on ν and Q^2 separately. The contributions carried by the flavor quarks u and d (called valence contributions) to the parton momentum distributions, xu(x) and xd(x), which are included in νW_2, each peak at about $x = \frac{1}{3}$. The analogous contributions carried by the other quarks and by the antiquarks (called sea contributions) peak at $x = 0$. Bjorken scaling is violated by about 20% over the experimentally studied (24) ranges of ν and Q^2.

Roughly speaking, the scaling violation can be described by an increase in the contribution of the sea and by a decrease in the contribution of the valence quarks, both for increasing Q^2. Nonetheless, the rough validity of scaling, and its simple derivation, provides support for the naive quark-parton model. The definition of the parton distribution functions leads to sum rules, which say, for example, that the number of u minus the number of $\bar{u}$ quarks in the proton is two, and, for another example, that the fraction of the nucleon momentum carried by the quarks and antiquarks is given by the sum of the first x moments of the quark and antiquark distribution functions, which can be expressed in terms of the x integrals of the structure functions for electron-nucleon and neutrino-nucleon scattering (25). Experimentally, this fraction is about one half; presumably the other half is carried by gluons. The fact that the same parton distribution functions describe both electromagnetic and weak inclusive processes leads to a relation between the structure functions for eN and νN scattering, which agrees well with data (25). The spin $\frac{1}{2}$ nature of the quark-partons leads to a relation (26) between two of the independent scaling structure functions for a given process, which is verified by data to within 10%. In addition, spin $\frac{1}{2}$ quark-partons, together with left-handed neutrinos and right-handed antineutrinos and the left-handed nature of the weak interaction currents for quarks, lead to the prediction of different kinematic dependences (on $y = v/E$) of the inclusive cross sections due to vq and $\bar{v}\bar{q}$ scattering on the one hand and due to $\bar{v}q$ and $v\bar{q}$ scattering on the other hand, with the result that quark-partons in neutrino scattering and antiquark-partons in anti-neutrino scattering both give y-independent contributions to their inclusive cross sections, and quark-partons in antineutrino scattering and antiquark-partons in neutrino scattering both contribute as $(1-y)^2$. After some conflicting data, the evidence now supports this prediction at the 10% level of accuracy. The parton model also predicts that the total neutrino and antineutrino cross sections rise linearly with the lab neutrino energy, and predicts $(\sigma^{vp} + \sigma^{vn})/(\sigma^{\bar{v}p} + \sigma^{\bar{v}n}) = 3$, in rough agree-ment with present data. These relations will be modified at energies where charm production is important, and at energies where momentum transfer effects of the $W^{\pm}$ and Z^0 gauge bosons that transmit the weak interactions must be taken into account.

The parton model, which had its origin in the theoretical study of electroproduction and was then applied to the analogous process of inclusive neutrino scattering, has also been applied to electron-positron annihilation to hadrons and to lepton-pair production in hadronic collisions, as well as to hadronic scattering processes. The prediction that $\sigma_{\text{tot}}(e^+e^- \to \text{hadrons}) \propto s^{-1}$ is verified to about 20%, and the coefficient is related to the sum of the squares of the quark charges with comparable accuracy. The presence of the factor of three coming from color in σ_{tot} is well verified. In hadronic production of e^+e^- and $\mu^+\mu^-$, the color degree of freedom results in a decrease of the cross section by a factor of three. Present data favor this prediction. Parton model studies of large momentum transfer inclusive hadronic cross sections using parton distributions found in inclusive lepton scattering together with QCD quark-quark, quark-gluon, and gluon-gluon scattering cross sections agree with present data.

The most naive predictions of the quark-parton model must be amended to allow $\sim 20\%$ scale breaking, and, as mentioned above, to allow for about half of the nucleon momentum to be carried by objects (color gluons) that do not participate in electromagnetic and weak interactions. Nonetheless, the model is useful in correlating a large body of inclusive scattering data using spin $\frac{1}{2}$, point-like color quarks with left-handed weak interactions. We point out below that the necessary amendments are compatible with QCD.

Quark model predictions for hadron spectroscopy and quark-parton model inclusive scattering predictions both give impressive support to the fundamental role of quarks in present elementary particle and high energy physics; however, we do not yet understand the precise relation between these two types of quark model, nor can we now derive them in all details from a basic theory of quark-gluon interactions, such as QCD.

How shall we characterize permanently confined quarks: Are they mathematical constructs without physical reality, or are they real physical particles? From the philosophical point of view, the notion of physical reality itself requires an act of faith; although it is not useful for prediction, solipsism is logically impeccable. As physicists we accept the notion of physical reality. To gain perspective concerning the reality and elementarity of quarks, we consider two examples with confined objects: phonons in a metal and north and south poles in a bar magnet. We do not consider phonons to be elementary particles because we understand phonons in terms of excitation and propagation of nodes in the metallic crystal lattice; the fact that phonons are confined to the metal is not decisive in excluding them from the list of elementary particles. We understand the fact that north and south poles necessarily occur together in terms of the generation of magnetic fields by currents of spin $\frac{1}{2}$ electrons (27). North and south poles are not elementary because we understand them in terms of smaller objects, not because they must occur together. Neither phonons nor north and south magnetic poles are less real because they are confined. The spin, flavor, and color quantum numbers of quarks lead to detailed predictions that agree with experiment. There is no present evidence for substructure underlying quarks; if a substructure is found in the future, quarks will remain as real as atoms are in view of their substructure of nuclei and electrons. We suggest that permanent confinement in no way compromises the status of quarks as real physical particles and that at present they appear to be elementary.

2   QUARK DEGREES OF FREEDOM

2.1   Spin

Spin $\frac{1}{2}$ quarks, together with orbital angular momenta, allow construction of all spin states of hadrons. More specifically, the successful $SU(6)_f$ analysis of baryon and meson spectroscopy, using spin $\frac{1}{2}$ flavor-triplet quarks, provides evidence based on strong interactions that quarks have spin $\frac{1}{2}$. The Callan-Gross sum rule (26) in the quark-parton model provides evidence that quarks have spin $\frac{1}{2}$, based on electromagnetic and weak interactions.

2.2   *Flavor*

Three flavors are necessary to construct the hadrons observed before the discovery of charm; the transformation of these flavors as a "unitary spin" triplet under $SU(3)_f$ leads to the observed $SU(3)_f$ structure of hadrons. The electromagnetic current

$$j_{em}^\mu = 3^{-1}(2\bar{u}\gamma^\mu u - \bar{d}\gamma^\mu d - \bar{s}\gamma^\mu s) \qquad 2.1$$

leads to the relative leptonic branching rates

$$\Gamma(\rho \to l^+l^-):\Gamma(\omega \to l^+l^-):\Gamma(\phi \to l^+l^-) = 9:1:2,$$

which agree with experiment up to uncertainties associated with mass effects. The triplet

$$\begin{pmatrix} u \\ d_c(\theta) \\ s_c(\theta) \end{pmatrix} = \begin{pmatrix} 1 & 0 & 0 \\ 0 & \cos\theta_C & \sin\theta_C \\ 0 & -\sin\theta_C & \cos\theta_C \end{pmatrix} \begin{pmatrix} u \\ d \\ s \end{pmatrix} \qquad 2.2$$

gives good agreement with weak processes using the charged ($\Delta Q = \pm 1$) Cabibbo currents

$$j_+^\mu = \bar{u}\gamma^\mu(1+\gamma_5)\, d_C(\theta), j_-^\mu = \bar{d}_C(\theta)\,\gamma^\mu(1+\gamma_5)\, u, \qquad 2.3$$

where u, d, s stand for (spin $\tfrac{1}{2}$) Dirac fields carrying quark quantum numbers. The Glashow-Iliopoulos-Maiani (GIM) (8) mechanism using the charm quark, c, cancels the essentially unavoidable $|\Delta S| = 1, \Delta Q = 0$ currents (which would lead to the decays $K^+ \to \pi^+\nu\bar{\nu}$, $K^+ \to \pi^+ e^+ e^-$, $K_L^0 \to \mu^+\mu^-$ at rates many orders of magnitude greater than experimental upper limits). Using the weak doublets

$$\begin{pmatrix} u_L \\ d_L(\theta) \end{pmatrix} \quad \text{and} \quad \begin{pmatrix} c_L \\ s_L(\theta) \end{pmatrix}, \qquad 2.4$$

where $q_L = (1+\gamma_5)q$, $\bar{q}_L = \bar{q}(1-\gamma_5)$, the GIM weak currents are

$$j_+^\mu = \bar{u}_L\gamma^\mu d_L(\theta) + \bar{c}_L\gamma^\mu s_L(\theta), j_-^\mu = (j_+^\mu)^\dagger \qquad 2.5$$

$$j_0^\mu = \bar{u}_L\gamma^\mu u_L + \bar{d}_L(\theta)\gamma^\mu d_L(\theta) + \bar{s}_L(\theta)\gamma^\mu s_L(\theta) + \bar{c}_L\gamma^\mu c_L$$

$$= \bar{u}_L\gamma^\mu u_L + \bar{d}_L\gamma^\mu d_L + \bar{s}_L\gamma^\mu s_L + \bar{c}_L\gamma^\mu c_L. \qquad 2.6$$

Note that the GIM neutral current conserves all hadronic quantum numbers. As mentioned in Section 1, more flavors may be needed.

2.3   *Color*

Precisely three colors are necessary to get agreement with baryon spectroscopy; the symmetric quark model provides detailed confirmation of the existence of the color degree of freedom. Other evidence for the color degree of freedom includes: (a) the saturation property of hadrons that low-lying states all have the quantum numbers of qqq for baryons and q$\bar{\text{q}}$ for mesons. The quark model with a single triplet fails to account for saturation (28); (b) the partial width $\Gamma(\pi^0 \to 2\gamma)$ in agreement with experiment as a consequence of the axial anomaly (29). The quark model with a single triplet gives one ninth of the observed partial width; (c) the cross section

$$\sigma(e^+e^- \to \text{hadrons}) = 4\pi\alpha^2(3q^2)^{-1} \sum_{i(\text{spin } \frac{1}{2})} Q_i^2; \qquad 2.7$$

the values of $\sum_i Q_i^2 = 2$ below charm threshold and $\sum_i Q_i^2 = \frac{10}{3}$ above charm threshold agree with experiment (away from resonances) to about 20%; without color, this cross section is $\frac{1}{3}$ of the above values; (d) with the color degree of freedom, the cross section $\sigma(\text{NN} \to l^+l^-X)$ is decreased by $\frac{1}{3}$ compared to the value without color. This prediction is favored but not yet confirmed by experiment.

3   QUARKS IN HADRON SPECTROSCOPY

3.1   *Baryons Without Charm*

Having introduced the spin, flavor, and color degrees of freedom and the SU(6)$_f$ combination of spin and flavor, we now discuss hadron spectroscopy in detail, starting with baryons. We assume that quarks and other color-carrying particles are permanently confined in color-singlet states. Since the color singlet formed from three color-triplet quarks is totally antisymmetric under permutations, each baryon composed of three quarks must be in a state totally symmetric in the combined SU(6)$_f$ and space degrees of freedom if we use the generalized Pauli principle to treat quarks with different flavors. This total symmetry requires that the SU(6)$_f$ and space states have matching permutation symmetries, so that the allowed SU(6)$_f$ space state is

$$\Psi(\text{tot}) = \sum \Psi[\text{SU}(6)_f] \times \Psi(\text{space}),$$

$$\qquad 3.1$$

where the Young tableaux give the permutation symmetry and, for the SU(6)$_f$ states, also label the irreducible representations. As stated in

Section 1, the $SU(6)_f$ representations, labeled by their multiplicities, are 56, 70, and 20 for the symmetric, mixed, and antisymmetric representations, respectively.

The use of $SU(6)_f$ for counting states, labeling supermultiplets, and analyzing mass operators is convenient on a phenomenological level. However, since $SU(6)_f$ is broken by the quark mass differences, we need not take states belonging to a single line of Equation 3.1 (in particular, with a space wave function belonging to a single irreducible representation of the permutation group S_3) as basis states in an analysis of baryons. Isgur & Karl (20) emphasize this point in their study of the lowest mass, negative-parity baryon supermultiplet, which we discuss below. Use of broken $SU(6)_f$ with the generalized Pauli principle is equivalent to treating quarks of different flavor as distinguishable; however, for a given calculation one approach may be simpler than the other.

Before discussing the orbitally excited states, we discuss the properties of the ground-state $SU(6)_f$ 56 supermultiplet. We assume the generalized Pauli principle and use a set of Bose annihilation and creation operators to carry the $SU(6)_f$ degrees of freedom and to take account of the fact that the 56 is symmetric. Let

$$a_\alpha^\dagger = a_{Aa}^\dagger, \ a^\alpha = a^{Aa}, \ \alpha = 1, 2, \ldots 6 \text{ of } SU(6)_f, \ A = 1, 2, 3 \text{ of } SU(3)_f,$$

$$a = 1, 2 \text{ of } SU(2)_S. \qquad 3.2$$

The generators of $SU(6)_f$, $SU(3)_f$, and $SU(2)_S$ are

$$F_\alpha^\beta = a_\alpha^\dagger a^\beta - \tfrac{1}{6}\delta_\alpha^\beta N, \quad N = a_\gamma^\dagger a^\gamma, \qquad 3.3$$

$$I_A^B = F_{Ac}^{Bc} = a_{Ac}^\dagger a^{Bc} - \tfrac{1}{3}\delta_A^B N, \quad \text{and} \qquad 3.4$$

$$S_a^b = F_{Ca}^{Cb} = a_{Ca}^\dagger a^{Cb} - \tfrac{1}{2}\delta_a^b N, \qquad 3.5$$

respectively. In particular, the spin, isospin, hypercharge, and electric charge operators are

$$S_+ = S_1^2, \quad S_- = S_2^1, \quad S_3 = \tfrac{1}{2}(S_1^1 - S_2^2), \qquad 3.6$$

$$I_+ = I_1^2, \quad I_- = I_2^1, \quad I_3 = \tfrac{1}{2}(I_1^1 - I_2^2), \qquad 3.7$$

$$Y = \tfrac{1}{3}(I_1^1 + I_2^2 - 2I_3^3), \quad \text{and} \qquad 3.8$$

$$Q = \tfrac{1}{3}(2I_1^1 - I_2^2 - I_3^3), \qquad 3.9$$

respectively. Relevant Casimir operators are

$$C_2^{(6)}(f) = F_\alpha^\beta F_\beta^\alpha, \quad C_2^{(3)}(f) = I_A^B I_B^A, \qquad 3.10$$

$$C_2^{(2)}(I) = I_{A'}^{B'} I_{B'}^{A'} - \tfrac{1}{2}(I_{C'}^{C'})^2 = 2I(I+1), \quad \text{and} \qquad 3.11$$

$$C_2^{(2)}(S) = S_a^b S_b^a = 2S(S+1), \qquad 3.12$$

Table 2 States in the nucleon octet with $J_3 = S_3 = \frac{1}{2}$

Baryon state	Quark wave function
$|p^+\uparrow\rangle$	$= 3^{-\frac{1}{2}} a_{11}^\dagger \varepsilon^{ab} a_{1a}^\dagger a_{2b}^\dagger |0\rangle$
$|n^0\uparrow\rangle$	$= 3^{-\frac{1}{2}} a_{12}^\dagger \varepsilon^{ab} a_{1a}^\dagger a_{2b}^\dagger |0\rangle$
$|\Lambda^0\uparrow\rangle$	$= 2^{-\frac{1}{2}} a_{31}^\dagger \varepsilon^{ab} a_{1a}^\dagger a_{2b}^\dagger |0\rangle$
$|\Sigma^+\uparrow\rangle$	$= 3^{-\frac{1}{2}} a_{1}^\dagger \varepsilon^{ab} a_{1a}^\dagger a_{3b}^\dagger |0\rangle$
$|\Sigma^0\uparrow\rangle$	$= 6^{-\frac{1}{2}} \varepsilon^{ab} [a_{11}^\dagger a_{2a}^\dagger + a_{21}^\dagger a_{1a}^\dagger] a_{3b}^\dagger |0\rangle$
$|\Sigma^-\uparrow\rangle$	$= 3^{-\frac{1}{2}} a_{21}^\dagger \varepsilon^{ab} a_{2a}^\dagger a_{3b}^\dagger |0\rangle$
$|\Xi^0\uparrow\rangle$	$= 3^{-\frac{1}{2}} a_{31}^\dagger \varepsilon^{ab} a_{1a}^\dagger a_{3b}^\dagger |0\rangle$
$|\Xi^-\uparrow\rangle$	$= 3^{-\frac{1}{2}} a_{31}^\dagger \varepsilon^{ab} a_{2a}^\dagger a_{3b}^\dagger |0\rangle$

where the sums over A', B', and C' run from one to two. The construction of the states is easy and compact. We construct the states[6] with $S_3 = S = \frac{3}{2}$ in the (10,4) with a monomial in the $a_{A1}^\dagger$. The construction of the states[6] in the (8,2) uses the fact that $O_{AB} \equiv \varepsilon^{ab} a_{Aa}^\dagger a_{Bb}^\dagger$ has $S = 0$ and the I belonging to an antisymmetric state of q_A and q_B. Then every state with $S_3 = S = \frac{1}{2}$ in the (8,2), except Σ^0, is the product of a single $a^\dagger$ and an O_{AB}, and Σ^0 is easily constructed using the isospin-lowering operator I_-. The proton and neutron with $S_3 = \frac{1}{2}$ are

$$|p^+\rangle = 3^{-\frac{1}{2}} a_{11}^\dagger O_{12} |0\rangle = 3^{-\frac{1}{2}} a_{11}^\dagger \varepsilon^{ab} a_{1a}^\dagger a_{2b}^\dagger |0\rangle \qquad 3.13$$

and

$$|n^0\rangle = 3^{-\frac{1}{2}} a_{21}^\dagger O_{12} |0\rangle = 3^{-\frac{1}{2}} a_{21}^\dagger \varepsilon^{ab} a_{1a}^\dagger a_{2b}^\dagger |0\rangle, \qquad 3.14$$

where the left-most $a^\dagger$ carries the spin and isospin of the nucleon in both cases. The states in the nucleon octet are given in Table 2.

The magnetic moments are

$$\mu_X = \langle X | \mu_3 | X \rangle, \qquad 3.15$$

where

$$\mu = \mu_0 Q_B^A a_{Aa}^\dagger a^{Bb} \sigma_b^a = 2\mu_0 (Q_u S_u + Q_d S_d + Q_s S_s), \qquad 3.16$$

$Q_B^A = (\frac{2}{3}, -\frac{1}{3}, -\frac{1}{3})$ is diagonal, and σ are the Pauli matrices. This leads directly to (30)

$$\mu_p = \mu_0 = 2.79 \; e\hbar/(2m_p c), \quad \mu_n = -(\tfrac{2}{3})\mu_p, \qquad 3.17$$

in agreement to 3% with the experimental μ_n/μ_p ratio. Predictions for the magnetic moments in the baryon octet are given in Table 3.

---

[6] See Equation 1.11.

Table 3 Magnetic moments in the baryon octet

Particle	$(\mu/\mu_p)^a$	$(\mu/\mu_p)^b$	$(\mu/\mu_p)_{exp}^c$
p	1^d	1^d	1
n	$-\frac{2}{3}$	-0.666	-0.685
Λ	$-\frac{1}{3}$	-0.215	-0.240 ± 0.021
Σ^+	1	0.956	0.938 ± 0.147
Σ^0	$\frac{1}{3}$	—	—
Σ^-	$-\frac{1}{3}$	-0.376	-0.530 ± 0.132
Ξ^0	$-\frac{2}{3}$	-0.500	—
Ξ^-	$-\frac{1}{3}$	-0.165	-0.662 ± 0.268

[a] Predicted from Equations 3.15 and 3.16.
[b] Predicted from Reference 19.
[c] Experimental value from Reference 31.
[d] Input.

Using the Dirac formula $\cdot \mu_0 = e\hbar/(2m_q c)$ for the magnetic moment together with Equation 3.17 leads to an effective constituent quark mass in the nucleon of $m_N/2.79 \approx 336$ MeV/c^2, i.e. about a third of the nucleon mass. When free quarks were thought to be massive, several authors (32) pointed out that the effective quark mass in a deep scalar potential could be much less than the free quark mass. With permanent quark confinement, quarks need not be massive in any case.

The result for G_A/G_V for the nucleon can be found in a similar way, using

$$G_V = \langle p, S_3 = \tfrac{1}{2} | I_+ | n, S_3 = \tfrac{1}{2} \rangle = 1 \qquad 3.18$$

$$G_A = \langle p, S_3 = \tfrac{1}{2} | I_+^5 | n, S_3 = \tfrac{1}{2} \rangle = \tfrac{5}{3}, \qquad 3.19$$

with $I_+^5 = a^\dagger_{1a} a^{2b}(\sigma_3)^a_b = u^{\uparrow\dagger}d^\uparrow - u^{\downarrow\dagger}d^\downarrow$, but disagrees with the experimental value $G_A/G_V = 1.2$.

Gürsey & Radicati (14) gave the first $SU(6)_f$ mass formula,

$$M = M_0 + M_1 Y + M_2[I(I+1) - \tfrac{1}{4}Y^2] + M_3 J(J+1), \qquad 3.20$$

for the 56. We derive this result here and outline a systematic approach to $SU(6)_f$ mass formulas (16) using the assumptions that one- and two-body operators and $SU(3)_f$ singlet and octet operators dominate. The one-body operators correspond to the quark masses. Neglecting flavor isospin symmetry breaking, the one-body operators are $m_n N_n$ and $m_s N_s$, where n and s stand for nonstrange and strange, respectively, and $N_n = a^\dagger_{A'a} a^{A'a}$ and $N_s = a^\dagger_{3a} a^{3a}$. We can write these two operators as a linear combination of the quark number operator, N, and the hypercharge operator, Y. From the $SU(6)_f$ point of view, these two operators are in

$$6 \times 6^* = 1 + 35, \qquad 3.21$$

where

$$1 \rightarrow (1,1) \text{ and } 35 \rightarrow (8,1) + (1+8,3), \qquad\qquad 3.22$$

under $SU(6)_f \rightarrow SU(3)_f \times SU(2)_S$. The available operators (with $I = Y = S = 0$) are then $(1,1)$, which transform as N, and the hypercharge operator Y in $(8,1)$. We label operators $^nT_a^{b,c}$, where n is the dimension of the one- or two-body $SU(6)_f$ representation on which T acts [it is zero on other $SU(6)$ representations], and T transforms as the representation a under $SU(6)_f$ and as (b,c) under $SU(3)_f \times SU(2)_S$. The most general two-body operator maps symmetric (antisymmetric) $SU(6)_f$ two-quark states into symmetric (antisymmetric) states, since parity is conserved, and the $SU(6)_f$ and space states are coupled (see Section 3.1). Using

$$6 \times 6 = 21 + 15 \text{ in } SU(6)_f, \qquad\qquad 3.23$$

$$\square \times \square = \square\!\square + \begin{array}{c}\square\\\square\end{array}$$

the most general two-body operator is in either

$$21 \times 21^* = 1 + 35 + 405, \qquad\qquad 3.24$$

or

$$15 \times 15^* = 1 + 35 + 189. \qquad\qquad 3.25$$

Only operators in $21 \times 21^*$ can be used for the s wave 56. Now

$$405 \rightarrow (1+8+27,1) + (8+8+10+10^*+27,3) + (1+8+27,5) \qquad 3.26$$

under $SU(6)_f \rightarrow SU(3)_f \times SU(2)_S$ so that only the $(1,1)$ and the hypercharge-like member of the $(8,1)$, both in the 405, are available. The two-body operators in the 1 and 35 are equivalent (for the baryons) to the one-body operators already discussed. The Gürsey-Radicati formula (Equation 3.20) results when the operators $^6T_1^{1,1}$, $^6T_{35}^{8,1}$, $^{21}T_{405}^{1,1}$, and $^{21}T_{405}^{8,1}$ are expressed in terms of I, Y, and $J = S$.

Without the restriction to one- and two-body operators, the most general operator acting on the 56 is in (33)

$$56 \times 56^* = 1 + 35 + 405 + 2695, \qquad\qquad 3.27$$

which contains eight spin zero operators in $(1,1)$, $(35,8)$, $(405,1)$, $(405,8)$, $(405,27)$, $(2695,8)$, $(2695,27)$, and $(2695,64)$, where $(a,b) = [\dim SU(6)_f, \dim SU(3)_f]$. There are eight isospin multiplets in the 56, so use of all eight spin zero operators places no constraint on the masses. Octet dominance alone leads to a five-parameter mass formula, as does two-body dominance alone. Octet and two-body dominance together lead to the four-parameter Gürsey-Radicati mass formula. Table 4 gives the predicted and observed masses of the $(56,0^+)$ baryon supermultiplet.

Table 4   The $(56, 0^+)$ baryon supermultiplet

Baryon/ state	$^4S, J^P = (\frac{3}{2})^+$			$^2S, J^P = (\frac{1}{2})^+$	
	Predicted mass $(\text{MeV}/c^2)^a$	Observed mass $(\text{MeV}/c^2)^b$		Predicted mass $(\text{MeV}/c^2)^a$	Observed mass $(\text{MeV}/c^2)^b$
Ω	1672	1672	Ξ	1325	1315–1321
Ξ	1528	1532–1535	Σ	1181	1189–1197
Σ	1384	1382–1387	Λ	1116	1116
Δ	1240	1230–1234	N	939	938–940

[a] From Jones, M. et al, Reference 17.
[b] From Reference 31.

To discuss higher supermultiplets with orbital excitation, we introduce the traceless coordinates

$$\rho = 2^{-\frac{1}{2}}(\mathbf{r}_1 - \mathbf{r}_2), \ \lambda = 6^{-\frac{1}{2}}(\mathbf{r}_1 + \mathbf{r}_2 - 2\mathbf{r}_3), \qquad 3.28$$

which are the basis vectors for the two-dimensional "mixed" irreducible representation of the permutation group on three objects, S_3. We associate momenta, $\mathbf{p}_\rho$ and $\mathbf{p}_\lambda$, angular momenta $\mathbf{l}_\rho$ and $\mathbf{l}_\lambda$, and annihilation and creation operators, $\mathbf{a}_\rho$, $\mathbf{a}_\rho^\dagger$, $\mathbf{a}_\lambda$, $\mathbf{a}_\lambda^\dagger$ with ρ and λ. Excitations associated with $\mathbf{a}_\rho^\dagger$ and $\mathbf{a}_\lambda^\dagger$ have one quantum of orbital angular momentum and odd parity. The total orbital angular momentum, parity, and total angular momentum are

$$\mathbf{L} = \mathbf{l}_\rho + \mathbf{l}_\lambda, \quad P = (-1)^{l_\rho + l_\lambda}, \quad \text{and} \quad \mathbf{J} = \mathbf{L} + \mathbf{S}, \qquad 3.29$$

where $l_\rho = |\mathbf{l}_\rho|$, etc.

The harmonic oscillator model is particularly convenient because the ρ and λ modes decouple, all calculations can be done analytically, and if desired, group theoretically, the harmonic potential provides permanent quark confinement and a fortiori good agreement with experimental results, both within a given $SU(6)_f$ supermultiplet, and for the equally spaced baryon Regge trajectories. The $SU(6)_f$ supermultiplets can be labeled $[\dim SU(6), L^P]_N$, where N is the total number of ρ and λ excitations. In general, further labels are needed. For a given N, the highest L and J will occur when $L = N$. For each pair of ρ and/or λ modes that couple to zero angular momentum, L will be reduced by two. Experimentally, resonances with the highest J at a given mass are easier to detect, in part because of a $2J + 1$ factor in their formation cross section. Among the states with $L = N$ is the sequence $[56,(2k)^+]$ and $[70,(2k+1)^-]$, $k = 0,1,2,\ldots$. This sequence is particularly prominent experimentally; however, we expect that all states of the symmetric quark model are present. Karl & Obryk gave a complete analysis of orbital excitations (34, 35).

The SU(6)$_f$ analysis of higher baryon supermultiplets, assuming one- and two-body operators and octet dominance initiated in Reference 16 for the (70,1$^-$) has been developed extensively by Dalitz and collaborators (11, 13, 17) among others. The reduction in the number of parameters (and thus the increased number of predictions) that this systematic approach allows is much greater in the higher supermultiplets. In particular, many, and sometimes all, of the reduced matrix elements necessary to derive mass formulas for higher supermultiplets are determined by fits to lower supermultiplets.

The phenomenological analysis of baryon spectra using the SU(6)$_f$ classification of states and mass operators has been very useful in establishing the necessity of the color degree of freedom and in accounting both qualitatively and, with few exceptions, quantitatively for spectra and decays. But this phenomenological analysis is just a single step toward a definitive theory. The next step, using guesses about the form of the interactions based on QCD, has already allowed more specific predictions (with fewer parameters) to be made. These guesses include: (a) the long-range confining interaction is spin and flavor independent, and leads to SU(6)$_f$ × O(3)$_L$ hadron supermultiplets; (b) SU(3)$_f$ breaking occurs only via quark masses; (c) short-range interactions are Coulombic with strength $-\frac{2}{3}\alpha_s$ for two quarks in an (antisymmetric) 3* and strength $-\frac{4}{3}\alpha_s$ for a quark-antiquark pair in a 1$_c$, and include Fermi contact and tensor interactions (19, 36). In particular, the Gürsey-Radicati mass formula for the (56,0$^+$) is recovered with the equivalent predictions (19)

$$\Delta > N \tag{3.30}$$

and

$$\Sigma^0 > \Lambda. \tag{3.31}$$

The relevant spin-dependent term of the Fermi contact interaction for baryons is

$$(\tfrac{16}{9}\pi\alpha_s) \sum_{i<j} (m_i m_j)^{-1} \mathbf{s}_i \cdot \mathbf{s}_j \delta(\mathbf{r}), \tag{3.32}$$

where α_s is the effective strong coupling constant, $\mathbf{s}_i$ and m_i are the quark spin and mass, and $\mathbf{r}$ is the separation between the quarks. Since $\mathbf{s}_i \cdot \mathbf{s}_j$ is positive (negative) for two quarks with parallel (antiparallel) spins, and all three quarks have parallel spins in the $J_z = \frac{3}{2}$ Δ states, Expression 3.30 follows. The Expression 3.31 follows directly from a more intricate argument; however, Expressions 3.31 and 3.30 are related by the form of the contact interaction.

This spin-spin Fermi contact interaction also leads to a calculation (37) of $\langle r_E^2(n)\rangle/\langle r_E^2(p)\rangle$, the ratio of the electric charge radii of the neutron and proton. The negative sign of $\langle r_E^2(n)\rangle$ follows because in the symmetric $SU(6)_f$ state for the nucleon, spin and isospin are coupled, $I = S = 1$ or $I = S = 0$, for a pair of nonstrange quarks. For the neutron, the dd pair must have $I = S = 1$, while the ud pairs have $I = S = 1$ with probability $\frac{1}{4}$ and $I = S = 0$ with $\frac{3}{4}$. Independent of the specific $\delta(\mathbf{r})$ potential in Equation 3.32, the repulsive interaction in the $S = 1$ state produces a greater separation between the two d's than between the ud pairs and thus makes $\langle r_E^2(n)\rangle < 0$. The calculation gives

$$\langle r_E^2(n)\rangle/\langle r_E^2(p)\rangle \approx -\tfrac{1}{2}\alpha_N' N(\Delta - N) \approx -0.14, \qquad\qquad 3.33$$

where N and Δ are the particle masses and α_N' is the nucleon Regge trajectory slope, in good agreement with the experimental value of -0.146 ± 0.005. The ratio of the magnetic radii has also been calculated.

Isgur & Karl (20) have reexamined the $(70,1^-)$ baryon supermultiplet using the harmonic oscillator potential for confinement, as in the phenomenological analyses. They (a) used both the contact and tensor terms in the color-gluon exchange hyperfine interaction to resolve degeneracies among states with the same J^P, I, and Y instead of the phenomenological $SU(6)_f$ mass operator; and they (b) took the different masses of the strange and nonstrange quarks into account for the $Y = 0$ and -1 states. They found that for the $Y = 0$ Λ and Σ states, for example, which have two nonstrange quarks and one strange quark, the harmonic oscillator Gaussian wave function factor for the ρ mode (nonstrange quarks excited in a relative P state) is more spread out than the corresponding factor for the λ mode (strange quark excited relative to the center of mass of the nonstrange quarks).

The contact interaction elevates the $S = \frac{3}{2}$ N states and $S = \frac{1}{2}$ Δ states relative to the $S = \frac{1}{2}$ N states. This pattern of mass shifts, due to the contact interaction, occurs because the wave functions of both the $S = \frac{3}{2}$ N states and $S = \frac{1}{2}$ Δ states are sums of products of flavor and spin factors that are symmetric and mixed: symmetric spin and mixed flavor for the $S = \frac{3}{2}$ N states, and mixed spin and symmetric flavor for the $S = \frac{1}{2}$ Δ states; so that for both cases the λ component of the space wave function symmetric in the coordinates of quarks one and two is multiplied by wave functions symmetric in both the flavor and spin of one and two. Thus, the contact interaction acts with probability $\frac{1}{2}$, and when it acts it sees the one, two pair with spin one. Since both Expression 3.32 and the flavor-spin-space baryon wave functions are symmetric, the matrix element of Expression 3.32 is proportional to the matrix element of $\mathbf{s}_1 \cdot \mathbf{s}_2$. Therefore the contact interaction shifts both the $S = \frac{3}{2}$ N states and the $S = \frac{1}{2} \Delta$ states by the same amount. For the $S = \frac{1}{2} N$ states, both the spin and flavor wave functions are mixed, and the λ component of the space wave function is multiplied by flavor and spin product wave functions that are both symmetric or both antisymmetric in one, two with equal probability. Thus the contact interaction that still acts with probability $\frac{1}{2}$ here sees with equal probability a one, two pair with spin one and with spin zero, and gives the $S = \frac{1}{2} N$ states a different shift than before. This discussion, together with the usual calculation of $s_1 \cdot s_2$ for spin zero and one states, leads to Isgur & Karl's result that the $S = \frac{3}{2} N$ and $S = \frac{1}{2} \Delta$ states are raised, and the $S = \frac{1}{2} N$ states are lowered by the same absolute amount. This effect is clearly visible in the $Y = 1 N$ and Δ states: the $S = \frac{3}{2}$ states $N(1675)\frac{1}{2}^-$, $N(1700)\frac{3}{2}^-$, and $N(1670)\frac{5}{2}^-$ and the $S = \frac{1}{2}$ states $\Delta(1655)\frac{1}{2}^-$ and $\Delta(1685)\frac{3}{2}^-$ are about 150 MeV/c^2 above the $S = \frac{1}{2}$ states $N(1515)\frac{1}{2}^-$ and $N(1520)\frac{3}{2}^-$. For the Λ's, Σ's, and Ξ's, singlet-octet, octet-decuplet, and octet-decuplet mixings, respectively, obscure this effect. The tensor part of the Fermi interaction

$$\frac{2}{3}\alpha_s \sum_{i<j} (m_i m_j)^{-1} \left[3(s_i \cdot \hat{r})(s_j \cdot \hat{r}) - (s_i \cdot s_j) \right] r^{-3}, \qquad 3.34$$

which is absent in the s wave $(56, 0^+)$, enters here for the quark pairs in a relative $L = 1$ state, and produces significant mixings between $S = \frac{3}{2}$ and $S = \frac{1}{2}$, but only small mass shifts. The α_s found from the Δ-N or Σ^0-Λ difference normalizes both of these effects absolutely. The agreement with experiment, particularly for the mixings that were found from decay data, is striking. For $N(1520)\frac{3}{2}^-$ the mixing angle between $S = \frac{3}{2}$ and $S = \frac{1}{2}$ is $6.3°$ vs $10°$ experimentally, and for $N(1530)\frac{1}{2}^-$ the angle is $-31.7°$ vs $-32°$ experimentally, where these angles are defined by,

$$|NJ^-\rangle = -\sin\theta \, |S = \tfrac{3}{2}, J\rangle + \cos\theta \, |S = \tfrac{1}{2}, J\rangle. \qquad 3.35$$

We emphasize that the a priori relation of contact to tensor interactions given by QCD leads to good agreement with data. The best constraint on this relation comes from the decay data, which leads to mixing among the two N states (with total quark spin $S = \frac{1}{2}$ and $\frac{3}{2}$) with $J^P = \frac{1}{2}^-$ or $\frac{3}{2}^-$ and the three Λ states (with $S = \frac{1}{2}$ in a flavor singlet and octet, and $S = \frac{3}{2}$ in a flavor octet) with $J^P = \frac{1}{2}^-$ or $\frac{3}{2}^-$, and the three Σ states (with $S = \frac{1}{2}$ in a flavor octet and decuplet, and $S = \frac{3}{2}$ in a flavor octet) with $J^P = \frac{1}{2}^-$ or $\frac{3}{2}^-$.

With different masses for the strange and nonstrange quarks, Isgur & Karl account for the reversal of order of $\Lambda \frac{5}{2}^-(1830)$ and $\Sigma\frac{5}{2}^-(1765)$ compared to the ground state $\Sigma \frac{1}{2}^+(1190)$ and $\Lambda \frac{1}{2}^+(1115)$ by showing that the $\Lambda \frac{5}{2}^-$ has the orbital angular momentum in the higher frequency ρ mode, while the $\Sigma \frac{5}{2}^-$ has it in the lower frequency λ mode. As mentioned above, there is no need to treat the nonstrange and strange quarks as indistinguishable via the generalized Pauli principle; the mass splitting of s from u and d makes it convenient to distinguish s from the other quarks.

Isgur & Karl find that the spin-orbit interaction that occurs in the Breit Hamiltonian can be present at most at one tenth of its expected strength, in agreement with indications from charmonium (38). The $SU(6)_f$ phenomenological analysis assumed that spin-orbit interactions were more important than the spin-spin interactions. Isgur & Karl speculate that, among other possibilities, Thomas precession may generate a spin-orbit interaction as a correction to the spin-independent long-range confining potential, and that this may cancel the Breit spin-orbit interaction. They point out that the $(70,1^-)$ spectrum supports the usual assumption that the confining potential is mass independent. Table 5 gives

Table 5   Reductions of the 70 and the $(70,1^-)$

Symmetry	Representations
$SU(6)_f$	70
$[SU(3)_f, SU(2)_S]$	$(1+8+10,2)+(8,4)$
$[SU(6)_f, O(3)^P]$	$(70,1^-)$
$[SU(3)_f, SU(2)_J]$	$(1+8+10,2+4)+(8,2+4+6)$

Table 6   The $(70,1^-)$ baryon supermultiplet

State	$J^P = (\tfrac{5}{2})^-$	$J^P = (\tfrac{3}{2})^-$	$J^P = (\tfrac{1}{2})^-$
^4P	Ξ 1930[a]	1985[a] 1900–1970?[b]	1900[a]
	Σ 1760 1766–1780[b]	1815	1750 1700–1790??[b]
	Λ 1815 1810–1840	1880	1800 (1825)
	N 1670 1660–1685	1745 (1710)	1655 1660–1690
^2P	Ω	2020[a]	2020[a]
	Ξ	1920	1930
	Σ	1805	1810 1700–1790??
	Δ	1685 1650–1720[b]	1685 1615–1695[b]
	Ξ	1800 1800–1850?	1780
	Σ	1675 1660–1680	1650 (1630)
	Λ	1690 1685–1695	1650 1660–1680
	N	1535 1510–1530	1490 1500–1530
	Λ	1490 1519	1490 1405

[a] Predicted masses in MeV/c^2 from Reference 20 are in the left columns. The states are mixed; see 20 for mixing angles. We placed each resonance in the state with the largest amplitude in the $[SU(3)_f, SU(2)_S]$ basis.

[b] Observed masses in MeV/c^2 from Reference 31 are in the right columns. Entries in parentheses are nominal masses for less well-established resonances. A single question mark means that J^P has not been measured. The double question mark means that the experimental mass range may contain two resonances.

the $[SU(3)_f, SU(2)_J]$ content of the $(70,1^-)$. Table 6 gives the predicted and observed masses of the $(70,1^-)$ baryon supermultiplet.

All parameters in the calculation of the masses and mixing angles in the $(70,1^-)$ are determined by the $(56,0^+)$. This is a striking success of QCD applied to the constituent quark model, and is important progress toward a fundamental theory of baryon spectroscopy.

3.2   Baryons With Charm

We now discuss baryons that contain charm quarks, using $SU(4)_f$ to replace $SU(3)_f$, and $SU(8)_f$ to replace $SU(6)_f$. For our discussion of the $SU(4)_f$ structure of baryons, we use the reduction $SU(4)_f \to SU(3)_f \times U(1)_c$, where here c stands for charm. The quark quartet is

$$4 \to 3_0 + 1_1, \text{ under } SU(4)_f \to SU(3)_f \times U(1)_c, \qquad 3.36$$
$$\square \longrightarrow (\square, \cdot) + (\cdot, \square)$$

where we use the notation $[\dim SU(3)_f]_C$, and C is the charm quantum number, for the $SU(3)_f \times U(1)_c$ reductions, and we give the Young tableaux for $[SU(3)_f, U(1)_c]$ underneath. The two-quark $SU(4)_f$ states are

$$4 \times 4 = 10 + 6; \qquad 3.37$$
$$\square \times \square = \square\square + \square\!\square$$

their reductions are

$$10 \to 6_0 + 3_1 + 1_2 \qquad 3.38$$
$$\square\square \longrightarrow (\square\square, \cdot) + (\square, \square) + (\cdot, \square\square)$$

and

$$6 \to 3^*_0 + 3_1 \qquad 3.39$$
$$\square\!\square \longrightarrow (\square\!\square, \cdot) + (\square, \square)$$

under $SU(4)_f \to SU(3)_f \times U(1)_c$. The three-quark $SU(4)_f$ states are

$$4 \times 4 \times 4 = 20_S + 20_M + 20_M + 4^*_A; \qquad 3.40$$
$$\square \times \square \times \square = \square\square\square + \square\square\!\square + \square\square\!\square + \square\!\square\!\square$$

their reductions are

$$20_S \to 10_0 + 6_1 + 3_2 + 1_3, \qquad 3.41$$
$$\square\square\square \longrightarrow (\square\square\square, \cdot) + (\square\square, \square) + (\square, \square\square) + (\cdot, \square\square\square)$$

$$20_M \to 8_0 + 6_1 + 3^*_1 + 3_2, \qquad 3.42$$
$$\square\square\!\square \longrightarrow (\square\square\!\square, \cdot) + (\square\square, \square) + (\square\!\square, \square) + (\square, \square\square)$$

and

$$4^*_A \rightarrow 1_0 + 3^*_1,$$
3.43

$$\square\!\!\!\!\square\!\!\!\!\square \rightarrow \left(\square\!\!\!\!\square\, , \cdot\right) + \left(\square\!\!\!\!\square\, , \square\right)$$

under $SU(4)_f \rightarrow SU(3)_f \times U(1)_c$. The 20_S, 20_M, and 4^*_A can occur as color-singlet baryons; it is worthwhile to have some feeling for them. Lichtenberg (39) introduced the notation given in Table 7 for charmed baryons; his rule is that a baryon with isospin I, strangeness S, and charm C is given the letter associated with an uncharmed baryon of isospin I and strangeness $S-C$ and a subscript C. The electric charge is given, as usual, as a superscript. We also give the notation of Reference 40 in Table 7. Gaillard et al (8) and Lichtenberg (5) give diagrams of the 20_S, 20_M, and 4^*_A. In particular, the 20_S includes the usual $C = 0$ $SU(3)_f$ decuplet; the 20_M includes the usual octet, and the 4^*_A includes the usual $C = 0$ singlet.

Taking the quark spin into account, the quark is an 8 under $SU(8)_f$ with the reduction

$$8 \rightarrow (4,2), \text{ under } SU(8)_f \rightarrow SU(4)_f \times SU(2)_S.$$
3.44

$$\square \longrightarrow (\square, \square)$$

Table 7 Properties and notations for charmed baryons

Quark content	I	I_3	S	Y	C	Z	Q	Symbol[a]	Symbol[b]
cuu	1	1	0	$\frac{2}{3}$	1	$-\frac{1}{4}$	2	Σ_1^{++}	C_1^{++}
cud	$\begin{cases}1\end{cases}$	0	0	$\frac{2}{3}$	1	$-\frac{1}{4}$	1	Σ_1^+	C_1^+
	$\begin{cases}0\end{cases}$	0	0	$\frac{2}{3}$	1	$-\frac{1}{4}$	1	Λ_1^+	C_0^+
cdd	1	-1	0	$\frac{2}{3}$	1	$-\frac{1}{4}$	0	Σ_1^0	C_1^0
cus	$\frac{1}{2}$	$\frac{1}{2}$	-1	$-\frac{1}{3}$	1	$-\frac{1}{4}$	1	Ξ_1^+	$\begin{matrix}A^{+c}\\S^+\end{matrix}$
cds	$\frac{1}{2}$	$-\frac{1}{2}$	-1	$-\frac{1}{3}$	1	$-\frac{1}{4}$	0	Ξ_1^0	$\begin{matrix}A^0\\S^0\end{matrix}$
css	0	0	-2	$-\frac{4}{3}$	1	$-\frac{1}{4}$	0	Ω_1^0	T^0
ccu	$\frac{1}{2}$	$\frac{1}{2}$	0	$\frac{1}{3}$	2	$-\frac{5}{4}$	2	Ξ_2^{++}	X_u^{++}
ccd	$\frac{1}{2}$	$-\frac{1}{2}$	0	$\frac{1}{3}$	2	$-\frac{5}{4}$	1	Ξ_2^+	X_d^+
ccs	0	0	-1	$-\frac{2}{3}$	2	$-\frac{5}{4}$	1	Ω_2^+	X_s^+
ccc	0	0	0	0	3	-2	2	Ω_3^{++}	Θ^{++}

[a] Notation from Reference 39.
[b] Notation from Gaillard, M. K. et al, Reference 8.
[c] A and S stand for states in the antisymmetric $SU(3)_f$ 3^* and the symmetric 6, respectively.

Here the notation is dim $SU(8)_f \to [\dim SU(4)_f, \dim SU(2)_S]$. The two-quark states are

$$8 \times 8 = 36 + 28;\qquad\qquad 3.45$$

$$\square \times \square = \square\square + \begin{array}{c}\square\\\square\end{array}$$

their reductions are

$$36 \to (10,3) + (6,1)\qquad\qquad 3.46$$

$$\square \to (\square\square, \square\square) + \left(\begin{array}{c}\square\\\square\end{array}, \begin{array}{c}\square\\\square\end{array}\right)$$

and

$$28 \to (6,3) + (10,1).\qquad\qquad 3.47$$

$$\begin{array}{c}\square\\\square\end{array} \to \left(\begin{array}{c}\square\\\square\end{array}, \square\square\right) + \left(\square\square, \begin{array}{c}\square\\\square\end{array}\right)$$

The three-quark states are

$$8 \times 8 \times 8 = 120 + 168 + 168 + 56;\qquad\qquad 3.48$$

$$\square \times \square \times \square = \square\square\square + \begin{array}{l}\square\square\\\square\end{array} + \begin{array}{l}\square\square\\\square\end{array} + \begin{array}{c}\square\\\square\\\square\end{array}$$

their reductions are

$$120 \to (20_S,4) + (20_M,2),\qquad\qquad 3.49$$

$$\square\square\square \to (\square\square\square, \square\square\square) + \left(\begin{array}{l}\square\square\\\square\end{array}, \begin{array}{l}\square\square\\\square\end{array}\right)$$

$$168 \to (20_M,4) + (20_S,2) + (20_S,2) + (4^*_A,2),\qquad\qquad 3.50$$

$$\begin{array}{l}\square\square\\\square\end{array} \to \left(\begin{array}{l}\square\square\\\square\end{array}, \square\square\square\right) + \left(\square\square\square, \begin{array}{l}\square\square\\\square\end{array}\right) + \left(\square\square\square, \begin{array}{l}\square\square\\\square\end{array}\right) + \left(\begin{array}{c}\square\\\square\\\square\end{array}, \begin{array}{l}\square\square\\\square\end{array}\right)$$

and

$$56 \to (4^*_A,4) + (20_M,2),\qquad\qquad 3.51$$

$$\begin{array}{c}\square\\\square\\\square\end{array} \to \left(\begin{array}{c}\square\\\square\\\square\end{array}, \square\square\square\right) + \left(\begin{array}{l}\square\square\\\square\end{array}, \begin{array}{l}\square\square\\\square\end{array}\right)$$

under $SU(8)_f \to SU(4)_f \times SU(2)_S$. Using the generalized Pauli principle, we find that the color-singlet baryons have the $SU(8)_f$ space state

$$\Psi(\text{tot}) = \sum \Psi[SU(8)_f] \times \Psi(\text{space})$$

$$= \square\square\square \quad \times \quad \square\square\square$$

$$+ \begin{array}{l}\square\square\\\square\end{array} \quad \times \quad \begin{array}{l}\square\square\\\square\end{array}$$

$$+ \begin{array}{c}\square\\\square\\\square\end{array} \quad \times \quad \begin{array}{c}\square\\\square\\\square\end{array}\qquad\qquad 3.52$$

The ground-state baryon supermultiplet will have a symmetric space state and will be in the 120 of $SU(8)_f$, which includes a spin $\frac{3}{2}$ 20_S and spin $\frac{1}{2}$ 20_M in $SU(4)_f$.

Since the charm quark mass is much larger than the u, d, and s masses, the reduction $SU(8)_f \to SU(6)_f \times SU(2)_{cs}$, where $SU(6)_f$ is associated with the uncharmed quarks and $SU(2)_{cs}$ is the charm quark spin group, is useful (41). The analogous reduction $SU(6)_f \to SU(4) \times SU(2)_{ss}$, where this $SU(4)$ is associated with the nonstrange quarks, and $SU(2)_{ss}$ is the strange quark spin group, has been used (42). These reductions of the three-quark states are

$$120 \to (56,1) + (21,2) + (6,3) + 1,4), \qquad\qquad 3.53$$

$$\square\square\square \longrightarrow (\square\square\square, \cdot) + (\square\square, \square) + (\square, \square\square) + (\cdot, \square\square\square)$$

$$168 \to (70,1) + (6,1) + (21,2) + (15,2) + (1,2) + (6,3), \qquad 3.54$$

$$\overset{\square\square}{\square} \to (\overset{\square\square}{\square}, \cdot) + (\square, \overset{\square}{\overset{\square}{\square}}) + (\square\square, \square) + (\overset{\square}{\overset{\square}{\square}}, \square) + (\cdot, \overset{\square\square}{\square}) + (\square, \square\square)$$

and

$$56 \to (20,1) + (6,1) + (15,2), \qquad\qquad 3.55$$

$$\overset{\square}{\overset{\square}{\square}} \to (\overset{\square}{\overset{\square}{\square}}, \cdot) + (\square, \overset{\square}{\overset{\square}{\square}}) + (\overset{\square}{\overset{\square}{\square}}, \square)$$

under $SU(8)_f \to SU(6)_f \times SU(2)_{cs}$, where the entries on the right-hand side are [dim $SU(6)_f$, dim $SU(2)_{cs}$]. To obtain the total spin angular momentum, S, the noncharmed spin and charmed spin must be coupled. For the 120, the states in the $SU(4)_f \times SU(2)_S$ and $SU(6)_f \times SU(2)_{cs}$ reductions are the same; however, in general the states in one reduction are linear combinations of those in the other.

We have given the analysis of the states of charmed baryons in some detail, both because it is relatively unfamiliar and because the classification of states is necessary before analyzing mass splittings using either the $SU(8)_f$ analog of the phenomenological mass analysis given earlier or the charm analog of the analysis using the hyperfine interaction guessed on the basis of QCD. The phenomenological analysis (41), using one- and two-body and octet dominance, requires nine parameters, four of which are determined by the $C = 0$ 56. There are not sufficient data to determine the remaining five parameters, although some conclusions, which agree with the QCD-based analysis, can be drawn.

DeRújula, Georgi & Glashow (19) initiated the analysis of mass splittings based on QCD before either charmed baryons or charmed mesons were found. Lee, Quigg & Rosner (43) updated this analysis using the neutrino event of Reference 44 and the photoproduction data of Reference 45. Lee, Quigg & Rosner assign the peak at 2250 MeV/c^2 in $\bar{\Lambda}\pi^+\pi^-\pi^-$ to the antiparticle of Λ_1^+, and the peak at 2500 MeV/c^2 in $\bar{\Lambda}\pi^+\pi^+\pi^-\pi^-$ as due to the combined effects of the antiparticles of $\Sigma_1^0\frac{1}{2}^+$ and $\Sigma_1^0\frac{3}{2}^+$, both decaying to $\bar{\Lambda}_1^-\pi^+$. They identify the neutrino-induced event with the antiparticles of $\Sigma_1^{++} \to \Lambda_1^+\pi^+ \to \Lambda\pi^+\pi^+\pi^+\pi^-$. They also discuss weak decays of charmed baryons, electromagnetic pair production, and strong decay widths, among other topics. Dalitz (17) pointed out that the neutrino excitation

$$\nu_\mu p \to \mu^- \Sigma_1^{++}\frac{3}{2}^+ \qquad\qquad 3.56$$

is analogous to

$$\nu_\mu p \to \mu^- \Delta^{++}, \qquad\qquad 3.57$$

which has a large cross section at lower energies. The papers of Reference 46 discuss Regge trajectories and Reference 47 discusses electromagnetic mass splittings for charmed baryons.

3.3 *Mesons*

The spin and flavor degrees of freedom are as important for meson spectroscopy as for baryon spectroscopy; however the spectrum of (color-singlet) meson states and the matrix elements involving a single current between a pair of meson states are independent of the color degree of freedom (9, 11). The color-singlet meson with color group SU(p)$_c$ has the wave function $p^{-\frac{1}{2}}\Sigma_1^p q_i\bar{q}_i$, and the color-singlet currents have the form $\Sigma_1^p \bar{q}_i\,\Gamma\,q_i$, where Γ is constructed from Dirac and flavor matrices, so that the factor of p coming from the sums in the currents cancels the two $p^{-\frac{1}{2}}$ normalization factors of the states.

The quark model mnemonic that a meson is a nonrelativistic q$\bar{q}$ bound state implies that the parity, P, charge conjugation, C, (defined only for neutral $Y = 0$ mesons), G-parity, G, relative orbital angular momentum, $\mathbf{L}$, q$\bar{q}$ spin, $\mathbf{S}$, total angular momentum, $\mathbf{J}$, and isospin, I, are related by

$$P = (-1)^{L+1}, C = (-1)^{L+S}, G = (-1)^{L+S+I}, \mathbf{J} = \mathbf{L} + \mathbf{S}. \qquad 3.58$$

The relation for C implies that the following J^{PC} values do not occur:

$$0^{--}; 0^{+-}, 1^{-+}, 2^{+-}, \ldots. \qquad\qquad 3.59$$

We exclude these J^{PC} values for relativistic q$\bar{q}$ states, because they have negative norm (48). The SU(6)$_f$ content of the ground-state mesons is $1 + 35$ [with charm, $1 + 63$ of SU(8)$_f$] with unitary spin and spin content given in Section 1.

For the lowest baryons, which are the nucleon $\frac{1}{2}^+$ octet and the $\frac{3}{2}^+$

Table 8 Ideal nonet vector meson states

Meson	Spin and flavor wavefunction[a]
ρ^+	$u^\uparrow \bar{d}^\uparrow$
ρ^0	$2^{-\frac{1}{2}}(u^\uparrow \bar{u}^\uparrow - d^\uparrow \bar{d}^\uparrow)$
ρ^-	$d^\uparrow \bar{u}^\uparrow$
K^{++}	$u^\uparrow \bar{s}^\uparrow$
K^{*0}	$d^\uparrow \bar{s}^\uparrow$
$\bar{K}^{*0}$ $-\Lambda$	$s^\uparrow \bar{u}^\uparrow$
$\bar{K}^{*-}$	$s^\uparrow \bar{d}^\uparrow$
ω^0	$2^{-\frac{1}{2}}(u^\uparrow \bar{u}^\uparrow + d^\uparrow \bar{d}^\uparrow)$
ϕ^0	$s^\uparrow \bar{s}^\uparrow$

[a] We give states with $J_3 = 1$.

decuplet, no two states have the same I, Y, and J, and thus mixing of $SU(3)_f$ multiplets cannot occur. For the lowest mesons (with $L = 0$ and no radial excitations) there are states, ω and ϕ with $I = Y = 0$, $J = 1$, and η and η' with $I = Y = 0$, $J = 0$, which are mixtures (really superpositions) of $SU(3)_f$ singlet, 1, and octet, 8, states with $I = Y = 0$. For the vector mesons ω and ϕ, this mixing is well accounted for by assuming that ω is almost entirely composed of the nonstrange quarks u and d (and their antiparticles) and that ϕ is almost a pure ss̄ state. Table 8 gives this "ideal" nonet mixing pattern for the vector mesons. The ideal mixing is

$$\omega = |1\rangle \cos\theta + |8\rangle \sin\theta, \quad \phi = -|1\rangle \sin\theta + |8\rangle \cos\theta, \qquad 3.60$$

with a mixing angle θ of arctan $2^{-\frac{1}{2}} \sim 35°$. Isgur (49) gave a qualitative argument for the observed singlet-octet mixing angle of about $-10°$ for η and η'. He observed that two simple cases of two-by-two mixings occur when the equal off-diagonal elements of the mass or mass squared matrix are either much smaller than or much larger than the difference of the diagonal elements. With strange and nonstrange $q\bar{q}$ states, $n = 2^{-\frac{1}{2}}(u\bar{u} + d\bar{d})$, $s = s\bar{s}$, as the basis, the former case leads to the ideal mixing of Equation 3.60 with pure s and n states and $\theta \sim 35°$, while the latter case leads to $45°$ mixing of the n and s states,

$$\eta = 2^{-\frac{1}{2}}(n - s), \quad \eta' = 2^{-\frac{1}{2}}(n + s), \qquad 3.61$$

and a mixing angle θ of arctan $[(1 - 2^{\frac{1}{2}})/(1 + 2^{\frac{1}{2}})] \sim -10°$ in good agreement with the experimental value of $-11°$. The mixing mechanism of $q\bar{q}$ annihilation via color gluons that produces J/ψ decay is the most likely

Table 9 Pseudoscalar meson states

Meson	Spin and flavor wave function
π^+	$2^{-\frac{1}{2}}(u{\uparrow}\bar{d}{\downarrow} - u{\downarrow}\bar{d}{\uparrow})$
π^0	$2^{-1}(u{\uparrow}\bar{u}{\downarrow} - u{\downarrow}\bar{u}{\uparrow} - d{\uparrow}\bar{d}{\downarrow} + d{\downarrow}\bar{d}{\uparrow})$
π^-	$2^{-\frac{1}{2}}(d{\uparrow}\bar{u}{\downarrow} - d{\downarrow}\bar{u}{\uparrow})$
K^+	$2^{-\frac{1}{2}}(u{\uparrow}\bar{s}{\downarrow} - u{\downarrow}\bar{s}{\uparrow})$
K^0	$2^{-\frac{1}{2}}(d{\uparrow}\bar{s}{\downarrow} - d{\downarrow}\bar{s}{\uparrow})$
$\bar{K}^0$	$2^{-\frac{1}{2}}(s{\uparrow}\bar{d}{\downarrow} - s{\downarrow}\bar{d}{\uparrow})$
$\bar{K}^-$	$2^{-\frac{1}{2}}(s{\uparrow}\bar{u}{\downarrow} - s{\downarrow}\bar{u}{\uparrow})$
η	$2^{-\frac{3}{2}}(u{\uparrow}\bar{u}{\downarrow} - u{\downarrow}\bar{u}{\uparrow} + d{\uparrow}\bar{d}{\downarrow} - d{\downarrow}\bar{d}{\uparrow}) - 2^{-1}(s{\uparrow}\bar{s}{\downarrow} - s{\downarrow}\bar{s}{\uparrow})$
η'	$2^{-\frac{3}{2}}(u{\uparrow}\bar{u}{\downarrow} - u{\downarrow}\bar{u}{\uparrow} + d{\uparrow}\bar{d}{\downarrow} - d{\downarrow}\bar{d}{\uparrow}) + 2^{-1}(s{\uparrow}\bar{s}{\downarrow} - s{\downarrow}\bar{s}{\uparrow})$

source of the mixing here (19); the mixing is large for two-gluon annihilation of pseudoscalar mesons. Table 9 gives these mesons.

The $L = 0$ to 2 mesons (apart from those related to charm) presently established are given in Table 10. The only established $L = 3$ meson is the $J^{PC} = 4^{++}$, $I^G = 0^+$ $h(2050)$.

Appelquist, Barnett & Lane (50) discuss meson resonances with hidden and explicit charm in this volume of the *Annual Review of Nuclear and Particle Science*; for this reason we do not discuss them here.

Table 10 Mesons[a] (without charm quarks)

$^{2S+1}L_J$	J^{PC}	(I,Y) (1,0)	$(\frac{1}{2}, \pm 1)$	(0,0)	(0,0)
$^1S_0$	0^{-+}	$\pi(136)$	$K(496)$	$\eta(549)$	$\eta'(958)$
$^3S_1$	1^{--}	$\rho(773)$	$K^*(892)$	$\omega(783)$	$\phi(1020)$
$^1P_1$	1^{+-}	$B(1228)$	$Q_1,Q_2(1300)^b$		
$^3P_0$	0^{++}	$\delta(976)$	$K(1250)$		
$^3P_1$	1^{++}	$A_1(1100)$	$Q_1,Q_2(1400)^b$	$D(1286)$	
$^3P_2$	2^{++}	$A_2(1310)$	$K^*(1421)$	$f(1271)$	$f'(1516)$
$^1D_2$	2^{-+}				
$^3D_1$	1^{--}				
$^3D_2$	2^{--}				
$^3D_3$	3^{--}	$g(1690)$	$K^*(1780)$	$\omega^*(1667)$	

[a] Data from Reference 31 with masses in MeV/c^2. Radial excitations exist, but are less well established; for data, see Reference 31.
[b] The Q_1 and Q_2 are mixtures of the $^1P_1$ and $^3P_1$ states.

3.4 Exotic States

The normal (nonexotic) states in the quark model are color-singlet states with the $SU(3)_f$ and J^{PC} quantum numbers of nonrelativistic qqq composites for baryons and $q\bar{q}$ composites for mesons. Restricting our discussion to color singlets, there are exotic states with (a) more quarks and antiquarks than the usual baryons and mesons, such as baryonium (51) $qq\bar{q}\bar{q}$, possible Z* baryon resonances (52), $qqqq\bar{q}$, six quark states, etc; (b) states composed only of color gluons (glueballs) (53), or with extra quarks and gluons; and (c) states with abnormal J^{PC} values such as occur in $q\bar{q}$ models of mesons analyzed with the Bethe-Salpeter equation (54). Jaffe (55) has analyzed baryonium through use of the MIT bag model with the color spin-spin $SU(6)_c$ symmetry [analogous to the unitary spin-spin $SU(6)_f$] suggested by QCD. He showed that the lightest baryonium states, which he called "cryptoexotic," will have normal $SU(3)_f$ quantum numbers. (See 56 for references to data on possible baryonium states.)

3.5 Appendix on Practical Group Theory Calculations

We collect practical rules to do group theory calculations relevant to hadron spectroscopy using $SU(n)$ groups (57). The irreducible representations of $SU(n)$ (and of the permutation group S_v, $v = \sum_i \alpha_i$) are uniquely labeled by the Young tableaux with j rows and α_i boxes in the ith row, subject to $\alpha_1 \geq \alpha_2 \geq \cdots \geq \alpha_j$, and $j < n$. Use $(\alpha_1, \alpha_2, \ldots, \alpha_j)$ as a symbol for both the Young tableau and the irreducible representation. The dimension of $(\alpha_1, \alpha_2, \ldots, \alpha_j)$ equals the number of standard arrangements of the tableau, where a standard arrangement has a number from one to n in each box and the numbers increase from top to bottom in a column and do not decrease from left to right in a row. Counting standard arrangements is tedious. A simpler way to find the dimension is to calculate the ratio N/D, where N is the product of the array obtained by inserting integers in the boxes starting with n in the upper left corner [for $SU(n)$], increasing by one going to the right, and decreasing by one going down. The number D is the product of the array obtained by counting for each box the number of boxes that the following line passes through: for the kth box in the jth row, start outside the diagram on the right, move horizontally to the left to this (row, column) $= (j,k)$ box, then move vertically down until the path leaves the diagram.

The complex conjugate representation to $(\alpha_1, \alpha_2, \ldots, \alpha_j)$ is the complement with respect to the rectangle $\alpha_1 \times n$:

$$(\underbrace{\alpha_1, \alpha_1, \ldots, \alpha_1}_{n-j \text{ times}}, \alpha_1 - \alpha_j, \alpha_1 - \alpha_{j-1}, \ldots, \alpha_1 - \alpha_2). \qquad 3.62$$

The fundamental representation with just one box, (1), has dimension n. The complex conjugate of (1) is $(1, 1, \ldots)$, where 1 occurs $n - 1$ times, and also has dimension n. The (self-conjugate) adjoint representation $(2, 1, \ldots, 1)$, where 1 occurs $n - 2$ times, has dimension $n^2 - 1$. The representation with no boxes at all is the identity representation or singlet of dimension one and is represented by a dot. A column with n boxes is equivalent to the singlet and can be omitted (it is not a proper tableau). A column with more than n boxes (also not a proper tableau) vanishes.

We have discussed treating a single quark as a 3 of $SU(3)_f$, as a 6 of $SU(6)_f$, as a 4 of $SU(4)_f$, etc, earlier in this section. We have given results for (a) the $SU(6)_f$ irreducible decomposition of three-quark states, which is the reduction of the outer product $6 \times 6 \times 6$ into irreducibles of $SU(6)$; (b) the $SU(3)_f \times U(1)_c$ irreducible decomposition of a given $SU(4)_f$ irreducible representation; and (c) the $SU(3)_f \times SU(2)_s$ irreducible decomposition of a given $SU(6)_f$ irreducible representation. Case (a) is an example of the reduction of the outer product of representations and occurs when we construct hadrons out of quarks and antiquarks where we use the generalized Pauli principle and treat the quarks and antiquarks as identical. Cases (b) and (c) occur when we want to know the content of a given hadron supermultiplet in terms of its quantum numbers and Casimir operators under a reduction of the type (b) $SU(m + n) \to SU(m) \times SU(n)$, or (c) $SU(mn) \to SU(m) \times SU(n)$. We discuss these three cases in turn.

Case (a): For most cases of interest, the reduction of the outer product of irreducibles R and S into irreducibles T_j can be guessed and checked using

$$(\dim R)(\dim S) = \sum_j (\dim T_j), \qquad\qquad 3.63$$

where a given T_j can occur more than once, without using all the fine details of the general rule.

The general rule is as follows: Draw the Young tableaux for R and for S. In one of the tableaux (say that for S) place the number i in each box in the ith row (for all rows). Take a box from the first row of S (with its number, one) and add it to R in all possible ways so that the new tableau is proper. Repeat this process with another box (if there is one) from the first row of S. Continue until the first row of S is exhausted. Repeat this for the second and lower rows of S until S is exhausted. Discard the resulting tableaux T_j in which the same number occurs twice in a column (i.e. in which boxes in a row of S occur in a column of a T_j). Discard T_j if, reading from right to left across the first row, then from right to left across the second row, etc, through T_j, the number of boxes passed with number i ever exceeds the number passed with $i - 1$. Finally, if more than one identical T_j has the identical pattern of numbers in it, count that T_j only once.

Case (b): For the reduction $SU(m + n) \to SU(m) \times SU(n)$, we give a procedure involving some guesswork, which works for most cases of practical interest. Consider

$$m + n \to (m,1) + (1,n) \qquad\qquad 3.64$$
$$\square \to (\square, \cdot) + (\cdot, \square)$$

under $SU(m + n) \to SU(m) \times SU(n)$. Build up the representation of interest in steps from properly symmetrized products of $(m + n)$, and use Expression 3.64 to perform the reduction. For example,

$$(m + n) \times (m + n) \to [(m,1) + (1,n)] \times [(m,1) + (1,n)]$$
$$\square \times \square \to [(\square,\cdot) + (\cdot,\square)] \times [(\square,\cdot) + (\cdot,\square)]$$

$$= (m,1) \times (m,1) + (m,1) \times (1,n) + (1,n) \times (m,1) + (1,n) \times (1,n)$$
$$(\square,\cdot) \times (\square,\cdot) + (\square,\cdot) \times (\cdot,\square) + (\cdot,\square) \times (\square,\cdot) + (\cdot,\square) \times (\cdot,\square)$$

$$= [\tfrac{1}{2}m(m+1),1] + [\tfrac{1}{2}m(m-1),1] + (m,n) + (m,n) +$$
$$(\boxminus\boxminus, \cdot) + \left(\boxminus, \cdot\right) + (\square,\square) + (\square,\square) +$$

$$+ [1,\tfrac{1}{2}n(n+1)] + [1,\tfrac{1}{2}n(n-1)] \qquad\qquad 3.65$$
$$+ (\cdot, \boxminus\boxminus) + \left(\cdot, \boxminus\right)$$

under $SU(m+n) \to SU(m) \times SU(n)$. We guess

$$\tfrac{1}{2}(m+n)(m+n+1) \to [\tfrac{1}{2}m(m+1),1] + (m,n) + [1,\tfrac{1}{2}n(n+1)] \qquad 3.66$$
$$\boxminus\boxminus \to (\boxminus\boxminus, \cdot) + (\square, \square) + (\cdot, \boxminus\boxminus)$$

and

$$\tfrac{1}{2}(m+n-1)(m+n) \to [\tfrac{1}{2}m(m-1),1] + (m,n) + [1,\tfrac{1}{2}n(n-1)], \qquad 3.67$$
$$\boxminus \to (\boxminus, \cdot) + (\square, \square) + (\cdot, \boxminus)$$

under $SU(m+n) \to SU(m) \times SU(n)$, and check that the dimensions add up correctly; i.e. $\tfrac{1}{2}(m+n)(m+n+1) = \tfrac{1}{2}m(m+1) + mn + \tfrac{1}{2}n(n+1)$, etc. To find reductions for three-particle states, which are relevant to baryons, use Expressions 3.66 and 3.67 and

$$\boxminus\boxminus \times \square = \boxminus\boxminus\boxminus + \boxminus\boxplus, \quad \boxminus \times \square = \boxminus\boxplus + \boxminus . \qquad 3.68$$

Case (c): For $SU(mn) \to SU(m) \times SU(n)$ we propose the procedure analogous to the case of $SU(m+n)$, and again illustrate it on the reduction of the two-particle state. We use

$$mn \to (m,n) \text{ under } SU(mn) \to SU(m) \times SU(n). \qquad 3.69$$
$$\square \to (\square, \square)$$

For

$$(mn) \times (mn) \to (m,n) \times (m,n)$$
$$\square \times \square \to (\square, \square) \times (\square, \square)$$
$$= (m \times m, n \times n)$$
$$(\square \times \square, \square \times \square)$$
$$= [\tfrac{1}{2}m(m+1)+\tfrac{1}{2}(m-1)m, \tfrac{1}{2}n(n+1)+\tfrac{1}{2}(n-1)n, \qquad 3.70$$
$$\left(\square\square + \square\!\!\!\!\square , \ \square\square + \square\!\!\!\!\square \right)$$

under $SU(mn) \to SU(m) \times SU(n)$. We now guess

$$\tfrac{1}{2}mn(mn+1) \to [\tfrac{1}{2}m(m+1), \tfrac{1}{2}n(n+1)] + [\tfrac{1}{2}(m-1)m, \tfrac{1}{2}(n-1)n] \qquad 3.71$$
$$\square\square \to (\square\square, \square\square) + \left(\square\!\!\!\!\square , \square\!\!\!\!\square \right)$$

and

$$\tfrac{1}{2}(mn-1)mn \to [\tfrac{1}{2}m(m+1), \tfrac{1}{2}(n-1)n] + [\tfrac{1}{2}(m-1)m, \tfrac{1}{2}n(n+1)], \qquad 3.72$$
$$\square\!\!\!\!\square \to (\square\square, \square\!\!\!\!\square) + \left(\square\!\!\!\!\square , \square\square \right)$$

under $SU(mn) \to SU(m) \times SU(n)$. We again check that the dimensions add up, i.e. $\tfrac{1}{2}mn(mn+1) = \tfrac{1}{4}m(m+1)n(n+1) + \tfrac{1}{4}(m-1)m(n-1)n$, etc. The procedure using Expression 3.68 as in case (b) leads to reductions for three-particle states.

In some cases, for example in analyzing mesons, it may be convenient to use products of the one-box fundamental representation and its complex conjugate to build the representation of interest.

4   QUARK-PARTONS IN INCLUSIVE COLLISIONS

4.1   *Inclusive Lepton-Hadron Scattering*

We discuss inclusive lepton-nucleon scattering as a prototype of processes that have been studied using the quark-parton model. Yan (58) reviewed the parton model in 1976; other recent reviews are given in Reference 59. We do not add to the brief discussions of kinematics and the basic assumptions of the naive parton model, nor to the list of processes included in $lN \to l'X$, given in Section 1. For an incident beam of charged leptons, the initial lepton energy E, the scattered lepton E', and the laboratory scattering angle θ are measured. We usually refer to this experimental situation. Assuming that a single photon, W boson, or Z boson mediates this process in lowest order, the matrix element factors into a known part describing the lepton vertex, the known boson propagator, and an unknown part describing the hadron vertex. The hadron vertex enters quadratically into the inclusive cross section. The contribution of the hadron vertex to this cross section depends on the scalar structure functions $W_{1,2}(v,Q^2)$ [and $W_3(v,Q^2)$ for neutrino and antineutrino processes], where

$$m_N v = P \cdot q, \, Q^2 = -q^2, \tag{4.1}$$

q is the four-momentum transfer at the lepton vertex and also the off-shell virtual boson momentum in the scattering of the boson from the nucleon, and P is the nucleon momentum. The measured quantities E, E', and θ determine v and Q^2. The dimensionless ratio

$$x = Q^2/(2m_N v), \tag{4.2}$$

the Bjorken scaling variable, plays an important role in our further discussion. The dimensionless variable

$$y = v/E \tag{4.3}$$

is also useful. In the Bjorken limit in which v and Q^2 both become large, and x is fixed, the two structure functions for electroproduction are found to be functions of x alone, rather than of v and Q^2 separately:

$$m_N W_1(v,Q^2) \to F_1(x), \, v W_2(v,Q^2) \to F_2(x). \tag{4.4}$$

If only spin $\frac{1}{2}$ partons contribute, the Callan-Gross (26) relation

$$F_2(x) = 2x F_1(x) \tag{4.5}$$

holds; this relation agrees with experiment for $x \gtrsim 0.3$, but deviates from experiment by about 20% for $x \lesssim 0.3$. [The low-x or "wee" parton region requires special treatment (21).] Assuming the Callan-Gross relation, the inclusive electroproduction cross section in the scaling limit is

$$d^2\sigma/dx \, dy = 4\pi\alpha^2 m_N E(q^2)^{-2} F_2(x)(2-2y+y^2). \tag{4.6}$$

We take the validity of the Callan-Gross relation as an indication that the active partons in electroproduction are quarks. Quarks contribute to inclusive lepton-hadron scattering via their flavor quantum numbers; in particular, via their electric charge for electroproduction. We label the parton distribution functions by the name of the quark in the proton. Then isospin symmetry implies that the distribution of, for example, the d (u) quark in the neutron is the same as the distribution of the u (d) quark in the proton, and that the strange and charm quarks to good approximation have the same distribution in the proton and neutron. Thus the structure functions for the proton and neutron can be expressed in terms of a single set of distributions u, d, s, and c and their antiquark distributions. In particular, the F_2 functions for inelastic electron-proton and electron-neutron scattering below charm threshold are

$$F_2^{ep}(x) = x\{\tfrac{4}{9}[u(x) + \bar{u}(x)] + \tfrac{1}{9}[d(x) + \bar{d}(x) + s(x) + \bar{s}(x)]\} \qquad 4.7$$

and

$$F_2^{en}(x) = x\{\tfrac{4}{9}[d(x) + \bar{d}(x)] + \tfrac{1}{9}[u(x) + \bar{u}(x) + s(x) + \bar{s}(x)]\}. \qquad 4.8$$

For neutrino- or antineutrino-nucleon inclusive scattering the charged current cross section in the scaling limit is

$$d^2 \sigma^{\nu N, \bar{\nu} N}/dx\, dy = (G^2 m_N E/\pi)\big[(1 - y + \tfrac{1}{2}y^2)F_2^{\nu N, \bar{\nu} N}(x)$$

$$\mp xy(1 - \tfrac{1}{2}y)\, F_3^{\nu N, \bar{\nu} N}(x)\big], \qquad 4.9$$

where the upper (lower) signs go with νN ($\bar{\nu} N$) and N is the proton p or neutron n. The proton structure functions below charm threshold are

$$F_2^{\nu p}(x) = 2x[d(x) \cos^2 \theta_C + s(x) \sin^2 \theta_C + \bar{u}(x)], \qquad 4.10$$

$$F_3^{\nu p}(x) = 2[-d(x) \cos^2 \theta_C - s(x) \sin^2 \theta_C + \bar{u}(x)], \qquad 4.11$$

$$F_2^{\bar{\nu} p}(x) = 2x[u(x) + \bar{d}(x) \cos^2 \theta_C + \bar{s}(x) \sin^2 \theta_C], \qquad 4.12$$

and

$$F_3^{\bar{\nu} p}(x) = 2[-u(x) + \bar{d}(x) \cos^2 \theta_C + \bar{\lambda}(x) \sin^2 \theta_C], \qquad 4.13$$

where θ_C is the Cabibbo angle. When the contributions from F_2 and F_3 are combined, the cross sections are

$$d\sigma^{\nu p}/dx\, dy = (G^2 m_N E/\pi)2x\,[d(x) \cos^2 \theta_C + s(x) \sin^2 \theta_C + \bar{u}(x)(1 - y)^2]$$

$$\qquad 4.14$$

and

$$d\sigma^{\bar{\nu} p}/dx\, dy = (G^2 m_N E/\pi)2x[u(x)(1 - y)^2 + \bar{d}(x) \cos^2 \theta_C + \bar{s}(x) \sin^2 \theta_C].$$

$$\qquad 4.15$$

The corresponding structure functions and cross sections for scattering by neutrons are given by interchanging u and d and $\bar{u}$ and $\bar{d}$ in the formulas above. Equations 4.14 and 4.15 illustrate the distinctive y dependences of neutrino scattering on quarks or antineutrino scattering on antiquarks:

$$d\sigma/dy \propto 1, \qquad 4.16$$

and neutrino-antiquark or antineutrino-quark scattering

$$d\sigma/dy \propto (1-y)^2. \hspace{4cm} 4.17$$

These y dependences follow from the left-handed nature of neutrinos and the chiral (left-handed) weak current of quarks. These cross sections agree well with experiment [with the demise of the "high-y" anomaly (60, 61)] and provide further support for the notion that the active partons are quarks.

Color, in contrast to flavor, does not directly manifest itself in these processes, because the electromagnetic and weak currents are color singlets, and therefore are blind to the color of both the quarks and the gluons. In particular, gluons that are flavor singlets and have only color do not directly contribute. We expect that gluons (and perhaps other inert partons as well) carry momentum. Llewellyn Smith (25) estimated the part of the momentum carried by inert partons by converting the sum rule

$$\int_0^1 [u(x)+d(x)+s(x)+\bar{u}(x)+\overline{d}(x)+\bar{s}(x)+g(x)]x \, dx = 1, \hspace{1cm} 4.18$$

where $g(x)$ is the distribution of gluons and any other inert partons, into a relation involving measurable quantities (taking $\theta_C \approx 0$)

$$\varepsilon = \int_0^1 g(x)x \, dx = 1 + \int_0^1 \left[\tfrac{3}{2} F_2^{\nu N}(x) - 9 F_2^{eN}(x)\right] dx, \hspace{1cm} 4.19$$

where here, and later, $F_2^{eN} = \tfrac{1}{2}(F_2^{ep} + F_2^{en})$ and $F_2^{\nu N} = \tfrac{1}{2}(F_2^{\nu p} + F_2^{\nu n})$. Present data give $\varepsilon \approx 0.44\text{--}0.46$, so that almost half of the nucleon momentum is carried by gluons or other inert partons. Still with $\theta_C \approx 0$, the strange parton distribution satisfies

$$0 \leq \tfrac{1}{9}x[s(x)+\bar{s}(x)] = F_2^{eN}(x) - \tfrac{5}{18} F_2^{\nu N}(x), \hspace{1cm} 4.20$$

so that, if strange partons can be neglected, the right-hand side of Equation 4.20 vanishes, and the sum rule

$$\int_0^1 F_2^{eN}(x) \, dx = \tfrac{5}{18} \int_0^1 F_2^{\nu N}(x) \, dx, \hspace{2cm} 4.21$$

in good agreement with experiment, follows.

A set of sum rules for the parton distribution functions state that the proton has a net number of two u quarks, one d quark and zero strange quarks (25):

$$\int_0^1 dx[u(x) - \bar{u}(x)] = 2, \hspace{3cm} 4.22$$

$$\int_0^1 dx[d(x) - \bar{d}(x)] = 1,$$ 4.23

$$\int_0^1 dx[s(x) - \bar{s}(x)] = 0.$$ 4.24

The sum rule for the number of quark partons minus the number of antiquark partons is (62)

$$\int_0^1 dx[u + d - \bar{u} - \bar{s}] = \int_0^1 dx\, F_3^{\nu N},$$ 4.25

where the strange quark- and antiquark-parton distributions are neglected. Experimentally, this has the value 3.2 ± 0.35, in agreement with the idea that there are three valence quarks in the nucleon.

We do not discuss neutral current neutrino and antineutrino inclusive processes except to remark that the cross sections for these processes depend on, and thus allow measurement of, the Weinberg angle (22). We also do not discuss processes involving charm production, since charm is discussed in the article by Appelquist et al (50).

We can see qualitatively how the observed structure functions come about by considering three stages in the quark-parton description of the nucleon (63). In the first stage, the nucleon is made of three quarks that share the nucleon's momentum equally. The structure functions F_2 are then

$$F_2^{ep}(x) = \tfrac{1}{3}\delta(x - \tfrac{1}{3}) \quad \text{and} \quad F_2^{en}(x) = \tfrac{2}{9}\delta(x - \tfrac{1}{3}).$$ 4.26

In the second stage, the three quarks no longer move freely but instead are bound together by gluon exchange. Then some of the nucleon momentum resides on the gluons, and therefore the structure functions, which are sensitive only to the part of the momentum that resides on the quarks, are spread out in x, still having a peak at about $x = \tfrac{1}{3}$. The gluon exchanges will give a tail near $x = 1$ of the form (63)

$$F_2(x) \sim (1 - x)^3.$$ 4.27

In the third stage, the quarks emit gluons that do not participate in the binding process and some of these gluons convert to quark-antiquark pairs. This process removes more momentum from the quarks and transfers it either to low-x gluons by bremsstrahlung or to low-x quark-antiquark pairs by gluon bremsstrahlung followed by pair production, and generates the component of the structure functions associated with the so-called "sea." The neutron-to-proton structure function ratio is $\tfrac{2}{3}$ in all three third stages. This ratio of $\tfrac{2}{3}$ follows from, among other things,

SU(6) symmetry. Experimental data do not support this ratio for all values of x, In particular, in the $x \to 0$ limit, where wee partons dominate, the ratio approaches one; and in the $x \to 1$ limit, the experimental ratio seems to approach approximately $\frac{1}{4}$, which is the lower limit of the allowed range implied by Equations 4.7 and 4.8:

$$\tfrac{1}{4} \leqq F_2^{en}(x)/F_2^{ep}(x) \leqq 4. \tag{4.28}$$

This situation requires a large breaking of SU(6) symmetry. This breaking can be interpreted with the idea that when $x \to 1$ one of the quarks carries all of the momentum of the nucleon, and the other two quarks are necessarily in a symmetric space state. If the leading quark, with $x \to 1$, carries the net spin and isospin of the nucleon, then the other pair of quarks must be an isospin 0 state, and in this case the ratio would approach $\frac{1}{4}$ for $x \to 1$ (64).

For $x \to 1$, the inelastic electron scattering structure functions are related to the squares of the elastic electromagnetic form factors, $G(Q^2)$, of the nucleon. This relation was first noted by Drell & Yan and by West (65) and has the form

$$F_2(x) \propto (1-x)^p, \quad G(Q^2) \propto Q^{-n}, \quad \text{with } 2n = p+1. \tag{4.29}$$

Experimental verification of this relation is difficult because it depends on measurements of the structure functions for x very close to 1, and these measurements are difficult to make. In addition, this relation is very sensitive to the exact choice of scaling variable.

Next we turn our attention to the deviations from the naive quark-parton model. In the naive model, the quark-parton distribution functions depend only on the scaling variable x; there is no dependence on Q^2. The structure functions depend on $2mv$ and Q^2. It is convenient to plot them (66) in a plane whose abscissa and ordinate respectively, are these two quantities. The physical region is the region between $Q^2 = 0$ and $Q^2 = 2mv$ in the first quadrant of this plane. The line $x = 0$ corresponds to $Q^2 = 0$ and is also related to real (on-shell) photon-nucleon scattering (proton Compton scattering), and the line $x = 1$ is related to elastic scattering from the nucleon. Generic x corresponds to a line from the origin with slope between 0 and 1. If the parton struck by the current has the momentum xP before being hit and absorbs the momentum q of the current, then after being struck it has momentum $xP + q$ and squared mass

$$M^2 = (xP + q)^2 = m^2 x^2 + 2mvx - Q^2. \tag{4.30}$$

If M^2 is taken to be small, say zero, and the x^2 term in Equation 4.30 is neglected, then one gets the Bjorken scaling variable defined above. If the x^2 term is not neglected, then one gets the light-cone scaling variable (67)

$$x_{LC} = \frac{Q^2}{m[(v^2 + Q^2)^{\frac{1}{2}} + v]}.$$
4.31

A third scaling variable, x' (68),

$$x' = \frac{Q^2}{2mv + m^2},$$
4.32

results if one considers the family of lines in the $2mv - Q^2$ plane that pass through the point $2mv = -m^2$ and $Q^2 = 0$. The light-cone scaling variable interpolates between the Bjorken x and the variable x': for x approaching 0, x_{LC} approaches x, and for x approaching 1, x_{LC} approaches x'. A scaling variable ξ (69) follows from operator product analyses of inelastic scattering; in particular ξ, takes account of changes in the quark-parton mass when heavy quark-partons, such as c, are excited. In the limit where the struck quark is light and the produced quark is heavy, ξ reduces to the variable (70)

$$z = x + (m_C^2/2m_N Ey), \quad m_C = \text{charm quark mass},$$
4.33

which has been used to analyze delayed rescaling in charm production.

Different scaling variables contain different implicit Q^2 dependences. QCD predicts explicit Q^2 dependences, which can be expressed in terms of quark-parton distributions that depend on both x and on Q^2. The electromagnetic and weak currents reveal a hierarchy of clusters as the momentum transfer Q is increased. At very small momentum transfers, the nucleon scatters as a single object and the current sees the electric or weak charge of the nucleon. At higher momentum transfers the current begins to see some structure in the nucleon and measures moments of the elastic form factors, such as the electric and magnetic charge radii of the proton and neutron. At still higher momentum transfers the current detects the virtual pion and other meson clouds surrounding the nucleon. At still higher momentum transfers where inelastic processes dominate, the current starts to see the quark-partons. With continued momentum transfer increase, the current sees that the quark itself has structure due to its emission of gluons and also due to the fact that gluons materialize into quark-antiquark pairs. Then higher order effects of QCD become relevant. If the quark is itself a constituent of more fundamental objects such as those discussed in Section 6, then the current will resolve these more fundamental objects. A good working hypothesis is that the entire quark structure is to be derived from QCD. One can then consider the valence quarks to be the quarks at relatively low momentum transfer, and the gluons and q$\bar{\text{q}}$ pairs emerge as the momentum transfer increases. These latter constitute the quark-parton sea. From this point of view, it seems intuitively reasonable that the quark-parton distribution functions require a Q^2 dependence in which the distribution functions grow for small x and decrease for large x as Q^2 increases, because gluons drain energy momentum out of the active quark-partons. This point of view is borne out by calculations in QCD (71, 72) and is consistent with present data. We conclude this section by mentioning that QCD seems to predict (73) smaller values of σ_L/σ_T, the longitudinal to transverse electroproduction cross-section ratio, than found experimentally. Effects of transverse momenta of partons might resolve this discrepancy (74).

4.2   Lepton-Pair Annihilation

The naive quark-parton model predicts

$$R = \frac{\sigma(e^+e^- \to \text{had})}{\sigma(e^+e^- \to \mu^+\mu^-)} \to \frac{1}{4}\sum_{\text{spin } 0} Q_i^2 + \sum_{\text{spin } \frac{1}{2}} Q_i^2, s \to \infty. \qquad 4.34a$$

Assuming that quark-partons have spin $\frac{1}{2}$, the second term on the right is the important term and, when evaluated in terms of the known quark charges, including the factor three from color, gives agreement with experiment within about 20% both for the range above the prominent photon-coupled resonances such as ρ, ω, and ϕ but below J/ψ production, and then above the resonance region associated with excitation of the charm quark, i.e. above about 5 GeV/c^2. The naive model also predicts the single-particle inclusive angular cross section

$$\frac{d\sigma}{d\Omega} \propto \frac{1}{2}\sum_{\text{spin } 0} Q_i^2(1-\cos^2\theta) + \sum_{\text{spin } \frac{1}{2}} Q_i^2(1+\cos^2\theta). \qquad 4.34b$$

More detailed jet structure, including the jet angular radius and its energy dependence have been derived from QCD (75), and new variables (75) have been introduced in order to test QCD predictions concerning jets. These predictions do not refer explicitly to the produced hadrons that occur in the laboratory; rather they infer statements about hadrons from QCD calculations for quarks and gluons.

4.3   Lepton-Pair Production

The Drell-Yan mechanism (76) for lepton-pair production views this process as due to quark-antiquark annihilation into lepton pairs via photons. Yan reviewed this process (58); here we confine ourselves to two remarks. First, because an antiquark is necessary, experiments involving meson-nucleon collisions in which antiquarks occur as valence partons in the meson should give a much larger yield of lepton pairs than nucleon-nucleon or nucleon-nucleus collisions, in which the antiquarks come only from the small $q\bar{q}$ sea of the nucleons. Second, unlike the case of lepton-pair annihilation, where color increases cross sections by a factor of three, here color suppresses the cross section by a factor of three because quark-antiquark annihilation can occur only if the quark and antiquark have the same color, and since there are three colors of quarks and three colors of antiquarks, the probability is only one third that the colors will match. The prediction of dimuon production from the Drell-Yan model with color fits present data better than without color; however, present data does not definitively establish the color factor. In addition, gluon and other QCD effects must be taken into account (77). We do not discuss more detailed electroproduction or leptoproduction or lepton annihilation processes such as $e^{+}e^{-} \rightarrow hX$ since Yan reviewed them.

4.4   Hadron-Hadron Scattering

4.4.1   COUNTING RULES   The counting rules (78) give a recipe for the energy dependence of hard scattering processes at fixed angle. The recipe is that such a process behaves as

$$\frac{d\sigma}{dt}(AB \rightarrow CD) \propto s^{2-n} f(t/s), \qquad 4.35$$

where n is the total number of fields in particles A, B, C, and D that participate in the hard scattering, i.e. that carry a finite fraction of the momentum, and s and t are the energy and momentum transfer, respectively. If not all of the quarks participate in the hard scattering, the energy and momentum transfer in Expression 4.35 are to be taken as the energy and momentum transfer of the participating constituents. According to the counting rules, the exponent n depends on the specific

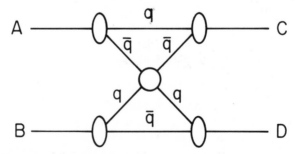

Figure 1   Quark-quark scattering contribution to meson-meson scattering.

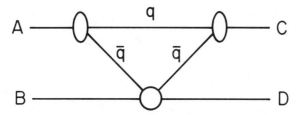

Figure 2 Quark-meson scattering contribution to meson-meson scattering.

subprocess involved, and different subprocesses will give different values of n for the same observed hadron-hadron scattering process. For example, in meson-meson scattering, the subprocess involving quark-quark scattering (see Figure 1) will give an exponent $n = 4$ and therefore an energy dependence of s^{-2} while the subprocess involving quark-meson scattering (see Figure 2) will give a value of $n = 6$ and therefore an energy dependence of s^{-4}. Analysis of Bethe-Salpeter equations shows (79) that these counting rules hold, provided that the short-distance behavior of bound states satisfies certain conditions, and that there is no internal mass scale for the hard scattering. These counting rules also give predictions for elastic scattering involving quarks, leptons, and photons and lead to the prediction for the energy dependence of the electromagnetic form factors

$$F(t) \propto t^{1-n},$$ 4.36

for example, where n here is the minimum number of fields in the hadron;

$$F^{\pi}(t) \propto t^{-1}, \quad F^{p}(t) \propto t^{-2}.$$ 4.37

Energy dependence of any process in hard scattering can be predicted by these counting rules. Reference 78 gives the net energy dependences that occur for various processes.

4.4.2 CONSTITUENT INTERCHANGE MODEL In the constituent interchange model (CIM) (80), the counting rules are used to determine the energy dependence of the fundamental hard scattering processes that are involved in high transverse momentum processes, and the properties of observed inclusive reactions are deduced from these. Since for a given hadronic collision, a number of constituent interchange processes can contribute, a rule is needed to decide which is the dominant process. The rule is that the dominant process is the one with the slowest decrease in transverse momentum according to the counting rules

$$E \frac{d\sigma}{d^3 p}(AB \to CX) = \sum_{a,b,c,d} (p_T^2 + m^2)^{2-n} f(M^2/s, \theta_{\text{c.m.}}),$$ 4.38

where n is the number of active elementary fields or equivalently the number of active quarks, leptons, photons, W's, Z's, etc. For particles emitted at $90°$, $M^2/s = 1 - x_T$. In that case, the function f of Equation 4.38 is approximately

$$f(M^2/s, \theta_{c.m.}) \sim (1 - x_T)^F f(\theta_{c.m.}), \qquad\qquad 4.39$$

for x_T close to one and possibly even down to one half. The number F is given in terms of the number of spectator quarks, that is to say the number of quarks in AB and the crossing of C that do not participate in the hard scattering. These rules would lead to the conclusion that quark-quark scattering is the dominant high momentum process at large energy; however the advocates of the CIM argue that the absolute normalization of this process is small and therefore that quark-meson scattering dominates. Depending on the details of the inclusive process, other dominant subprocesses may be relevant; for example in pp → K⁻X, because there is no strange quark in either of the particles in the initial state, a "fusion" subprocess is taken to be dominant. There is an element of choice in the selection of the dominant subprocess; however, a number of processes have been successfully fitted with the constituent interchange model.

4.4.3 FEYNMAN-FIELD-FOX MODEL   Feynman, Field & Fox (FFF) (81) initially adopted quark-quark scattering as the fundamental primary process involving large momentum transfer events. Until recently, they found that they had to assume an s^{-4} dependence of the fundamental quark-quark scattering cross section rather than the s^{-2} dependence expected from QCD or simple scaling arguments. With this ad hoc assumption, a number of processes could be accounted for; however, the assumption's ad hoc nature was a severe difficulty of this approach. Stimulated by the necessity to introduce transverse momentum distributions in both the initial hadrons and in the quark-to-hadron fragmentation functions for the outgoing quark jets, Field reevaluated their approach, first introducing relatively large transverse momenta, and then redoing the entire calculation staying as close as possible to the ideas of quantum chromodynamics (82). He found that the experimental p_T^{-8} behavior could be obtained with a fundamental quark-quark cross section going as p_T^{-4}, provided several effects, no one of which is dominant, were taken into account:

1. The asymptotic freedom effect that the strong interaction coupling constant $\alpha_s(Q^2)$ decreases logarithmically with Q^2, for example (83),

$$\alpha_s(Q^2) = 2\pi/[25 \log (Q^2/\Lambda^2)], \quad \Lambda = 0.4 \text{ GeV}. \qquad 4.40$$

2. The parton distribution functions in the proton do not scale, but rather they depend on Q^2 in a way suggested by QCD (4).

3. The quark-to-hadron fragmentation functions also are Q^2 dependent. Here, in contrast to the parton distribution functions, there are no directly relevant data; the Q^2 dependence of the fragmentation functions is estimated by using renormalization group arguments and by using the fragmentation functions from semi-inclusive lepton-hadron processes to give input at $Q^2 = 4\ \text{GeV}^2$.

4. The incoming partons are given a transverse momentum of 848 MeV/c, much larger than before; in the quark-to-hadron fragmentation process a larger transverse momentum, 439 MeV/c, is used than previously.

5. The hard quark-quark scattering process and the associated crossed processes $\bar{q}q \to \bar{q}q$ and $\bar{q}\bar{q} \to \bar{q}\bar{q}$ are augmented by other processes, necessary in QCD (4):

$$\text{gq} \to \text{gq}, \text{g}\bar{q} \to \text{g}\bar{q}, \text{gg} \to \bar{q}q, \bar{q}q \to \text{gg}, \text{ and gg} \to \text{gg}. \qquad 4.41$$

One cannot say that this recalculation represents a fundamental derivation from bedrock principles of QCD, because a number of additional assumptions had to be made. For example, there is no direct information about the fragmentation function for gluons to hadrons. A number of other necessary inputs had to be parametrized using the best judgment. The results are sensitive to the treatment of soft gluons (84). Nonetheless, the net result of these changes is to allow the QCD processes that go as p_T^{-4} to fit data that appear to go as p_T^{-8}. This seems to remove a major difficulty of the FFF approach.

It is important to distinguish between the CIM and FFF·models. Certain phenomena are predicted by both models. For example, the ratio of $\text{pp} \to \pi^+ X$ to $\text{pp} \to \pi^- X$ depends primarily on the quark content of the proton; in particular the fact that the proton has two u quarks and one d quark. Thus there will be more $\pi^+ X$ than $\pi^- X$. Brodsky (85) and Field (86) emphasize conflicting predictions of these two models; however, the changes in the FFF model must be incorporated to make a proper comparison. If QCD is correct, then at very large s and p_T, the p_T^{-4} behavior should manifest itself unambiguously.

4.5   Relation of Constituent Quarks, Quark-Partons, and Current Quarks

Although there is no definitive theory of quarks based on fundamental physical ideas, such as those of QCD, and thus there is no definitive theory of the relation between the constituent quarks of hadron spectroscopy and the quark-partons of high energy inclusive scattering, there are proposals concerning the relation of constituent quarks, quark-partons, and also "current" quarks, the quark fields with associated parameters such as masses that enter the QCD Lagrangian (87).

Weinberg (88), using current algebra, estimated the current quark mass ratios to be

$$m_d/m_u = (K^0 + \pi^+ - K^+)/(K^+ + 2\pi^0 - K^0 - \pi^+) = 1.80 \qquad 4.42$$

and

$$m_s/m_d = (K^0 + K^+ - \pi^+)/(K^0 + \pi^+ - K^+) = 20.1, \qquad 4.43$$

where the meson symbol stands for the square of the meson mass. This surprising nondegeneracy of the current quark masses raises the question of why isospin and flavor SU(3) are rather good symmetries; the answer is that all three current quark masses are small compared to a typical hadron mass, such as the nucleon mass. Weinberg estimates the renormalized current quark masses to be

$$m_u = 4.3 \text{ MeV}, m_d = 7.5 \text{ MeV, and } m_s = 150 \text{ MeV}. \qquad 4.44$$

The constituent quark masses, estimated from the proton magnetic moment (9)

$$m_u \approx m_d = m_p/2.79 = 336 \text{ MeV}/c^2, \qquad 4.45$$

and from the Λ magnetic moment,

$$m_s = 429\text{--}511 \text{ MeV}/c^2, \qquad 4.46$$

are quite different.

Such light constituent quarks do not move nonrelativistically; rather their velocities are large enough that relativistic corrections are important and must be taken into account. This leads to a number of effects (87). First, there is a Lorentz contraction of the wave function of the quarks in the nucleon; this Lorentz contraction will also be reflected in the parton model (89). In particular, the relativistic quark motion in the nucleon rest frame is associated with a transverse momentum of the partons in the infinite momentum frame. Second, the spin $\frac{1}{2}$ quarks moving relativistically must be described by Dirac wave functions rather than by Pauli spinors. These Dirac wave functions will have a small component whose squared value will give corrections to the G_A/G_V ratio calculated in the nonrelativistic quark model (87). Finally, there will be configuration mixing associated with breaking of SU(6) symmetry, which not only changes the pattern of masses within a single SU(6) super-multiplet, but also mixes different SU(6) supermultiplets. Configuration mixing has been considered phenomenologically; however, it also follows from QCD. The types of configuration mixing most widely considered are those between $(56,0^+)_0$ and $(70,1^-)_1$ (90) and between $(56,0^+)_0$ and $(70,0^+)_2$ (87). Configuration mixing plays a crucial role in understanding the nonzero value of the electric charge radius of the neutron as discussed in Section 3, and in explaining the strong deviation from SU(6) predictions and inequalities for the ratio F_2^{en}/F_2^{ep} for $x \rightarrow 1$. The proton-to-neutron magnetic moment ratio and the F-to-D ratio are very little affected by these considerations, which is fortunate since the naive model gives good results for these quantities. A further issue, discussed above, is the quark structure expected from QCD. This quark structure will likely consist, in large part, of soft gluons and relatively low momentum quark-antiquark pairs, which on the average are likely to be in an $SU(6)_f$ singlet and therefore will not have an important effect in changing hadron spectroscopy. However, as we noted above, in the quark-parton model such effects are connected with the fact that gluons carry of the order of half of the energy momentum of the nucleon. Thus quark structure effects associated with QCD will have a decisive importance in the quark-parton model of high energy and momentum transfer inclusive processes, but very little importance in the constituent quark model at rest. There are some exceptions, in particular the charge radius of the proton probably derives an important contribution from quark structure (91).

5 BOUND STATES OF QUARKS, SATURATION, AND PERMANENT QUARK AND COLOR CONFINEMENT

In this section we discuss several related topics: (a) bound states of quarks considered in terms of principle, rather than phenomenology; (b) saturation; that is, the property that the observed low-lying hadrons can all be constructed from $q\bar{q}$ for mesons and qqq for baryons; and (c) permanent quark and color confinement, that is the notion that isolated single quarks or antiquarks are never observed, but rather are seen only bound into hadrons. With the advent of QCD, where the color degree of freedom plays the essential role in confinement, the fundamental degree of freedom being confined is really color, including both (colored) quarks and colored gluons, so that confinement of quarks is incidental to confinement of color.

We discuss attempts to show that strongly bound heavy quarks can have nonrelativistic motion in the bound state (although these attempts are largely superseded by the idea of permanent confinement of light quarks), since they may become relevant if isolated single quarks are ever found. The main idea is that if quarks are bound in a deep potential that is flat at the origin, then they can have strong binding together with nonrelativistic motion (92). This idea encounters the following difficulties: (a) The flat bottom well is hard to get from meson exchange; more likely one would get a Coulomb or Yukawa potential, and for such potentials strong binding implies relativistic motion (93). (b) Dalitz (17) has pointed out that the simultaneous requirements of nonrelativistic motion and strong binding do not lead to a normal Schrödinger equation as is usually assumed. An equation of the Blankenbecler-Sugar (94) type might be appropriate. More likely one should use a fully relativistic equation such as the Bethe-Salpeter equation (54). Joos and collaborators (95), using the Bethe-Salpeter equation together with certain phenomenological assumptions, have given an extensive treatment of hadrons. Their approach is successful in relating the Regge slope and intercept to the range parameter that governs strong decays of hadrons and to the leptonic decay constants for decays of radially excited vector mesons. Their phenomenological assumptions include using a harmonic oscillator approximation for the potential and in studying general spin structure for the quark-binding interaction. They require a type of saturation in which the quark interaction approaches a finite limit as the quark mass goes to infinity. They find that the saturating potential is in conflict with a partially conserved axial current (PCAC) and chiral symmetry. This seems to be a significant difficulty of this approach. This difficulty might arise because of assumptions concerning the Bethe-Salpeter kernel; in particular, that it has convolution type.

Preparata (96) proposed a heavy quark model in which the quark mass is taken to infinity. This approach implies that quarks are not real objects, but simply things that provide a space-time description of extended hadrons.

Nambu (97) suggested a mechanism using color gluon exchange that produces saturation in the context of heavy quarks. Let the quark mass in the absence of gluon exchange be $m_q \gg 1$ GeV/c^2. Let the effect of color gluon exchange be described by the $SU(3)_c$ generators g_α^i,

Table 11 Casimir operator $C = C_2^{(3)}$ for some irreducible representations[a] of SU(3)

Representation	1	3	6	8	10	27
C	0	$\frac{4}{3}$	$\frac{10}{3}$	3	6	8

[a] The general irreducible has the Young tableau (α_1, α_2) with α_1 boxes in row i, $i = 1, 2$. $C = (\frac{1}{3})(\alpha_1^2 - \alpha_1\alpha_2 + \alpha_2^2) + \alpha_1$. C is the same for a representation and its complex conjugate.

$\alpha = 1, 2, \ldots 8$, of the ith quark. If the gluon interaction strength U is large, $U \gg 1$ GeV; then the mass of a system of N quarks and antiquarks will be (28, 97)

$$M = Nm_q + U \sum_{i<j} \sum_{\alpha=1}^{8} g_\alpha^i g_\alpha^j = N\tilde{m}_q + \tfrac{1}{2}UC, \qquad 5.1$$

where $\tilde{m}_q = m_q - \tfrac{1}{2}UC_0$ ($C_0 = \tfrac{4}{3}$) is the effective quark mass in the bound state, and C is the quadratic Casimir operator of SU(3)$_c$. Table 11 gives C for several SU(3) representations. If m_q and U are chosen so that $\tilde{m}_q \ll m_q$ is about one third of the nucleon mass, then color-singlet states will have typical hadronic masses and color-nonsinglet states will have much larger masses. The interaction energy of two clusters μ and v (more clusters can be treated similarly) into which a set of quarks and antiquarks are divided will be

$$U(\mu, v) = U \sum_{i\in\mu} \sum_{j\in v} \sum_{\alpha=1}^{8} g_\alpha^i g_\alpha^j = \tfrac{1}{2}U[C - C(\mu) - C(v)]. \qquad 5.2$$

If, say, a cluster μ is a color singlet, then $C(\mu) = 0$, and $C = C(v)$, so $U(\mu, v) = 0$. Thus color-singlet systems coupled by a universal color-octet gauge interaction are neutral or inert if all pairs of quarks and antiquarks interact with the same strength, for example if all qq or q$\bar{\text{q}}$ pairs are in the same relative space state. In particular, states of zero triality with more quarks and antiquarks than the usual baryons and mesons will not be bound strongly against dissociation into baryons and mesons. However, analysis at this qualitative level does not exclude weakly bound or resonant states of this kind. For example, although Equation 5.1 shows that the mass of a six-quark color-singlet state is twice the mass of two baryons, the approximate nature of Equation 5.1 does not indicate whether or not the six-quark state is bound as a single-centered system. A two-centered six-quark bound state does exist: the deuteron. The existence of the deuteron and other nuclei as bound states means that the notion that a state can be either single-centered or many-centered is crucial. Saturation would rule out single-centered bound $3n$ quark states for $n > 1$, and single-centered bound n quark–n antiquark states for $n > 1$, but does not rule out n-fold-centered bound $3n$ quark states, since these exist as nuclei at least up to $n \sim 250$.

The parton model and the large value of the proton magnetic moment are most easily understood in terms of light quarks, for example, with mass about a third the mass of the nucleon. Such models require permanent quark confinement, otherwise the light quarks will emerge and should already have been detected. A simple approach to such models is to assume a confining effective potential between quarks and quarks and between quarks and antiquarks such as a harmonic oscillator or linear potential. Most models of this kind now use the color degree of freedom as an essential ingredient in the permanent binding, and assume that quarks are attractive in an antisymmetric color state and that quark and antiquark attract in the singlet-color state, while the other quark-quark and quark-antiquark color states have repulsive interactions. For a given hadron, this procedure works perfectly well at the phenomeno-logical level; however when one considers several separated hadrons, there is the problem of cancelling out the long-range potential so that it does not act between the separated hadrons. In such models it is necessary to have repulsive interactions of exactly the same long-distance space dependence (but opposite sign) as the attractive ones to provide cancella-tion. However this cancellation, in general, is likely to reduce the radial dependence of the potential by only two powers. Thus a harmonic oscillator potential is changed to one that is constant, independent of separation, and a linear potential is changed to one that is Coulombic. Since, in nature, hadrons do not seem to have such long-range interactions, this model has a serious difficulty.

The MIT bag model (98) has objects carrying color, such as quarks and gluons, enclosed in finite regions of space by introducing into the Lagrangian a pressure characterized by a bag constant B. This model treats each hadrons, in lowest approximation, as a collection of free quarks and antiquarks inside the bag, and in next approximation takes into account perturbations due to the interactions of the quarks and antiquarks via gluon exchange. In this model, the color fields are all confined to the bag, and therefore the difficulty concerning interactions between separated hadrons is eliminated by fiat. A number of phenomeno-logical results have been well described by this model; however, it cannot be considered a fundamental approach to quark confinement. The SLAC bag (99–101) achieves confinement using spontaneously broken symmetry involving a scalar field. This model is in many ways similar to the MIT bag, except that in this model the quark densities are highest near the boundary of the region. This feature leads to some phenomenological difficulties. Friedberg & Lee (102) have studied nontopological solitons that also provide quark confinement via spontaneous symmetry breaking using a scalar field. The solitons act like a "gas bubble" immersed in a "medium," and the quarks are confined in the bubble. This model reduces in certain limits to the MIT and SLAC bags.

QCD is the most promising approach to quark confinement. In QCD, the quarks and other color-carrying objects emit flux lines mediated by gluons that attract themselves. Thus one can consider that a separated pair of quarks or quark and antiquark are connected by an attracting line of gluons, which form themselves into a string (103). A potential that is linear with the distance might emerge in this model (104). In addition, if the colored flux lines are quantized, then all the flux from a given particle with color might land on another such particle, so that there would be no residual long-range interaction between separated hadrons. Attempts to achieve confinement using QCD include the use of lattice models (105), vacuum degeneracy in the nonabelian gauge theory (106), and instantons and merons (107). A phase transition may play an important role in the confinement problem (108). In particular, 't Hooft (109) recently proposed a confinement mechanism from a domain structure associated with singular gauge transformations. In a qualitative way, this solution of QCD resembles the bag models.

The most abstract approach to confinement involves algebraic ideas. Günaydin & Gürsey (110) analyzed octonionic representations of the Poincaré group, and found that new representations, not included in Wigner's list, exist precisely for the states corresponding to mesons and baryons, assuming that the quarks are octonion-valued fields. These ideas seem to require significant modification in the usual quantum-mechanical theory of measurement. Domokos & Kövesi-Domokos (111) introduce an algebra that leads to the color rules for the couplings of quarks and antiquarks.

6   SPECULATIONS ABOUT POSSIBLE COMPOSITENESS OF QUARKS

The quark model has evolved a great deal from the initial suggestion in 1964 that there are three types of quarks u, d, and s in terms of which all hadrons can be constructed. Our present point of view requires at least four flavors (and probably more) and three colors. This escalation in the number of quarks, together with the associated gluons for the strong color interaction, has made the quark point of view start to become rather complicated. Theory also requires at least six leptons and their gauge bosons. The unified models of all interactions (112) require even more fundamental objects. To reduce the number of fundamental objects, several authors (113) consider composite models in which, for example, the flavor and color degrees of freedom reside in different fields. Then the quark would effectively be a composite of a flavor piece (probably spin $\frac{1}{2}$, since the flavor is the quantum number that participates in lepton-hadron processes where the quark-partons in good first approximation do have spin $\frac{1}{2}$) and a spin 0 or 1 piece that carries the color. There are modifications of these speculations in which the quark is a bound state of three or possibly even more objects. Perhaps the most bold speculation of this type, is that of Nowak, Sucher & Woo (114), who speculate that the leptons themselves may be the flavor-carrying part of the quark. These authors showed that this idea, which a priori seems entirely unlikely, cannot yet be ruled out. We emphasize that there is at present no direct evidence that requires assuming that quarks are composites of still more fundamental objects. (Of course quarks are composites of quarks and gluons from the standpoint of QCD.)

7   SEARCHES FOR QUARKS

Jones (115) recently made a comprehensive survey of quark search experiments; we rely heavily on this survey, and refer to it for all references to quark search experiments that were available at the time it was published. Searches for quarks have been done using accelerators, cosmic rays, and stable matter.

7.1   Searches with Accelerators

Quark searches with accelerators look for fractionally charged particles with a given momentum at a given angle. Given a particle flux, one needs to assume a mechanism by which the quarks are produced and a model of this production in order to deduce the cross section that follows from the measured particle flux.

Most of the quark search experiments rely on detection of the fractional charge of the quarks, in particular on the fact that the ionization produced by a particle of charge Q is proportional to Q^2, so that quarks and antiquarks with $Q = \pm e/3$ or $\pm 2e/3$ would produce less than minimum ionization for a particle with $Q = \pm e$.

For hadronic quark production we consider two main mechanisms. We write these in terms of nucleon-nucleon collisions; however they can easily be converted to processes involving nucleon-nucleus or nucleus-nucleus collisions. We also allow the possibility that additional particles, X, are produced in the final state. These processes are diffraction dissociation,

$$NN \to qqq\ NX, \qquad\qquad\qquad 7.1$$

and hadronic pair production,

$$NN \to q\bar{q}\ NNX \text{ or } N\bar{N} \to q\bar{q}X. \qquad\qquad 7.2$$

Roughly speaking, the accelerator experiments set upper limits on production cross sections below about 10^{-34} cm^2 for masses up to 20 GeV, and about 10^{-38} cm^2 for masses below 10 GeV. These conclusions depend sensitively on the production mechanism; in particular using the statistical model for production reduces the lower limits on quark masses to about 5 GeV.

Electromagnetic pair production of quarks

$$e^- Z \rightarrow q\bar{q}X \qquad\qquad 7.3$$

provides a better limit on the existence of free point-like quarks with low masses, in particular, with masses close to one third the nucleon mass. Since for point-like quarks the electromagnetic production mechanism is completely known, these experiments can be interpreted unambiguously, and rule out free point-like quarks with low mass.

7.2 Searches in Cosmic Rays

Cosmic-ray searches provide an upper limit for the flux of quarks, ϕ (cm^2 sr sec)$^{-1}$, incident upon the detection apparatus.

7.2.1 SINGLE-PARTICLE SEARCHES The simplest type of experiment involves searches for single particles with less than minimum ionization. Results are usually reported in terms of 90% confidence level of detection. Typical upper limits at sea level are

$$\phi \leqq \sim 10^{-10} \, (cm^2 \, sr \, sec)^{-1}. \qquad\qquad 7.4$$

Some experiments searched for particles of charge $\pm 4e/3$, rather than of charge $\pm e/3$ and $\pm 2e/3$. In general the experiments searching for charge $\pm e/3$ are more sensitive than those for the higher charges.

7.2.2 EXTENSIVE AIR SHOWERS Cosmic-ray events often produce at sea level electromagnetic cascades that contain many charged particles per unit area near the center or core of the cascade. In such an event, a finite-sized detector would necessarily see several charged particles (of integer charge) along with any fractionally charged quark, and thus would fail to detect a (below minimum ionization) fractionally charged quark in isolation. More sophisticated experiments take account of this possibility, and insure that fractionally charged quarks are not missed because of their close association with integrally charged particles. These experiments give an upper limit on the quark flux of the same order as from single-particle searches,

$$\phi \leqq \sim 10^{-10} \, (cm^2 \, sr \, sec)^{-1}. \qquad\qquad 7.5$$

7.2.3 TIME DELAY EXPERIMENTS If quarks produced in the upper atmosphere are very massive, then they will arrive at the ground after less massive particles produced in the same event, since the latter (electrons, muons, and nucleons) move at close to light velocity. Massive particles will have a typical time delay of the order of nanoseconds after the rest of the electromagnetic shower. Experiments of this kind do not directly detect the charge of the delayed particle; in addition these experiments are less well developed than the two types just discussed above and do not now give better limits.

Given a measurement of the flux, ϕ, deduction of a limit on the production cross section of quarks requires a model, and this production cross section in general will depend upon the mass of the quark. Jones (115) gives limits derived from three different models.

The discussion of cosmic-ray events containing quarks given so far tacitly assumes that these quarks are produced in the upper atmosphere. Another possibility is that primordial quarks are present from the big bang. Zel'dovich, Okun & Pikel'ner (116) studied this possibility, and estimated a concentration of 10^{-9}–10^{-18} quarks per nucleon, depending upon various assumptions. Feinberg (117) argued that observations of the three-degree black body radiation constrain this estimate to 10^{-10}–10^{-13} quarks per nucleon. Assuming that primary cosmic-ray quarks are not attenuated in the atmosphere, the present upper limits on the flux of quarks in cosmic rays correspond to an upper limit for the quark concentration among primary cosmic rays of about 10^{-10}–10^{-13} quarks per nucleon.

7.3 Searches in Stable Matter

7.3.1 DESCENDANTS OF THE OIL DROP EXPERIMENT The descendants of the oil drop experiment involve studying a sample of bulk matter to see if it has a fractional charge or if it has a nonzero residual charge when it is irradiated with x rays or other ionizing radiation, or electrons or positrons, to try to bring its charge to zero. These experiments require no assumptions about nonelectromagnetic interactions of quarks. The most popular experiments at present involve magnetic suspension of the sample and a study of its oscillations or lack of oscillations when an alternating electric field is applied. Jones (115) gives a table of experiments of this type, with upper limits for recent experiments in the range 5×10^{-19} to 3×10^{-21} quarks per nucleon. Some positive results in such experiments in the past have been discounted by their authors; however a recent positive result by La Rue et al (118) has been advanced as evidence for the existence of fractionally charged particles. In this experiment superconducting niobium spheres of mass about 9×10^{-5} g were magnetically suspended. Some of these spheres were annealed on a tungsten substrate and some were not. Those spheres annealed on the tungsten substrate appeared to have fractional charge, which might therefore be associated with tungsten having been deposited on these spheres. This interpretation, however, would require a very large quark density in the thin tungsten film on the spheres. So far this experimental result has not been reproduced; on the contrary other experiments of this general type, performed since, have yielded negative results (119). No final conclusion concerning this experiment can be made at present; however, standing by itself this experiment does not give convincing evidence that isolated fractionally charged particles exist.

7.3.2 SPECTROSCOPIC SEARCHES   Searches in the spectra of terrestrial material and in the solar spectrum have yielded negative results. Experiments originally designed to check the exact equality of the magnitudes of the electron and proton electric charges were made using tanks of gas together with an electrometer. These experiments gave very good upper limits on the difference between the magnitudes of the electronic and protonic charges; after the fact, they can also be interpreted as searches for fractional residual charges.

Attempts to concentrate particles with fractional charges have been made with natural mechanisms such as oysters and other scavengers in the ocean, or with boiling or electromagnetic enrichment procedures. The resulting concentrated material can then be examined with either optical or mass spectrometry. Jones (115) tabulates experiments that give upper limits from 10^{-15}–10^{-33} quarks per nucleon. Another technique is the use of molecular beams. The negative experiments of this type can give upper limits on the number of quarks per nucleon of the order of 10^{-20} for the bulk matter experiments and as small as 10^{-23} or even 5×10^{-27} for the concentration experiments.

7.4   Relation of Cosmic-Ray Experiments to Stable Matter Experiments

If a flux of quarks in cosmic rays has been incident on the earth's surface over the lifetime of the earth and if the quarks are absolutely stable (which we assume because of their fractional charge) and further do not diffuse away or become eroded by weather, etc, then quarks accumulate in some thickness, y (g cm^{-2}), of matter at the surface of the earth's crust. The relation between the cosmic-ray flux and the quark-per-nucleon ratio is given by

$$\rho = \frac{\pi t \phi_Q}{N_0 y},$$ 7.6

where t is the time over which the quarks are accumulating, N_0 is Avogadro's number, and ϕ_Q is the quark flux defined earlier. This relation is sensitive to the choice of y; with a reasonable value of y, the upper limit on cosmic-ray flux of 10^{-11} (cm^2 sr sec)$^{-1}$ implies that the quark-per-nucleon ratio is of the order of 10^{-25}–10^{-23}. This value is at least two orders of magnitude below the quark-to-nucleon ratio implied in the experiment of La Rue et al (118).

7.5   Conclusion

Here, as throughout this review, we consider only fractionally charged quarks, at least one of which is absolutely stable because of charge conservation. Thus our discussion is irrelevant to the unstable integrally charged quarks in models such as that of Pati & Salam (12). There is no convincing evidence at present for the existence of isolated particles with fractional charge.

8   ASTROPHYSICAL AND COSMOLOGICAL IMPLICATIONS

8.1   Quark Stars

The possibility that stars formed of degenerate quark matter exist depends on the nature of quark matter at high densities and pressures. Many authors studied the equation of state of quark matter under these conditions, starting with heavy noninteracting quarks (120). Recent work mainly assumes that quarks are light and confined, and that they are asymptotically free at short distances (121). The main idea is that if the matter in the central core of a star has a density greater than the density of a neutron, which is about 8×10^{14} g cm^{-3}, the baryons in the core overlap, the clustering of three quarks per baryon disappears, and the matter consists of a gas of quarks, rather than a gas of nucleons. Some authors (122) find it likely that stable quark stars with masses greater than the Oppenheimer-Volkoff (123) limit can exist. This raises the possibility of an explanation for objects such as that in Cygnus X-1 as quark stars rather than black holes. Other authors consider the possibility of a neutron gas to quark gas transition in the core without addressing the question of the star mass (124). Still others find it unlikely that quark matter will occur in stars because the neutron to quark-matter transition will occur at densities exceeding that of the core (125). References (126) give general discussions of dense quark matter. Zel'dovich argued that causality implies that the $p = \varepsilon$ (pressure = energy) equation of state is the stiffest possible (127); Reference 128 presents the argument that stiffer equations of state can occur. Brecher (129) gave an observational test based on the gravitation mass vs redshift to distinguish neutron stars from quark stars.

8.2   Element Production in the Early Universe

The number of quarks and antiquarks is frozen in at an early time (130), if quarks are massive and can travel freely. In this case, Zel'dovich and others (116, 131) have shown that the quark-to-nucleon ratio would be much larger than upper limits established by observations on earth. On the other hand if quarks are permanently confined, at least below some high temperature, then there is no contradiction between the failure to observe free quarks on earth and the possibility of the existence of quarks.

8.3   Entropy Production in the Early Universe

The specific entropy of the universe, that is the entropy per baryon, S_b, approximately equal to the number of photons per baryon, is an important cosmological parameter because it is conserved in an adiabatically expanding universe (130). The standard hot big bang model (131) assumes that S_b has been close to its present value since the early universe, in particular since $T \gtrsim 10^{10\circ}$K. Reference 132 consider the contrary possibility that entropy has been generated during the dissipation of anisotropies and inhomogenities in the early universe. In particular the possibility of large entropy generation was studied with (a) a model with asymptotically free quarks as the dominant material constituent of the early universe, and (b) a model of nuclear matter interacting via vector meson exchange. The former gives a soft equation of state and is unlikely to provide significant entropy generation; the latter, with a stiff equation of state, might account for significant entropy production.

9   CONCLUSIONS

Spin $\frac{1}{2}$ quarks with flavor and color, and their associated spin one, flavor-less gluons with color, are the fundamental objects in the description of hadrons. The local color-SU(3) gauge invariant field theory of quarks and gluons, QCD, is a good candidate for the theory of hadrons and their interactions. The replacement of naive guesses for quark dynamics by calculations based on QCD[7] seems likely to transform the quark model, already successful in a large range of areas in elementary particle physics, into a theory that can unify the many disparate applications of the quark model, foremost of which are the applications to hadron spectroscopy via the constituent quark model and applications to high energy inclusive collisions via the quark-parton model.

---

[7] We emphasize that all quark model estimates must be replaced by calculations that include gluons and other QCD effects.

ACKNOWLEDGMENTS

We acknowledge the hospitality of Dr. Frank B. McDonald at NASA/ Goddard, helpful discussions concerning Section 8 with Dr. Floyd W. Stecker and concerning spin-orbit interactions with Dr. Howard J. Schnitzer. We thank Dr. Gabriel Karl, Dr. Joseph Sucher, Dr. Ching Hung Woo, and Dr. Gaurang B. Yodh for reading and commenting on the manuscript.

Literature Cited

1. Gell-Mann, M. 1964. *Phys. Lett.* 8:214
2. Zweig, G. 1964. *CERN 8182/TH.401*; *8419/TH.412*
3. Yang, C. N., Mills, R. L. 1954. *Phys. Rev.* 96:191; Abers, E. S., Lee, B. W., 1973. *Phys. Rep. C* 9:1; Taylor, J. C. 1976. *Gauge Theories of Weak Interactions*, Cambridge Engl.: Cambridge Univ. Press. 167 pp.
4. Weinberg, S. 1973. *Phys. Rev. Lett.* 31:494; *Phys. Rev. D* 8:4482; Gross, D. J., Wilczek, F. 1973. *Phys. Rev. D* 8:3633; Fritzsch, H., Gell-Mann, M. Leutwyler, H. 1973. *Phys. Lett. B* 47: 365; Marciano, W., Pagels, H. 1978. *Phys. Rep. C* 36:137
5. Hendry, A. W., Lichtenberg, D. B. 1978. *Rep. Prog. Phys.* In press; Lipkin, H. J. 1973. *Phys. Rep C* 8:173; Morpurgo, G. 1970. *Ann. Rev. Nucl. Sci.* 20:105–46; Kokkedee, J. J. J. 1969. *The Quark Model*, New York: Benjamin. 239 pp.; Close, F. E. 1978. *Introduction to Quarks and Partons.* London: Academic. In press
6. Herb, S. W., Hom, D. C., Lederman, L. M., Sens, J. C., Snyder, H. D., Yoh, J. K., Appel, J. A., Brown, B. C., Brown, C. N., Innes, W. R., Ueno, K., Yamanouchi, T., Ito, A. S., Jöstlein, H., Kaplan, D. M., Kephart, R. D. 1977. *Phys. Rev. Lett.* 39:252
7. Gell-Mann, M. 1961. *Cal. Inst. Tech. Synchrotron Lab. Rep. CTSL-20*; Ne'eman, Y. 1961. *Nucl. Phys.* 26:222; Gell-Mann, M., Ne'eman, Y. 1964. *The Eightfold Way*, New York: Benjamin
8. Glashow, S. L., Iliopoulos, J., Maiani, L. 1970. *Phys. Rev. D* 2:1285; Appelquist, T., Barnett, M., Lane, K. 1978. *Ann. Rev. Nucl. Part. Sci.* 28:387– 499; Gaillard, M. K., Lee, B. W., Rosner, J. 1975. *Rev. Mod. Phys.* 47:277
9. Greenberg, O. W. 1964. *Phys. Rev. Lett.* 13:598
10. Han, M. Y., Nambu, Y. 1965. *Phys. Rev.* 139:B1006

11. Greenberg, O. W., Nelson, C. A. 1977. *Phys. Rep. C* 32:69
12. Pati, J. C. 1977. GIFT VII *Int. Semin. Theor. Phys., Univ. Md. Tech. Rep. 78-065*; Pati, J. C. 1977. In *Fundamentals of Quark Models*, ed. I. M. Barbour, A. T. Davies, p. 89. Edinburgh: Scottish Univ. Summer Sch. Phys. 588 pp.
13. Greenberg, O. W. 1977. In *Color Symmetry and Quark Confinement*, ed. J. Tran Thanh Van, p. 51. Orsay: Inst. Natl. Phys. Nucl. Phys. Part 230 pp.
14. Gürsey, F., Radicati, L. A. 1964. *Phys. Rev. Lett.* 13:173
15. Sakita, B. 1964. *Phys. Rev.* 136:B1756
16. Greenberg, O. W. 1967. *Univ. Md. Tech. Rep. 680*; Greenberg, O. W., Resnikoff, M. 1967. *Phys. Rev.* 163:1844
17. Horgan, R. R., Dalitz, R. H. 1973. *Nucl. Phys. B* 66:135; Horgan, R. R. 1974. *Nucl. Phys. B* 71:514; Jones, M., Horgan, R. R., Dalitz, R. H. 1977. *Nucl. Phys. B* 129:45; Dalitz, R. H., Horgan, R. R., Reinders, L. J. 1977. *Oxford Univ. Rep. 52/77*; Dalitz, R. H., 1977. In *Fundamentals of Quark Models*, ed. I. M. Barbour, A. T. Davies, p. 151. Edinburgh: Scottish Univ. Summer Sch. Phys. 588 pp.
18. Divgi, D. R. 1968. *Phys. Rev.* 175:2027
19. DeRújula, A., Georgi, H., Glashow, S. L. 1975. *Phys. Rev. D* 12:147
20. Isgur, N., Karl, G. 1977. *Phys. Lett. B* 72:109; 1978. *Phys. Lett. B* 74:353; *Phys. Rev. D.* In press
21. Feynman, R. P. 1969. In *High-Energy Collisions*, ed. C. N. Yang, J. A. Cole, M. Good, R. Hwa, J. Lee-Franzini, p. 237. New York: Gordon & Breach. 525 pp; Feynman, R. P. 1969. *Phys. Rev. Lett.* 23:1415; Bjorken, J. D., Paschos, E. A. 1969. *Phys. Rev.* 185:1975; 1970. *Phys. Rev. D* 1:3151; Feynman, R. P. 1972. *Photon-Hadron Interactions*, Reading, PA: Benjamin. 282 pp. Kuti, J., Weisskopf, V. F. 1971. *Phys. Rev. D* 4:3418

22. Weinberg, S. 1967. *Phys. Rev. Lett.* 19: 1264; Salam, A. 1968. In *Elementary Particle Theory*, ed. N. Svartholm, p. 367. Stockholm: Almquist & Wiksells. 399 pp.; Abers, E. S., Lee, B. W. 1973. See Ref. 3; Taylor, J. C. 1976. See Ref. 3; Bég, M. A. B., Sirlin, A. 1974. *Ann. Rev. Nucl. Sci.* 24: 379–449; Bernstein, J. 1974. *Rev. Mod. Phys.* 46: 7; Weinberg, S. 1974. *Rev. Mod. Phys.* 46: 255

23. Bjorken, J. D. 1969. *Phys. Rev.* 179: 1547

24. Taylor, R. E. 1976. In *Proc. 1975 Int. Symp. Lepton Photon Interactions*, ed. W. T. Kirk, p. 679. Stanford: SLAC. 1072 pp. (Abbreviated as SLAC); Riordan, E. M., Bodek, A., Breidenbach, M., Dubin, D. L., Elias, J. E., Friedman, J. I., Kendall, H. W., Poucher, J. S., Sogard, M. R., Coward, D. H. 1974. *Phys. Lett. B* 52: 249; Atwood, W. B. 1975. PhD thesis: *SLAC Rep. 185*; Mo, L. W. SLAC, p. 651; Watanabe, Y., Hand, L. N., Herb, S. et al. 1975. *Phys. Rev. Lett.* 35: 898; Chang, C., Chen, K. W., Fox, D. J. et al. 1975. *Phys. Rev. Lett.* 35: 901; Anderson, H. L., Bharadwaj, V. K., Booth, N. E. et al. 1977. *Phys. Rev. Lett.* 38: 1450

25. Llewellyn Smith, C. H. 1970. *Nucl. Phys. B* 17: 277; 1971. *Phys. Rev. D* 4: 2392; 1974. *Phys. Rep. C* 3: 264

26. Callan, C. G., Gross, D. J. 1969. *Phys. Rev. Lett.* 22: 156

27. Drell, S. D. 1977. *Daedalus* 106(3): 15

28. Nambu, Y. 1966. In *Preludes in Theoretical Physics*, ed. A. de Shalit, H. Feshbach, L. Van Hove, p. 133. Amsterdam: North-Holland. 351 pp.; See Ref. 10; Greenberg, O. W., Zwanziger, D. 1966. *Phys. Rev.* 150: 1177; Lipkin, H. J. 1973. *Phys. Lett. B* 45: 267; 1975. *Phys. Lett. B* 58: 97; Capps, R. H. 1974. *Phys. Lett. B* 49: 178; Dolgov, A. D., Okun, L. B., Zakharov, V. I. 1974. *Phys. Lett. B* 49: 455

29. Adler, S. L. 1969. *Phys. Rev.* 177: 2426; 1970. In *Lectures on Elementary Particles and Quantum Field Theory*, Vol. 1, ed. S. Deser, M. Grisaru, H. Pendleton, p. 1. Cambridge, Mass.: MIT. 592 pp. Bell, J. S., Jackiw, R. 1969. *Nuovo Cimento A* 60: 47; Jackiw, R. 1972. In *Lectures on Current Algebra and Its Applications*, ed. S. B. Trieman, R. Jackiw, D. J. Gross, p. 97. Princeton Univ. Press, NJ; Bardeen, W. A., Fritzsch, H., Gell-Mann, M. 1973. In *Scale and Conformal Symmetry in Hadron Physics*, ed. R. Gatto, p. 139. New York: Wiley

30. Bég, M. A. B., Lee, B. W., Pais, A. 1964.

Phys. Rev. Lett. 13: 514; Sakita, B. 1964. *Phys. Rev. Lett.* 13: 643

31. Particle Data Group. 1976. *Rev. Mod. Phys.* 48, Part II: S1

32. Bogolyubov, N. N., Struminskiĭ, B., Tavkhelidze, A. N. 1965. *JINR Preprint D-1968*; Lipkin, H. J., Tavkhelidze, A. 1965. *Phys. Lett.* 17: 331; Greenberg, O. W. 1965. *Phys. Lett.* 19: 423; Bogolyubov, P. N. 1967. *Sov. J. Nucl. Phys.* 5: 321; 1968. *Ann. Inst. H. Poincaré* 8: 163

33. Harari, H., Rashid, M. A. 1966. *Phys. Rev.* 143: 1354

34. Karl, G., Obryk, E. 1968. *Nucl. Phys. B* 8: 609

35. Horgan, R. 1976. *J. Phys. G.* 2: 625

36. DeGrand, T., Jaffe, R. L. Johnson, K., Kiskis, J. 1975. *Phys. Rev. D* 12: 2060; DeGrand, T., Jaffe, R. L. 1976. *Ann. Phys. NY* 100: 425; DeGrand, T. A. 1976. *Ann. Phys. NY* 101: 496

37. Carlitz, R. D., Ellis, S. D., Savit, R. 1977. *Phys. Lett. B* 68: 443

38. Henriques, A. B., Kellett, B. H., Moorhouse, R. G. 1976. *Phys. Lett. B* 64: 85; Schnitzer, H. J. 1978. *Brandeis Preprint*

39. Lichtenberg, D. B. 1975. *Lett. Nuovo Cimento* 13: 346; Hendry, A. W., Lichtenberg, D. B. 1975. *Phys. Rev. D* 12: 2756

40. Gaillard, M. K., Lee, B. W., Rosner, J. See Ref. 8

41. Dunbar, I. H. 1977. *J. Phys. G* 3: 765, 1025

42. Bég, M. A. B., Singh, V. 1964. *Phys. Rev. Lett.* 13: 418

43. Lee, B. W., Quigg, C., Rosner, J. L. 1977. *Phys. Rev. D* 15: 157

44. Cazzoli, E. G., Cnops, A. M., Connolly, P. L. et al. 1975. *Phys. Rev. Lett.* 34: 1125

45. Knapp, B., Lee, W., Leung, P. et al. 1976. *Phys. Rev. Lett.* 37: 882

46. Finkelstein, J., Tuan, S. F. 1976. *Phys. Rev. D* 15: 902; Igi, K. 1977. *Phys. Rev. D* 16: 196; Boal, D. H. 1977. *Phys. Lett. B* 69: 237

47. Lichtenberg, D. B. 1977. *Phys. Rev. D* 16: 231

48. Böhm, M., Joos, H., Krammer, M. 1973. *Nucl. Phys. B* 51: 397

49. Isgur, N. 1975. *Phys. Rev. D.* 12: 3770; 1976. *Phys. Rev. D* 13: 122

50. Appelquist, T., Barnett, R. M., Lane, K. 1978. See Ref. 8

51. Rosner, J. L. 1968. *Phys. Rev. Lett.* 21: 950; 1974. *Phys. Rep. C* 11: 189; Freund, P. G. O., Waltz, R., Rosner, J. L. 1969. *Nucl. Phys. B* 13: 237

52. Miller, R. C., Novey, T. B., Yokosawa, A., Cutkosky, R. E., Hicks, H. R., Kelly,

R., Shih, C. C., Burleson, G. 1972. *Nucl. Phys. B* 37:401; Cutkosky, R. E., Hicks, H. R., Kelley, R. L. et al. 1973. In *Baryon Resonances—73*, p. 175. West Lafayette, IN: Purdue; Arndt, R. A., Hackman, R. H., Roper, L. D., Steinberg, P. H. 1974. *Phys. Rev. Lett.* 33:987; Giacomelli, G., Lugaresi-Serra, P., Mandrioli, G. et al. 1973. *Nucl. Phys. B* 56:346; Jaffe, R. L. 1977. In *Proc. Topical Conf. Baryon Resonances*, ed. R. T. Ross, D. H. Saxon, p. 455. Chilton, Engl.: Rutherford

53. Fritzsch, H., Gell-Mann, M. 1972. In *Proc. Int. Conf. High Energy Phys., 16th*, ed. J. D. Jackson, A. Roberts, Vol. 2, p. 135. Batavia, IL: Natl. Accel. Lab. 476 pp.; Kogut, J., Sinclair, D. K., Susskind, L. 1976. *Nucl. Phys. B* 114: 199

54. Lipkin, H. J. 1969. *Rev. Nuovo Cimento* Spec. No. 1:134; Llewellyn Smith, C. H. 1969. *Ann. Phys. NY* 53:521; See Ref. 48

55. Jaffe, R. L. 1977. *Phys. Rev. D* 15:267, 281

56. Protopopescu, S. D. 1978. Brookhaven *Preprint BNL-23612*; See Ref. 31

57. Boerner, H. 1963. *Representations of Groups.* Amsterdam: North-Holland. 325 pp.; Gilmore, R. 1974. *Lie Groups, Lie Algebras and Some of Their Applications.* New York: Wiley. 587 pp.; Hamermesh, M. 1962. *Group Theory.* Reading, PA: Addison-Wesley. 509 pp.; Itzykson, C., Nauenberg, M. 1966. *Rev. Mod. Phys.* 38:95; Miller, W. 1972. *Symmetry Groups and Their Applications.* New York: Academic. 432 pp.; Pais, A. 1966. *Rev. Mod. Phys.* 38:215

58. Yan, T. M. 1976. *Ann. Rev. Nucl. Sci.* 26:199–238

59. Perkins, D. H. 1977. *Rep. Prog. Phys.* 40:409; West, G. B. 1975. *Phys. Rep. C* 18:263; Roy, P. 1975. *Theory of Lepton-Hadron Processes at High Energies.* Oxford: Clarendon. 172 pp.; DeRújula, A., Georgi, H., Glashow, S. L., Quinn, H. R. 1974. *Rev. Mod. Phys.* 46:391; Friedman, J. I., Kendall, H. W. 1972. *Ann. Rev. Nucl. Sci.* 22:203–54; See Ref. 21; See Ref. 25

60. Aubert B., Benvenuti, A., Cline, D. et al. 1974. *Phys. Rev. Lett.* 33:984; Fermilab-IHEP-ITEP-Michigan Neutrino Group. 1977. *Phys. Rev. Lett.* 39:382

61. Holder, M., Knobloch, J., May, J. et al. 1977. *Phys. Rev. Lett.* 39:433

62. Gross, D. J., Llewellyn Smith, C. H. 1969. *Nucl. Phys. B* 14:337

63. Zakharov, V. I. 1977. In, *Proc. Int. Conf. on High Energy Physics, 18th*, ed. N. N. Bogolubov, V. P. Dzhelepov, V. G.

Kadyshevsky et al, Vol. II, p. B69. Dubna: JINR

64. Feynman, R. P. 1972. *Photon-Hadron Interactions.* See Ref. 21; Close, F. E. 1973. *Phys. Lett. B* 43:422; Carlitz, R. 1976. *Phys. Rev. Lett.* 36:1001; 1975. *Phys. Lett. B* 58:345

65. Drell, S. D., Yan, T. M. 1970. *Phys. Rev. Lett.* 24:81; West, G. 1970. *Phys. Rev. Lett.* 24:1206; Bloom, E., Gilman, F. J. 1970. *Phys. Rev. Lett.* 25:1140

66. Nachtmann, O. 1971. *Nucl. Phys. B* 28: 283

67. Domokos, G. 1971. *Phys. Rev. D* 4: 3708; Greenberg, O. W., Bhaumik, D. 1971. *Phys. Rev. D* 4:2048; Gürsey, F., Orfanidis, S. 1972. *Nuovo Cimento A* 11:225; Nachtmann, O. 1973. *Nucl. Phys. B* 63:237; 1974. *Nucl. Phys. B* 78:455

68. Bloom, E., Gilman, F. J. 1970. See Ref. 65

69. Georgi, H., Politzer, H. D. 1976. *Phys. Rev. Lett.* 36:1281; 37:68 (erratum); *Phys. Rev. D* 14:1829; Barbieri, R., Ellis, J., Gaillard, M. K., Ross, G. G. 1976. *Phys. Lett. B* 64:171; *Nucl. Phys. B* 117: 50

70. Barnett, R. M. 1976. *Phys. Rev. Lett.* 36:1163; *Phys. Rev. D* 14:70

71. Parisi, G. 1973. *Phys. Lett. B* 43:207; Altarelli, G., Parisi, G. 1977. *Nucl Phys.* 298; Buras, A. J. 1977. *Nucl. Phys. B* 126:298; Johnson, P. W., Tung, W. 1977. *Nucl. Phys. B* 121:270

72. DeRújula, A., Georgi, M., Politzer, H. D. 1977. *Ann. Phys. NY* 103:315; Politzer, H. D. 1977. *Nucl. Phys. B* 122:237

73. Riordan, E. M. et al. 1974. See Ref. 24; Buras, A. J., Floratos, E. G., Ross, D. A., Sachrajda, C. T. 1977. *CERN Report TH2340*

74. Nachtmann, O. 1977. In *Proc. 1977 Int. Symp. on Lepton and Photon Interactions at High Energies*, ed. F. Gutbrod, p. 811. Hamburg: DESY. 1043 pp.

75. Sterman, G., Weinberg, S. 1977. *Phys. Rev. Lett.* 39:1436; Farhi, E. 1977. *Phys. Rev. Lett.* 39:1587; Georgi, H., Machacek, M. 1977. *Phys. Rev. Lett.* 39: 1237

76. Drell, S. D., Yan, T. M. 1970. *Phys. Rev. Lett.* 25:316; 1971. *Ann. Phys. NY* 66: 578

77. Kaplan, D. M., Fisk, R. J., Ito, A. S., Jöstlein, H., Appel, J. A., Brown, B. C., Brown, C. N., Innes, W. R., Kephart, R. D., Ueno, K., Yamanouchi, T. Herb, S. W., Hom, D. C., Lederman, L. M., Sens, J. C., Snyder, H. D., Yoh, J. K. 1978. *Phys. Rev. Lett.* 40:435; Quigg, C. 1978. Private communication

78. Brodsky, S. J., Farrar, G. R. 1972. *Phys. Rev. Lett.* 31:1153; 1975. *Phys. Rev. D* 11:1309; Matveev, V., Muradyan, R., Tavkhelidze, A. 1973. *Lett. Nuovo Cimento* 7:719; Sivers, D., Brodsky, S. J., Blankenbecler, R. 1976. *Phys. Rep. C* 23:1; Ellis, S. D., Stroynowski, R. 1977. *Rev. Mod. Phys.* 49:753

79. Brodsky, S. J., Farrar, G. R. 1975. See Ref. 78

80. Blankenbecler, R., Brodsky, S. J., Gunion, J. F. 1972. *Phys. Lett. B* 39:649; 1973. *Phys. Lett B* 42:461; 1975. *Phys. Rev. D* 12:3469

81. Field, R. D., Feynman, R. P. 1977. *Phys. Rev. D* 15:2590; Feynman, R. P., Field, R. D., Fox, G. C. 1977. *Nucl. Phys. B* 128:1

82. Field, R. D. 1978. *Phys. Rev. Lett.* 40:997

83. Politzer, H. D. 1974. *Phys. Rep. C* 14:129

84. Halzen, F., Ringland, G. A., Roberts, R. G. 1978. *Phys. Rev. Lett.* 40:991

85. Brodsky, S. J. 1977. In *Deep Scattering and Hadronic Structure*, ed. J. Tran Thanh Van, p. 321. Orsay: Inst. Natl. Phys. Nucl. Phys. Part. 485 pp.

86. Field, R. D. 1977. See Ref. 85, p. 207

87. Le Yaouanc, A., Oliver, L., Pene, O., Raynal, J. C. 1977. In *Quarks and Hadronic Structure*, ed. G. Morpurgo, p. 167. New York: Plenum. 318 pp. Close, F. E., p. 137; Altarelli, G., Cabibbo, N., Maiani, L. Petronzio, R. 1974. *Nucl. Phys. B* 69:531; Nachtmann, O. See Ref. 74

88. Weinberg, S. 1977. *Trans. NY Acad. Sci. Ser.* II 38:185

89. Licht, A. L., Pagnamenta, A. 1970. *Phys. Rev. D* 2:1150; LeYaouanc, A. et al. See Ref. 87

90. Gilman, F. J., Harari, H. 1968. *Phys. Rev.* 165:1803; Altarelli, G. et al. See Ref. 87

91. Chanowitz, M. S., Drell, S. D. 1973. *Phys. Rev. Lett.* 30:807; 1974. *Phys. Rev. D* 9:2078; See Ref. 87

92. Morpurgo, G. 1965. *Physics* 2:95; 1966. *Phys. Lett.* 20:684; Nambu, Y. 1967. In *Symmetry Principles at High Energy*, ed. B. Kursunoğlu, A. Perlmutter, I. Sakmar. London: Freeman; Katz, A. 1969. *J. Math. Phys.* 10:2215; Lipkin, H. J. See Ref. 5

93. Greenberg, O. W. 1966. *Phys. Rev.* 147:1077; Dalitz, R. H. 1977. See Ref. 17

94. Blankenbecler, R., Sugar, R. 1966. *Phys. Rev.* 142:1051

95. See Ref. 48; Böhm, M., Joos, H., Krammer, M. 1973. *Acta Phys. Austriaca* Suppl. XI:3; 1974. *Nucl. Phys.*

B 69:349; Joos, H. See Morpurgo, G., Ref. 87, p. 203

96. Preparata, G. 1976. In *Current Induced Reactions*, ed. J. G. Körner, G. Kramer, D. Schildknecht, p. 522. Heidelberg: Springer. 553 pp.

97. See Ref. 10; See Ref. 28; See Ref. 92

98. Chodos, A., Jaffe, R. L., Johnson, K. et al. 1974. *Phys. Rev. D* 9:3471; 10:2599; De Grand, T., Jaffe, R. L., Johnson, K., Kiskis, J. 1975. *Phys. Rev. D* 12:2060

99. Bardeen, W. A., Chanowitz, M. S., Drell, S. D. et al. 1975. *Phys. Rev. D* 11:1094

100. Vinciarelli, P. 1972. *Lett. Nuovo Cimento* 4:905

101. Creutz, M. 1974. *Phys. Rev. D* 10:1749; 12:3126; Creutz, M., Soh, K. S. 1975. *Phys. Rev. D* 12:443

102. Friedberg, R., Lee, T. D. 1977. *Phys. Rev. D* 15:1694; 16:1096

103. Jacob, M., ed. 1974. *Dual Theory*. Amsterdam: North-Holland

104. Tryon, E. P. 1972. *Phys. Rev. Lett.* 28:1605

105. Wilson, K. 1974. *Phys. Rev. D* 10:2445; Kogut, J., Susskind, L. 1975. *Phys. Rev. D* 11:395; Drell, S. D. 1977. In *Quark Spectroscopy and Hadron Dynamics*, ed. M. C. Zipf, p. 81. Stanford: SLAC

106. Belavin, A. A., Polyakov, A., Schwartz, A., Tyupkin, Y. 1975. *Phys. Lett. B* 59:85; Callan C., Dashen R., Gross, D. 1976. *Phys. Lett. B* 63:334; Jackiw, R., Rebbi, C. 1976. *Phys. Rev. Lett.* 37:172

107. Callan, C., Dashen, R., Gross, D. 1977. *Phys. Lett. B* 66:375; Glimm, J., Jaffe, A. 1978. *Phys. Rev. Lett.* 40:277

108. Marciano, W., Pagels, H. See Ref. 4

109. 't Hooft, G. 1977. *Utrecht Preprint*

110. Günaydin, M., Gürsey, F. 1974. *Phys. Rev. D* 9:3387

111. Domokos, G., Kövesi-Domokos, S. 1978. *Johns Hopkins Preprint JHU-HET 7801*

112. Fritzsch, H. See Ref. 13, p. 215

113. See Ref. 11, p. 83; Pati, J. C., Rajpoot, S., Salam, A. 1977. *Imperial Coll. Preprint*

114. Nowak, E., Sucher, J., Woo, C. H. 1977. *Phys. Rev. D* 16:2874

115. Jones, L. W. 1977. *Rev. Mod. Phys.* 49:717

116. Zel'dovich, Y. B., Okun, L. B., Pikel'ner, S. B. 1966. *Sov. Phys. Usp.* 8:702

117. Feinberg, E. L. 1967. *Sov. Phys. Usp.* 10:256

118. La Rue, G. S., Fairbank, W. M., Hebard, A. F. 1977. *Phys. Rev. Lett.* 38:1011

119. Gallinaro, G., Marinelli, M., Morpurgo,

G. 1977. *Phys. Rev. Lett.* 38:1255; Bland, R., Bocobo, D., Eubank, M., Royer, J. 1977. *Phys. Rev. Lett.* 39:369

120. Ivanenko, D., Kurdgelaidze, D. F. 1969. *Lett. Nuovo Cimento* 2:13; Itoh, N. 1970. *Prog. Theor. Phys.* 44:291

121. Collins, J. C., Perry, M. J. 1975. *Phys. Rev. Lett.* 34:1353

122. Brecher, K., Caporaso, G. 1976. *Nature* 259:377

123. Oppenheimer, J. R., Volkoff, G. 1939. *Phys. Rev.* 55:374

124. Keister, B. D., Kisslinger, L. S. 1976. *Phys. Lett.* B 64:117; Kislinger, M. B., Morley, P. D. 1977. *Phys. Lett.* B 67: 371; Baluni, V. 1978. *Phys. Lett. B* 72:381; 1978. *Phys. Rev. D* 17:2092

125. Baym, G., Chin, S. A. 1976. *Phys. Lett.* B 62:241; Chapline, G., Nauenberg, M. 1976. *Nature* 264:235; 1977. *Phys. Rev. D* 16:450; Bowers, R. L., Gleeson, A. M., Pedigo, R. D. 1977. *Astrophys. J.* 213:840

126. Kisslinger, M. B., Morley, P. D. 1976. *Phys. Rev. D* 13:2765, 2771; Freedman,

B. A., McLerran, L. D. 1977. *Phys. Rev. D* 16:1130, 1147, 1169; 17:1109

127. Zel'dovich, Y. B. 1962. *Sov. Phys. JETP* 14:1143

128. Caporaso, G., Brecher, K. 1978. *MIT Preprint*

129. Brecher, K. 1977. *Astrophys. J.* 215: L17

130. Schramm, D. N., Wagoner, R. V. 1977. *Ann. Rev. Nucl. Sci.* 27:37–74; Weinberg, S. 1977. *The First Three Minutes.* New York: Basic. 188 pp.; *Gravitation and Cosmology. Principles and Applications of the General Theory of Relativity.* New York: Wiley; Harrison, E. R. 1973. *Ann. Rev. Astro. Astrophys.* 11:115

131. Frautschi, S., Steigman, G., Bahcall, J. N. 1972. *Astrophys. J.* 175:307; Zel'dovich, Y. B. 1970. *Comments Astrophys. Space Phys.* 2:12; Chapline, G. F. 1976. *Nature* 261:550

132. Zel'dovich, Y. B. 1973. *Sov. Phys. JETP* 37:33; Chapline, G. F. 1976. *Nature* 261: 550; Liang, E. P. T. 1977. *Phys. Rev. D* 16:3369

Ann. Rev. Nucl. Part. Sci. 1978. 28 : 387–499

CHARM AND BEYOND[*]   ✕5598

Thomas Appelquist[1]

J. Willard Gibbs Laboratory, Yale University, New Haven, Connecticut 06520

R. Michael Barnett[2]

Stanford Linear Accelerator Center, Stanford University, Stanford, California 94305

Kenneth Lane[3]

Lyman Laboratory of Physics, Harvard University, Cambridge, Massachusetts 02138

CONTENTS

[*] The US Government has the right to retain a nonexclusive, royalty-free license in and to any copyright covering this paper.

[1] Alfred P. Sloan Foundation Fellow. Research supported in part by the Department of Energy under contract EY-76-C-02-3075.

[2] Research supported in part by the Department of Energy.

[3] Research supported in part by the National Science Foundation under contract PHY77-22864.

388   APPELQUIST, BARNETT & LANE

1 INTRODUCTION

The physics of elementary particles has changed dramatically during the 1970s, especially during the last three or four years. A new quark carrying a new quantum number called charm has been discovered and there is mounting evidence for the existence of yet a heavier quark. A great deal has been learned about the structure of the weak interactions, and there is considerable optimism that we are beginning to understand strong interaction dynamics at a fundamental level. Indeed, some visionaries are already attempting grand syntheses of the strong, weak, and electromagnetic interactions. In this paper, we describe these theoretical and experimental developments, emphasizing the role of the new heavy quarks. It is largely a review for nonspecialists but specialists will find some new results or at least some new perspectives.

Hadrons, that is mesons and baryons, are made of quarks (1); after the events of the last three years there are no longer any skeptics. Since many of the details of the quark model of hadrons are discussed by O. W. Greenberg in a paper appearing in this issue of the Annual Review (2), we recall only some of the basic properties of the old quarks to prepare for our discussion of the new ones. Until 1974, all the known mesons and baryons could be understood as quark-antiquark and three-quark bound states respectively, where the quarks come in three varieties or "flavors," commonly called u, d, and s. The u and d (up and down) quarks form an isotopic spin SU(2) doublet and are the constituents of the nucleons and mesons of nuclear physics. The heavier s (strange) quark can bind with the others or with itself to produce the so-called strange particles that fill out the SU(3) multiplets of Gell-Mann (1). The relative heaviness of the strange quark means, of course, that this SU(3) symmetry is rather badly broken. All of these quarks have spin $\frac{1}{2}$ and carry fractional electric charge. The u quark has charge $\frac{2}{3}$ and the d and s quarks have charge $-\frac{1}{3}$.

1.1   The Need For Charm

With the success of SU(3) in the early 1960s, many people soon considered the possibility that yet heavier quarks might exist (3). They would presumably be constituents of hadrons too heavy and too short lived to have been seen at the time. It was partly a matter of "why not?" and partly motivated by primitive notions of quark-lepton symmetry. There were four leptons—the electron, the muon, and their respective neutrinos—so why shouldn't there be four quarks? Quark-lepton symmetry continues to be an important guiding principle in attempting to unify the fundamental forces of nature, but it seems likely that it will ultimately take a more subtle form than equal numbers of each.

It was a problem with weak interaction phenomenology that led Glashow, Iliopoulos, & Maiani (GIM), in their classic paper of 1970, to provide a genuine *raison d'être* for a fourth quark (4). It had been known for decades that the weak interactions, such as neutron β decay and $\mu^- \to e^- + \bar{v}_e + v_\mu$, could all be described by the interaction of two charge-changing currents with an interaction strength $G_F \approx 10^{-5}/M_P^2$. It is now known that neutral currents also play a role in the weak interactions. Processes like $v_\mu + p \to v_\mu + p$ have been observed and require both vector and axial-vector neutral current interactions with strength of order G_F. In 1970, the neutral weak currents had not been seen experimentally but, for the most part, neither had they been ruled out at the level G_F. It was expected by some people that they would appear at this level and this expectation was given a sound basis with the proof, a year later, of the renormalizability of gauge theories of the weak and electromagnetic interactions (see below).

There was one embarrassing problem with this state of affairs. One class of neutral current interactions, those involving strangeness-changing hadronic weak currents, was known to be tremendously suppressed. The branching ratios (5)

$$\frac{\Gamma(K_L^0 \to \mu^+\mu^-)}{\Gamma(K_L^0 \to \text{all})} \sim 10^{-8} \quad \text{and} \quad \frac{\Gamma(K^\pm \to \pi^\pm e^+ e^-)}{\Gamma(K^\pm \to \text{all})} = (2.6 \pm 0.5) \times 10^{-7}$$

$$1.1$$

exhibit the problem. That a strangeness-changing neutral current, coupling with strength G_F, might have been expected can be seen by examining the structure of the charge-changing hadronic weak current

$$J_\pm^\mu = \bar{q}\gamma^\mu \tfrac{1}{2}(1-\gamma_5)T_\pm q$$

$$\equiv \bar{q}_L \gamma^\mu T_\pm q_L. \qquad\qquad 1.2$$

The quark spinor q contains $4 \times N$ components where N is the number of flavors and $T_\pm$ is an $N \times N$ matrix that changes the electric charge by ± 1 unit. Here $q_L \equiv \frac{1}{2}(1-\gamma_5)q$ is the left-handed part of the quark field[4]. It was expected by some and unified theories demanded, that there should exist a neutral partner of these two currents

$$J_0^\mu = \bar{q}\gamma^\mu \tfrac{1}{2}(1-\gamma_5)T_0 q \qquad \qquad 1.3$$

coupling with the same strength G_F, where

$$T_0 = [T_+, T_-]. \qquad \qquad 1.4$$

The charged current was known to have the Cabibbo form (6)

$$J_+^\mu = \bar{u}\gamma^\mu \tfrac{1}{2}(1-\gamma_5)(d\cos\theta_C + s\sin\theta_C) \qquad \qquad 1.5$$

where $\sin\theta_C = 0.23$. Therefore, the neutral current must contain a piece of the form

$$J_0^\mu = \bar{s}\gamma^\mu \tfrac{1}{2}(1-\gamma_5)\,d\sin\theta_C \cos\theta_C + \ldots, \qquad \qquad 1.6$$

which gives rise to $\Delta S = 1$ neutral interactions of order G_F.

The GIM solution (4) was to introduce a new, heavy, charge $\frac{2}{3}$ quark c with a left-handed weak coupling to the orthogonal Cabibbo combination $s\cos\theta_C - d\sin\theta_C$. The c quark was postulated to carry a new quantum number charm, conserved by the strong interactions. It can then easily be checked that with

$$J_+^\mu = \bar{u}\gamma^\mu \tfrac{1}{2}(1-\gamma_5)(d\cos\theta_C + s\sin\theta_C) + \bar{c}\gamma^\mu \tfrac{1}{2}(1-\gamma_5)(s\cos\theta_C - d\sin\theta_C),$$
$$1.7$$

the $\Delta S = 1$ piece of J_0^μ in Equation 1.6 is cancelled and the neutral current is, in fact, diagonal in all flavors. This current couples to itself and to leptonic neutral currents with strength G_F and, to lowest order in this interaction, the reactions in Equation 1.1 are forbidden. Higher order corrections might, of course, induce such reactions and because the branching ratios (Expression 1.1) are so small, this must be looked at carefully. It is only possible to do this in a renormalizable theory, so we turn next to a discussion of gauge theories of weak and electromagnetic interactions.

1.2   The Weinberg-Salam Model

Unified gauge theories provide the general framework for most modern work on weak and electromagnetic interactions. The prototype for all

---

[4] The spinor field q_L is not, strictly speaking, left-handed. The operator $\frac{1}{2}(1-\gamma_5)$ projects out the left-handed (negative helicity) part of the Dirac spinor $u(p)$ only in the zero mass limit. Thus for the large momentum $(p \gg m_q)$ components of $q_L(x)$, $\frac{1}{2}(1-\gamma_5)$ can be thought of as a covariant version of a left-handed projection operator.

such models is based on the group SU(2) × U(1) and was first written down in detail by Weinberg in 1967 (7). It remains viable today in the face of a large amount of experimental data and will survive as at least a subgroup of the ultimate weak and electromagnetic theory.

Unified gauge theories are constructed by generalizing what is known about electromagnetic interactions to include the weak interactions. The inclusion of strong interactions in this framework is described in the next section. The forces are all mediated by the exchange of spin one bosons corresponding to vector gauge fields that are present to maintain the local gauge invariance of the Lagrangian. In electrodynamics, this means invariance under a space-time-dependent phase transformation $\psi(x) \rightarrow e^{iQ\theta(x)}\psi(x)$ on each matter field of charge Q, along with the transformation $A_\mu(x) \rightarrow A_\mu(x) + \partial_\mu\theta(x)$ on the electromagnetic field. Invariance is insured if A_μ enters the Lagrangian only in the gauge covariant derivative $D_\mu\psi = (\partial_\mu - iQA_\mu)\psi$ and if derivatives of A_μ enter only through the electromagnetic field tensor $F_{\mu\nu} = \partial_\mu A_\nu - \partial_\nu A_\mu$. By using these ingredients in a minimal way (excluding nonrenormalizable couplings), one obtains the Maxwell-Dirac theory for photons interacting with charged, spin $\frac{1}{2}$ particles of mass m

$$\mathscr{L} = \bar{\psi}[i\gamma^\mu D_\mu - m]\psi - \tfrac{1}{2}F_{\mu\nu}F^{\mu\nu}. \qquad 1.8$$

The phase transformations of electrodynamics form the group U(1), the group of unitary 1 × 1 matrices. Since both neutral and charge-changing currents play a role in the weak interactions, the corresponding group must be larger, but the gauge principle can be carried over by associating a separate gauge field $A_\mu^a(x)$ with each infinitesimal parameter of the group (8). The minimal possibility is SU(2) and, since this is a three-parameter group, there will be three gauge fields forming an isotopic triplet. The Lagrangian will take the form of Equation 1.8 except that now $F_{\mu\nu}$ will contain a term quadratic in A_μ^a and a coupling matrix will multiply the charge Q in $D_\mu\psi$. Such a theory can be proven to be renormalizable[5] (9).

The problem is that the weak bosons must be very massive—they have not yet been seen and the lower limit on the mass is about 25 GeV (10). The addition of a mass term $-\tfrac{1}{2}M^2 A_\mu^a A^{a\mu}$ to the Lagrangian, however, destroys the local gauge invariance and the renormalizability. The solution is to preserve the local gauge invariance of the Lagrangian, but allow it to be spontaneously broken, that is, not respected by the physical states. There is a general theorem that whenever this happens, massless

---

[5] This means that the infinities that appear in higher order computations can all be absorbed into the physical masses and coupling constants of the theory. The S matrix can then be computed to any order in terms of these few parameters.

scalar particles, known as Goldstone bosons, must be present (11). However, in gauge theories these quanta get mixed in with the longitudinal parts of the gauge fields. They allow the gauge bosons to become massive by providing the zero helicity degree of freedom forbidden to a massless vector boson. This is called the Higgs mechanism (12). A particularly simple way of realizing this is to introduce into the Lagrangian a multiplet of elementary spinless fields. After spontaneous symmetry breaking, some of these become the Goldstone bosons absorbed by the gauge fields, while the remaining members of the multiplet survive as physical, massive scalar particles known as Higgs bosons.

A unified model of this sort must incorporate both a weak gauge group and the electromagnetic gauge group. The $SU(2) \times U(1)$ model involves four gauge bosons. The spontaneous breakdown is arranged to preserve a local $U(1)$ subgroup so that one boson stays massless and is identified with the photon, while the other three ($W^{\pm}$ and Z^0) become the intermediate bosons of the weak interactions. In the original form of the model (7), the left-handed leptons and quarks are put into $SU(2)$ doublets and the right-handed components are in $SU(2)$ singlets. The fermions are massless in the Lagrangian; their mass, along with the weak boson masses, arises from the spontaneous symmetry breakdown. There are two independent coupling constants g and g' for the $SU(2)$ and $U(1)$ groups respectively. The electric charge is given by $e = gg'/(g^2 + g'^2)^{\frac{1}{2}}$ and since $W^{\pm}$ exchange describes the familiar process of β decay, the Fermi coupling constant G_F is given by

$$\frac{G_F}{2^{\frac{1}{2}}} = \frac{g^2}{8M_W^2},$$

1.9

It follows that

$$M_W = \left(\frac{e^2}{32^{\frac{1}{2}}G_F}\right)^{\frac{1}{2}} \frac{1}{\sin \theta_W} = 37.3 \text{ GeV}/\sin \theta_W,$$

where θ_W is the Weinberg angle defined by $\tan \theta_W = g'/g$. In the simplest version of the theory, there is one complex $SU(2)$ doublet of Higgs fields, and the Z^0 mass is given by $M_Z = M_W/\cos \theta_W$.

In this model, there remains one massive physical Higgs scalar. Its mass is a free parameter theoretically, but its presence in the theory is absolutely crucial for renormalizability. It has by now been proven that such theories are, in fact, renormalizable to all orders (9). Spontaneously broken gauge theories of weak and electromagnetic interactions are theories in the same sense that quantum electrodynamics itself is a theory.

Let us now return to the problem of $\Delta S = 1$ neutral currents in the four-quark GIM model. A process that is $O(G_F)$ in Born approximation will be

$O(G_F e^2)$ at the one-loop level in a renormalizable gauge theory. If a process is forbidden to lowest order, however, it is not hard to see that it will also vanish to $O(G_F e^2)$. Consider the decay $K_L^0 \to \mu^+ \mu^-$ for example. It can proceed through an intermediate state consisting of two charged bosons, but the graph with an exchanged c quark is exactly cancelled by the graph with an exchanged u quark in the limit $m_u = m_c$. The amplitude is of order $G_F^2(m_c^2 - m_u^2)$ as long as m_c, $m_u \ll M_W$. This can be made consistent with rates and limits such as those in Expression 1.1 with $m_c \approx 1\text{–}2$ GeV. The mass difference between the K_L^0 and K_S^0 mesons arises from a similar interaction and led Gaillard & Lee (13) in 1973 to a similar estimate for m_c. These estimates are only approximate because of uncertainties associated with strong interaction corrections, but they later were found to be qualitatively correct.

1.3 Experimental Indications

The first direct experimental evidence for the existence of the charmed quark came from e^+e^- annihilation into hadrons. It was widely expected that $R(W) \equiv \sigma_{tot}(e^+e^- \to \text{hadrons})/\sigma(e^+e^- \to \mu^+\mu^-)$ would become constant above center-of-mass energy $W = 1\text{–}2$ GeV. The theoretical basis for this expectation and the value of the constant is given in Section 2. When this ratio was first measured above $W = 3$ GeV at the Cambridge Electron Accelerator (CEA) in 1972 (14), it was found to be well above the expected value—and apparently rising with W! These measurements, eventually confirmed at the Stanford Linear Accelerator Center (SLAC) (15) and the Deutsches Elektronen Synchrotron (DESY) (16), gave hope to the small band of charm enthusiasts.

The next piece of evidence for charm came from the neutrino scattering experiments at Fermilab. In the collision of muon-type neutrinos with matter, events were discovered that contained a $\mu^+ \mu^-$ pair in the final state (17). These events, both in character and rate, could be naturally explained by charm excitation. The hadronic current (Equation 1.7) can couple to the muonic current $\bar{\mu}\gamma^\mu \frac{1}{2}(1 - \gamma_5)\nu_\mu$ through W^+ exchange leading to a muon and c quark in the final state. The c quark can decay by this same interaction into a lepton-neutrino pair and preferentially a strange quark. This source of dimuon pairs requires the presence of K mesons in the final state in most events, and that was indeed verified somewhat later (17).

The case was strengthened considerably by the spectacular events of November 1974. The simultaneous discovery of the ψ/J resonance[6] by

---

[6] Although the $^3S_1$ c$\bar{c}$ ground state was simultaneously dubbed ψ (at SLAC) and J (at Brookhaven), the excited states have all been discovered at SLAC and DESY in e^+e^- annihilation. We refer throughout this paper to the c$\bar{c}$ states that can be directly produced in e^+e^- annihilation as ψ, ψ', ψ'', etc.

groups at SLAC (18) and Brookhaven (19) and the subsequent measurement of excited states (20) and radiative transitions (21) could only be understood in terms of the existence of a new quark of mass around 1.5 GeV. In the ψ it was bound to its own antiquark but, before long, particles made of one heavy and one or more old quarks were discovered (22), and their weak decay properties made it clear that the new quark was precisely the charmed quark predicted by GIM.

The discovery of the upsilon Υ in July 1977 (23) will perhaps be the beginning of a similar story. At least two and perhaps three states in the mass range 9.5–10.5 GeV have been discovered in the reaction $p + Be \rightarrow \Upsilon + \cdots \rightarrow \mu^+ \mu^- + \cdots$, and it is irresistible to guess that they are the ground state and two radial excitations of yet another quark and its antiquark. In the case of the ψ/J, it is the $e^+ e^-$ colliding beam machines that have produced the most important discoveries. Unfortunately, the highest energy presently available with these machines is about 8 GeV, which puts the upsilon just out of range. However, the next generation of higher energy colliding beam machines at SLAC, DESY, and Cornell will be operating in the near future. They will, no doubt, produce a great deal of information about the upsilon and even heavier quark-antiquark systems if they exist.

It is important to mention one more experimental development here. In 1975, the SLAC–Lawrence Berkeley Laboratory (LBL) group announced the discovery (24) of a new particle τ with $m_\tau \approx 1.8$ GeV. The initial evidence came in the form of the "anomalous" events $e^+ e^- \rightarrow e^\pm + \mu^\mp +$ missing energy, which indicates a primary process $e^+ e^- \rightarrow \tau^+ \tau^-$ followed by weak decay of the τ^+ and τ^-. With further study, it now seems clear that the τ is a lepton (25). These experiments and the τ properties are not discussed here, but it is important to keep the τ's existence in mind throughout. On the theoretical side, it must loom large in any attempts at grand synthesis or considerations of quark-lepton symmetry. Experimentally, there is the curious fact that its mass is so close to that of the charmed quark. In $e^+ e^-$ annihilation, charm threshold and $\tau^+ \tau^-$ threshold come nearly on top of each other, so that the experimental study of each is complicated by the other.

2   QUANTUM CHROMODYNAMICS

2.1   Background

Even though charm was invented in response to problems with the weak and electromagnetic interactions, its discovery has had a considerable impact on strong interaction physics. The strong interactions have been a notoriously difficult problem for several decades and one might ask what impact the discovery of new strongly interacting quarks could have on our understanding of these forces. The answer is that early in 1973, a little more than a year before the experimental discovery of the ψ/J, a remarkable theoretical discovery was made (26), which has led to the growing consensus among theorists that we may finally have in hand a fundamental theory of the strong interactions. This has not yet been proven, but the measured properties of heavy quarks have reinforced this view and if yet heavier quarks are discovered, the reinforcement should become even stronger. The interplay of charm and this candidate theory of strong interactions called, quantum chromodynamics (QCD), is discussed throughout the paper.

2.2   The Hidden Color Hypothesis

The first ingredient in QCD is a new, completely hidden, quantum number known as color (27). This notion is explained in some detail in the accompanying article by Greenberg (2) and so we only summarize the essentials. Several problems with quark phenomenology have suggested that the number of quarks should be tripled so that each type of quark u, d, s, c, ... comes in three so-called colors. In the most popular version of the color model, all the fundamental forces respect the symmetry under color interchange (28). This symmetry can be viewed as a new SU(3) symmetry and the observed hadrons are all singlets with respect to this new SU(3).

There are several reasons for believing in the color hypothesis. With color, the ground states of baryons can be simply understood in the quark model without abandoning Fermi-Dirac statistics. Consider the Ω^- for example; it consists of three s quarks in an orbital S state with spins aligned, leading to an overall symmetric wave function. With three colors, however, overall antisymmetry can be restored if the Ω^- is a color singlet. Color also plays a sort of counting role, raising the value of certain amplitudes to agree with experiment. One example is the decay $\pi^0 \to \gamma\gamma$ (29), and another is the value of $\sigma_{tot}(e^+ e^- \to \text{hadrons})$ within the free quark (parton) approximation (30). There, neglecting fermion masses,

$$R(W) \equiv \frac{\sigma_{tot}(e^+ e^- \to \text{hadrons})}{\sigma(e^+ e^- \to \mu^+ \mu^-)} = \sum_i Q_i^2 \qquad 2.1$$

where Q_i is the electric charge of a quark of type i in units of e. The inclusion of color in this sum raises the prediction from $\frac{2}{3}$ to 2 below charm threshold. This is a good first approximation to the experimental value of R (31–34) and within QCD it also turns out to be a good first approximation theoretically.

2.3   The Color SU(3) Gauge Theory

QCD is constructed by extending the global $SU(3)_c$ color symmetry to a local gauge symmetry. This requires the introduction of massless vector gauge fields $A_\mu^a(x)$, $a = 1, 2, \ldots, 8$, transforming according to the adjoint representation of $SU(3)_c$. The eight gauge fields are called colored gluons. The Lagrangian density is

$$\mathscr{L}(x) = \bar{q}(x)\left[i\gamma^\mu D_\mu - M_0\right]q(x) - \tfrac{1}{4}F_{\mu\nu}^a(x)\,F^{a\mu\nu}(x), \qquad 2.2$$

where the quark field $q(x)$ contains $4 \times 3 \times N$ components, corresponding to the three colors and the unknown number N of quark flavors; M_0 is the bare mass matrix, a product of Dirac and color unit matrices and a diagonal flavor matrix M_0^{AB}. The gauge field covariant derivatives are

$$D_\mu = \partial_\mu - ig\,T^a A_\mu^a$$

and   2.3

$$F_{\mu\nu}^a = \partial_\mu A_\nu^a - \partial_\nu A_\mu^a + gf^{abc}A_\mu^b A_\nu^c,$$

and we normalize the T matrices in the conventional way:

$$T_a = \tfrac{1}{2}\lambda_a, \qquad [\lambda_a, \lambda_b] = 2if_{abc}\lambda_c, \qquad \text{Tr } \lambda_a\lambda_b = 2\delta_{ab}. \qquad 2.4$$

The theory is invariant under the gauge transformation

$$A_\mu(x) \equiv A_\mu^a(x)T^a \to U(x)A_\mu(x)U^{-1}(x) + \frac{i}{g}U(x)\partial_\mu U^{-1}(x)$$

$$q(x) \to U(x)q(x), \qquad 2.5$$

where $U(x) = \exp\{ig\theta^a(x)T^a\}$.

While QCD bears a superficial resemblance to quantum electro-dynamics (QED), it is, in fact, considerably more complicated. It is a renormalizable theory, like QED, so that the ultraviolet divergences that appear in perturbation theory can be handled in the traditional way. Field theory Green's functions can be computed to any order in a power series expansion in a renormalized coupling constant. One expects this to be useful only if the coupling constant is small, however, and that depends on how the theory is renormalized. It is here that QCD and QED part company.

In quantum electrodynamics, the renormalized coupling constant α can be defined at zero momentum transfer (infinite distances), since the non-zero electron mass prevents infrared divergences in the photon propagator in this limit. The potential energy between two charged particles separated by a distance $r \gg 1/m_e$ is αr^{-1} with α determined experimentally to be approximately $\frac{1}{137}$. As a consequence of the r^{-1} behavior, the asymptotic states of the theory are the charged particles and photons corresponding to the fields in the Lagrangian. At distances $r \ll 1/m_e$, the charged shielding effect of vacuum polarization begins to disappear and the effective coupling strength begins to increase. The familiar one-loop result is

$$V(r) = \frac{\alpha}{r}\left[1 + \frac{\alpha}{3\pi}\log{(rm_e)^{-1}}\right] \qquad 2.6$$

and at distances $r \simeq m^{-1}e^{-3\pi/\alpha}$, the increase becomes substantial. Since the perturbation expansion breaks down at this point, the asymptotic behavior of QED as $r \to 0$ is unknown. Fortunately, physics at laboratory distance scales is insensitive to this asymptotic ignorance.

QCD differs from QED in many ways, in particular in the way the effective coupling strength as computed in perturbation theory varies with distance. This different behavior is the important feature of QCD called asymptotic freedom (26). We shall describe this property with an eye toward our later discussion of its role in heavy quark physics. At the end of this section some of the other known and conjectured properties of QCD will be summarized.

The Feynman rules for QCD can be generated using either functional methods or the canonical Hamiltonian formalism (35). In either case, a gauge must be chosen as in QED. In a general class of covariant gauges, the gluon propagator is

$$D_{\mu\nu}^{ab}(k) = \delta^{ab}\left(g_{\mu\nu} - \xi\frac{k_\mu k_\nu}{k^2}\right)\frac{1}{k^2 + i\varepsilon}, \qquad 2.7$$

where ξ is an arbitrary parameter specifying the gauge. The remaining Feynman rules, consisting of the quark propagator and the various vertices, can be read off directly from the Lagrangian (35). The only exception to this is that a careful treatment of unitarity demands the inclusion of fictitious scalar particles called Fadde'ev-Popov ghosts, which propagate only around closed loops (36).

For some purposes, it is convenient to adopt the physical Coulomb gauge (35). The propagator then consists of an instantaneous potential part and a transverse gluon part

$$\begin{aligned} D_{\mu\nu}^{ab}(k) &= \delta^{ab}i/\mathbf{k}^2 \qquad &\mu = \nu = 0 \\ &= \delta^{ab}\frac{i}{k^2 + i\varepsilon}\left(\delta_{ij} - \frac{k_i k_j}{\mathbf{k}^2}\right) \qquad &\mu = i, \nu = j. \end{aligned} \qquad 2.8$$

A Fadde'ev-Popov ghost must again be included, but it is instantaneous and it couples only to transverse gluons. In either covariant or Coulomb gauge, Ward identities can be established and renormalizability proven.

Now consider again the potential energy between two charge sources separated by a distance r (37, 38). The charge is now color and the sources can be taken to be a very heavy quark and antiquark. If they are in a quantum mechanical color singlet state, then the Born approximation to the potential is $-C_F \alpha_s / r$, where $T_{ik}^a T_{kj}^a = C_F \delta_{ij}$ and $\alpha_s = g^2/4\pi$ is the strong coupling constant. For SU(N), $C_F = (N^2 - 1)/2N$. The one-loop corrections to the potential are most conveniently computed in Coulomb gauge and, if light quarks are neglected for the moment, the two relevant diagrams are shown in Figure 1 (37). The necessary renormalization subtractions cannot be performed at infinite separation because the loops are infrared divergent at this point. Some other arbitrary point, say $r = \mu^{-1}$, must be chosen, and this necessarily brings in a new dimensional parameter μ. In QED, such a parameter *could* be introduced; in QCD, because of the self-coupling of the gauge field, it *must* be introduced. The static potential through one loop is

$$V(r) = -C_F \frac{\alpha_\mu}{r}\left[1 + \frac{5}{6}\frac{\alpha_\mu}{\pi}C_A \ln(r\mu)^{-1} - \frac{16}{6}\frac{\alpha_\mu}{\pi}C_A \ln(r\mu)^{-1}\right], \qquad 2.9$$

where $\alpha_\mu = g_\mu^2/4\pi$ is the coupling strength corresponding to the scale $r = \mu^{-1}$ and $f_{acd}f_{bcd} = C_A \delta_{ab} = N\delta_{ab}$ for SU(N). The two logarithmic modification terms correspond to Figures 1b and 1a respectively. The first term is vacuum polarization of transverse gluon pairs, a charge-shielding effect similar to electron-positron vacuum polarization in QED. As expected, this contribution tends to make the effective coupling strength increase as $r \to 0$. The other contribution, unique to Yang-Mills theories, is a self-energy of the Coulomb field. It comes with the opposite sign and is larger in magnitude than vacuum polarization. The net result is that the effective coupling strength decreases as r decreases. This is the property called asymptotic freedom.

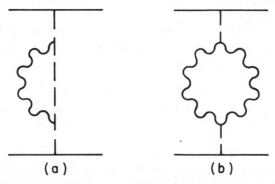

(a)   (b)

Figure 1 Contributions to coupling-constant renormalization for the Yang-Mills theory in Coulomb gauge.

It is not difficult to understand the physical mechanism behind asymptotic freedom (37). We do not describe it in detail except to say that the Coulomb field self-energy produces a collimation of the color electric field lines connecting the quark to the antiquark. The collimation becomes more pronounced as $r \to \infty$, increasing the energy stored in the field relative to the pure Coulomb potential. The decrease of the collimation as r decreases is the explanation of asymptotic freedom.

The behavior of the effective coupling strength as $r \to 0$ can be described to all orders in perturbation theory by a consideration of its scaling properties[7] (39). The potential can be written quite generally in the form

$$V(r) = -C_F \frac{\alpha(r\mu, \alpha_\mu)}{r}, \qquad \qquad 2.10$$

where $\alpha(r\mu, \alpha_\mu) \equiv \alpha_s(r)$ is the effective coupling strength, given through one loop by Equation 2.9. The potential must be independent of the choice of μ and the condition $(d/d\mu)\alpha_s(r) = 0$ can be used to show that

$$r \frac{\partial}{\partial r} \alpha_s(r) = \beta[\alpha_s(r)], \qquad \qquad 2.11$$

where

$$\beta(x) \equiv r \frac{\partial}{\partial r} \alpha(r\mu, x)\Big|_{r = \mu^{-1}}. \qquad \qquad 2.12$$

From Equation 2.9 we find

$$\beta(x) = \frac{11}{2} \frac{x}{\pi} + O(x^2) \qquad \qquad 2.13$$

and, therefore, to lowest order, Equation 2.11 can easily be integrated. The result is

$$\alpha_s(r) = \frac{\alpha_\mu}{1 + \dfrac{11}{2} \dfrac{\alpha_\mu}{\pi} \ln(r\mu)^{-1}}, \qquad \qquad 2.14$$

which agrees with Equation 2.9 to lowest order and which goes to zero as r decreases. As r increases Equation 2.14 shows that the theory becomes strongly coupled and the perturbation expansion breaks down. In this sense, QCD and QED are oppositely behaved.

---

[7] In a renormalizable field theory, the study of scaling properties is not just a simple matter of dimensional analysis. A new dimensional parameter is introduced through renormalization, and that complicates the scaling behavior. The formalism for studying this behavior is called the renormalization group.

2.4   *Properties of Quantum Chromodynamics*

2.4.1   COVARIANT FORMULATION   For most purposes, it is best to work in a covariant gauge. The asymptotic form Equation 2.14 of the running coupling constant is gauge invariant, but its Feynman diagrammatic breakdown is completely different in a covariant gauge (26). As a result, the physical mechanism behind asymptotic freedom is not as transparent as in Coulomb gauge. The momentum space effective coupling constant $\alpha_s(-q^2)$ is defined in terms of both propagator and vertex functions at a Euclidean momentum $q^2 = q_0^2 - q^2 < 0$. If there are f quark flavors for which $m \ll q$, then

$$\alpha_s(-q^2) = \frac{\alpha_\mu}{1 + \left(\frac{11}{4} - \frac{1}{6}f\right)\frac{\alpha_\mu}{\pi}\ln\left(-\frac{q^2}{\mu^2}\right)}. \qquad 2.15$$

It is convenient to re-express $\alpha_s(-q^2)$ in the form

$$\alpha_s(-q^2) = \frac{\pi}{\left(\frac{11}{4} - \frac{1}{6}f\right)\ln\left(-\frac{q^2}{\Lambda^2}\right)}, \qquad 2.16$$

where $\Lambda = \mu\exp\left[-\pi/(11/2 - f/3)\alpha_\mu\right]$. This explicitly exhibits the fact that $\alpha_s(-q^2)$ depends only on one parameter (α_μ and μ are not independent): Λ must be determined experimentally, and then $\alpha_s(-q^2)$ is completely specified.

2.4.2   APPLICATIONS OF ASYMPTOTIC FREEDOM   It is possible to directly confront asymptotic freedom with experiment only in those few situations where the measured quantity is sensitive to the short-distance behavior of QCD alone. The most important example is deep inelastic electroproduction, where approximate Bjorken scaling (40a; in the context of QCD, see 40b) at momentum transfers larger than 1–2 GeV indicates that the theory is nearly free. In order to make this precise, that is, to truly isolate short-distance behavior from the data, one must make use of the Wilson operator product expansion (41). An explanation of this formalism falls beyond the purview of this paper and we simply summarize the situation: Short-distance physics can indeed be isolated and the experimental data suggests that $\alpha_s(-q^2) < 1$ for $(-q^2)^{\frac{1}{2}} > 1$–2 GeV. Analyses of electroproduction typically suggest a value of Λ around 500 MeV. This gives $\alpha_s(-q^2) \approx 0.5$ at $(-q^2)^{\frac{1}{2}} \approx 2$ GeV.

Another important application of asymptotic freedom is the total cross section for $e^+e^- \to$ hadrons (42). Moments of the cross section are related by dispersion relations to hadronic vacuum polarization at space-like q^2. This is sensitive only to short-distance behavior and is therefore computable in perturbation theory for $-q^2$ large enough. It has been conjectured that asymptotic freedom can be applied directly to $\sigma_{\text{tot}}(e^+e^- \to \text{hadrons})$ for $q^2 = W^2 > 0$ as though the final particles are quarks and gluons (43). This is true, at best, away from important thresholds such as charm-anticharm, and there the prediction for $R(W)$, Equation 2.1, is

$$R(W) = \sum_i Q_i^2 \left[1 + \frac{\alpha_s(W)}{\pi} \right]. \qquad \qquad 2.17$$

This result can be stated in a somewhat more solid form by averaging R over an interval in W (44). But where R shows no rapid variation with W, this should not be necessary and Equation 2.17 can be used reliably. The total-cross-section data is still rather poor for these purposes (31–34). Above charm threshold, the errors are large and the presence of the heavy τ lepton (24) complicates matters. Below charm threshold ($1 < W < 3$ GeV), the experimental value of R is somewhat above the parton model value of $\Sigma_i Q_i^2 = 2$. About the best that can be said now is that the data are consistent with $\Lambda \sim 500$ MeV.

There are many other situations where asymptotic freedom plays at least some role. Some examples are high momentum transfer hadron-hadron scattering (45) and the $\Delta I = \frac{1}{2}$ in nonleptonic weak decays (46). However, the short- and long-distance behavior of QCD cannot be clearly separated in these problems and they are perhaps less important for testing asymptotic freedom. There is one other possible application of asymptotic freedom that is very important for heavy quark physics. The ψ particle is very long lived and it has been suggested that this can be understood as a consequence of asymptotic freedom (47). This possibility will be discussed in detail in Section 3, but we emphasize here that if the narrow width of the ψ can be explained in this way, a value of $\alpha_s(M_\psi^2) \approx 0.2$ is required. Whether the analysis of electroproduction allows such a small effective coupling is not yet clear.

2.4.3 BEYOND PERTURBATION THEORY A widespread hope is that the spectrum of QCD consists only of the color singlet hadrons observed in the laboratory. The underlying quarks and gluons would then not be among the asymptotic states, and would exist only as the constituents of hadrons. To demonstrate this "confinement" and to compute the masses and other properties of hadrons starting from the QCD Lagrangian is an extraordinarily difficult and completely unsolved problem. A discussion of the many approaches to this problem is beyond the scope of this review and we only make a few comments to provide some perspective for the next sections.

None of these conjectured features of QCD can be seen in perturbation theory. As the distance scale r increases, the effective coupling (Equation 2.14 or Equation 2.16) increases, and perturbation theory is no longer directly useful. One might hope that general properties could still be extracted from the perturbation expansions or that the series could be summed to deal with large-distance effects, but even this seems unlikely. On the one hand, it can be shown that to any finite order of perturbation theory, there is no indication of confinement (48). The self-coupling of the gauge field suggests an infrared divergence structure much worse than QED, and one speculation was that this structure might have something to do with confinement. However, it has been shown (48) that properly defined transition probabilities are infrared finite order by order in a renormalized coupling constant α_μ. The situation is not unlike QED.

As far as summing the expansion is concerned, there is every indication that the series is not convergent. In fact, it would appear that the series is not even Borel summable (49) so that the perturbation expansion cannot be used to define the theory except for weak coupling. QCD is known to contain essential singularities at $\alpha_\mu = 0$ even at the classical level. The most important examples of this at the present are the instantons and other Euclidean field configurations with nontrivial topological structure (50). The ultimate role of these features of QCD in confinement is not yet known but they surely point up the inadequacy of perturbation theory. Perhaps the most ambitious attack on confinement and hadron structure has been the strong coupling expansion for QCD pioneered by Wilson (51). A short-distance cut-off in the form of a spatial or space-time lattice eliminates the weakly coupled sector of the theory. A linear confining potential $V(r) \sim \alpha r$ between quarks appears naturally as a consequence of the fact that the color electric flux is quantized on the lattice (52). The problem of taking the lattice spacing to zero is, however, unsolved and this is an important ingredient in computing the properties of hadrons on the lattice. The linear potential suggested by the lattice strong coupling expansion has been applied rather successfully to charmonium spectrum computations; an example of the interplay between charm and (in this case) suggested properties of QCD.

3   CHARMONIUM AND BEYOND

The existence of the narrow resonances discovered in November 1974 had been anticipated theoretically (47) prior to the experiments. If charm was real, then narrow $c\bar{c}$ bound states should exist below the threshold for charm production and the name charmonium was suggested, in analogy to positronium.

At the time it was, in fact, thought that the resemblance to positronium might be more than just an analogy (47, 53). Asymptotic freedom says that at short distances (less than about $\frac{1}{5}$ fermi) the strong interactions should behave nearly like α_s/r with $\alpha_s = g^2/4\pi \ll 1$ (but of course $\gg 1/137$). If the c and $\bar{c}$ were to spend most of their time within $\frac{1}{5}$ fermi of each other, then the similarity with positronium would become almost complete. It became clear, very quickly after the initial discoveries, that things were not going to be so simple. The radius of charmonium was closer to 1 fermi than $\frac{1}{5}$ fermi and the binding potential had a structure completely unlike the Coulomb potential of electrodynamics. Nevertheless, the qualitative resemblance to positronium is undeniable and the name charmonium has caught on.

This section reviews in some detail the major areas of theoretical research into this charmonium system. After a brief survey of the experimental situation in Section 3.1, we lay the groundwork for what we shall call the charmonium model (Section 3.2). This is the atomic model of heavy (c) quarks bound in a static, confining potential that, together with asymptotic freedom applied to short-distance processes, describes the spectrum and decay widths of states in the $c\bar{c}$ system.

The simplest version of the model is discussed in Section 3.3, with special emphasis on the choice of a phenomenological potential and the spectrum of states and transition probabilities resulting therefrom. At this naive level, the model already provides a fairly good description of the charmonium system with only two glaring exceptions—the even-charge-conjugation (even-C) states at 2.83 and 3.45 GeV. The main attempts to go beyond the basic model, by incorporating effects of quark spin dependence and of coupling $c\bar{c}$ states to charmed hadron decay channels, are reviewed critically in Sections 3.4 and 3.5. In both cases, we have tried to motivate the directions research has taken, evaluate the outcomes, and, by stressing the shortcomings of existing work, hopefully point the way for future improvement. Finally, Section 3.6 summarizes the straightforward applications of the potential model to bound systems of still heavier quarks. Here, the data is still too limited for critical evaluation of the theory, but we can look forward to rigorous experimental tests in just a year or two.

Lack of space prevents our discussing other charmonium research topics. Most notable are the attempts to compute charmonium properties without recourse to a potential model. These include Regge-trajectory analysis of the spectrum (e.g. 54); a topological S-matrix approach to understanding the small hadronic widths of charmonium (55); and the use of dispersion relations to (a) derive sum rules between leptonic widths of charmonium levels and integrals over the charm contribution to R, and (b) estimate two-photon widths of appropriate states by using sum rules derived from $\gamma\gamma$ scattering (56).

Table 1 The properties of charmonium. All data is taken from Reference 59, which cites original work; question marks indicate unknown values

Particle	$I^G(J^{PC})$	Mass (MeV)	Full width (MeV)	Decay mode	Fraction (%)
X(2830)	$?^?(?^{?+})$	2830 ± 30	?	$\gamma\gamma$	> 0.8
ψ/J(3095)	$0^-(1^{--})$	3098 ± 3	0.069 ± 0.015	e^+e^-	7.3 ± 0.5
				$\mu^+\mu^-$	7.3 ± 0.5
				Direct hadrons	86 ± 2
				γX(2830)	< 1.7
				γX(2830) $\to 3\gamma$	0.013 ± 0.004
				γX(2830) $\to \gamma p\bar{p}$	< 0.004
χ(3415)	$0^+(0^{++})$	3415 ± 5	?	$\gamma\psi$(3095)	3 ± 3
				Hadrons	
P_c/χ(3510)	$0^+(1^{++})$	3508 ± 4	?	$\gamma\psi$(3095)	35 ± 16
				Hadrons	
χ(3455)	$?^?(?^{?+})$	3454 ± 7	?	$\gamma\psi$(3095)	$> 24 \pm 16$
χ(3550)	$0^+(2^{++})$	3552 ± 6	?	$\gamma\psi$(3095)	14 ± 9
				Hadrons	
ψ'(3684)	$0^-(1^{--})$	3684 ± 4	0.228 ± 0.056	e^+e^-	0.88 ± 0.13
				$\mu^+\mu^-$	0.88 ± 0.13
				$\psi\pi^+\pi^-$	33.1 ± 2.6
				$\psi\pi^0\pi^0$	15.9 ± 2.8
				$\psi\eta$	4.1 ± 0.7
				$\gamma\chi$(3415)	7.3 ± 1.7
				$\gamma\chi$(3510)	7.1 ± 1.9
				$\gamma\chi$(3550)	7.0 ± 2.0
				$\gamma\chi$(3455)	< 2.5
				$\gamma\chi$(3455) $\to \gamma\gamma\psi$	0.6 ± 0.4
				γX(2830)	< 1.0
				Direct hadrons	~ 9
ψ''(3772)	$0^-(1^{--})$	3772 ± 6	26 ± 5	e^+e^-	0.0010 ± 0.0005
				$D^0\bar{D}^0$	56 ± 3
				D^+D^-	44 ± 3
ψ(4028)	$?(1^{--})$	4028 ± 20	~ 50	e^+e^-	~ 0.002
				Charmed mesons	~ 100
ψ(4414)	$?(1^{--})$	4414 ± 7	33 ± 10	e^+e^-	0.0013 ± 0.0003
				Charmed mesons	~ 100

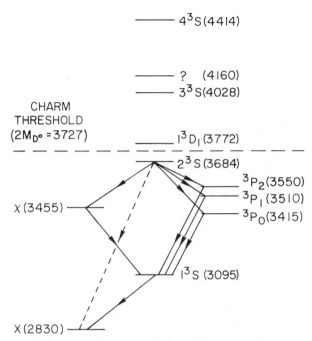

Figure 2 The observed charmonium levels with the notation $n^{2S+1}L_J$ and masses in MeV. The correct identification for X(2830) and χ(3455) is not known. Observed radiative transitions are shown as solid directed lines. The dotted line represents an M1 transition not yet observed. Splittings are approximately to scale.

3.1 *Experimental Overview*

There are several excellent, up-to-date reviews (57–59) on the production and decay characteristics of the charmonium states (58 and 59 reference experimental work not cited explicitly here). We content ourselves here with a brief survey of what has been seen, with special emphasis on production via e^+e^- annihilation; for other production mechanisms of ψ/J, see Section 6. Table 1 summarizes the known properties of charmonium. A level diagram[8] is shown in Figure 2.

Except for the simultaneous discovery of ψ/J in e^+e^- annihilation at SLAC (18) and in proton-beryllium collisions at Brookhaven (19), all the charmonium levels—including narrow states below charm threshold ($W_c = 2M_{D^0} = 3.727$ GeV) and broad ones above—have been discovered at the SLAC and DESY e^+e^- storage rings (SPEAR and DORIS). The

---

[8] We use the spectroscopic notation $n^{2S+1}L_J$, where n is the number of radial nodes plus one, S is the total q + q̄ spin (0 or 1), L is the orbital angular momentum of the qq̄, and J is the total angular momentum of the state.

experimental situation is incredibly beautiful and simple: States with $J^{PC} = 1^{--}$, the quantum numbers of the photon, are produced directly and copiously as resonances in e^+e^- collisions; narrow states lying below $\psi'(3684)$ and having even charge conjugation are observed in radiative transitions, $\psi' \to \gamma\chi$ and $\psi \to \gamma X$. It should be added that there remain additional charmonium levels to be discovered. An important one is the 1^1P_1 state with $J^{PC} = 1^{+-}$, expected near 3.45 GeV. In e^+e^- annihilation, this could be observed (with difficulty) only in the reaction chain

$$
\begin{aligned}
e^+e^- \to \psi'&(3684; 2^3S_1) \\
&\hookrightarrow \eta'_c(?; 2^1S_0) + \gamma \\
&\qquad \hookrightarrow \chi(?; 1^1P_1) + \gamma \\
&\qquad\quad \hookrightarrow \text{hadrons.}
\end{aligned}
$$

(3.1)

The most striking feature of states lying below W_c is their very small total widths. In particular, first and second order electromagnetic decays such as $\psi' \to \gamma\chi$ and $\psi \to e^+e^-$ are competitive with strong interaction decays such as $\psi' \to \psi\pi\pi$ and $\psi \to$ hadrons. If we assume that these high-mass mesons are $c\bar{c}$ bound states, it follows that decay to ordinary hadrons, not containing charmed quarks, must proceed by mutual annihilation of the $c\bar{c}$ pair. The reluctance of any $q\bar{q}$ pair within a single hadron to annihilate, known as the Okubo-Zweig-Iizuka (OZI) rule (60), has been observed in the light hadrons for a long time. The most notable example is the suppressed decay ϕ ($s\bar{s}$) $\to \pi^+\pi^-\pi^0$ (all containing u, d quarks). This empirical rule could be a simple consequence of asymptotic freedom in the case of heavy quarks.

Two features of e^+e^- annihilation make it ideal for discovery and study of the myriad of charmonium levels: First, the energy in a single beam, $E_b = \frac{1}{2}W$ (W = total center-of-mass energy), is known very precisely, to within 1–2 MeV. Second, the $J^{PC} = 1^{--}$ states are produced at rest in the center of mass. Together, these allow very precise measurement of the mass of the directly produced states. The even-C states below ψ' are produced by "sitting on" ψ' and ψ, i.e. setting $E_b = \frac{1}{2}M_{\psi'}$ or $\frac{1}{2}M_\psi$, and observing their radiative decays. For this, three methods have been used:

1. Measurement of the invariant mass of the charged hadrons in decays such as

$$\psi' \to \pi^+\pi^-\pi^+\pi^- + \text{missing neutrals,} \tag{3.2}$$

corresponding, perhaps, to $\psi' \to \chi + \gamma$, $\chi \to 4\pi$. If a peak is found in the charged hadron mass spectrum and if the mass of the missing neutral is consistent with its being a photon (rather than a π^0), a fairly precise determination of the χ mass results from constraining the mass of the parent to $M_{\psi'}$ and that of the neutral to zero. The X(2830) was found by a similar method (61), but with all neutral particles; in particular, this state has been seen only in the decay chain

$$\psi \to \gamma + X(2830); \qquad X(2830) \to \gamma\gamma. \qquad\qquad 3.3$$

Therefore, all we know about this meson is that it has even C and cannot have $J^P = 1^{\pm}$.

2. Measurement of the inclusive photon energy distribution in $\psi' \to \gamma +$ anything and $\psi \to \gamma +$ anything. Here, the photon is definitely identified and its energy measured, usually with an energy resolution of $(5\text{–}10\%)/E_\gamma^{\frac{1}{2}}$ (in GeV). Peaks in this distribution correspond to ψ or $\psi' \to \gamma + a$ narrow $C = +1$ state. To date, only the states $\chi(3415)$, $\chi(3510)$, and $\chi(3550)$ have been detected by this method (as well as by the other two).

3. Observation of the double-cascade process,

$$\psi' \to \gamma\chi; \qquad \chi \to \gamma\psi. \qquad\qquad 3.4$$

Here, the ψ is identified by its decay to $\mu^+\mu^-$, and one photon is detected by its conversion to e^+e^-. The missing neutral energy is determined to be consistent with zero mass. There is a potential ambiguity in determining the χ mass because one does not know which photon came first in a given event of this type. The ambiguity is resolved neatly by plotting the two possible values of the $\psi\gamma$ invariant mass. The wrong solution shows a characteristic Doppler broadening induced by the motion of the χ. In addition to the well-established states at 3414, 3508, and 3552 MeV, this method has revealed the existence of a fourth intermediate state, $\chi(3455)$ (21). Seen in no other way, the only known decay mode of this even-C state is $\chi(3455) \to \gamma\psi$.

The comparative ease of detecting and identifying ψ and χ decay products in e^+e^- annihilation makes it also the best method for determining their spin-parities and branching fractions to individual final states. An outstanding example of this is the determination $J^{PC} = 1^{--}$ for ψ and ψ', assignments that are not obvious a priori. This was done by observing the characteristic destructive interference, just below $W = M_{\psi,\psi'}$, between

$$e^+e^- \to \gamma_v \to \mu^+\mu^- \qquad (\gamma_v = \text{virtual photon}) \qquad 3.5a$$

and

$$e^+e^- \to \gamma_v \to \psi,\psi' \to \gamma_v \to \mu^+\mu^-. \qquad\qquad 3.5b$$

The assignments of $J^P = 0^+$ to $\chi_0 = \chi(3415)$, $J^P = 1^+$ to $\chi_1 = \chi(3510)$, and $J^P = 2^+$ to $\chi_2 = \chi(3550)$ are based on the following considerations

(62): (a) χ_0 and χ_2 decay to $\pi^+\pi^-$ and K^+K^- and, therefore, both states have natural spin-parity; these modes have not been observed for χ_1, consistent with the 1^+ assignment; and (b) the angular distribution of the photon in $\psi' \to \gamma\chi_0$ is well fitted by $1 + \cos^2\theta$, which is expected for $J = 0$. The angular distributions $1 - \frac{1}{3}\cos^2\theta$ for $J = 1$ and $1 + \frac{1}{13}\cos^2\theta$ for $J = 2$ are consistent with the rather meager measurements for χ_1 and χ_2, respectively.

Finally, the normal hadronic widths, on the order of 10–100 MeV, of the directly produced resonances above $\psi'(3684)$ are further dramatic confirmation of the OZI rule. Here, the charmed quarks need no longer annihilate since it is energetically possible for them to emerge (together with light quarks) as the charmed mesons D, D*, F, F*, and so on. All this is discussed in Section 4. Suffice it to say that these broad charmonium resonances were solely responsible for the unambiguous isolation of charmed mesons.

3.2   Foundations of the Charmonium Model

Perhaps the most important feature of the charmonium spectrum in Figure 2 is the fact that the level spacings are very small compared to the overall mass scale of the system. This suggests, at least for the states below charm threshold, that the system is nonrelativistic, with excitation energies small compared to the masses of the constituents. This is something completely new in strong interaction physics, and a great deal of theoretical work has gone into analyzing this system using a nonrelativistic Schrödinger equation formalism (63, 64).

In retrospect, this approach is somewhat too naive, especially with regard to the assumption of spin independence of the dominant $c\bar{c}$ interaction. The hyperfine splitting is not much smaller than the radial and orbital excitation energies. Nevertheless, the model has, at the very least, been a powerful predictive guide to the qualitative features of charmed quark physics.

The other aspect of the charmonium model is the attempt to understand the narrowness of the states below charm threshold as a consequence of asymptotic freedom (see Section 2.4). We now discuss the theoretical basis for this possibility in some detail, but it is important to keep in mind that it is largely a separate issue from the energy level structure. There, it is clear from Figure 2 that without a solution to QCD in the strong coupling regime, some phenomenological input is necessary. In the case of the decay widths, there is the possibility that perturbation theory may be directly relevant.

The idea is basically that with $c\bar{c}$ annihilation into light hadrons proceeding through gluons, the decay will be inhibited, since the effective

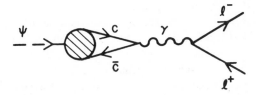

Figure 3   The electromagnetic decay of the $\psi(1\,^3S_1)$.

gluon coupling constant should be small at high energies (47, 65). The dominant contribution will come from the minimum number of intermediate gluons, which depends upon the quantum numbers of the charmonium state. Some rather striking experimental predictions can be made on the basis of this "gluon counting"; these are discussed in Section 3.3.

Consider the decay of the $\psi(^3S_1)$ state. Its dominant electromagnetic decay is shown in Figure 3. The $c\bar{c}$ pair must come together to annihilate into the virtual photon, and if the bound state is nonrelativistic, then, to first approximation, the decay width will be given by

$$\Gamma(\psi \to l^+ l^-) = \frac{4\alpha^2 (2/3)^2}{M^2} |\Psi(0)|^2. \qquad 3.6$$

The charge of the charmed quark is $\frac{2}{3}|e|$, and M is the charmonium mass. The nonrelativistic radial wave function at the origin is $\Psi(0)$, and one cannot expect to be able to compute it in perturbation theory. The reason for this is that the mean radius $\langle r \rangle$ of charmonium is on the order of one fermi, a distance scale at which the effective coupling strength for the binding has become large. Thus $\Psi(0)$ will be determined in part by the nonperturbative, long-range part of the potential. The hadronic decay of the ψ must proceed through a minimum of three gluons. If this is indeed the dominant contribution, that is to say, if perturbation theory is truly relevant to this problem, the decay will proceed as shown in Figure 4. The $c\bar{c}$ annihilation will be essentially local—on the order of $1/m_c \, (\ll \langle r \rangle)$. The computation of the decay matrix element is then done

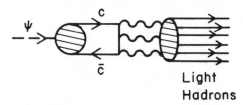

Figure 4   The hadronic decay of the $\psi(1\,^3S_1)$, assuming the minimal gluon mechanism.

in analogy to the parton model computation of $\sigma_{tot}(e^+e^- \to$ hadrons) as if the final state consisted of three on-mass-shell gluons. This amounts to the statement that the transition from the three-gluon state into physical hadrons takes place with unit probability. A more satisfying theoretical justification of the three-gluon mechanism can be given and we turn to it shortly. If the mechanism just described is correct, then the total hadronic width of the ψ is given by (65)

$$\Gamma(\psi \to \text{hadrons}) = \frac{40}{81\pi}(\pi^2 - 9)\frac{\alpha_s^3(M^2)}{M^2}|\Psi(0)|^2. \qquad 3.7$$

The strong coupling constant is defined at the ψ mass and, as before, $\Psi(0)$ is the nonrelativistic wave function at the origin.

Before proceeding to the comparison of these expressions with experiment, we sketch the analysis that underlies Equation 3.7. A necessary condition for the applicability of lowest order perturbation theory is that no large dynamical factors enter in higher orders to make the expansion break down. One must analyze the quantity $\Sigma_n |M(c\bar{c} \to n)|^2$ where n is some quark-gluon final state; M is the decay matrix element and is defined to be two-particle ($c\bar{c}$) irreducible in the decay channel. Two-particle reducible contributions are absorbed into the definition of the wave function. If it can be shown that $\Sigma_n |M(c\bar{c} \to n)|^2$ is free of infrared singularities for the $c\bar{c}$ pair at rest, then there can be no large dynamical factors. This is because the result will involve no small energy or momentum factors, only the (large) charmed quark mass and the (large) renormalization scale. The infrared finiteness through order α_s^4 (the lowest order is α_s^3) for ψ decay has been checked and it is conjectured to be true to all orders. It is technically simpler to use the Coulomb gauge rather than covariant gauges in this analysis.

We make one last point before proceeding. The infrared analysis is necessary but not sufficient. It could well be that threshold singularities in high orders, or even completely nonperturbative effects prevent the use of perturbation theory in this simple way. The use of asymptotic freedom to explain the OZI rule is speculative. It is on much less solid footing than the conventional deep Euclidean application or even the direct computation of $\sigma_{tot}(e^+e^- \to$ hadrons), since there the production is truly local, coming from the off-shell photon.

3.3   The Basic Charmonium Model

To go beyond qualitative predictions, a model of the $c\bar{c}$ interaction is necessary, and it is natural to realize this in terms of an instantaneous, central potential, $V_0(r)$. Such a model presupposes that, to a first approximation, one may neglect the influences of spin-dependent forces and of nearby open decay channels on spectroscopy and decay rates. The attempts to incorporate these effects in the basic model are described in Sections 3.4 and 3.5.

A simple possibility for V_0, motivated by asymptotic freedom at short $c\bar{c}$ separations and quark confinement at large ones, is (63, 64, 66)

$$V_0(r) = -\frac{\kappa}{r} + \frac{r}{a^2}. \qquad 3.8$$

Asymptotic freedom tells us to expect a rather small short-distance coupling, say $\kappa \sim 0.2$. The choice of a linear confining term is suggested by the lattice gauge theory (51, 52) and the dual string model (for a review, see 67). Then $1/a^2$ is related to the Regge slope and is $1/a^2 \approx 1$ GeV/fm ≈ 0.2 (GeV)^2. The third parameter of this model is the charmed quark mass, which is expected to be $m_c \simeq \frac{1}{2}M_\psi \simeq 1.5$ GeV.

We emphasize that these parameters are purely phenomenological. For example, m_c^{-1} is really the strength of the kinetic energy term in the $c\bar{c}$ Hamiltonian, and not necessarily equal to half the difference between the mass of a state and its energy eigenvalue.[9] Thus, within the general guidelines set by the above expectations, the three parameters will be determined by fitting to selected pieces of data. When this model is applied to bound systems of heavier quarks, the Υ for example, it is presumed that only the quark mass will change substantially; κ may decrease slightly, while the linear potential strength a^{-2} is thought to be independent of the quark mass (37).

A final word about this choice of V_0: From a purely phenomenological point of view, other forms may be equally reasonable and give as acceptable an account of the data. The advantages of V_0 are that it is well motivated and that it contains the minimum number of parameters needed to reach agreement with the observed spectra and decay rates in charmonium. There is evidence that V_0 does not adequately describe the Υ spectrum (23, 57), and some possible modifications are discussed in Section 3.6.

To obtain a "best" overall description of charmonium data, Eichten et al (63, 68) chose parameters κ, a, and m_c by (a) fitting to the $\psi' - \psi$ mass difference; (b) taking the electronic width of ψ to be 5.3 keV, one standard deviation above the measured value of 4.8 ± 0.5 keV; and (c) constraining $1.5 \lesssim m_c \lesssim 2.0$ GeV and $0.1 \lesssim \kappa \lesssim 0.4$. These constraints are consistent with the notion of heavy constituents moving nonrelativistically and

---

[9] The zero of energy of a system bound in an infinitely rising potential cannot be defined unambiguously.

with weak short-range interactions. Acceptable results are obtained with a range of parameters, and their preferred choice is

$$m_c = 1.65 \text{ GeV}, \qquad a = 2.07 \text{ (GeV)}^{-1}, \qquad \kappa = 0.132. \qquad\qquad 3.9$$

A check on the self-consistency of the nonrelativistic approximation is provided by the mean-squared quark velocity in the 1S and 2S states; these are $\langle v^2 \rangle = 0.17$ and 0.28, respectively. It is worth emphasizing that the Coulomb part of the potential, although the most certain feature according to QCD, may not be very meaningful for charmonium. For this value of κ, it only becomes important below distance scales $\sim m_c^{-1}$, where a nonrelativistic potential picture ceases to be sensible.

3.3.1   CHARMONIUM SPECTRUM   The most important consequence of the existence of a confining potential (of almost any shape) between c and c̄ is that there will be ^{3}P states lying between the ^{3}S states, ψ and ψ' (63, 64, 69, 70). The reason for this is simple. Recalling that, in a purely Coulomb potential, the 2S and 1P states are degenerate, it is clear that the presence of a confining term will impart a greater kinetic energy to the 2S state, with its one radial node, than to the 1P, which has only an orbital node, thus lifting the degeneracy. The same remark applies to the relative ordering of 3S, 2P, and 1D states, and so only the amount of this splitting depends on details of the potential and quark mass. In charmonium, we are in an extremely fortunate position to observe this consequence of quark confinement because, lying below the OZI-allowed decay threshold, $2M_D$, the 1^3P states, like ψ and ψ', will be unusually narrow. Furthermore, they have even-charge conjugation ($C = \pm 1$), and so may be detected by radiative transitions from ψ' and, if narrow enough,

Table 2 Predicted (68) and observed spin-triplet charmonium levels; assignment of $\psi(4028)$ to 3^3S_1, and existence of $\psi(4160)$ with its assignment to 2^3D_1 are open to question

State $n^3L_J(J^{PC})$	Predicted mass (MeV)	Candidate (measured mass)
$1^3S_1(1^{--})$	3095 (Input)	$\psi(3095)$
$1^3P_J(0^{++},1^{++},2^{++})$	3457	$\chi_{0,1,2}(3522 \pm 5)$
$2^3S_1(1^{--})$	3684 (Input)	$\psi'(3684)$
$1^3D_J(1^{--},2^{--},3^{--})$	3755	$\psi''(3772) = 1^3D_1$
$2^3P_J(0^{++},1^{++},2^{++})$	3957	Above charm threshold; difficult to produce
$3^3S_1(1^{--})$	4157	$\psi(4028)$
$2^3D_J(1^{--},2^{--},3^{--})$	4204	$\psi(4160) = 2^3D_1$
$4^3S_1(1^{--})$	4567	$\psi(4414)$

down to ψ. Again, branching ratios will depend on details of the model, but the existence of these states and their mode of observation is inescapable.

Using the potential V_0 with parameters in Equation 3.9, the Schrödinger equation may be numerically integrated to obtain the spectrum of low-lying spin-triplet states shown in Table 2 (68), together with the most likely candidates for these states. A number of explanatory remarks are in order: (a) For brevity only those states even remotely accessible to existing experimental techniques have been included. (b) With the neglect of spin-dependent forces, spin-singlet states such as $\eta_c = 1^1S_0$ and 1^1P_1 would be degenerate with the predicted center of gravity (c.o.g.) of the corresponding triplet states. And, by adjusting the constant zero of energy to give the correct ψ mass, that constant subsumes some of the hyperfine interaction ($\propto \mathbf{S}_1 \cdot \mathbf{S}_2$) for the low-lying ^3S states. (c) As noted previously, there is fairly strong evidence for the assignment of $\chi_0 = \chi(3415)$, $\chi_1 = \chi(3510)$, and $\chi_2 = \chi(3550)$ to $1^3P_{0,1,2}$ respectively; the c.o.g. of these levels is 3522 ± 5 MeV, somewhat higher than predicted. (d) Most model calculations of the splitting among the 1^3D_J levels expect it to be smaller than that observed for the 1^3P_J (see the next section), so that all three of these $L = 2$ states may be fairly close to the mass of $\psi(3772) = 1^3D_1$. (e) The region between c.m. energy $W = 4.0$ and 4.3 GeV in e^+e^- annihilation is quite complicated (see Section 3.5) and difficult to interpret. It is obvious that the peak $\psi(4028)$ is a resonance and, within the charmonium model, is assigned to the 3^3S_1 level. If it should become equally clear that the enhancement at $W \approx 4.16$ GeV is a resonance, the candidate assignment is 2^3D_1. That both these states and the $\psi(4414)$, assigned to 4^3S_1, appear ~ 150 MeV lower than predicted shows that the approximations underlying the model are breaking down. Another sign of this is the sizeable electronic width, ~ 0.2–0.4 keV, of $\psi(3772)$. In the nonrelativistic potential model, $\Gamma_e \propto |\Psi(0)|^2$ vanishes for all but $^3S_1$ states. The observed electronic width is fairly well understood as a coupled-channel effect, and is discussed in Section 3.5.

3.3.2 CHARMONIUM TRANSITION RATES   A great deal can be said about the decay rates and branching ratios of the charmonium levels (below $D\bar{D}$ threshold), most of which requires some knowledge of the bound-state wave functions. The rates that can be computed in the nonrelativistic model fall into two classes: those that depend on the wave function at very short $c\bar{c}$ separations (electronic, two-photon, and the hadronic widths obtained from gluon counting), and those that involve overlaps of radial wave functions (E1 and M1 radiative widths). The reader should be aware of what is being neglected in these calculations. Those involving the wave function $\Psi_{nL}(r)$ near $r = 0$ certainly ignore possibly important relativistic and quantum effects such as spin dependence and pair creation, as well as the mixing among states (e.g. 2^3S_1 and 1^3D_1) induced by coupling to decay channels. The second type of calculation does not include almost certain reductions in the overlap integral due to differing gluon distributions in the initial and final states, nor does it take account of the spin-dependent and mixing effects noted above. Finally, the charmonium model has very little to say on the important question of $\psi' \to \psi\pi\pi$ (and $\Upsilon' \to \Upsilon\pi\pi$, $\Upsilon\eta$) because the rather low momentum imparted to the $\pi\pi$ or η makes gluon counting especially dubious. See, however, Gottfried (71) for scaling rules for these decays.

The leptonic decay width is readily computed from the graph in Figure 3. If Q is the charge of the quark q, M_{n0} the mass of the $J^{PC} = 1^{--}$ states $n\,^3S_1$, $\Psi_{n0}(0)$ its radial wave function at the origin, and m_l the mass of the charged lepton, the result is (72)

$$\Gamma(n^3S_1 \to l^+l^-) = 4Q^2\alpha^2 \frac{|\Psi_{n0}(0)|^2}{M_{n0}^2}\left(1 + \frac{2m_l^2}{M_{n0}^2}\right)\left(1 - \frac{4m_l^2}{M_{n0}^2}\right)^{\frac{1}{2}}. \qquad 3.10$$

The terms involving m_l are relevant for decays such as $\psi' \to \tau^+\tau^-$ and $\Upsilon \to \tau^+\tau^-$.

The rates for the OZI-forbidden direct hadronic decays of heavy quark systems via annihilation to the minimum possible number of gluons can be computed from graphs as in Figure 4. The results are (to lowest order in quark velocity) (65, 73)

$$\Gamma(n^3S_1 \to \text{hadrons}) = \frac{40}{81\pi}(\pi^2 - 9)\alpha_s^3 \frac{|\Psi_{n0}(0)|^2}{M_{n0}^2}, \qquad 3.11$$

$$\Gamma(n^1S_0 \to \text{hadrons}) = \frac{8}{3}\alpha_s^2 \frac{|\Psi_{n0}(0)|^2}{M_{n0}^2}, \qquad 3.12$$

$$\Gamma(n^3P_0 \to \text{hadrons}) = 96\alpha_s^2 \frac{|\Psi_{n1}'(0)|^2}{M_{n1}^4}, \qquad 3.13$$

$$\Gamma(n^3P_1 \to \text{hadrons}) = \frac{128}{3\pi}\alpha_s^3 \frac{|\Psi_{n1}'(0)|^2}{M_{n1}^4} \ln\left(\frac{4m_q^2}{4m_q^2 - M_{n1}^2}\right), \qquad 3.14$$

$$\Gamma(n^3P_2 \to \text{hadrons}) = \frac{128}{5}\alpha_s^2 \frac{|\Psi_{n1}'(0)|^2}{M_{n1}^4}, \qquad 3.15$$

$$\Gamma(n^1P_1 \to \text{hadrons}) = \frac{320}{9\pi}\alpha_s^3 \frac{|\Psi_{n1}'(0)|^2}{M_{n1}^4} \ln\left(\frac{4m_q^2}{4m_q^2 - M_{n1}^2}\right). \qquad 3.16$$

Here α_s is evaluated at the bound-state mass M_{nL}, and $\Psi'_{nL}(0)$ is the slope of the radial wave function at $r = 0$. The rate for ψ_{nL} to decay directly to a photon plus hadrons may also be obtained from gluon counting; the formula is (69, 74)

$$\Gamma(n^3S_1 \to \gamma + \text{hadrons}) = \frac{32}{9\pi} (\pi^2 - 9)\, \alpha_s^2\, \alpha Q^2 \frac{|\Psi_{n0}(0)|^2}{M_{n0}^2}. \qquad 3.17$$

Two more direct decay widths, not involving gluon counting but of great importance nonetheless, are

$$\Gamma(n^3S_1 \to \gamma \to \text{hadrons}) = R_{\text{bkgd}}\, \Gamma(n^3S_1 \to e^+e^-) \quad \text{and} \qquad 3.18$$

$$\Gamma(n^1S_0 \to \gamma\gamma) = 12Q^4\alpha^2 \frac{|\Psi_{n0}(0)|^2}{M_{n0}^2} \qquad 3.19a$$

$$= \frac{9}{2}\, Q^4 \left(\frac{\alpha}{\alpha_s}\right)^2 \Gamma(n^1S_0 \to \text{hadrons}), \qquad 3.19b$$

where R_{bkgd} is the value of R just off the resonance peak.

Note that Equation 3.19a involves only the short-distance (positronium-like) assumption, while Equation 3.19b involves the much stronger gluon-counting assumption. The two-photon width of $^3P_0$ and $^3P_2$ states is related to their hadronic (two-gluon) widths by the same factor, $9Q^4\alpha^2/2\alpha_s^2$, as for $^1S_0$ states. To the extent that wave functions are independent of total quark spin and angular momentum $\mathbf{J} = \mathbf{L} + \mathbf{S}$ (for fixed n, L), we have

$$\Gamma(n^3S_1 \to \text{hadrons})/\Gamma(n^1S_0 \to \text{hadrons}) = \frac{5(\pi^2 - 9)\alpha_s}{27\pi} \qquad 3.20$$

and

$$\Gamma(n^3P_0 \to \text{hadrons}) : \Gamma(n^3P_1 \to \text{hadrons}) : \Gamma(n^3P_2 \to \text{hadrons})$$

$$= 1 : \frac{4\alpha_s}{9\pi} \ln\left(\frac{4m_q^2}{4m_q^2 - M_{n1}^2}\right) : \frac{4}{15}. \qquad 3.21$$

Equations 3.14 and 3.16 for $^3P_1$, $^1P_1 \to$ hadrons deserve some comment, In the spirit of gluon counting, a spin one state cannot decay to two massless, on-shell gluons, and so we expect these rates to be $O(\alpha_s^3)$, not $O(\alpha_s^2)$ (73, 75). In particular, χ_1 should be the narrowest of the ^3P_J levels. In an actual calculation of these rates (73), the dominant contribution involves the exhibited logarithm, due to a threshold singularity at $2m_c = M_{\chi_1}$. Such a large logarithm is always worrisome in QCD calculations, since it may signal the breakdown of perturbation theory. So it is

Table 3 Direct decays in charmonium, calculated from Equations 3.10 to 3.21: $\Gamma(\chi_J \to \text{all}) = \Gamma(\chi_J \to \text{hadrons}) + \Gamma_\gamma(\chi_J \to \psi), \Gamma(\eta'_c \to \text{all}) = \Gamma(\eta'_c \to \text{hadrons}) + \Gamma(\eta'_c \to \eta_c \pi\pi)$, and $\Gamma(\eta'_c \to \eta_c \pi\pi) = \Gamma(\psi' \to \psi \pi\pi)$ are assumed

Mode	Rate (keV)	Branching ratio (%)
$\eta_c \to \text{hadrons}$	5.1×10^3	100
$\eta_c \to \gamma\gamma$	7.1	0.14
$\psi \to \gamma + \text{hadrons}$	6.1	8.0
$\chi_0 \to \text{hadrons}$	1.8×10^3	90
$\chi_0 \to \gamma\gamma$	2.5	0.13
$\chi_1 \to \text{hadrons}$	105	21
$\chi_2 \to \text{hadrons}$	480	48
$\chi_2 \to \gamma\gamma$	0.66	6.6×10^{-2}
$\eta'_c \to \text{hadrons}$	3.3×10^3	97
$\eta'_c \to \gamma\gamma$	4.5	0.13
$\psi' \to \text{hadrons}$	31	14
$\psi' \to e^+ e^-$	3.4	1.5
$\psi' \to \gamma + \text{hadrons}$	3.9	1.7

difficult to take these results too seriously beyond the reasonable (and conservative) guess that $\Gamma(^3P_1 \to \text{hadrons})$ is $\sim \alpha_s \Gamma(^3P_{0,2} \to \text{hadrons})$.

The predictions of these formulae for the charmonium system are listed in Table 3. Experimental comparisons are best delayed until the discussions of radiative transition rates. The value of α_s used here is determined by fitting to the total width of ψ,

$$\Gamma(\psi \to \text{all}) \cong \Gamma(\psi \to e^+ e^-)[2 + R_{\text{bkgd}}] + \Gamma(\psi \to \gamma + \text{hadrons})$$
$$+ \Gamma(\psi \to \text{hadrons})$$
$$= 69 \pm 15 \text{ keV}, \qquad\qquad 3.22$$

where the small width for $\psi \to \gamma X(2830)$ has been ignored. Using $R_{\text{bkgd}} \approx 2.2$ (58, 59) and $\Gamma(\psi \to e^+ e^-) = 5.3$ keV gives[10]

$$\alpha_s(M_\psi) \approx 0.19. \qquad\qquad 3.23$$

The wave functions for states other than ψ are determined using the parameters in Equation 3.9.

The important qualitative features of these calculations are: (a) The ground-state pseudoscalar η_c is expected to have a total width in the MeV range, about 100 times greater than the width of ψ, and its branching ratio to two photons should be $\sim 10^{-3}$. Similar remarks apply to its first radial excitation η'_c, which may also decay to $\eta_c + \pi\pi$ [presumably, $\Gamma(\eta'_c \to \eta_c \pi\pi)$

---

[10] This value of $\alpha_s(M_\psi)$ is about a factor of two smaller than that deduced from analyses of electroproduction data.

$\approx \Gamma(\psi' \to \psi\pi\pi)$]. (b) As discussed above, gluon counting implies that $^3P_1$ and $^1P_1$ have considerably less hadronic width than $^3P_0$ and $^3P_2$. An immediate consequence is that the branching ratio $B(\psi' \to \chi_J\gamma) B(\chi_J \to \psi\gamma)$ should be largest for the 1^{++} state χ_1. (c) The potential model predicts

$$\Gamma(\psi' \to e^+e^-) = 3.4 \text{ keV} \quad \text{and} \quad \Gamma(\psi' \to 3 \text{ gluons} \to \text{hadrons}) = 31 \text{ keV},$$

both in fair agreement with the measured values (59), 2.0 keV and ~ 20 keV respectively. These results lend some support to both the presence of a linear confining term in V_0 (since for $\kappa = 0$, $|\Psi_{10}(0)| = |\Psi_{20}(0)|$) and to the gluon-counting calculation of the direct hadronic width.

We turn now to the radiative decays. The E1 transition rate between S- and P-wave states having the same total quark spin is

$$\Gamma_\gamma(\text{S} \leftrightarrow \text{P}) = \frac{4}{9}\left(\frac{2J_f+1}{2J_i+1}\right) Q^2\alpha |E_{if}|^2 \omega^3, \qquad 3.24$$

where ω is the photon energy, J_i (J_f) the total angular momentum of the initial (final) state, $Q = \frac{2}{3}$ for charmed quarks, and E_{if} is the transition dipole matrix element,

$$E_{if} = \int_0^\infty r^2 \, dr \Psi_i(r) \, \Psi_f(r) \, r. \qquad 3.25$$

Here, $\Psi_{i,f}$ are initial (final) state radial wave functions. The E1 rates between $^3D_1$ and ^3P_J states are given by

$$\Gamma_\gamma(^3D_1 \leftrightarrow {}^3P_J) = \frac{4}{9}\left(\frac{2J_f+1}{2J_i+1}\right) D_{if} Q^2\alpha |E_{if}|^2 \omega^3, \qquad 3.26$$

where $D_{if} = 1, \frac{1}{4}, \frac{1}{100}$ for $J = 0, 1, 2$ respectively. For the charmonium system, the 1^3D_1 state lies above charm threshold and so it has a very small branching ratio to $\chi_J + \gamma$. Therefore, Equation 3.26 is useful only indirectly, through the mixing between 2^3S and 1^3D_1 (see Section 3.5). In the Υ and heavier quark bound systems, however, there is an excellent chance for direct observation of the $^3D_1 \leftrightarrow {}^3P_J$ E1 transitions.

The M1 transition rate between $^3S_1$ and $^1S_0$ states is taken to be

$$\Gamma_\gamma(^3S_1 \leftrightarrow {}^1S_0) = \frac{16}{3}\left(2J_f+1\right)\left(\frac{Q}{2m_q}\right)^2 \alpha |M_{if}|^2 \omega^3, \qquad 3.27$$

where a Dirac moment is assumed for the quark, and

$$M_{if} = \int r^2 \, dr \, j_0(\tfrac{1}{2}\omega r) \, \Psi_i(r) \, \Psi_f(r); \qquad 3.28$$

where j_0 is the spherical Bessel function of order zero.

For the "allowed" M1 transitions between hyperfine partners, M_{if} is very nearly unity because $\Psi_i = \Psi_f$ and $\frac{1}{2}\omega\langle r\rangle_{if} \ll 1$, so that $j_0 \approx 1$. For the same reason, M1 transitions between S states corresponding to different radial quantum numbers $n_i \neq n_f$ are strongly suppressed. It is still true that $\frac{1}{2}\omega\langle r\rangle_{if} \ll 1$, and

$$M_{if} \approx -\frac{\omega^2}{24}\int r^2\, dr\, \Psi_i(r)\, \Psi_f(r)\, r^2 \ll 1 \qquad (i \neq f). \qquad 3.29$$

The immediate consequence is that the "hindered" M1 transitions $\psi'(2^3S_1) \to \eta_c(1^1S_0)+\gamma$ and $\eta'_c(2^1S_0) \to \psi(1^3S_1)+\gamma$ are expected to be very rare compared to allowed M1 and E1 transitions.

The predictions of the potential model for E1 rates and branching fractions are compared with experimental observations in Table 4. Following the custom of traditional spectroscopy, experimental values of the χ masses are used, so what is being tested here is the theoretical strength Γ_γ/ω^3.

Given the naiveté of the simple potential model, the agreement is rather good, with theory lying within 1–2 standard deviations of experiment. Especially noteworthy are: (a) The normalized experimental rates are (with large errors) (59)

$$\frac{\Gamma(\psi' \to \gamma\chi_2)}{5\omega_2^3} : \frac{\Gamma(\psi' \to \gamma\chi_1)}{3\omega_1^3} : \frac{\Gamma(\psi' \to \gamma\chi_0)}{\omega_0^3} = 1:0.7:0.6, \qquad 3.30$$

with unity expected for pure E1 transitions from $^3S_1$ to ^3P_J states. (b) The measured branching ratios for $\chi_J \to \gamma\psi$ are quite consistent with predictions based on both the potential model and gluon counting for

Table 4 E1 decays in charmonium for theory (68) and experiment (59); predicted total widths for $\chi_{0,1,2}$ of 2.0, 0.5, and 1.0 MeV, respectively have been assumed

Datum	Theory (keV for Γ_γ, % for B_γ)	Experiment (keV for Γ_γ, % for B_γ)
$\Gamma_\gamma(\psi' \to \chi_2)$	27	16 ± 9
$\Gamma_\gamma(\psi' \to \chi_1)$	38	16 ± 8
$\Gamma_\gamma(\psi' \to \chi_0)$	44	16 ± 9
$\Gamma_\gamma(\chi_2 \to \psi)$	525	—
$B_\gamma(\chi_2 \to \psi)$	52	14 ± 9
$\Gamma_\gamma(\chi_1 \to \psi)$	395	—
$B_\gamma(\chi_1 \to \psi)$	79	35 ± 16
$\Gamma_\gamma(\chi_0 \to \psi)$	190	—
$B_\gamma(\chi_0 \to \psi)$	9.5	3 ± 3

the χ hadronic widths, with χ_1 considerably more narrow than χ_0 and χ_2. These facts strengthen the J^P assignments discussed in Section 3.1.

The M1 transition rates are compared with experiment in Table 5, where we have made the tentative assignments of $X(2830) = \eta_c(1^1S_0)$ and $\chi(3455) = \eta_c'(2^1S_0)$. If these identifications are found to be correct, they will represent a serious failure of the charmonium model:

1. From Table 1,

$$B[\psi \to X(2830)\gamma]\, B(X \to \gamma\gamma) = 1.3 \pm 0.4 \times 10^{-4} \qquad \text{3.31a}$$

$$B[\psi \to X(2830)\gamma] < 0.017 \quad (90\% \text{ confidence level}). \qquad \text{3.31b}$$

From these, one infers

$$B[X(2830) \to \gamma\gamma] \gtrsim 8 \times 10^{-3}, \qquad \text{3.31c}$$

which is at least a factor of five larger than the predicted value (Table 3). This is to be contrasted with the apparent success of calculated direct decay rates for the ^3P states and for ψ'.

2. The model fails by at least an order of magnitude in predicting $\psi \to \gamma\eta_c$. This is especially puzzling when one recalls that M1 transitions among the light mesons are fairly well described by the nonrelativistic quark model (72, 76). Particularly relevant to $\psi \to \gamma\eta_c$ are the predictions $\Gamma(\phi \to \gamma\eta) = 70$ keV and $\Gamma(\phi \to \gamma\pi^0) = 6.9$ keV, to be compared to the measured widths of 82 ± 16 keV and 5.7 ± 2.0 keV, respectively. To add to the puzzle, there is the apparently successful prediction of a strongly suppressed $\psi' \to \gamma\eta_c$ transition.

Table 5 M1 decays in charmonium; observed total widths of ψ and ψ' (Table 1) and predicted total widths of η_c and η_c' (Table 2) are used in determining branching ratios

Datum	Theory (keV for Γ_γ, % for B_γ)	Experiment (keV for Γ_γ, % for B_γ)
$\Gamma_\gamma(\psi \to \eta_c)$	26	< 1.2
$B_\gamma(\psi \to \eta_c)$	37	< 1.7
$B_\gamma(\psi \to \eta_c)B(\eta_c \to \gamma\gamma)$	0.052	0.013 ± 0.004
$\Gamma_\gamma(\psi' \to \eta_c)$	1.9	< 2.3
$B_\gamma(\psi' \to \eta_c)$	0.83	< 1.0
$B_\gamma(\psi' \to \eta_c)B(\eta_c \to \gamma\gamma)$	1.2×10^{-3}	$< 3.4 \times 10^{-2}$
$\Gamma_\gamma(\psi' \to \eta_c')$	17	< 5.7 .
$B_\gamma(\psi' \to \eta_c')$	7.5	< 2.5
$B_\gamma(\psi' \to \eta_c')B(\eta_c' \to \gamma\gamma)$	1.0×10^{-2}	$< 3.1 \times 10^{-2}$
$\Gamma_\gamma(\eta_c' \to \psi)$	0.53	—
$B_\gamma(\eta_c' \to \psi)$	1.6×10^{-2}	$> 24 \pm 16$
$B_\gamma(\psi' \to \eta_c')B_\gamma(\eta_c' \to \psi)$	1.2×10^{-3}	0.6 ± 0.4

3. While the prediction $B(\psi' \to \gamma\eta'_c) = 0.075$ is only three times larger than the observed branching ratio limit, the inferred lower limit, $B(\eta'_c \to \gamma\psi) \gtrsim 0.15$, is 2–3 orders of magnitude greater than what one expects theoretically for this hindered M1 transition.

If X(2830) really is the η_c, the resolution to these difficulties must lie partly in correcting the assumption of identical radial wave functions for ψ and η_c, i.e. that gluons play an important role in suppressing both the M1 overlap integral and $\Gamma(\eta_c \to 2 \text{ gluons})/\Gamma(\eta_c \to \gamma\gamma)$. On the other hand, given the successes of gluon counting for direct decays of spin-triplet states, the verified suppression of the hindered M1 transition $\psi' \to \gamma\eta_c$, and the experimental fact that $\Gamma(\psi \to \gamma\eta_c) \gtrsim 1$ keV, there is no way to understand the identification $\chi(3455) = \eta'_c$ with such a large branching ratio to $\gamma\psi$.

It is always possible, of course, that η_c and η'_c have not been discovered yet and that they lie $\lesssim 100$ MeV below their hyperfine partners, as originally expected (47, 53). In that case, theoretical estimates of the M1 rates are greatly reduced and, in fact, lie within experimental limits for states at such masses.

The natural question then is: What are these two states? Various conjectures abound including:

1. They are four-quark ($c\bar{c}\, q\bar{q}$) or molecular states (η_c and π^0 bound in an S wave, say) (77) or, perhaps $c\bar{c}$ states with a gluon excitation (78). There are no convincing models for such relatively low-mass systems that are not pure $c\bar{c}$, nor is there even the ability to make convincing estimates of transition rates. Such techniques are sorely needed.

2. Another interesting speculation is that they are Higgs mesons H (79). If this is the case, H certainly does not have the "conventional" coupling to quarks $\sim G_F^{\frac{1}{2}} m_q$, for then one would estimate (with $m_c \approx \frac{1}{2} M_{\psi,\psi'}$)

$$\frac{\Gamma(\psi,\psi' \to \gamma H)}{\Gamma(\psi,\psi' \to e^+ e^-)} \approx \frac{G_F M_{\psi,\psi'}^2}{2^{\frac{1}{2}}\pi\alpha} \frac{M_{\psi,\psi'}^2(M_{\psi,\psi'} - M_H^2)}{(M_{\psi,\psi'}^2 + M_H^2)^2} \approx 10^{-4} \qquad 3.32$$

Not only does one need the coupling of H to charmed quarks to be anomalously large, but the coupling to light quarks must be anomalously small if $H \to \gamma\gamma$ is to be a sizable decay mode (80). [Actually, $\chi(3455) = H \to \psi\gamma$ might be a dominant decay mode of a more-or-less conventional Higgs meson.]

3. One final possibility, suggested by Harari (81), is that $\chi(3455)$ is the $^1D_2$ level of charmonium. Because of the strongly hindered nature of $^3S_1 \to {}^1D_2$ radiative transitions, such a possibility is viable only if ψ' and ψ contain a sizable admixture (~ 5–10%) of $^3D_1$ and if $\Gamma(^1D_2 \to \text{hadrons}) \approx \Gamma(\psi \to \text{hadrons})$ (82). Perhaps the most serious objection to this identification, based as it is on the presumed large splitting between

all triplet and singlet states of given L, is that $\Gamma(\psi' \to \gamma\eta_c')$ is expected to be $\sim 10\,\Gamma(\psi' \to \gamma^1D_2)$ and yet, on this hypothesis, η_c' has not been seen yet.

To summarize, while the basic model gives a very good qualitative and creditable quantitative description of the spectrum and transition rates of the spin-triplet charmonium levels, it fails to account for practically all observed features of the proposed singlet levels. Either something very important (and largely unknown) is missing from the model or, more happily, the model is telling us that new degrees of freedom—which it was never intended to handle—have been discovered.

3.4 *Including Spin Dependence*

Even before the states between ψ and ψ' were discovered and their splittings measured, many people began trying to incorporate quark spin dependence into the charmonium model. The very earliest work utilizing a Coulomb model (47, 53, 69) was much too naive and was soon abandoned. However, the nonrelativistic character of the low-lying $c\bar{c}$ states suggested that the Bethe-Salpeter equation would be a useful formalism and that some analog of the Breit-Fermi Hamiltonian for positronium would continue to be relevant. Almost nothing was known about the structure of the Bethe-Salpeter kernel and some educated guesses were needed to make the computation of splittings possible. Some perturbation-theoretic analyses of QCD already suggested that the spin-dependent part of the Hamiltonian would be strongly modified away from a Coulomb form just as the spin-independent confining part was (83). These same investigations further indicated that the modifications of the two pieces might not be simply related.

One early guess, however, was that the spin-dependent interaction would have only a short-range Coulomb-type structure (84). Very small fine structure splittings were predicted, such as $M_\psi - M_{\eta_c} \approx 30$ MeV and $M(^3P_2) - M(^3P_1) \approx 5$–10 MeV. This is in sharp disagreement with experiment, and such an approach now seems inadequate. In particular, Johnson (85) has recently emphasized that at least one part of the spin effect is necessarily long range, namely that part of the spin-orbit interaction arising from Thomas precession (see Equation 3.37 below). Indeed, almost all treatments of the spin forces in charmonium have focused on the long-range part of the $c\bar{c}$ interaction. One exception is the model of Celmaster et al (86), mentioned below in Section 3.6. They assume an r-dependent short-distance coupling $\kappa(r)$, whose form is suggested by asymptotic freedom, and they use $V_{AF}(r) = -\kappa(r)/r$ to generate the Breit-Fermi interaction. Although they obtained much larger splittings than those in Reference 84 (see Table 6), they have neglected to take account of the long-range contribution from Thomas precession.

The most popular approach, pioneered by Schnitzer (87) and by Pumplin, Repko & Sato (88), and since generalized by many authors (89), has been to assume that heavy quark binding is effectively due to "single gluon exchange with renormalization group improvement," summarized by an instantaneous Bethe-Salpeter kernel consisting of vector and scalar interaction terms:

$$V_{\text{coul}}(\mathbf{k}^2)\gamma_1^\mu\gamma_{2\mu} + V_v(\mathbf{k}^2)\Gamma_1^\mu(k)\Gamma_{2\mu}(k) + V_s(\mathbf{k}^2)\mathbf{1}_1\mathbf{1}_2. \qquad 3.33$$

The subscripts 1 and 2 refer to the c and c̄ quarks, $\mathbf{1}$ is a unit matrix in Dirac space, k is the four-momentum carried by the exchanged gluon, and (with $\sigma_{\mu\nu} = (2i)^{-1}[\gamma_\mu, \gamma_\nu]$)

$$\Gamma_\mu(k) = \gamma_\mu - \frac{i\lambda}{2m_c}\sigma_{\mu\nu}k^\nu, \qquad 3.34$$

where λ is the color magnetic moment of the quark—an adjustable parameter. In the spin-independent, nonrelativistic limit of this interaction, the potential is

$$V_0 = V_{\text{coul}} + V_v + V_s. \qquad 3.35$$

In most discussions, it has been assumed that

$$V_v = \eta V_{\text{lin}} \qquad V_s = (1-\eta)V_{\text{lin}}, \qquad 3.36$$

where V_{lin} is the linear confining potential r/a^2 in coordinate space, and η is another adjustable parameter.

Using Equations 3.33 to 3.36, the authors obtained the spin-dependent potential by following the same steps used to convert the kernel for positronium into the Breit interaction. The result is:

$$V_{\text{spin}}(r) = \frac{1}{2m_c^2}\left[\frac{4\kappa}{r^3} + 4(1+\lambda)\frac{1}{r}\frac{dV_v}{dr} - \frac{1}{r}\frac{dV_0}{dr}\right]\mathbf{L}\cdot\mathbf{S}$$

$$+ \frac{2}{3m_c^2}\left[4\pi\kappa\delta(\mathbf{r}) + (1+\lambda)^2\,\nabla^2 V_v(r)\right]\mathbf{S}_1\cdot\mathbf{S}_2$$

$$+ \frac{1}{3m_c^2}\left[\frac{3\kappa}{r^3} + \frac{1}{r}\frac{dV_v}{dr} - \frac{d^2V_v}{dr^2}\right]S_{12}, \qquad 3.37a$$

where

$$S_{12} = 3\mathbf{S}_1\cdot\hat{\mathbf{r}}\,\mathbf{S}_2\cdot\hat{\mathbf{r}} - \mathbf{S}_1\cdot\mathbf{S}_2. \qquad 3.37b$$

The last term in the spin-orbit part of V_{spin} is the Thomas precession contribution, and contains the only influence of the scalar interaction on spin dependence. For the potentials in Equation 3.36, V_{spin} is given by

$$V_{\text{spin}}(r) = \frac{1}{2m_c^2} \left\{ \frac{3\kappa}{r^3} + \frac{1}{ra^2} \left[\eta(3+4\lambda) - (1-\eta) \right] \right\} \mathbf{L} \cdot \mathbf{S}$$

$$+ \frac{2}{3m_c^2} \left[4\pi\kappa\delta(\mathbf{r}) + \frac{2\eta}{ra^2}(1+\lambda)^2 \right] \mathbf{S}_1 \cdot \mathbf{S}_2$$

$$+ \frac{1}{3m_c^2} \left[\frac{3\kappa}{r^3} + \frac{\eta}{ra^2}(1+\lambda)^2 \right] S_{12}. \tag{3.38}$$

When used perturbatively, this interaction generates the following mass formulae (below, M_L = bare mass of the level ψ_{nL} determined by V_0 and $\langle r^{-c} \rangle_L = \langle \psi_{nL} | r^{-c} | \psi_{nL} \rangle$):

$$M(^3S_1) = M_0 + \frac{1}{6m_c^2} \left[4\pi\kappa |\Psi_{n0}(0)|^2 + 2(1+\lambda)^2 \eta a^{-2} \langle r^{-1} \rangle_0 \right]$$

$$M(^1S_0) = M_0 - \frac{1}{2m_c^2} \left[4\pi\kappa |\Psi_{n0}(0)|^2 + 2(1+\lambda)^2 \eta a^{-2} \langle r^{-1} \rangle_0 \right]$$

$$M(^3P_2) = M_1 + \frac{1}{5m_c^2} \left\{ 7\kappa \langle r^{-3} \rangle_1 + \left[\left(9 + 13\lambda + \frac{3}{2}\lambda^2 \right)\eta - \frac{5}{2}(1-\eta) \right] \right.$$
$$\left. a^{-2} \langle r^{-1} \rangle_1 \right\}$$

$$M(^3P_1) = M_1 + \frac{1}{m_c^2} \left\{ -\kappa \langle r^{-3} \rangle_1 + \left[\left(-1 - \lambda + \frac{1}{2}\lambda^2 \right)\eta + \frac{1}{2}(1-\eta) \right] \right.$$
$$\left. a^{-2} \langle r^{-1} \rangle_1 \right\}$$

$$M(^3P_0) = M_1 - \frac{1}{m_c^2} \left\{ 4\kappa \langle r^{-3} \rangle_1 + \left[(3+4\lambda)\eta - (1-\eta) \right] a^{-2} \langle r^{-1} \rangle_1 \right\}$$

$$M(^1P_1) = M_1 - \frac{(1+\lambda)^2\eta}{m_c a^2} \langle r^{-1} \rangle_1. \tag{3.39}$$

For completeness, we include formulae for the splitting of D levels and mixing of $^3D_1$ with $^3S_1$:

$$M(^3D_3) = M_2 + \frac{1}{7m_c^2} \left\{ 20\kappa \langle r^{-3} \rangle_2 + \left[(23+32\lambda+2\lambda^2)\eta - 7(1-\eta) \right] \right.$$
$$\left. a^{-2} \langle r^{-1} \rangle_2 \right\}$$

$$M(^3D_2) = M_2 + \frac{1}{m_c^2}\left\{-\kappa\langle r^{-3}\rangle_2 + \left[\left(-1-\lambda+\frac{1}{2}\lambda^2\right)\eta + \frac{1}{2}(1-\eta)\right]\right.$$
$$\left. a^{-2}\langle r^{-1}\rangle_2\right\}$$

$$M(^3D_1) = M_2 + \frac{1}{6m_c^2}\left\{-30\kappa\langle r^{-3}\rangle_2 + \left[(-26-34\lambda+\lambda^2)\eta + 9(1-\eta)\right]\right.$$
$$\left. a^{-2}\langle r^{-1}\rangle_2\right\}$$

$$M(^1D_2) = M_2 - \frac{(1+\lambda)^2\eta}{m_c^2 a^2}\langle r^{-1}\rangle_2$$

$$\langle n^3S_1|V_{spin}|m^3D_1\rangle = \frac{8^{\frac{1}{2}}}{12m_c^2}\left[3\kappa\langle nS|r^{-3}|mD\rangle + (1+\lambda)^2\right.$$
$$\left. \eta a^{-2}\langle nS|r^{-3}|mD\rangle\right]. \qquad 3.40$$

The splittings among the P and S states determined by various authors from Equation 3.39 are listed in Table 6, together with the measured mass differences; for comparison purposes, we assumed $X(2830) = 1^1S_0$ and

Table 6 Spin-dependent splittings in charmonium measured in MeV. The numbers in brackets are values of κ, $1/a^2$ (GeV^2), m_c (GeV), η, and λ used by various authors. In Reference 86 the Coulomb parameter is not constant (see Equation 3.70) and only this short-range part of the potential is kept in computations of spin dependence

Author parameters	$M_{\chi_2}-M_{\chi_1}$	$M_{\chi_1}-M_{\chi_0}$	$M_\psi-M_{\eta_c}$	$M_{\psi'}-M_{\eta_c'}$
Experiment	44±6	95±5	265±14	230±7
Schnitzer (87) [0.2, 0.19, 1.6, 1.0, 0.0]	87	63	70	58
Pumplin et al (88) [0.0, 0.30, 1.5, 1.0, 0.0]	152	117	119	92
Henriques et al (89) [0.8, 0.18, 1.6, 0.0, 0.0]	40 (input)	80	95	—
Schnitzer (87) [0.2, 0.19, 1.6, 1.0, 1.1]	182	170	268 (input)	225
Chan (89) [0.2, 0.19, 1.6, 0.12, 5.0]	40 (input)	98	262 (input)	225
Carlson & Gross (89) [0.27, 0.20, 1.37, 0.08, 4.4]	41 (input)	98	265 (input)	181
Celmaster et al (86) [—, —; 1.98, 1.0, 0.0]	92	100	150	80

$\chi(3455) = 2^1S_0$. For the D states we content ourselves with two remarks. First the 1^3D_1 level comes out about 40 MeV below its unperturbed mass[11] (which is 3.755 GeV in Table 2). Thus, the agreement between the coupled-channel calculation (in Section 3.5) and the observed mass 3.772 GeV of ψ'' is perhaps fortuitous. Second, the predicted mixing angle between 1^3D_1 and 2^3S_1 is

$$\varepsilon = \frac{1}{2}\tan^{-1}\frac{2\langle 2^3S_1 | V_{\text{spin}} | 1^3D_1\rangle}{M(2^3D_1) - M(2^3S_1)} \approx 6°. \qquad 3.41$$

This is much less than one infers from measurements of the ψ'' electronic width. Using

$$\varepsilon = \tan^{-1}\left\{\frac{M_{\psi''}}{M_{\psi'}}\left[\frac{\Gamma(\psi'' \to e^+e^-)}{\Gamma(\psi \to e^+e^-)}\right]^{\frac{1}{2}}\right\}, \qquad 3.42$$

these are $\varepsilon = (26 \pm 3)°$ for $\Gamma_e(\psi'') = 0.37 \pm 0.09$ keV (90), and $\varepsilon = (19 \pm 3)°$ for $\Gamma_e(\psi'') = 0.18 \pm 0.05$ keV (91).

The lessons of this attack on the problem of spin dependence may be summarized thus: Insofar as one is willing to extend the hypothesis of an instantaneous interaction between heavy quarks beyond the realm of the simple spin-independent potential V_0, the ansatz Equation 3.33 is the basis of a quite reasonable first effort. Clearly, the assumption of a purely short-range origin for spin forces is inadequate. Economy of parameters then demands that one attribute spin forces to the long-range part, here assumed to be r/a^2. As we have seen, this still leaves some freedom in the Dirac structure of the kernel, and the work of Henriques et al (89) first suggested that the P-state splittings are best fit if the long-range interaction is scalar. The reason for this is that the Thomas precession term, which is the most important part of the spin-orbit interaction in this case (η small), orders the ^3P_J levels oppositely from the other terms in V_{spin}. This feature is needed to explain the unexpectedly small ratio $[M(^3P_2) - M(^3P_1)]/[M(^3P_1) - M(^3P_0)] \approx 0.42$.

If one now assigns X(2830) to $\eta_c(1^1S_0)$, the splitting from ψ may be fit, as Schnitzer found (87), by assuming a rather large quark color anomalous moment, $\lambda(\eta)^{\frac{1}{2}} \approx 1$. All this was put together by Chan and by Carlson & Gross (89), who combined the successful features of the last two named pieces of work to obtain excellent agreement with the data.

But, in this flush of success, one must not lose sight of two important facts. First, all that has been accomplished so far is to fit four mass splittings with two parameters, λ and η. In fact, as Carlson & Gross point out, $M(^3P_1) - M(^3P_0)$ is far less sensitive to parameters than is

---

[11] For this, we have used the parameters of Carlson & Gross (89).

$M(^3P_2) - M(^3P_1)$. Therefore, in a fit to the latter plus the $\psi - X(2830)$ splitting, the former is almost automatically correct. And, it is not surprising that a large $1^3S_1 - 1^1S_0$ implies a comparable $2^3S_1 - 2^1S_0$ splitting. The real test of this phenomenology will come when the multitude of intramultiplet splittings in the Υ system are measured—a potentially difficult task if they are ~ 5–10 times smaller, as expected. Second, given the difficulty of accommodating the X(2830) and $\chi(3455)$ in the simple charmonium scheme, it may well be that it has been a great mistake all along to use these states in determining the parameters in V_{spin}. Again, only time will tell.

From a theoretical standpoint, the procedure outlined above suffers from the lack of a "first principles" justification for its starting point, the Bethe-Salpeter kernel defined in Equations 3.33 and 3.36. More ambitious approaches are in progress in a number of dynamical models that incorporate gluon degrees of freedom more or less explicitly. These include lattice gauge theories (51, 52) the MIT bag model (85, 92), and the quark-string models (78, 93). In each of these models, the long-range spin-independent potential between heavy quarks is shown to be linear. Further there is a reasonably well-defined procedure for extracting the spin-dependent interaction. While it is too early to assess these approaches, it is hoped that they will provide insight to this most puzzling aspect of charmonium dynamics.

3.5   Coupling Charmonium to Its Decay Channels

It was recognized very early (63) that the quantum mechanical coupling between charmonium states and their OZI-allowed decay channels could modify the predictions of the naive potential model. The development of a model for this coupling and the resulting predictions have been presented in a series of papers by Eichten et al (68, 75, 94; see also 95).[12] The main issues one wants to address in such a model are these:

1. Renormalization (shifts) of the bare spectrum generated by $V_0(r)$ and the widths of $c\bar{c}$ states above charm threshold, $W_c = 2M_{D^0} = 3.727$ GeV.
2. Renormalizations of the wave functions deduced from V_0. This includes leakage from the $c\bar{c}$ sector to the charmed hadron sector, as well as the mixing among charmonium levels having the same J^{PC}. Both of these will affect rates for all the transitions discussed in Section 3.3. Of special interest is the fact that decays forbidden or strongly suppressed in the potential model, such as $^3D_1 \rightarrow e^+e^-$ and $2^3S_1 \rightarrow 1^1S_0 + \gamma$,

---

[12] Kogut & Susskind (95) implement the coupling to decay channels in a quite different way from Eichten et al. In particular, charmed mesons never appear explicitly.

can be enhanced through mixing between 1^3D_1 and 2^3S_1, and between 2^3S_1 and 1^3S_1 (or 2^1S_0 and 1^1S_0), respectively.

3. A description of e^+e^- annihilation in the charm threshold region, $W_c \leq W \lesssim 4.4$ GeV. In particular, one wants to interpret the structure of $R = \sigma(e^+e^- \rightarrow \text{hadrons})/\sigma(e^+e^- \rightarrow \mu^+\mu^-)$ in this region, and to discuss the general (and sometimes peculiar) features of the exclusive channel cross sections $\sigma(e^+e^- \rightarrow D^0\bar{D}^0)$, $\sigma(e^+e^- \rightarrow D^+D^-)$, $\sigma(e^+e^- \rightarrow D^0\bar{D}^{*0})$, etc.

3.5.1 FORMALISM AND DESCRIPTION OF THE MODEL   The description of what happens to a discrete set of states in one part of Hilbert space when it is immersed in and coupled to a continuum of states belonging to another subspace is a classic problem in quantum mechanics and was, formally, solved long ago (96a; more recent discussions are given in 96b).

Let the total Hamiltonian of this system be

$$\mathscr{H} = \mathscr{H}_0 + \mathscr{H}_I, \qquad 3.43$$

where $\mathscr{H}_0$ is responsible for the binding of the discrete states $|n\rangle$ and the continuum states $|v\rangle$:

$$\mathscr{H}_0|n\rangle = \varepsilon_n|n\rangle, \qquad \langle n|m\rangle = \delta_{nm}, \qquad 3.44$$

$$\mathscr{H}_0|v\rangle = \omega_v|v\rangle, \qquad \langle v|\mu\rangle = \delta(v-\mu),$$

$$\langle v|n\rangle = 0,$$

while $\mathscr{H}_I$ is responsible for their coupling:

$$\langle v|\mathscr{H}|n\rangle = \langle v|\mathscr{H}_I|n\rangle. \qquad 3.45$$

In the present shorthand notation, $|n\rangle$ stands for any of the pure $c\bar{c}$ levels $|\psi_{nLJ}\rangle$, and $|v\rangle$ for any state of charmed hadrons having zero net charm, zero total momentum, and total energy ω_v. (States with more than one c and one $\bar{c}$ are ignored.)

The problem we face is to describe the eigenstates $|N\rangle$ of $\mathscr{H}$ with eigenvalue E_N. These states include the observed bound states (ψ, X_J, ψ', etc) below charmed threshold as well as the continuum states and resonances in e^+e^- annihilation above W_c. (Because $\mathscr{H}$ allows for decay, E_N need not be real—and certainly won't be for the resonances.) We begin by expanding $|N\rangle$ in the complete basis formed by the eigenstates $|n\rangle$ and $|v\rangle$ of $\mathscr{H}_0$:

$$|N\rangle = \sum_n a_{nN}|n\rangle + \int dv\, a_{vN}|v\rangle,$$

$$a_{nN} = \langle n|N\rangle, \qquad a_{vN} = \langle v|N\rangle. \qquad 3.46$$

From Equations 3.44 to 3.46, the expansion coefficients and the energy E_N are solutions of the eigenvalue problem

$$\sum_m \left[(E_N - \varepsilon_n)\,\delta_{nm} - \Omega_{nm}(E_N)\right] a_{mN} = 0 \quad \text{and} \qquad 3.47$$

$$\det\left[W - \varepsilon - \Omega(W)\right] = 0 \quad \text{(roots } E_N\text{),} \qquad 3.48$$

where

$$\Omega_{nm}(W) = \int \frac{dv}{W - \omega_v + i0} \langle n|\mathscr{H}_I|v\rangle \langle v|\mathscr{H}_I|m\rangle$$

$$\equiv \Delta_{nm}(W) - \frac{i}{2}\Gamma_{nm}(W). \qquad 3.49$$

For $E_N < W_c$, the solutions of Equations 3.47 and 3.48 correspond to the bound charmonium levels, the matrix $\Delta_{nm}(E_N)$ describes the shift in the mass of these levels from the "bare" masses ε_n, and the width matrix $\Gamma_{nm}(E_N < W_c)$ vanishes. For $|E_N| > W_c$, the state $|N\rangle$ is a resonance that decays almost exclusively to charmed hadrons, having mass M_N, width Γ_N, and $E_N = M_N - \frac{1}{2}i\Gamma_N$. Given the coefficients a_{nN}, one may determine a_{vN} from

$$a_{vN} = \frac{1}{E_N - \omega_v + i0}\sum_m a_{mN}\langle v|\mathscr{H}_I|m\rangle \qquad 3.50$$

Finally, the continuum eigenstates $|N\rangle$ may be determined from integral equations similar in structure to Equations 3.47 and 3.48. For reasons that will become clear shortly, the model makes little use of these. Rather, the recalculation of the transition rates of the bound states and of

$$\Delta R \equiv \frac{\sigma(e^+e^- \to \text{charmed hadrons})}{\sigma(e^+e^- \to \mu^+\mu^-)} \qquad 3.51$$

requires only a knowledge of Ω, which in turn requires a model for $\mathscr{H}_I$.

The assumptions and approximations defining the model for $\mathscr{H}_I$ and Ω are (68, 75):

1. The Hamiltonian is taken to be

$$\mathscr{H} = -\frac{3}{8}\sum_{a=1}^{8}\int d^3x\,d^3y\,\rho_a(\mathbf{x},t)\,V(\mathbf{x}-\mathbf{y})\,\rho_a(\mathbf{y},t) \qquad 3.52$$

+ quark kinetic energy terms,

where

$$\rho_a(x) = \sum_{\text{flavors}(i)} q_i^\dagger(x)\frac{\lambda_a}{2}q_i(x)$$

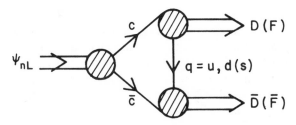

Figure 5   The transition amplitude (Equation 3.53) for $\psi_{nL} \to D\bar{D}$ or $F\bar{F}$.

is the octet of color charge densities and $V(\mathbf{x}-\mathbf{y}) = |\mathbf{x}-\mathbf{y}|/a^2$ is the instantaneous confining potential in Equation 3.8. The Coulomb piece has been dropped for simplicity.[13]

2. As the form of V implies, calculations with this Hamiltonian necessarily are carried out in the nonrelativistic approximation. Therefore, $\mathscr{H}$ is explicitly spin independent. The only dependence on quark total spin that enters the computation of Ω is through the use of "spin-split thresholds" for the continuum states $|v\rangle$, i.e. the mass difference between D* and D, F* and F, is put in by hand.

3. When decomposed into creation and annihilation operators, $\mathscr{H} = \mathscr{H}_0$ (binding of $q_i\bar{q}_j$)$+\mathscr{H}_I$ (pair emission: $q_i \to q_i + \bar{q}_j + q_j$) + other terms (e.g. emission of two pairs from the vacuum) that are discarded. While the nonrelativistic binding mechanism is presumably consistent only for $c\bar{c}$ states (where it reproduces the results of the spin-independent potential model described in Section 4.3), it is also used to generate bound-state wave functions for charmed meson states $c\bar{q}$ and $\bar{c}q$ (q = u,d,s from now on).

4. The model assumes that the transition $\psi_{nLJ} \to$ charmed mesons is a sequential quasi-two-body process,[14] e.g. $\psi_{nLJ} \to \bar{D}D^*; D^* \to D\pi$. Accordingly, the only terms kept in $\mathscr{H}_I$ are those describing light-pair emission, $c \to c+q+\bar{q}$ and $\bar{c} \to \bar{c}+q+\bar{q}$, which governs $\psi_{nLJ}(c\bar{c}) \to D(c\bar{q})+\bar{D}(\bar{c}q)$, as depicted in Figure 5. There, the shaded circles denote bound-state vertex functions (simply related to the wave functions in the nonrelativistic limit) and they emphasize that the model incorporates the extended nature of the parent and its decay products. While certain features (to be mentioned below) of the transition amplitudes computed with $\mathscr{H}_I$ may be model independent, the nonrelativistic approximation used in the computation is very questionable.

[13] Note that the Hamiltonian in Equation 3.52 corresponds to the nonrelativistic limit of a pure vector kernel in Equation 3.33.

[14] The experimental justification for the quasi-two-body hypothesis and for the neglect of charmed baryons is presented in Section 4.

5. The final approximation made in References 68 and 75 is a drastic truncation of the continuum states $|v\rangle$ to include only the ground-state charmed mesons D, D*, F, and F*. Consequently, the model is reliable (even semiquantitatively) only where the effects of higher thresholds (e.g. charmed P states) may be ignored. For the calculation of ΔR, the breakdown due to neglect of higher thresholds is already apparent at $W \approx 4.1$ GeV.

The general form of the transition amplitude for $\psi_{nLJ} \to$ pair of ground state charmed mesons is

$$\langle c\bar{q}(J_1; E_1, \mathbf{p}), \bar{c}q(J_2; E_2, -\mathbf{p}) | \mathcal{H}_1 | c\bar{c}(nLJ; W, \mathbf{0}) \rangle$$

$$= \left(m_q \frac{m_c m_q}{m_c + m_q} \right)^{-1} \times \text{spin factor } (J_i) \times \text{form factor } (n, L; |\mathbf{p}|; m_c, m_q, a),$$

3.53

where $(E_i, \mathbf{p}_i)$ are the four-momenta of the outgoing pair, $W = E_1 + E_2$, and J_i are total angular momenta. The parameters entering the calculation of this amplitude are the quark masses m_c, m_q and the linear potential strength a^{-2}. Because of the spin independence of the $q\bar{q}$ production mechanism, all dependence on quark spin appears in a Clebsch-Gordon coefficient, the second factor in Equation 3.53. The first factor, which implies a suppressed production of F = $c\bar{s}$ relative to D = $c\bar{u}$, arises from the S-wave nature of the production mechanism and from the charmed meson wave function. The P-wave form factors for the first

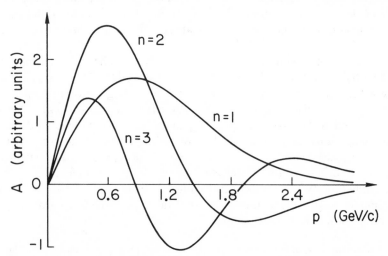

Figure 6   The P-wave form factor in Equation 3.53 for $\psi_{n0} \to D\bar{D}$ ($n = 1, 2, 3$) as a function of the momentum of the outgoing charmed meson, $p = \frac{1}{2}(W^2 - 4M_D^2)^{\frac{1}{2}}$ (from 68). See Table 7 for parameters used in this calculation.

three $L = 0$ cc̄ levels to go to a DD̄ pair are plotted in Figure 6 as a function of the D momentum, $|\mathbf{p}|$. The high-momentum fall-off and the oscillations reflect the finite extent of the bound states and the nodes in the wave functions of radial excitations. These zeroes in the form factor play an important role in the behavior of the exclusive cross sections $\sigma(e^+e^- \to$ DD̄, DD̄*, D*D̄*) as a function of center-of-mass energy W.

The calculation of level positions, radiative transition rates, and ΔR in this model proceeds as below.

3.5.2 RENORMALIZATION OF SPECTRUM   The parameters of the model are m_c and a, the light quark masses $m_u = m_d$ and m_s, and the charmed meson masses M_D, M_{D*}, M_F, and M_{F*}. The last named are now chosen to have their measured values. Both m_c and a generate a bare cc̄ spectrum, which is renormalized by the coupling to the continuum. For fixed $m_u \approx \frac{1}{3}$ GeV and $m_s \approx \frac{1}{2}$ GeV, m_c and a are adjusted until the Green's function

$$\mathcal{G}_{mn} = [W - \varepsilon - \Omega(W)]_{mn}^{-1} \qquad 3.54$$

has $J^{PC} = 1^{--}$ poles at 3.095 and 3.684 GeV (corresponding to ψ and ψ'), and a residue at 3.095 GeV such that the ψ electronic width is 5.3 keV. The computed ψ' electronic width is 3.4 keV, which is the same as found in the basic model (Table 3) and about 1.7 times the observed value. Table 7 contains a list of the bare and renormalized masses of the spin-triplet states below charm threshold. Especially noteworthy are the large downward shifts in masses and the very small splitting of the ^3P_J states. The first effect shows that $\mathcal{H}_I$ is by no means a weak perturbation. The second is a consequence of the fact that the only spin dependence comes from the split thresholds and will be sizable only for states very near (within ~ 50–100 MeV) these thresholds.

Table 7   Mass shifts in the coupled-channel model (68); parameters used are $m_c = 1.69$ GeV, $a = 1.80$ GeV^{-1}, $m_u = m_d = 0.33$ GeV, and $m_s = 0.50$ GeV. D and D* masses are taken from experiment, while $M_F = 2.00$ GeV and $M_{F*} = 2.14$ GeV are assumed

State	Bare mass (MeV)	Mass shift (MeV)	Renormalized mass (MeV)
ψ	3191	-96	3095
ψ'	3893	-209	3684
ψ''	3976	-208	3768
χ_2	3622	-170	3451
χ_1	3622	-180	3442
χ_0	3622	-191	3431

3.5.3 RADIATIVE TRANSITIONS With the parameters determined by renormalization, the radiative transition rate for $\psi' \to \chi_J + \gamma$, say, is computed as follows: Using the (oversimplified) notation of Equation 3.46, let $|N\rangle = |\psi'\rangle$, $|M\rangle = |\chi_J\rangle$, and $j_\lambda = \frac{2}{3}(\bar{c}\gamma_\lambda c + \bar{u}\gamma_\lambda u) - \frac{1}{3}(\bar{d}\gamma_\lambda d + \bar{s}\gamma_\lambda s)$ be the electromagnetic current. The E1 transition amplitude is (68)

$$\langle M|j_\lambda|N\rangle = \sum_{m,n} a^*_{mM} a_{nN} \langle m|j_\lambda|n\rangle + \int d\mu \, d\nu \, a^*_{\mu M} a_{\nu N} \langle \mu|j_\lambda|\nu\rangle$$

$$+ \sum_n \int d\nu \left(a^*_{\nu M} a_{nN} \langle \nu|j_\lambda|n\rangle + a^*_{nM} a_{\nu N} \langle n|j_\lambda|\nu\rangle \right). \qquad 3.55$$

The first term on the right in Equation 3.55 includes only the parts of ψ' and χ_J in the discrete (c$\bar{\text{c}}$) sector, with $\langle m|j_\lambda|n\rangle$ computed just as in the potential model without coupling to the continuum. The second term, involving the continuum components of ψ' and χ_J, contains electromagnetic transition matrix elements of charmed mesons; these are taken from standard quark model calculations. The third (cross) term involves a transition between the discrete and continuum sectors under the action of j_λ.

In lieu of some long and not-very-illuminating formulae for the terms in Equation 3.55, a few remarks on their relative importance are offered. The most important contribution to the discrete-sector terms is obviously the diagonal one: $|n\rangle = |2^3S_1\rangle$ and $|m\rangle = |1^3P_J\rangle$. The next single most important contribution to this set comes from $|n\rangle = |1^3D_1\rangle$, i.e. the mixing of $^3S_1$ and $^3D_1$ states due to nearby spin-split thresholds, and this is rather sensitive to the precise position of the D$\bar{\text{D}}$ and D$\bar{\text{D}}^*$ thresholds. The S-D mixing is most important for $\psi' \to \chi_0 + \gamma$ because of a large Clebsch-Gordan coefficient for $1^3D_1 \to 1^3P_0 + \gamma$ (see Equation 3.26). Because of energy denominators, the continuum-sector terms are dominated by the nearest threshold accessible to both ψ' and χ_J. Thus, the continuum is considerably more important (roughly a factor of two in amplitude) for χ_0 than for χ_1 and χ_2 because $\chi_0 \to D\bar{D}$ in an S wave, while $\chi_1 \to D\bar{D}$ is forbidden and $\chi_2 \to D\bar{D}$ is suppressed by a D-wave factor. In amplitude, the continuum contribution to χ_0 is about half the diagonal contribution and of the same sign. Finally, the mixed terms are practically negligible.

For the M1 transitions, only the discrete-sector terms have been computed so far. They are obtained from the standard formula (68, 72)

$$\Gamma(\psi_N \to \eta_{c,M} + \gamma) = \frac{16}{27} \alpha \frac{\omega^3}{m_c^2} \left| \sum_n a_{nN} a^*_{nM} \right|^2, \qquad 3.56$$

i.e. only nonhindered terms (same principal quantum number) are kept.

Table 8 Radiative transition rates in the coupled-channel model (68); parameters used are given in Table 7

Mode	Width (keV)
$\psi' \to \gamma\chi_2$	19
$\psi' \to \gamma\chi_1$	28
$\psi' \to \gamma\chi_0$	37
$\psi \to \gamma\eta_c$	21
$\psi' \to \gamma\eta_c$	12
$\psi' \to \gamma\eta'_c$	8
$\eta'_c \to \gamma\psi$	0.1

The overlap factor $\sum_n a_{nN} a_{nM}^*$ is 0.7 for $\psi' \to \eta'_c$, -0.13 for $\psi' \to \eta_c$, -0.05 for $\eta'_c \to \psi$, and 0.9 for $\psi \to \eta_c$, where η_c and η'_c were taken to lie at 2.8 and 3.45 GeV for the purpose of this calculation. For the hindered M1 transition amplitudes, one may reasonably expect the neglected terms to be comparable to those so far computed.

The final results for E1 and M1 transition rates are displayed in Table 8. Compared with the results of the potential model (Tables 4, 5), the E1 rates show a modest improvement, though they are still one to two standard deviations from experiment. Once again the M1 rates bear no resemblance to those observed for $\psi' \to \gamma\chi(3455)$, $\chi(3455) \to \gamma\psi$, and $\psi \to \gamma X(2830)$. Taking this together with the unexpectedly large hyperfine splittings, there can no longer be any doubt that something very important is missing from the charmonium model *or* that the identification of these states as hyperfine partners of ψ' and ψ is wrong.

3.5.4 CHARMED MESON PRODUCTION IN e^+e^- ANNIHILATION The essence of the model for ΔR is that charmed meson production is a quasi-two-body process mediated by those $c\bar{c}$ states that couple to the photon. It is thus in the spirit of vector-meson dominance (97), generalized to include coupled-channel mixing. The quasi-two-body hypothesis, which has proven to be correct for c.m. energy $W \lesssim 4.4$ GeV, implies that charmed meson spectroscopy can be readily and accurately carried out by studying the invariant mass recoiling against D (or F) in the reaction $e^+e^- \to D(F) + \text{anything}$.

For a charmed quark charge of $\frac{2}{3}$, ΔR is given by (68)

$$\Delta R(W) = \frac{32\pi}{W^2} \sum_{m,n} \Psi_m(0) \left[\frac{\mathscr{G}^\dagger(\Omega^\dagger - \Omega)\mathscr{G}}{2i} \right]_{mn} \Psi_n(0), \qquad 3.57$$

where the quantity in brackets is the absorptive part of the Green's function in Equation 3.54. The $\Psi_n(0)$ are the wave functions at zero $c\bar{c}$ separation of the discrete-sector states. Since Ω is really a sum over the allowed continuum channel types, $v = D^0\bar{D}^0, D^+D^-$, etc, ΔR may be written as a sum over exclusive-channel ratios, R_v. Since $\Psi(0) \neq 0$ only for S-wave states in this model,

$$R_v(W) = \frac{32\pi}{W^2} \sum_{m,n} \sum_{l,l'=0,2} \Psi_{m0}(0)\, \mathscr{G}^*_{m0,m'l}\Gamma^v_{m'l,n'l'}\, \mathscr{G}_{n'l',n0}\Psi_{n0}(0), \qquad 3.58$$

where the orbital quantum number $l,l' = 0,2$ has been made explicit. Equation 3.49 defines $\Gamma^v_{m'l,n'l'}$, and it is the only factor in Equation 3.58 that varies from one channel to the next—through its dependence on the momentum p_v, the intrinsic angular momenta, and the constituent quark masses of the outgoing charmed mesons (see Equation 3.53).

The reason for including $^3D_1$ levels, l or $l' = 2$, in the orbital sum in R_v is this: As mentioned, the only dependence on total quark spin in the calculation of Ω and $\mathscr{G}$ enters through the use of spin-split thresholds. This induces a mixing between nearby S and D states that can emerge as a D-state resonance pole in off-diagonal elements such as $\mathscr{G}_{20,12}$. This mixing is strongest when a $^3D_1$ pole sits in the middle of a set of spin-

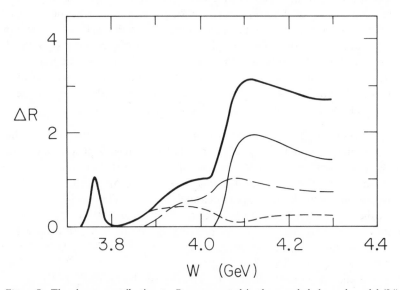

Figure 7 The charm contribution to R as computed in the coupled-channel model (94). The heavy solid curve is the sum of the contributions from D$\bar{\text{D}}$ (*short-dashed curve*), D$\bar{\text{D}}$*+D*$\bar{\text{D}}$ (*long-dashed*), and D*$\bar{\text{D}}$* (*light solid*); F-meson production makes a negligible contribution. The thresholds used are given in the text; other parameters are very similar to those in Table 7.

split thresholds (e.g. at $W \approx 3.8$ GeV), and is considerably weaker when it is far from such thresholds (so that, for example, D and D* look degenerate).

Figure 7 shows a graph of ΔR taken from Reference 94. Completed some time before charmed meson masses were accurately measured, it assumed the thresholds $W_{DD} = 3.730$, $W_{DD*} = 3.885$, $W_{D*D*} = 4.040$, $W_{FF} = 4.00$, $W_{FF*} = 4.15$, and $W_{F*F*} = 4.30$ GeV. For comparison, the most recent data from the various collaborations at SPEAR (31, 32) and at DORIS (33, 34) is shown in Figures 8–11.

The prediction of the parameters of $\psi''(3772)$ more than a year before its discovery must be regarded as the greatest success of the coupled-channel model, especially in view of the fact that all attempts so far to understand the spin-dependent forces in charmonium have failed to give the requisite $2^3S_1 - 1^3D_1$ mixing by more than an order of magnitude.

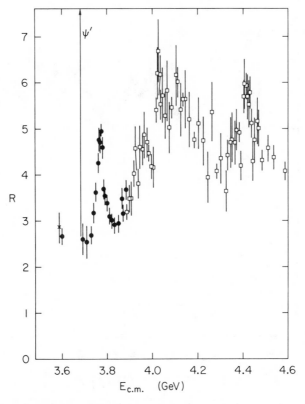

Figure 8 R in the charm threshold region as measured by the SLAC-LBL collaboration (31) at SLAC. Radiative corrections have been applied.

Typical predictions of Γ_e for this state based on the tensor force in a Breit Hamiltonian are ~ 20 eV. The predicted mass and hadronic width of ψ'' agree, within errors, with the measured values (see Table 1). The predicted electronic width of about 150 eV is 2.5 times smaller than that reported by the SLAC-LBL collaboration (370 ± 90 eV) (90) while nearly the same as that measured by the DELCO group (180 ± 45 eV) (91). As we discuss in Section 4, the most important feature of ψ'' is that it decays exclusively to $D\bar{D}$, providing a unique, high-precision setting in which to study these mesons.

Comparison of the theoretical curve with R data for energies W between 3.8 and 4.2 GeV shows only qualitative agreement between the two.

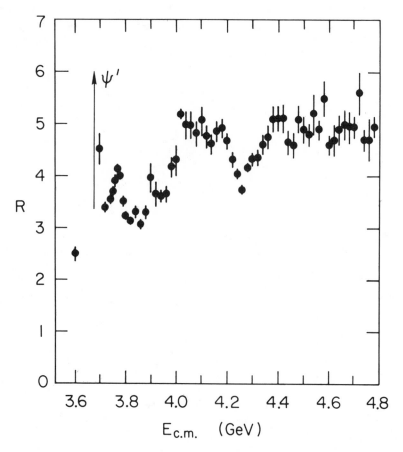

Figure 9 R as measured by the DELCO collaboration (32) at SLAC. Errors shown are statistical and the vertical scale may have an overall systematic uncertainty of $\pm20\%$.

Points of agreement include: (*a*) The dip in ΔR to zero near 3.8 GeV due, in the model, to the vector-meson dominated production; (*b*) The rise in ΔR near 3.95 GeV. This is the $D\bar{D}^*$ threshold in the model calculation,

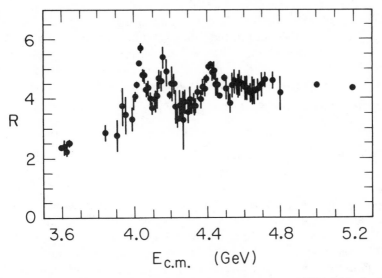

Figure 10   R as measured by the DASP collaboration (33) at DESY. The absolute normalization is estimated to have a $\pm 10\%$ systematic error.

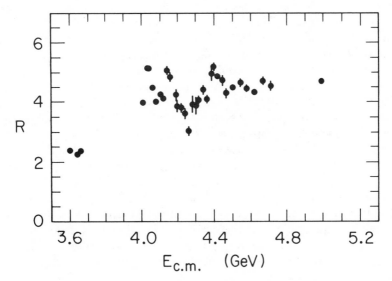

Figure 11   R as measured by the PLUTO collaboration (34) at DESY. Errors shown are statistical only.

but there ΔR (≈ 1) is only about one half the measured value;[15] (c) The sharp rise shown in Figure 7 is due to the concurrence of the important $D^*\bar{D}^*$ threshold with the 3^3S_1 charmonium level—the corresponding rise in the data, culminating in $\psi(4028)$, is considerably sharper ($\Delta R \approx 3$ in about 30 MeV); and (d) The dip in exclusive $D\bar{D}$ production at this resonance is due to a zero in the $3^3S_1 \to D\bar{D}$ form factor near $p_D \approx 750$ MeV. This striking prediction of the model (68, 94, 98) is confirmed experimentally. Study of the mass distribution recoiling against observed D's at 4.028 GeV gives the relative exclusive-channel ratios as (99):

$$R(D^{*0}\bar{D}^{*0}): R(D^{*0}\bar{D}^0 + D^0\bar{D}^{*0}): R(D^0\bar{D}^0)$$

$$= 1.00 \pm 0.10: 0.85 \pm 0.09: 0.10 \pm 0.06 \qquad 3.59$$

This preference of $\psi(4028)$ for $D^*\bar{D}^*$, despite the limited phase space, has been interpreted by some authors (77, 100) as an indication that $\psi(4028)$ is an almost bound state of these two mesons—a $D^*\bar{D}^*$ molecule. It is difficult to test this rather ad hoc hypothesis because no model of such objects exists that can be relied upon for further predictions. In the meantime, the existence of a near zero in $D\bar{D}$ production near 4.028 GeV can be tested by careful study of this region, and will further establish the notion of quarks through the observation of a node in their bound-state wave functions.

Above $W \approx 4.1$ GeV, the model calculation breaks down badly, and bears little resemblance to the data. In particular, the enhancement near 4.15 GeV, the dip at 4.3 GeV, and the obvious resonance $\psi(4414)$ are all beyond the reach of the model as presently constituted. If $\psi(4028)$ is indeed the (highly distorted) 3^3S_1 charmonium level, then the spectroscopy of the naive potential model would lead one to interpret the enhancement at 4150 MeV as the 2^3D_1 state and the resonance at 4414 MeV as the 4^3S_1 state. But this is perhaps pushing the naive model too far. Most of its assumptions are questionable for such high excitations, and even the nonrelativistic spectroscopic notation may be meaningless.

Finally, it should be mentioned that several other models predict states in this region beyond those expected in the linear potential model. To mention two examples: (a) The model of Giles & Tye (78), in which the $c\bar{c}$ pair is bound by a string with dynamical degrees of freedom, expects a number of levels corresponding to vibrational excitations of the string. No prediction is made for the leptonic width of these new states, so that their observability is an open question. (b) The (essentially) logarithmic potential proposed by a number of authors (86, 101, 102), has a greater

---

[15] To extract ΔR from experiment, one should subtract from R a constant background of about 2.2 due to production of noncharmed hadrons plus the contribution of τ production, which is approximately $(1 + 2m_\tau^2/W^2)(1 - 4m_\tau^2/W^2)^{\frac{1}{2}}$ at c.m. energy W.

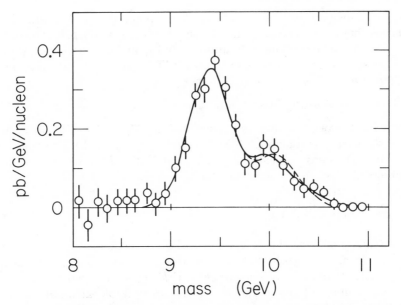

Figure 12   The $\mu^+\mu^-$ invariant mass spectrum (background subtracted) in the Υ region (23). The solid curve is a fit to two zero-width resonances (smeared by resolution); the dashed curve is a fit to three resonances.

level density than does the linear model, and predicts the 4^3S_1 and 5^3S_1 levels at 4.25 and 4.41 GeV, respectively.

3.6   *Beyond Charmonium*

The recent discovery at Fermilab of enhancements in the μ-pair invariant mass near $M_{\mu^+\mu^-} \approx 10$ GeV (23) is widely interpreted as solid evidence for the existence of a new quark, Q, with mass $m_Q \simeq 5$ GeV. The data (Figure 12) shows clearly two, and possibly three, resonances, called Υ, Υ', and Υ''. Assuming these to have zero width (consistent with the experimental resolution of ~ 200 MeV), a fit to the data gives their masses as

$$M_\Upsilon = 9.40 \pm 0.013 \text{ GeV}$$
$$M_{\Upsilon'} = 10.01 \pm 0.04 \text{ GeV}$$
$$M_{\Upsilon''} = 10.40 \pm 0.12 \text{ GeV}. \qquad\qquad 3.60$$

While the statistical significance of Υ'' is still not large, the obvious interpretation is that these are the ground and first two radially excited $^3S_1$ states of a $Q\bar{Q}$ system.[16]

---

[16] We do not discuss here the possibility that Υ and Υ' are the ground states of two distinct, but nearly degenerate, $Q\bar{Q}$. See, for example, Cahn & Ellis (103).

Since they are narrow enough to have appreciable $\mu^+\mu^-$ branching ratios, they must all lie below the threshold for decay into a pair of mesons $Q\bar{q} + \bar{Q}q$ containing one new and one old quark $(q = u,d)$ each. Two other striking features of these states readily inferred from the data are these: (a) The $2^3S - 1^3S$ mass difference

$$M_{\Upsilon'} - M_{\Upsilon} = 610 \pm 50 \text{ MeV} \qquad 3.61$$

is, within errors, the same as in charmonium, $M_{\psi'} - M_{\psi} = 589$ MeV. As we shall see, this is about 150 MeV larger than expected if one adheres to the "standard" potential V_0 in Equation 3.8. (b) The ratio of the observed $\mu^+\mu^-$ signals at Υ' and Υ is

$$R_\mu = \frac{B(\Upsilon' \to \mu^+\mu^-) \, d\sigma/dy}{B(\Upsilon \to \mu^+\mu^-) \, d\sigma/dy}\bigg|_{y=0} = 0.37 \pm 0.04. \qquad 3.62$$

The corresponding value of R_μ for ψ' and ψ production is about 0.02 (59). This strongly suggests (103) that $B(\Upsilon' \to \mu^+\mu^-) \gg B(\psi' \to \mu^+\mu^-)$ and, therefore, that $\Gamma(\Upsilon' \to \Upsilon + \text{anything}) \ll \Gamma(\psi' \to \psi + \text{anything}) \approx 130$ keV. These features of the Υ system receive considerable attention in the following discussion.

It hardly need be emphasized that bound systems of quarks heavier than charm will provide critical tests of the foundations of the charmonium model—gluon counting and the use of a nonrelativistic potential. Furthermore, the observed spectrum and branching ratios for radiative and direct decays will sharpen our knowledge of the form of this potential, since it is expected to be largely independent of m_Q. And, of course, the relative strength of radiative and leptonic decays to purely hadronic ones will help determine the new quark's charge.

These issues and more have already sparked considerable theoretical interest in the Υ system where, as we just mentioned, the prediction (104) of the standard potential for $M_{\Upsilon'} - M_{\Upsilon}$ appears to have failed. But a complete test of the form of the potential requires a comparison with experiment of its expectations for the myriad of branching ratios and absolute widths, as well as the details of the spectrum accessible only to e^+e^- storage ring experiments. And preliminary to making meaningful predictions, one must decide the relative positions of the ground state $Q\bar{Q}$ and the threshold for OZI-allowed decays. Only then can one know in a given model how many states are bound (i.e. narrow) and what transitions among these should be observed.

Eichten & Gottfried (104) have addressed the question of the m_Q dependence of the threshold and ground-state energies. While their arguments are, strictly speaking, valid only in the $m_Q \to \infty$ limit, errors should be small so long as $m_Q \gg m_q = m_{u,d,s}$, which is already true for

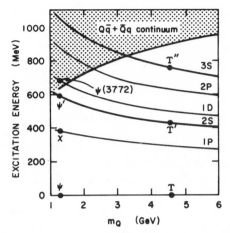

Figure 13   Predicted excitation spectrum of $Q\overline{Q}$ levels and the threshold for OZI-allowed decays, as a function of the heavy quark mass, m_Q, in the linear + Coulomb potential model (104) with $m_c = 1.37$ GeV.

the charmed quark. In the $Q\bar{q}$ system, for example, the reduced mass $\mu = m_Q m_q/(m_Q + m_q) \approx m_q$, and the dynamics is essentially independent of m_Q. The mass of the $Q\bar{q}$ ground-state pseudoscalar is then the sum of $m_Q + m_q$, the binding energy (a function of m_q only), and a correction due to the $^3S - {^1S}$ hyperfine splitting. This last term decreases like m_Q^{-1} in a heavy-light system, so that the threshold energy W_Q may be written as

$$W_Q = 2\left[M_D + m_Q - m_c + \frac{3}{4}\left(1 - \frac{m_c}{m_Q}\right)\left(M_{D^*} - M_D\right)\right].$$ 3.63

In the $Q\overline{Q}$ system, the mass M_1 of the ^3S ground state is

$$M_1 = E_1(Q\overline{Q}) + E_0(m_Q),$$ 3.64

where E_1 is the ground state eigenvalue and E_0 is the zero of energy, whose definition is not completely obvious in an infinitely rising potential such as V_0. Writing

$$E_0 = 2m_Q + \Delta(m_Q),$$ 3.65

all that is known about Δ is (*a*) $\Delta(m_c) \approx -205$ MeV and (*b*) $m_Q^{-1}\Delta(m_Q) \to 0$ as $m_Q \to \infty$. Eichten & Gottfried interpolate[17] with $\Delta(m_Q) = \Delta(m_c)m_c/m_Q$. Using the potential V_0 in Equation 3.8, with essentially the same parameters as in Equation 3.9, Eichten & Gottfried computed the excitation spectrum shown in Figure 13. Using Equations 3.63 to 3.65, the quark

[17] Quigg and Rosner (105) have recently argued that the number of bound $^3S_1$ levels grows approximately as $2(m_Q/m_c)^{\frac{1}{2}}$.

mass appropriate to the Υ system is $m_Q = 4.6$ GeV, and the threshold for OZI-allowed decays is 900 MeV above the Υ. Thus, they predicted three bound ^3S states, as seems to be the case, with masses

$$M_{\Upsilon'} - M_\Upsilon \approx 450 \text{ MeV}, \qquad M_{\Upsilon''} - M_\Upsilon \approx 750 \text{ MeV},$$

$$M(Q\bar{u}; 1^1S_0) \approx 5.16 \text{ GeV} \qquad (\text{for } m_c = 1.37 \text{ GeV}),$$

$$M(Q\bar{u}; 1^3S_1) - M(Q\bar{u}; 1^1S_0) \approx 100 \text{ MeV}. \qquad 3.66$$

One possible explanation for the apparent failure of the highly motivated linear + Coulomb model to predict correctly the $\Upsilon' - \Upsilon$ and $\Upsilon'' - \Upsilon$ separations is this: While good arguments exist for both the small and large r behavior of the potential, it may well be that the systems under investigation, ψ and Υ, see mainly an intermediate-range portion of the $Q\bar{Q}$ potential. This may be neither linear nor Coulomb and, in any case, no arguments exist that give a clue to its shape.

Whatever the reason for failure, the large $\Upsilon' - \Upsilon$ mass difference has caused renewed interest in alternative forms of the potential. One choice, the logarithmic potential, studied some time ago by Machacek & Tomozawa (101) and more recently by Quigg & Rosner (102), is currently in vogue because of its peculiar property that the $Q\bar{Q}$ excitation spectrum is independent of the Q mass. Thus, by fitting to the ψ system, Quigg & Rosner find

$$V_1(r) = 0.733 \text{ GeV} \ln (r/r_0) \qquad 3.67$$

with r_0 an arbitrary constant, and $m_c = 1.1$ GeV. The spectrum of the first few excited $Q\bar{Q}$ levels may be found from Figure 13 by drawing horizontal lines through the dots corresponding to members of the charmonium family. The predictions of the two models for charmonium start to deviate around the 3^3S_1 state; in the logarithmic potential,

$$M(c\bar{c}; 3^3S_1) = 4.03 \text{ GeV}, \qquad M(c\bar{c}; 4^3S_1) = 4.25 \text{ GeV} \qquad 3.68$$

compared to 4.17 and 4.6 GeV, respectively, in the V_0 model. The separation between the ground state and OZI threshold will not be the same as in the Eichten-Gottfried calculation because quark masses appropriate to V_1 will differ somewhat from those for V_0. In particular, Quigg & Rosner find $m_Q \cong 4.5$ GeV so that, using Equation 3.63, they predict

$$M(Q\bar{u}; 1^1S_0) = 5.33 \text{ GeV} \qquad (\text{for } m_c = 1.1 \text{ GeV}) \qquad 3.69$$

and that three to four ^3S states will be bound.

Celmaster, Georgi & Machacek (86) have proposed still another potential inspired by the linear + Coulomb model:

$$V_2 = \frac{-8\pi}{27r}\left(\ln\frac{1}{r\Lambda e^\gamma}\right)^{-1} + \frac{(1-e^{-Ar})r}{2\pi} + E_0;$$

$$\Lambda = 0.50\ \text{GeV}, \qquad \gamma = 0.577, \qquad A = 0.16\ \text{GeV}, \qquad E_0 = 0.39\ \text{GeV}.$$

3.70

The logarithm and coefficient in the short-range part of V_2 is motivated by appeal to asymptotic freedom (however, the argument of the logarithm has been modified). The linear potential strength, $(2\pi)^{-1}$ $(\text{GeV})^2$, is taken from the slope of the Regge trajectory. The new, intermediate-range part, $(r/2\pi)\exp(-Ar)$, is chosen arbitrarily. The parameters in Equation 3.68 are determined by fitting to the spectra of light as well as heavy ($c\bar{c}$) mesons—a very questionable procedure for any nonrelativistic potential. The charmed quark mass that results is $m_c = 1.98$ GeV. With all parameters determined (including the zero of energy, E_0, which is assumed m_Q independent), the Υ mass fixes $m_Q = 5.4$ GeV. This, in turn, leads to

$$M(Q\bar{u}; 1^1S_0) = 5.35\ \text{GeV} \qquad (\text{for } m_c = 1.98\ \text{GeV})$$

3.71

and the prediction that three to four ^3S levels will be bound below OZI threshold.

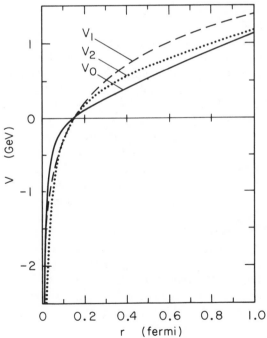

Figure 14   The $Q\bar{Q}$ potentials V_0 (68), V_1 (102), and V_2 (86). The zeros of energy have been chosen to make them cross at the same value of r.

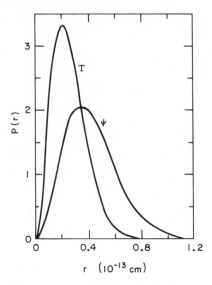

Figure 15 Radial probability for the ψ and Υ ground states, computed from V_0 (68), as a function of the $Q\bar{Q}$ separation r. We thank K. Gottfried for the use of his figures.

The potentials V_0, V_1, V_2 are plotted in Figure 14, and the radial probability $P(r) \, dr$ for the $\psi(1\,^3S_1)$ and $\Upsilon(1\,^3S_1)$ states as determined by V_0 is shown in Figure 15. Several remarks immediately follow from these figures: (*a*) V_1 and V_2 are practically congruent, and so the level spacing determined by V_2 is almost independent of m_Q and fits well what is known about the Υ spectrum. There will be small differences in the wave functions, hence the rates, predicted by the two models because of the different values of m_Q. (*b*) All three potentials are nearly congruent in the region in which the ψ wave function is large, $0.2 \lesssim r \lesssim 0.6$ fm. Thus, it is not surprising that all three give the same spectrum of low-lying $c\bar{c}$ levels and roughly comparable strengths for radiative decays. (*c*) Over most of the region in which the Υ wave function is concentrated, $0.1 \lesssim r \lesssim 0.3$ fm, there is considerable difference between the shape of V_0 and of $V_{1,2}$. It is not surprising, therefore, that the V_0 spectrum is quantitatively different from the V_1, V_2 spectra here.

The predictions in References 86, 102, and 104 for Υ' and Υ branching ratios and total widths are listed in Table 9. To estimate $\Upsilon' \to \Upsilon\pi\pi + \Upsilon\eta$, we have used Gottfried's scaling relation (71)

$$\Gamma(\Upsilon' \to \Upsilon\pi\pi + \Upsilon\eta) \approx \left(\frac{m_c}{m_Q}\right)^2 \Gamma(\psi' \to \psi\pi\pi + \psi\eta) \qquad 3.72$$

Table 9 Predicted Υ and Υ' transition rates (86, 102, 104)

Particle	Mode	V_0		V_1		V_2	
		$\lvert e_Q \rvert = \tfrac{1}{3}$	$\tfrac{2}{3}$	$\tfrac{1}{3}$	$\tfrac{2}{3}$	$\tfrac{1}{3}$	$\tfrac{2}{3}$
Υ'	$l^+ l^- (l = e, \mu, \tau)$	0.45	1.8	0.5	1.9	1	4
	$\sum_J (\gamma + 1\,^3P_J)$	6.8	27	10	40	8.8	35
	$\Upsilon \pi\pi + \Upsilon \eta$	~12	~12	~12	~12	~12	~12
	Hadrons (direct)	8.6	8.6	9.1	9.1	19	19
	Hadrons (2nd order E.M.)	1.9	7.6	2.0	8.0	4.2	17
	All	~31	~60	~35	~75	~47	~95
Υ	$l^+ l^- (l = e, \mu, \tau)$	0.7	2.7	1.1	4.3	2.5	10
	Hadrons (direct)	14.5	14.5	23	23	52	52
	Hadrons (2nd order E.M.)	2.8	11.4	4.5	18	10	40
	All	~19	~34	~31	~54	~70	~122

with $(m_c/m_Q)^2 \approx 1/10$. Since cascade radiative decays will constitute only a very small part of $\Upsilon' \to \Upsilon$ transitions, the suppression in Equation 3.72 goes a long way toward explaining the unexpectedly large value of R_μ in Equation 3.62.

While very little is known of the details of the Υ system at present, we can look forward in the next few years to a flood of data from the new $e^+ e^-$ storage rings at Cornell, DESY, and SLAC. It can only be hoped that out of all this will come a clearer picture of the "correct" phenomenological potential for heavy quark binding and, indeed, of the foundations of the charmonium model. Beyond this, we need a better understanding of the corrections to the naive model due to quark spin and the inevitable presence of light quark and gluonic degrees of freedom. The experimental study of $Q\bar{Q}$ systems is essential, but equally so is progress in understanding the structure of quantum chromodynamics itself—a theory still very much in its infancy.

4 CHARMED HADRONS

Direct evidence for the existence of charmed particles was announced in the summer of 1976, a year and a half after discovery of the ψ/J (22). For an early discussion of charm phenomenology that anticipated many of the recent discoveries, the reader is referred to the paper of Gaillard, Lee & Rosner (106). The properties of the charmed hadrons observed so far are shown in Table 10. Several other recent theoretical and experimental reviews are available (59, 107, 108).

4.1 *Theory*

4.1.1 SPECTROSCOPY

Since all the charmed particles contain one or more light quarks, relativistic motion will very likely make them much more difficult to treat with a potential model than charmonium. A detailed discussion of attempts at light quark dynamics falls outside the scope of our paper. The point we want to make here is that nothing terribly surprising seems to be going on. The experiments have largely confirmed the qualitative expectations of the charm model.

Table 10 Charmed particle properties: All data is taken from Reference 108, which cites original work; quark content is shown below particle name; D^{*+} branching ratios are estimated (108) as described in the text for $m_u/m_c = m_\rho/m_\psi$ (parenthetical numbers are for $m_u/m_c = 0$)

Particle	$I(J^P)$	Mass (MeV)	Decay mode	Branching fraction (%)
D^0 (cū)	$\frac{1}{2}(0^-)$	1863.3 ± 0.9	$K^- \pi^+$	2.2 ± 0.6
			$\bar{K}^0 \pi^+ \pi^-$	4.0 ± 1.3
			$K^- \pi^+ \pi^0$	12 ± 6
			$K^- \pi^+ \pi^- \pi^+$	3.2 ± 1.1
			$e^+ \nu_e + \text{hadrons}$	~ 10
D^+ (cd̄)	$\frac{1}{2}(0^-)$	1868.3 ± 0.9	$\bar{K}^0 \pi^+$	1.5 ± 0.6
			$K^- \pi^+ \pi^+$	3.9 ± 1.0
			$e^+ \nu_e + \text{hadrons}$	~ 10
		$\delta = M_{D^+} - M_{D^0}$ $= 5.0 \pm 0.8$		
D^{*0}	$\frac{1}{2}(1^-)$	2006 ± 1.5	$D^0 \pi^0$	55 ± 15
			$D^0 \gamma$	45 ± 15
D^{*+}	$\frac{1}{2}(1^-)$	2008.6 ± 1.0	$D^0 \pi^+$	$68 \pm 8 \ (63 \pm 9)$
			$D^+ \pi^0$	$30 \pm 8 \ (27 \pm 7)$
		$\delta^* = M_{D^{*+}} - M_{D^{*0}}$ $= 2.6 \pm 1.8$ $\delta - \delta^*$ $= 2.4 \pm 2.4$	$D^+ \gamma$	$2 \pm 1 \ (10 \pm 5)$
F^+ (cs̄)	$0(0^-)$	2039.5 ± 1.0	$\eta \pi^+$	?
			$K^+ K^- \pi^+$	?
			$K^+ \bar{K}^0$	?
			$K^+ K^- \pi^+ \pi^- \pi^+$	?
F^{*+}	$0(1^-)$	2140 ± 60	$F^+ \gamma$	100
Λ_c^+ [c(ud)$_A$]	$0(\frac{1}{2}^+)$	2260 ± 10	$\Lambda \pi^+ \pi^+ \pi^-$	?
Σ_c^{*++} [c(ud)$_S$]	$1(\frac{3}{2}^+)$	2426 ± 12	$\Lambda_c^+ \pi^+$	?

The masses in Table 10 can, in fact, be qualitatively understood in the most naive form of the nonrelativistic quark model by adding the appropriate masses:

$$m_c = \tfrac{1}{2}M_\psi \approx 1.55 \text{ GeV} \qquad m_u = m_d \approx 0.33 \text{ GeV} \qquad m_s \approx 0.46 \text{ GeV}$$

4.1

This reproduces the pseudoscalar and $J^P = \tfrac{1}{2}^+$ baryon masses reasonably well and the vector mesons require roughly an extra 150 MeV hyperfine energy. Given the observed masses $M_D \approx 1.865$ GeV and $M_{D*} \approx 2.005$ GeV, a better estimate is

$$M_{F*} = M_{D*} + M_\phi - M_{K*} = 2.13 \text{ GeV}$$
$$M_F = M_{F*} - (M_{D*} - M_D) = 1.99 \text{ GeV}.$$

4.2

The reported F and F* masses in Table 10 are reproduced nicely by this sum rule.

Mass splittings between D^0 and D^+ and between D^{*0} and D^{*+} have been measured (108) and are also interesting theoretically. In fact, because the $D^* - D$ mass difference leads to an extremely small Q value for $D^* \to D\pi$, the $D^0 - D^+$ and $D^{*0} - D^{*+}$ splittings have important experimental consequences in sorting out the spectroscopy of the D's. In the nonrelativistic quark model, the splitting is the sum of the down-up mass difference and a contribution from single-photon exchange:

$$M_{D^+} - M_{D^0} = m_d - m_u + \frac{2}{3}\alpha \left[\left\langle \frac{1}{r} \right\rangle_D + \frac{2\pi}{m_c m_u} \left| \Psi_D(0) \right|^2 \right]$$

4.3

$$M_{D^{*+}} - M_{D^{*0}} = m_d - m_u + \frac{2}{3}\alpha \left[\left\langle \frac{1}{r} \right\rangle_D - \frac{2\pi}{3m_c m_u} \left| \Psi_D(0) \right|^2 \right].$$

These expressions can be evaluated using a current algebraic extraction of $m_d - m_u$ (109) and an atomic quark model analogous to charmonium for the D mesons. The result is $M_{D^+} - M_{D^0} \approx 7.0$ MeV and $M_{D^{*+}} - M_{D^{*0}} \approx 6.5$ MeV to be compared with the experimental values of 5.0 ± 0.8 MeV and 2.6 ± 1.8 MeV, respectively. An alternative estimate of these splittings using the Massachusetts Institute of Technology (MIT) Bag model (92, 110) gives essentially the same result.

In addition to the S-wave charmed mesons, P states should exist as well. Their masses have been estimated by Eichten et al (68) to be

$$M_D(1^3P_0) = 2.44 \text{ GeV} \qquad M_D(1^3P_1) = 2.58 \text{ GeV}$$
$$M_D(1^1P_1) = 2.45 \text{ GeV} \qquad M_D(1^3P_2) = 2.58 \text{ GeV}.$$

4.4

The evidence for these states (as well as the beautiful measurements of the D and F masses) are discussed shortly.

4.1.2 DECAYS In the standard WS-GIM model (4, 7), the hadronic current Equation 1.7 leads to the selection rule $\Delta C = \Delta S = \pm 1$ for charm decays. Ignoring QCD renormalization corrections, the complete effective Hamiltonian for charm decays in this model is

$$\mathcal{H}_{\Delta C} = 2^{-\frac{1}{2}} G_F \left[\cos \theta_C \, \bar{c} \gamma_\lambda (1 - \gamma_5) s - \sin \theta_C \, \bar{c} \gamma_\lambda (1 - \gamma_5) d \right]$$
$$\times \left[\cos \theta_C \, \bar{d} \gamma^\lambda (1 - \gamma_5) u + \sin \theta_C \, \bar{s} \gamma^\lambda (1 - \gamma_5) u \right. \qquad 4.5$$
$$\left. + \sum_{l = e, \mu} \bar{l} \gamma^\lambda (1 - \gamma_5) v_l \right] + \text{h.c.}$$

If one naively assumes that charmed hadron decays are processes in which only the constituent c quark participates, one estimates from Equation 4.5 the following relative rates (ignoring questions of phase space):

$$c \to s u \bar{d} = 3 \cos^4 \theta_C$$
$$c \to d u \bar{d} = 3 \cos^2 \theta_C \sin^2 \theta_C$$
$$c \to s u \bar{s} = 3 \cos^2 \theta_C \sin^2 \theta_C$$
$$c \to d u \bar{s} = 3 \sin^4 \theta_C$$
$$c \to s l^+ v_l = \cos^2 \theta_C \qquad (l = e, \mu)$$
$$c \to d l^+ v_l = \sin^2 \theta_C \qquad (l = e, \mu). \qquad 4.6$$

The factor of 3 is due to a sum over the color of the quarks in the non-charmed piece of the hadronic current. Ignoring all but $\Delta C = \Delta S$ transitions, one expects D (and F) decays will be 60% nonleptonic and 40% semileptonic, divided equally between e and μ. All of these will involve a single kaon, which provides the outstanding signal for the presence of charm. Note, in particular, that $\Delta C = \Delta S$ implies that one should see a D^+ signal in the exotic channel $K^- \pi^+ \pi^+$, but not in the nonexotic $K^+ \pi^+ \pi^-$. Finally, the D lifetime is estimated in this model to be

$$\tau(D \to \text{all}) = \tfrac{1}{5} (m_\mu / M_D)^5 \, \tau(\mu \to e v \bar{v}) \approx 10^{-13} \text{ sec.} \qquad 4.7$$

Thus visual observation of D decays can be made only with high resolution techniques using emulsions or streamer chambers (111).

More detailed studies of charmed meson decays have been made by Einhorn & Quigg (112) and by Ellis et al (113), as well as by several other groups (114). These are based on the analyses of the operator structure of the nonleptonic weak Hamiltonian carried out by Gaillard & Lee (46) and by Altarelli & Maiani (46). The nonleptonic Hamiltonian is found to consist of two pieces, one transforming as the [20] representation under SU(4) (flavor), the other as [84]. When decomposed with respect to SU(3) subgroups (the symmetry group of u, d, s quarks), one finds

$$[20] = 6 \oplus 8 \oplus 6^*$$

$$[84] = 6 \oplus \{3 \oplus 15_M\} + 6^* \oplus \{3^* \oplus 15_M^*\} + \{1 \oplus 8 \oplus 27\}$$

4.8

with square brackets used to distinguish representations of SU(4) from those of SU(3), and the subscript M denoting a representation of mixed symmetry. The octet in the decomposition of $[20]$ is the $\Delta C = 0, |\Delta S| = 1$ operator responsible for nonleptonic K decay; its matrix elements are enhanced relative to the octet and 27-plet in $[84]$.

On this basis, Einhorn & Quigg argued that the $\Delta C = \pm 1$ pieces, 6 and 6^* in $[20]$ should have enhanced matrix elements relative to the $\Delta C = \pm 1$ parts of $[84]$, namely $3 \oplus 15_M + 3^* \oplus 15_M^*$. (Actually, only $15_M \oplus 15_M^*$ appear in the Hamiltonian.) Now, part of this octet enhancement is due to the sign of anomalous dimensions appearing in the operator product expansion (41), while an appreciable further enhancement is due to incalculable matrix elements of the operator, i.e. it is of uncertain origin. Furthermore, some of the octet enhancement arises from the choice of renormalization point in the evaluation of the anomalous dimensions of the operators; this was taken to be 1 GeV for K decay (46), and assumed to be the same by Einhorn & Quigg (112). Ellis et al argue that this renormalization point should be taken higher when dealing with decays of charmed hadrons, say $m_c \sim 2$ GeV. This, they claim, diminishes sextet enhancement of the $|\Delta C| = 1$ Hamiltonian.

Now, all of this has measurable consequences. Under the reasonable assumption that decays of high-mass states such as D and F are quasi-two-body, Einhorn & Quigg point out that sextet enhancement implies that D^+ has no Cabibbo-enhanced decays ($\propto \cos^4 \theta_c$) to a pair of pseudoscalars (such as $\bar{K}^0 \pi^+$) or a pair of vectors (like $\bar{K}^{*0} \rho^+$). The only Cabibbo-enhanced decays of D^+ then would be to a pseudoscalar plus a vector, say $D^+ \to \bar{K}^0 \rho^+ \to \bar{K}^0 \pi^+ \pi^0$ and $D^+ \to \bar{K}^{*0} \pi^+ \to K^- \pi^+ \pi^+$ or $\bar{K}^0 \pi^0 \pi^+$.

To the contrary, Ellis et al find $\Gamma(D^+ \to \bar{K}^0 \pi^+) \approx \Gamma(D^0 \to K^- \pi^+)$, a "sextet enhanced" rate. Using a variety of techniques, they estimate the following ranges of branching ratios for charmed meson decay:

$$B(D, F \to l + v_l + \text{hadrons}) = 0.1–0.25$$

$$B(D \to l + v_l + K) = 0.03–0.08$$

$$B(F \to l + v_l + \eta) = 0.02–0.05$$

$$B(D^0 \to K^- \pi^+ + \bar{K}^0 \pi^0) = 0.03–0.18$$

$$B(D^+ \to \bar{K}^0 \pi^+) = 0.02–0.10$$

$$B(F^+ \to \eta \pi^+ + K^+ \bar{K}^0) = 0.02–0.12.$$

4.9

Arguments over operator enhancement aside, all authors (112–114) agree that, because of the large number of modes available for decay, no single branching ratio is expected to be more than a few percent.

One other interesting aspect of D decays has to do with the possibility of $D^0\bar{D}^0$ mixing (115). If this is induced by charm-changing neutral currents such as $\bar{c}\gamma_\lambda\frac{1}{2}(1\pm\gamma_5)u$ coupled to the Z^0 weak boson, then

$$\Delta M(D^0,\bar{D}^0) \sim G_F M_D^3 \gg \Gamma(D^0 \to \text{all}) \sim G_F^2 M_D^5$$

and $D^0\bar{D}^0$ mixing will be complete. One then will see $D^0 \to K+\dots$ as often as $D^0 \to \bar{K}+\dots$. If $|\Delta C| = 2$ transitions are mediated by second order (G_F^2) processes or by neutral Higgs bosons, mixing may be less than complete but still appreciable. Thus, a measurement of $\Gamma(D^0 \to K+\dots)/\Gamma(D^0 \to \bar{K}+\dots)$ gives us important information about the structure of weak currents (both charged and neutral) as well as constraints on the couplings of Higgs mesons to quarks.

Charmed baryon decays are considerably more complicated and correspondingly uncertain. The reader is referred to the above papers (and enclosed references) for details beyond the gross estimates one can make from Equations 4.6 and 4.7.

As we hinted earlier, the masses of D's and D*'s are so delicately arranged that they cause an unprecedented complication in sorting out D spectroscopy. The problem is that the Q values for $D^* \to D\pi$ are so small that the electromagnetic (M1) decay $D^* \to D\gamma$ is competitive with the strong one. Therefore, when studying the invariant mass recoiling against D^0 produced in e^+e^- annihilation at 4.028 GeV, say, one sees a very rich structure corresponding to

$$e^+e^- \to D^0\bar{D}^0 \qquad \text{(Recoil mass } M_x = M_{\bar{D}^0})$$

$$e^+e^- \to D^0\bar{D}^{*0} \qquad (M_x = M_{\bar{D}^{*0}})$$

$$e^+e^- \to D^{*0}\bar{D}^0, \quad D^{*0} \to D^0\pi^0 \text{ or } D^0\gamma \qquad (M_x = M_{\bar{D}^0\pi^0} \text{ or } M_{\bar{D}^0\gamma})$$

$$e^+e^- \to D^{*0}\bar{D}^{*0}, \quad D^{*0} \to D^0\pi^0 \text{ or } D^0\gamma \qquad (M_x = M_{\bar{D}^{*0}\pi^0} \text{ or } M_{\bar{D}^{*0}\gamma})$$

$$e^+e^- \to D^{*+}D^-, \quad D^{*+} \to D^0\pi^+ \qquad (M_x = M_{D^-\pi^+})$$

$$e^+e^- \to D^{*+}D^{*-}, \quad D^{*+} \to D^0\pi^+ \qquad (M_x = M_{D^{*-}\pi^+}). \qquad 4.10$$

The decay $D^{*0} \to D^+\pi^-$ is energetically forbidden. The relative strength of each component of the recoil distribution is determined by the product of the exclusive-channel cross section and the appropriate D* branching ratio. So these branching ratios are of great importance in charmed meson spectroscopy.

In addition to their model for calculating exclusive-channel cross sections, Eichten et al (68) have estimated the various D* branching ratios as follows: For the M1 decays they use the naive quark model formula

$$\Gamma[D^*(c\bar{q}) \to D(c\bar{q}) + \gamma] = \frac{4}{3}\alpha \left(\frac{e_c}{2m_c} + \frac{e_q}{2m_q}\right)^2 p^3, \qquad 4.11$$

where $e_c = e_u = \frac{2}{3}$, $e_d = -\frac{1}{3}$, $p = (M_{D^*}^2 - M_D^2)/2M_{D^*}$, and they use quark masses of $m_u = m_d = 0.33$ GeV and $m_c = 1.87$ GeV [determined from $\Gamma(\psi \to e^+e^-)$ and $M_{\psi'} - M_\psi$ in the pure linear potential model].

The $D^* \to D\pi$ width is obtained by assuming a form suggested by their model calculation of $\psi_{nL} \to D\bar{D}$; it is

$$\Gamma(D^* \to D\pi) = \frac{p^3}{72\pi M_{D^*}^2} C^2 |(M_{D^*} E_D E_\pi)^{\frac{1}{2}} A|^2, \qquad 4.12$$

where $E_{D,\pi} = (p^2 + M_{D,\pi}^2)^{\frac{1}{2}}$, C is an isospin Clebsch-Gordan coefficient, and A is an amplitude depending only on m_u in the limit that heavy quark mass $m_c \to \infty$. Assuming further that m_s is large enough so that A can be estimated from $K^* \to K\pi$ decays gives

$$A = 47.8 \text{ GeV}^{-\frac{3}{2}}. \qquad 4.13$$

Using the measured (108) D and D* masses, they obtain the widths and branching ratios listed in Table 11. As we will see shortly, the results in Table 11 are in remarkable agreement with those determined from experiment under much less model-dependent assumptions.

To conclude this discussion, we mention first that the expected (and measured) $F^* - F$ mass difference ($\lesssim M_\pi$) implies the $F^* \to F\gamma$ is the only decay mode of this $C = S = 1$ vector meson. Using $m_s = 0.50$ GeV in Equation 4.11 gives

$$\Gamma(F^{*+} \to F\gamma) = 0.2 \text{ keV}. \qquad 4.14$$

Table 11 Predicted D* widths and branching ratios (68)

Mode	Width (keV)	Branching ratio (%)
$D^{*0} \to D^0\pi^0$	39.7	53.0
$D^{*0} \to D^0\gamma$	35.2	47.0
$D^{*0} \to$ all	74.9	—
$D^{*+} \to D^+\pi^0$	22.2	28.4
$D^{*+} \to D^0\pi^+$	53.4	68.4
$D^{*+} \to D^+\gamma$	2.5	3.2
$D^{*+} \to$ all	78.0	—

Second, the apparent success of the M1 formula, Equation 4.11, for D*
decays stands in sharp contrast to its apparent failure in the charmonium
system.

Finally, it is unfortunate that the formula, Equation 4.12, is unlikely to
be testable in bound systems of a still heavier quark Q with u, d, s. With
the hyperfine splitting decreasing like M_Q^{-1}, the only energetically allowed
decays of possible new mesons M*(Qū) to M(Qū), will be the radiative
ones. Looking on the bright side, this situation will make M, M* spectroscopy a little easier, and—if the M* width can be measured—provide
further tests of the M1 formula.

4.2   Experiment

What follows is a brief discussion of the properties summarized in Table
10. For more detail and reference to experimental work not cited explicitly, the reader is referred to Feldman's recent review (108).

To date, charmed mesons have been identified directly only in e^+e^-
annihilation experiments, where their production cross sections are $\sim 50\%$
of the total and the kinematics is especially simple. The D and D* mesons
have been positively identified by the SLAC-LBL collaboration (22).
We hasten to add that before and since their discovery, there has been
plenty of indirect evidence for charmed mesons in neutrino experiments
(see Section 6) in e^+e^- annihilation at CEA (14), at SLAC (31, 32), and
DESY (33, 34), and in photoproduction at SLAC (see Section 6.3). The
F^+ and F^{*+} (decaying to $F^+\gamma$) were discovered by the double-arm
spectrometer (DASP) collaboration (116) at DESY at c.m. energy 4.414
GeV. And the F^+ is tentatively confirmed by the SLAC-LBL collaboration in data taken at 4.16 GeV (108). All of these discoveries and all of
the precision data on D^0 and D^+ were obtained at the peaks in the
annihilation cross section—3.772, 4.028, 4.16, and 4.414 GeV—which
are charmonium resonances above threshold. [There is one exception:
the beautiful measurement (108) of $M_{D^{*+}} - M_{D^0}$ required high energy,
6.8 GeV, to detect the π^+ in $D^{*+} \to D^0\pi^+$.]

Evidence for charmed baryons comes from two sources. The first is a
single neutrino event in the Brookhaven 7-ft bubble chamber (117):

$$\nu_\mu p \to \mu^- \Lambda \pi^+ \pi^+ \pi^+ \pi^-. \qquad\qquad 4.15$$

This is interpreted as $\nu_\mu p \to \mu^- \Sigma_c^{*++}$; $\Sigma_c^{*++} \to \Lambda_c^+ \pi^+$; $\Lambda_c^+ \to \Lambda \pi^+ \pi^+ \pi^-$
because the event violates the $\Delta S = \Delta Q$ rule. The second comes from a
peak in the inclusive $\bar{\Lambda}\pi^-\pi^-\pi^+$ spectrum at 2.26 GeV in a photoproduction experiment at Fermilab (118). This group also reports evidence for
the sequence $\overline{\Sigma_c^{*0}} \to \Lambda_c^- \pi^+$, with a Σ_c^* mass of 2.43 GeV. These masses
are exactly those determined in the Brookhaven experiment and expected

in the quark model. Very indirect evidence for charmed baryons in e^+e^- annihilation comes from the sharp rise in proton and Λ production in the 4.4–5.0 GeV region (108). But upper limits on cross section times branching ratio to observable modes (σB) are typically an order of magnitude lower than for D production at the same energies. Most interesting in these studies is that Λ production is consistently only 10–15% of proton production at all energies, which suggests that charmed baryons preferentially decay to K + nucleon + ... rather than to strange baryons.

The isospin assignments in the table are made purely on theoretical grounds; no experimental information exists other than the fact that $D^* \to D\pi$ precludes $I = 0$ for both these charmed mesons.[18] Similarly, no measurements of J^P exist for F, F*, and charmed baryons. Assuming that D^0 and D^+ have the same J^P, observation of $D^0 \to K^-\pi^+$ and study of the Dalitz plot for $D^+ \to K^-\pi^+\pi^+$ proves that parity is violated in their decays (119), which suggests that they decay via weak interactions. This fact of parity violation is now obvious (without the assumption of equal J^P) from the observed decay modes of D^0 and D^+. Assuming that the parity of D is -1, observing that $D^* \to D\pi$ and $e^+e^- \to D\bar{D}^*$, and measuring the angular distributions for

$$e^+e^- \to D\bar{D}^*$$
$$\mathrel{\rule[0.5ex]{0pt}{0pt}}\!\longrightarrow \bar{K}\pi \; , \qquad\qquad 4.16$$

the SLAC-LBL collaboration is able to rule out $J_D^P = J_{D^*}^P$ for $J_D^P = 0^\pm$ and $J_D^P = 1^-$, $J_{D^*}^P = 0^-$, whereas they find high confidence levels for the hypothesis $J_D^P = 0^-$, $J_{D^*}^P = 1^-$.

The remarkably precise measurements (108) of the D^0 and D^+ masses come from data taken at the peak of $\psi'' = \psi(3772)$, the 1^3D_1 charmonium level. This accuracy is possible because: (a) ψ'' decays exclusively to $D^0\bar{D}^0$ and D^+D^- ($D\bar{D}^*$ is energetically forbidden), and (b) it lies just about 40 MeV above threshold, so that the D's are moving very slowly. Thus, small errors in the measurement of p_D are unimportant in determining M_D from

$$M_D = [(M_{\psi''}/2)^2 - p_D^2]^{\frac{1}{2}} = (E_b^2 - p_D^2)^{\frac{1}{2}}, \qquad\qquad 4.17$$

where the beam energy E_b is very well measured. Of course, p_D is determined from the momenta of the D-decay products. Figure 16 shows the

---

[18] Of course, the fact that only D^0 and D^+ (with no D^- or D^{++}, say) are observed in the decays $D^{*+} \to D^+\pi^0$, $D^0\pi^+$, and the fact that the D^- is observed to decay to states of positive strangeness, are rather convincing evidence in favor of $I = \frac{1}{2}$ as well as $Q = +\frac{2}{3}$ for the charmed quark.

invariant mass spectra for D^0 and D^+. The clearly visible $\delta = M_{D^+} - M_{D^0} = 5.0 \pm 0.8$ MeV is only slightly less than predicted in References 109 and 110.

The D^{*0} mass is measured by a similar trick at $W = 4.028$ GeV. Here, $e^+e^- \to D^{*0}\overline{D}^{*0}$ is picked out and $p_{D^{*0}}$ is measured as follows: The momentum spectrum of the D^0's detected at $\psi(4.028)$ is measured. This is shown in Figure 17. The small Q value for $D^{*0} \to D^0\pi^0$ makes these

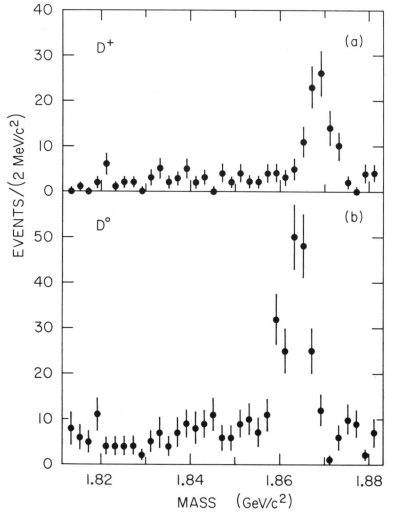

Figure 16 Invariant mass spectra for the sum of all observed (*a*) D^+ and (*b*) D^0 decay modes yielding all-charged-particle final states (108).

components of the distribution (curves B and E in the figure) rather sharply peaked. The lower peak (B) obviously corresponds to $D^{*0}\bar{D}^{*0}$, and a simple kinematical exercise gives the center of this peak as

$$p_D^{ctr}(D^{*0}\bar{D}^{*0}; D^{*0} \rightarrow D^0\pi^0) = p_{D^{*0}}E_{D^0}/M_{D^{*0}}.$$ 4.18

Here, $E_{D^0} = M_{D^0}$, and $M_{D^{*0}} = W/2$ to a very good approximation, so that

$$M_{D^{*0}} = [(W/2)^2 - (p_D^{ctr})^2]^{\frac{1}{2}}$$ 4.19

determines the mass quite accurately. With the $D^{*+} - D^0$ mass difference accurately determined from high energy data as noted above, there results

$$\delta^* = M_{D^{*+}} - M_{D^{*0}} = 2.6 \pm 1.8 \text{ MeV, and } \delta - \delta^* = 2.4 \pm 2.4 \text{ MeV.}$$
4.20

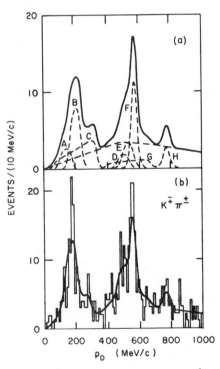

Figure 17 The D^0 momentum spectrum at 4.028 GeV, for $D^0 \rightarrow K^{\mp}\pi^{\pm}$ (from 99). The solid curves represent an isospin-constrained fit to the data. Part (*a*) shows the various contributions to the fit in (*b*). Curves A, B, C are from $e^+e^- \rightarrow D^*\bar{D}^*$ with (A) $D^{*+} \rightarrow D^0\pi^+$, (B) $D^{*0} \rightarrow D^0\pi^0$, and (C) $D^{*0} \rightarrow D^0\gamma$. Curves D, E, F, G are from $D^*\bar{D} + \bar{D}^*D$ production with (D) $D^{*+} \rightarrow D^0\pi^+$, (E) $D^{*0} \rightarrow D^0\pi^0$, (F) direct D^0, and (G) $D^{*0} \rightarrow D^0\gamma$. Curve H is the contribution from $D^0\bar{D}^0$ production.

This is purely an electromagnetic hyperfine splitting and is expected to be ~1 MeV in most theoretical estimates. Finally, the Q values used in constructing Table 11 are shown in Figure 18.

Masses of the other charmed mesons are determined by similar techniques, with the most precise measurements always coming from $e^+e^- \to M_c \bar{M}_c$ where M_c is a charmed meson, $\bar{M}_c$ its antiparticle. These masses are determined in standard ways dictated by the experimental arrangement. And, finally, at $W = 4.4$ GeV, there is some evidence for peaking in the recoil mass distribution against D^0 (99). The peak occurs near $M_x = 2.4$ GeV, possibly corresponding to the quasi-two-body production of charmed P states with D or D*. If we use the masses in Equation 4.4, the process is

$$e^+e^- \to D \, \bar{D}(1^1P_1) + D(1^1P_1) \, \bar{D}. \qquad 4.21$$

Before the discovery of ψ'', it was possible to measure with certainty only σB for the various D-meson decay modes. With this pure source of $D^0\bar{D}^0$ and D^+D^- comes the ability to measure absolute branching fractions. This, in turn, permits absolute determination of the charmed component of the total annihilation cross section. The main features of the nonleptonic fractions in Table 10 are: (a) they are all only a few

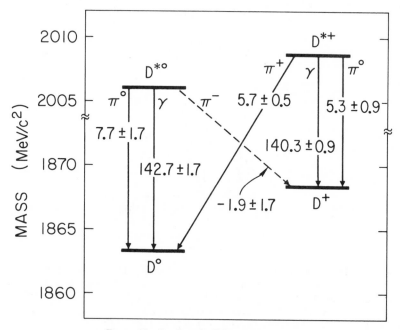

Figure 18 Q values for D* → D transitions.

percent, as expected; (b) the decay $D^+ \to K^0 \pi^+$ does not seem suppressed, so that the nonleptonic $\Delta C = \Delta S$ Hamiltonian may not be as simple (or as mysterious—depending on one's point of view) as the one governing K decay; and (c) so far there is little evidence (119) for quasi-two-body decay, e.g. $D^0 \to (K^{*-} \pi^+, \bar{K}^0 \rho^0) \to \bar{K}^0 \pi^+ \pi^-$. There is no evidence for Cabibbo-suppressed decays; present limits are somewhat above theoretical expectations of order $\tan^2 \theta_c$ (59).

The semielectronic decay fractions measured to date are really an average determined from measurement of

$$\sum_{M_c = D^0, D^+, F^+, \ldots} \sigma(e^+ e^- \to M_c + \ldots) \cdot B(M_c \to e^+ \nu_e + \ldots). \qquad 4.22$$

To reduce contamination from F's and charmed baryons (so that their semileptonic fractions can be unfolded in future experiments), one wants data taken at the lowest possible energies. This still gives an average over D^0 and D^+, a problem that can be resolved using "tagged" D's from ψ'' decays (108). At any rate, three experiments have now measured the average semielectronic branching ratio at low energies: the DASP collaboration at DESY found $\langle B_e \rangle = 0.10 \pm 0.03$ at $W = 3.99$–4.08 GeV (120); an LBL-SLAC-Stanford-Northwestern-Hawaii (LSSNH) collaboration found $\langle B_e \rangle = 0.072 \pm 0.028$ from running at ψ'' (121); and, at the same energy, the DELCO collaboration at SLAC has measured 0.11 ± 0.03 (122). It is worth remarking that DELCO has an order of magnitude more solid angle (60%) for electron detection than do either of the other two experiments. The average of these three is

$$\langle B_e \rangle = 0.093 \pm 0.017. \qquad 4.23$$

In Table 10 we arbitrarily took $B_e \approx 10\%$ for both D^0 and D^+. This result is a factor of two smaller than expected in a naive quark model calculation, but within the wide range estimated by Ellis et al (113).

The DASP, LSSNH, and DELCO collaborations also have measured the electron momentum spectrum in multiparticle events, presumably corresponding to $D \to e^{\pm} + X$ (rather than $\tau \to e + X$). Figure 19 shows the DELCO results; the results of the other two groups are quite similar but with poorer statistics. All of the spectra show the characteristic shape expected from $D \to K e \nu$ and $K^* e \nu$, with good fits obtained by assuming sizable fractions of these two modes. None of the experiments were sensitive to the V, A structure of the amplitude for the decay $D \to K^* e \nu$. Certainly, this is one of the most important questions for future study.

The question of $D^0 \bar{D}^0$ mixing has been studied by two methods. The first (99) is to observe $D^0 \to K^- \pi^+$ and look at the charge of the kaon resulting from decay of the $\bar{D}$ in recoil. The second (108) is to tag D^0 by observing the π^+ in $D^{*+} \to D^0\pi^+$ and then count the number of times D^0 decays to $K^+\pi^-$ instead of $K^-\pi^+$. In both cases, the apparent $\Delta C = -\Delta S$ decays are consistent with what is expected from $\pi - K$ misidentification; and the violation of the $\Delta C = \Delta S$ rule is $\lesssim 17\%$ at the 90% confidence level. This certainly rules out maximal $D^0\bar{D}^0$ mixing, i.e. $|\Delta C| = 2$ currents coupled to Z^0, but not necessarily a small mixing due to second order weak or Higgs boson effects.

D* branching ratios may be determined as follows (108): Fitting the relative contributions of curves B and C in Figure 17, there results (99)

$$B(D^{*0} \to D^0\gamma) = 0.45 \pm 0.15.$$ 4.24

Hence, $B(D^{*0} \to D^0\pi^0) = 0.55 \pm 0.15$, and D^{*+} branching ratios may be obtained under more general assumptions than those made in Equation 4.12, namely (a) isospin conservation in $D^* \to D\pi$ decays; (b) $\Gamma(D^* \to D\pi)$

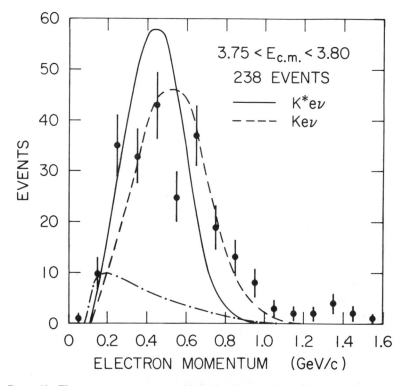

Figure 19 The momentum spectrum of inclusive electrons in multiparticle decays of the ψ'' as measured by the DELCO collaboration (32) at SLAC. Solid and dashed curves are theoretical spectra expected for $D \to K^*e\nu$ and $D \to Ke\nu$. The dot-dashed curve indicates the estimated background remaining in the data.

proportional to p_D^3; (c) Equation 4.11 for M1 rates. It is then clear that if $I = \frac{1}{2}$ for D and D* and $m_u/m_c \lesssim \frac{1}{4}$, the resulting widths and branching fractions are nearly indistinguishable from the predicted ones in Table 11.

All in all, the gross characteristics of charmed hadrons are just what were expected on theoretical grounds. But the theory is a long way from being well tested, and a great deal more experimental study of the details of charmed particle weak interactions is needed. It is to be hoped that this "bread and butter" physics (best carried out at the ψ'') will not be overlooked in the rush for new physics at the Υ and still higher energies.

5 BEYOND CHARM

5.1 Motivations

The existence and characteristics of the quarks u, d, s, and c are well established. However, there are reasons for considering the existence of further quarks, even heavier than the c. The quarks to be considered here are those of charge $\frac{2}{3}$ (called t quarks) and of charge $-\frac{1}{3}$ (called b quarks). While there is no known reason that quarks of charges $-\frac{4}{3}$, $+\frac{5}{3}$, etc. are forbidden, there is no theoretical or experimental motivation for them; they present no essentially different features and are not discussed here.

The theoretical motivation for quarks beyond charm comes from gauge theories for the weak and electromagnetic interactions. Within the group SU(2) × U(1) there is little purely theoretical motivation for more than four quarks. However, if for aesthetic or other reasons, one required the existence of both left- and right-handed charged currents (for quarks), then no SU(2) × U(1) model with only four quarks is even remotely consistent with the data. One six-quark model (123) with the coupling[19] $(ub)_R$ (and $m_b > 11$ or 12 GeV) is possible; all other SU(2) × U(1) models (e.g. see 124) with right-handed couplings for u and/or d quarks appear to be inconsistent with the data. Of course, there is absolutely no experimental evidence requiring any right-handed charged currents among quarks.

Many models based on higher gauge groups require six (or more) quarks. If an SU(3) of "flavor" is contained in the group, then triplets of quarks with charges $\frac{2}{3}$, $-\frac{1}{3}$, $-\frac{1}{3}$ can be found.[20] Some argue that the noninteger nature of quarks can be understood in a natural manner in

---

[19] Our notation $(q_1 q_2)_L$ [or $(q_1 q_2)_R$] refers to the charge-changing weak current, Equation 1.2, with $(1 - \gamma_5)$ [or $(1 + \gamma_5)$] implied. The $W^\pm$ boson couples the two fermions, q_1 and q_2.

[20] The SU(3) used in such models refers to the weak interactions: a triplet might contain u, d, and b quarks. It is quite distinct from the old SU(3) associated with the light quarks (u, d, and s).

such theories where the sum of quark charges is zero. In theories involving SU(3), the fifth and sixth quarks are both expected to have charges $-\frac{1}{3}$. It should be mentioned that some higher groups contain SU(2) × U(1) as a physically relevant subgroup and are able to reproduce all of the WS-GIM predictions for neutrino-hadron interactions. SU(3) generalizations of SU(2) × U(1) models are discussed in Section 5.2.

There have been attempts to find theories that unify the strong interactions with the electromagnetic and weak interactions. Among such theories are those based on the exceptional groups E_6 and E_7, studied by Gürsey, Ramond, and Sikivie (125). The group E_6 has SU(3) × SU(3) × SU(3)$_c$ as a maximal compact subgroup, while E_7 has SU(6) × SU(3)$_c$. The E_6 theory can be reduced to a model very similar to the WS-GIM model, and E_7 to the SU(2) × U(1) model with (ub)$_R$. Both models require six quarks (two with charges $\frac{2}{3}$ and four with $-\frac{1}{3}$). The quarks found in these theories based on exceptional groups automatically have fractional charges.

There are further motivations for new quarks. Following the discovery of the charmed quark, four quarks and four leptons were known to "exist." Later a fifth lepton τ (of mass 1.8 GeV and charge ± 1) was discovered (24). Horn & Ross (126) showed that in the WS-GIM model, existing data require the existence of a neutral lepton coupled to τ (as ν_e is to e). Some have speculated that these additional leptons may indicate the need for additional quarks. Within the WS-GIM model, it is necessary to have equal numbers of quark doublets and lepton doublets in order to cancel VVA triangle anomalies (127), which otherwise prevent renormalizability of the gauge theory. In any unified gauge theory, the proof of renormalizability makes strong use of current conservation through the associated Ward identities (9). The formal conservation of axial currents, however, is not necessarily true in the presence of fermions. Triangle diagrams with one axial and two vector vertices will destroy the axial conservation unless cancellations are arranged among the different fermions that can circulate in the loop. In the WS-GIM model, the anomaly cancels between the $(\nu_e\ e)_L$ and $(u\ d)_L$ doublets, and between the $(\nu_\mu\mu)_L$ and $(cs)_L$ doublets. Therefore the presence of a doublet associated with τ requires a quark doublet (tb)$_L$ in that model. It may be relevant to mention that the present, limited data show the branching ratio (128) for the decay $\tau \to \nu\pi$ to be substantially lower than expected; this apparent failure of a firm theoretical prediction clouds the interpretation of τ, but more data are needed before taking this result seriously. In other models the triangle anomalies are cancelled by different means so that conclusions concerning new quarks can be different.

An early experimental motivation for new quarks was the report of anomalous energy dependencies for antineutrino scattering cross sections and distributions (129). However, more recent experiments (130) with higher statistics find no large anomalies. Another relevant observation in neutrino scattering has been the discovery (131) of events with three outgoing muons ($\mu^-\mu^-\mu^+$). The number of events reported at this time is quite small, and it is impossible to determine their origin now. Three possible sources involving new heavy particles have been suggested; however, two "background" sources could also provide a significant rate of "trimuon" production. One source of trimuons in neutrino scattering could be the production of a charged heavy lepton (heavier than the τ lepton) that has a sequential decay (involving another new heavy lepton) into three muons and other particles (132). Another source (133) involves the simultaneous production of a neutral lepton (which decays into $\mu^-\mu^+\nu$) and a quark b (which decays into a negative muon and other particles). Finally, a heavy quark t could be produced and decay sequentially through either a quark b or a neutral heavy lepton (134). Alternatively, trimuon events could simply be the result of ρ, ω, ϕ, and ψ production and decay to $\mu^+\mu^-$, or muon-pair bremsstrahlung (135) off quarks or the muon. At this time one cannot, therefore, say whether or not trimuon events are an indication of the existence of new heavy quarks, but the amount of data should increase sharply in the near future.

There is, of course, one substantial motivation for quarks beyond charm. Upon its discovery (23) in pp scattering, the $\Upsilon(9.4)$ was immediately interpreted as a $\bar{q}q$ meson (103). Analyses indicated that the associated quark had charge $-\frac{1}{3}$; however, these analyses involve significant assumptions, and it should be emphasized that one cannot reach reliable conclusions concerning the charge in hadronic collisions.

In e^+e^- annihilation it should be easy to determine the nature of $\Upsilon(9.4)$ and the charge of its constituent quark. The charge will be evident by determination of the leptonic width, $\Gamma(\Upsilon \to \mu^+\mu^-)$, [about 0.7 keV for b quarks and 2.8 keV for t quarks of mass 5 GeV, according to Eichten & Gottfried (104)]. Use of the branching ratio of Υ to $\mu^+\mu^-$ is not completely reliable, since theoretical calculation of the total width is difficult. The cross sections expected for Υ are much smaller than those for ψ (see for example 136). The integrated area under a resonance in e^+e^- annihilation is given by

$$\sum = \frac{6\pi^2}{m_\Upsilon^2} B_{had} \Gamma(\psi \to e^+e^-), \qquad\qquad 5.1$$

where $\Gamma(\psi \to e^+e^-) = \Gamma(\psi \to \mu^+\mu^-)$. For a $-\frac{1}{3}$ charge quark one finds $\Sigma \approx 150$ nb-MeV compared with 10^4 nb-MeV for ψ, and the signal-to-background ratio may be only 2 to 1. The maximum ratio of the cross section in the resonance Υ (for charge $-\frac{1}{3}$) to background would be approximately 10 compared to 300 for ψ. While Υ will not be as dramatic as ψ, it will be quite noticeable in $e^+ e^-$ experiments, and its discovery there will be an important confirmation of a new quark.

5.2 Extending the Standard Model

The simplest extension of the WS-GIM model within $SU(2) \times U(1)$ is the addition of a new left-handed doublet with t and b quarks (137), which, together with a new doublet for τ leptons, would cancel triangle anomalies:

$$\binom{u}{d}_L \quad \binom{c}{s}_L \quad \binom{t}{b}_L$$

$$\binom{v_e}{e}_L \quad \binom{v_\mu}{\mu}_L \quad \binom{v_\tau}{\tau}_L \qquad 5.2$$

with all right-handed components in singlets. These new doublets have little impact on the phenomenology of the lighter particles. The d, s, and b quarks in these doublets are actually mixtures similar to the Cabibbo mixture for the four-quark model (see the discussion in Section 8). As discussed in Sections 6 and 7 and elsewhere, there are almost no data in conflict with this expanded WS-GIM model.

Within the gauge group $SU(2) \times U(1)$ it is also possible to construct models with right-handed charged currents. The relevant couplings are those to u and d quarks. Some models (123, 124) have $(u\, b)_R$ or $(t\, d)_R$ or both. If one is willing to consider quarks of charge $-\frac{4}{3}$ or $+\frac{5}{3}$, then models with $(d\, v)_R$ or $(r\, u)_R$ can be obtained. Of models with such right-handed doublets, only one (123) is consistent with present neutral current data (see Section 7):

$$\binom{u}{d}_L \quad \binom{c}{s}_L \qquad\qquad \binom{u}{b}_R \quad \binom{c}{g}_R$$

$$\binom{v_e}{e}_L \quad \binom{v_\mu}{\mu}_L \quad \binom{v_\tau}{\tau}_L \quad \binom{N_e}{e}_R \quad \binom{N_\mu}{\mu}_R \quad \binom{N_\tau}{\tau}_R, \qquad 5.3$$

where $m_b > 11$ or 12 GeV (see Section 6) but the g quark could be the constituent of Υ. The N_e, N_μ and N_τ are heavy neutral leptons. For this model, the cancellation of triangle anomalies occurs separately within the quark sector and within the lepton sector.

One can modify such models to include a coupling $(c\, s)_R$ but not $(c\, d)_R$; this has been discussed by Golowich & Holstein and others (138).

There is no reason, a priori, that quarks (or leptons) must be in doublets. However, SU(2) triplets (or higher representations) require quarks of charge $-\frac{4}{3}$ or $+\frac{5}{3}$, and will not be considered here.

We have mentioned a variety of theoretical reasons for considering other weak and electromagnetic gauge groups beyond SU(2) × U(1). Furthermore it is possible that future data will rule out SU(2) × U(1) as the full group. However the present phenomenological success of the WS-GIM model indicates that SU(2) × U(1) will be a good subgroup of any larger group.

Various authors (139) have noted that there are models within SU(2)^L × SU(2)^R × U(1) [where L = left and R = right and SU(2)^L is the same SU(2) as above] that reproduce virtually all the neutrino-hadron scattering results of the WS-GIM model. Georgi & Weinberg (140) have generalized these results and shown that at zero momentum transfer, the neutral current interactions of neutrinos in an SU(2) × G × U(1) gauge theory are the same as in the corresponding SU(2) × U(1) theory if neutrinos are neutral under G. They also noted that one of the neutral gauge bosons in the expanded group must have a mass below that of the Z^0 (80 GeV) of the SU(2) × U(1) model.

In SU(2)^L × SU(2)^R × U(1) there are seven gauge bosons, $W_L^\pm$, $W_R^\pm$, Z_1^0, Z_2^0, γ, in contrast to the four in SU(2) × U(1) ($W^\pm$, Z^0, γ). It can be arranged so that Z_1^0 has purely axial-vector couplings to all particles (except neutrinos) and that Z_2^0 has purely vector couplings; this assures the absence of parity violation in neutral current interactions (see Section 7). One version of the model has the couplings:

$$
\begin{pmatrix} u \\ d \end{pmatrix}_L \quad \begin{pmatrix} c \\ s \end{pmatrix}_L \quad \begin{pmatrix} t \\ b \end{pmatrix}_L \qquad (u\ b)_R \quad (c\ s)_R \quad (t\ d)_R
$$

$$
\begin{pmatrix} \nu_e \\ e \end{pmatrix}_L \quad \begin{pmatrix} \nu_\mu \\ \mu \end{pmatrix}_L \quad \begin{pmatrix} \nu_\tau \\ \tau \end{pmatrix}_L \qquad (N_e\ e)_R \quad (N_\mu\ \mu)_R \quad (N_\tau\ \tau)_R,
$$

5.4

where column doublets are coupled by W_L (the usual W) and row doublets by W_R. Since $(u\ b)_R$ is coupled by W_R, which has no direct couplings to ν_μ, the usual lower limits on the mass of b do not apply (and $\Upsilon \equiv \bar{b}b$ is possible).

Another gauge group that has received considerable attention is SU(3) × U(1). For the models (141) considered, extreme values of the parameters can be chosen, which will reduce these models to conventional SU(2) × U(1) models. For intermediate values of the parameters, the phenomenological results are somewhat different. One version of the models resembles the WS-GIM model, while another resembles the SU(2) × U(1) model with $(u\ b)_R$.

One extension (125, 142) of the standard WS-GIM model, which has SU(2) × U(1) as a good subgroup in a fairly natural way, is based on the group SU(3) × SU(3). The neutrino-hadron scattering results are essentially the same as for the WS-GIM model although the value $\sin^2 \theta_W = \frac{3}{8}$ predicted in this model seems somewhat larger than present experimental indications. There are 16 gauge bosons, including the usual $W^\pm$, Z^0, and γ. Many of these bosons must be three (or more) times as heavy as the $W^\pm$ for phenomenological purposes. Among these are the right-handed equivalents of $W^\pm$ and most of the bosons carrying diagonal neutral currents. In this model the leptons are placed in two $(\bar{3},3)$ representations. The quarks are in triplets such as

$$\begin{pmatrix} u \\ d \\ b \end{pmatrix}_L \quad \begin{pmatrix} c \\ s \\ g \end{pmatrix}_L \quad (u\,s\,b)_R \quad (c\,d\,g)_R, \qquad 5.5$$

where the first two quarks in each column triplet are coupled by $W^\pm$ and all other quarks are coupled by different bosons. One of the most interesting features of this model is that the lightest new quark, b, always decays semileptonically, including modes such as $b \to u l^- \bar{\nu}$ and $b \to d \nu \bar{\nu}$. Thus, this model predicts a large amount of missing neutral energy in $e^+ e^- \to (b\,\bar{q}) + (\bar{b}\,q)$.

One of the questions in constructing new models concerns the weak coupling of the b quark where $\Upsilon(9.4) \equiv b\bar{b}$. While the standard assumption places the b quark in a left-handed doublet with a t quark, there are several other couplings that are consistent with all data (see Sections 6 and 7). In SU(2) × U(1) models, the couplings $(t\,b)_R$ and $(c\,b)_R$ are allowed. In a model such as SU(2)^L × SU(2)^R × U(1) the b quark can even have a right-handed coupling to u quarks since that interaction is mediated by W_R, which does not couple to ν_μ. Those models with quarks in SU(3) triplets can have couplings such as $(u\,d\,b)_L$, $(t\,b\,d)_R$, or $(c\,b\,d)_R$. There certainly is no evidence that b quarks have left-handed couplings to t quarks. In fact, some of the models mentioned here have no t quarks.

Not all theories involve quarks with fractional charge. Pati & Salam and others (143) have proposed models with quarks of integer charge [following Han & Nambu (144)], which nonetheless reproduce many of the results of conventional gauge theories. In this theory, however, quarks and gluons can exist as free particles (before decaying). In the basic model, there are 16 fermions:

$$\begin{pmatrix} u_R & u_Y & u_B & \nu_e \\ d_R & d_Y & d_B & e^- \\ s_R & s_Y & s_B & \mu^- \\ c_R & c_Y & c_B & \nu_\mu \end{pmatrix} \quad \text{with charges} \quad \begin{pmatrix} 0 & 1 & 1 & 0 \\ -1 & 0 & 0 & -1 \\ -1 & 0 & 0 & -1 \\ 0 & 1 & 1 & 0 \end{pmatrix}, \qquad 5.6$$

where R, Y, B are the "colors" red, yellow, blue, and the leptons are considered to be the fourth color. The model can be expanded to include other quarks and leptons. One of the problems with this model is that it predicts free, massive gluons that have not been observed. This and other aspects of the Pati-Salam model are discussed critically in Reference 145.

6 PRODUCTION BY NEUTRINOS, HADRONS, AND PHOTONS

Although most experimental information about new quarks has come from e^+e^- annihilation, other methods of production have played an extremely important role. The ψ/J was produced hadronically at the same time that it appeared in e^+e^- annihilation, and charmed baryons have been observed only in neutrino and photoproduction. In this section, some of these other methods are discussed. They can yield important information about the weak and electromagnetic theory and about QCD as a possible strong interaction theory.

6.1 *Production by Neutrinos*

In neutrino scattering in the WS-GIM model, where u quarks have a left-handed coupling to d quarks $(u\,d)_L$, one expects $\nu_\mu d \to \mu^- u$ or $\bar{\nu}_\mu u \to \mu^+ d$ to be the usual charged-current processes. Most results are consistent with this hypothesis, and one must look at rare processes in order to learn more.

In the scattering of neutrinos off nucleons, it is possible to produce single charmed mesons (or baryons). However, since there is no large coupling of valence (u or d) quarks to c quarks, this additional cross section is not large. The coupling $(\bar{c}d \sin \theta_c)$ with $\sin^2 \theta_C \approx 0.05$ leads to a 5% rise in the expected cross section for neutrinos (above the threshold energy). There is no similar Cabibbo-suppressed $(\sin \theta_C)$ process possible for antineutrinos. The coupling $(\bar{c}s \cos \theta_C)$ leads to an increase in both neutrino and antineutrino cross sections; however, the amount of strange quarks in the sea (i.e., of $s\bar{s}$ pairs in the nucleon) is quite small, of order 5% (146), so that resulting effects are small. Since 5% effects are difficult to measure experimentally and since comparable or larger QCD effects may occur, little evidence for charm is found in total cross sections. Similarly, little effect is seen in y distributions $[y \equiv (E_\nu - E_\mu)/E_\nu]$.

It might be helpful to give a simplified description of several of the features of QCD that should result in similar effects in neutrino scattering (147). With increasing $Q^2 (\equiv -q^2 = 4E_\nu E_\mu \sin^2 \theta_{lab}/2)$ one expects that: (*a*) the x distributions of quarks [where $x \equiv Q^2/2M_N(E_\nu - E_\mu)$] will shrink (i.e. become more peaked toward zero); (*b*) the fraction of the struck nucleon's momentum carried by valence quarks will decrease slowly; (c) the fraction carried by sea quarks ($u\bar{u}$, $d\bar{d}$ and $s\bar{s}$ pairs in the nucleon) will increase. There are helicity arguments that show $\sigma(\nu q_1 \to \mu^- q_2) = 3\sigma(\nu\bar{q}_2 \to \mu^- \bar{q}_1) = 3\sigma(\bar{\nu}q_2 \to \mu^+ q_1) = \sigma(\bar{\nu}\bar{q}_1 \to \mu^+ \bar{q}_2)$, where $\bar{q}$ indicates antiquark. In neutrino reactions then, scattering off valence quarks is enhanced relative to that off sea quarks, while in antineutrino reactions scattering off sea quarks is enhanced (although most momentum is always carried by valence quarks). As a result, one expects neutrino cross sections to decrease with increasing E_ν (which is proportional to $\langle Q^2 \rangle$) and antineutrino cross sections to increase slightly. A related effect is the increase of $\langle y \rangle$ for antineutrinos with increasing E_ν (for neutrinos there is little effect).

Although charm is difficult to detect in total cross sections and distributions, evidence for charm is quite clear in other aspects of neutrino experiments. Charmed particles decay into muons and into electrons 10–20% of the time, and these leptons can be detected. If charm production is 5–10% of the total, and the branching ratio to muons (electrons) is 10–20%, then 0.5–2% of all neutrino-induced events should contain an extra muon (electron). This rate of "dilepton" production (17) is in fact roughly what is observed (because of experimental cuts and efficiencies the exact rate is not easy to determine directly). Furthermore, in distributions of variables such as y, E_μ, and various angles, one finds (148) strong evidence for the additional lepton coming from the decay of a produced heavy quark (with mass of approximately 2 GeV).

Neither the rate nor the distributions show that this heavy quark is charm. However, since charm usually decays to s quarks, in bubble chamber experiments one can see if events with two leptons also have a K meson or a Λ^0 baryon. When neutrinos change d quarks into c quarks, one strange particle should result. However, when an s quark in the sea is changed into a c quark, there is always the remaining $\bar{s}$ quark from the pair in addition to the s quark from c quark decay, so that two strange particles result. Since antineutrino scattering lacks a Cabibbo-suppressed mode of charm production, the number of strange particles (two) is expected to be greater than for neutrinos (roughly 1.5). At present, results have been reported only for neutrinos. Two experiments (149) have reported about 3.5 K mesons per $\mu^- e^+$ event, while one other (150) with much higher statistics has reported about 1.0 K mesons per event. The latter corresponds closely to the predictions for charm.

In all of these features (cross sections, y distributions, dilepton rates, presence of strange particles) little room remains for significant production of any heavier quarks. Of course, for sufficiently massive quarks, all production would be deferred until higher energies. Present data

(129, 130) indicate that any b quark (charge $-\frac{1}{3}$) that has a right-handed coupling to u quarks (through W bosons) must have $m_b \gtrsim 11$ or 12 GeV, certainly excluding the quark in $\Upsilon(9.5)$. If given that $m_b = 5$ GeV, then the coupling squared for $(u\,b)_R$ must be 0.1 (or less) of that for $(u\,d)_L$. Any t quark (charge $\frac{2}{3}$) that has a right-handed coupling to d quarks (through W bosons) must have $m_t \gtrsim 5$ or 6 GeV. For the left-handed couplings $(u\,b)_L$ and $(t\,d)_L$, the limits from analysis of the data are $m > 8$ GeV in both cases if the couplings are full strength. For 5-GeV t or b quarks, the (left-handed) couplings squared must be 0.3 (or less) of that for $(u\,d)_L$.

In the WS-GIM model, the additional coupling $(t\,b)_L$ (see Section 5) would lead to little t or b quark production, because the mixing angles between heavy quarks and light quarks must be small (see Section 8.1). From the universality of quark and lepton couplings and from the $K_L^0 - K_S^0$ mass difference, one finds (151) that the $\bar{t}d$ coupling ($\bar{u}d$ coupling) is not likely to be more than 10% (5%) of the $\bar{u}d$ coupling [i.e. rates at the 1% (0.3%) level]. Clearly even at high energies there will be little impact on cross sections.

There are many other couplings possible for t and b quarks besides $(t\,b)_L$, and some of these also have few observable consequences in neutrino physics. Among such couplings are the right-handed couplings $(t\,b)_R$ and $(c\,b)_R$. Also the couplings $(u\,b)_R$ and $(t\,d)_R$ are possible (even for relatively light t and b) in models such as $SU(2)^L \times SU(2)^R \times U(1)$, where those couplings occur not through the usual W boson but through a new boson that does not couple ν_μ to μ and is heavier than W.

One can look for signals for new quarks in multilepton events, although in the WS-GIM model the rates will not be high. Given the coupling squared ($\lesssim 0.01$), the branching ratio to muons (~ 0.2), and the phase space suppression F (see Table 12), the rate for dilepton events from new heavy quarks is less than 0.2-F times the rate for dileptons from charm in this model. Clearly then, detection of t and b quarks will be difficult. One possible method of distinguishing dileptons for t or b

Table 12 The phase space suppression F (relative to zero mass quarks) for production in neutrino scattering of quarks of given mass

Quark mass (GeV)	F for Fermilab Quad triplet flux		F for CERN Wide-band flux	
	All E	$E > 100$ GeV	All E	$E > 100$ GeV
5	10^{-1}	3×10^{-1}	10^{-1}	2×10^{-1}
10	6×10^{-3}	3×10^{-2}	10^{-3}	10^{-2}
15	10^{-4}	6×10^{-4}	10^{-5}	10^{-4}

decay from dileptons for charm decay involves the energy of the secondary (decay) lepton. Dileptons from heavy quark decay should be significantly more energetic (152).

Another means of finding evidence for the production of t or b quarks comes from examination of events in which three leptons are produced. In the WS-GIM model, b quarks are likely to decay into c quarks unless $m_t < m_b$ (see Section 8.1). In some cases, both the b and c quark decays would involve leptons. In antineutrino scattering, b quarks can then be produced (along with the usual μ^+) and decay sequentially into two muons (or electrons). The resultant "trilepton" events would occur at less than $10^{-4} F$ (see Table 12) of the total rate. If $m_b > m_t$, it is clear that trilepton events can also occur at the same rate, but with b decay to t instead.

If $m_t > m_b$, then (in this model) t quarks can decay into b quarks, which decay into c quarks. In neutrino scattering t quarks can be produced and decay into two, three, or more leptons. For such trilepton events the rate would be less than $8 \times 10^{-4}F$ of the total rate (counting both $\mu^- \mu^- \mu^+$ and $\mu^- \mu^+ \mu^+$ events).

Trimuon events have been reported by three groups (131). At Fermilab, one group (131) reports a rate of 10^{-4} of the total rate, while at CERN a rate of 5×10^{-5} has been reported (131) (in both cases $E_\nu > 100$ GeV and $E_\mu > 4$ GeV is required). Presumably some (and possibly all) of these events come from background sources (such as ρ decay). The rates given above for the WS-GIM model are upper bounds (since upper bounds on t$\bar{\text{d}}$ and u$\bar{\text{b}}$ couplings are used) and even then appear to be lower than these reported rates, but the question of backgrounds should be resolved. Various studies (132–135) have been made concerning the expected characteristics of such trilepton events; more data are needed before conclusions can be reached.

The multilepton events found in b or t quark decays are expected (in the WS-GIM model) to be accompanied by the presence of strange particles. If $m_b < m_t$, then b quarks decay into c quarks, which decay into strange particles; and the t quarks decay into b quarks. If $m_t < m_b$, then t quarks decay dominantly into s quarks directly; and the b quarks decay into t quarks.

While the above remarks were taken in the context of the WS-GIM model, similar conclusions follow in many other models. There are a large variety of left- and right-handed couplings possible for b and t quarks in both SU(2) × U(1) and other models, and frequently these result in detectable signals in neutrino experiments as described above.

6.2   Hadroproduction of Heavy Particles

Heavy vector mesons such as ψ and Υ, which decay to $\mu^+\mu^-$ or e^+e^-, have been observed in pp, pN, p$\bar{\text{p}}$, and πp scattering experiments (153).

Several different approaches to calculating the cross sections have been advocated. Some of them have been modified as further data were reported, and here most attention will be given to the later versions.

One obvious way to produce ψ mesons is by the fusion of a c quark from one of the incoming hadrons with a $\bar{c}$ quark from the other hadron, where the c and $\bar{c}$ quarks come from the "sea" of their respective hadrons (154). The magnitude of such a process is difficult to estimate since assumptions are needed about the validity of SU(4) and/or about the manner of decay of the ψ (used to estimate the fusion coupling). However, this process has an unavoidable consequence that is easy to test: All ψ production should be accompanied by the simultaneous production of two charmed particles. The experiments done report (155) no evidence for this process, and it must be quite suppressed compared to other processes.

One modification of such an approach is to argue that the ψ is produced by the fusion of light quarks and antiquarks (154). However, since there are not many antiquarks in a proton compared to an antiproton, one expects ψ production in p$\bar{p}$ scattering to be at least twenty times that in pp scattering (depending on the shape of quark and antiquark distributions). Experiment indicates a factor of about seven. While this approach appears to be inadequate, such diagrams are included in two other (more successful) methods.

In one of these other approaches it is argued that the P-wave states (χ) of the ψ family are produced much more frequently than are ψ's, and that the observed ψ's are primarily decay products of those P-wave states (156a; for Υ production, see 130, 156b). The production of ψ's (relative to χ's) is said to be suppressed because ψ couplings require at least three gluons, so that the resulting effective coupling is much smaller than that for χ, which requires only two gluons. In such a picture, the χ is produced by two processes: (a) the fusion of two gluons, one from each of the colliding hadrons, and (b) the fusion of a quark and an antiquark from the colliding hadrons. The latter process can be assumed negligible in pp scattering (where there are few antiquarks), but is important in p$\bar{p}$ scattering. This approach can obtain the correct ratio (157) for ψ production in pp relative to p$\bar{p}$ scattering, which is 0.15 ± 0.08 at $s^{\frac{1}{2}} = 8.75$ GeV. The cross section (via gluons, which are labeled g below) can be written as:

$$\sigma(A + B \to \psi + X) = \int dx_1 dx_2 f^A(x_1) f^B(x_2) \sigma(gg \to \chi) B(\chi \to \psi + \gamma)$$

$$= \frac{8\pi}{m^3} \Gamma(\chi \to gg) B\tau \int_\tau^1 \frac{dx}{x} f^A(x) f^B\left(\frac{\tau}{x}\right), \qquad 6.1$$

where $f(x)$ are gluon distribution functions, $\tau \equiv m^2/s$ and m is the mass of χ. Using

$$\Gamma(\chi \to gg)B \approx \Gamma(\chi \to \psi + \gamma) = \tfrac{4}{3}\alpha e_Q^2 \omega^3 \left| \langle \psi | \mathbf{r} | \chi \rangle \right|^2 \propto e_Q^2 m^{-\frac{5}{3}}, \qquad 6.2$$

where e_Q is the charge of the quark ($\tfrac{2}{3}$ for charm), ω is the $\chi - \psi$ mass difference, and the last proportionality holds for a linear potential model, then one finds

$$\sigma(A + B \to \psi + \chi) \propto e_Q^2 m^{-\frac{14}{3}} F(\tau). \qquad 6.3$$

In this approach both the form and magnitude of $F(\tau)$ can be calculated, and the results shown in Figure 20 are in reasonable agreement with the data. Even if $F(\tau)$ was not calculable, one could take $F(\tau)$ from the data for ψ and apply Equation 6.3 to the production of Υ and heavier particles.

One motivation for this approach is that it provides an obvious explanation for the observed large suppression of ψ' relative to ψ production (153). Since there are no P-wave states that can decay into ψ', it can only be produced directly; but direct production was assumed to be suppressed. If ψ production is really an indirect process involving $\chi \to \psi + \gamma$, then the observation of γ's associated with ψ production is a

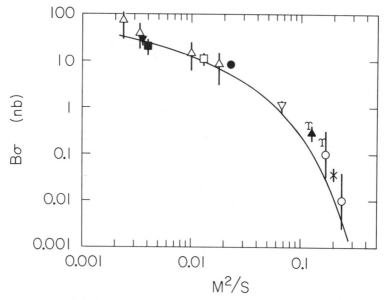

Figure 20   The cross section for ψ production in pp scattering times the branching ratio to muons as a function of τ (where $\tau \equiv m_\psi^2/s$). The data is from Reference 153. The curve is a theoretical prediction (156a) of Carlson & Suaya. The two symbols Υ are the data for Υ production adjusted according to Equation 6.3.

crucial test of this approach [a recent experiment may in fact see such γ's (158)].

In another approach (159a; for Υ production, see 159b), the production of a pair of quarks, c and $\bar{c}$, is calculated. When the invariant mass of the pair is less than two times the mass of D mesons, it is assumed that ψ's (or other ψ family members) can be produced. There are three types of diagrams that contribute to $c\bar{c}$ production in this approach: (a) a quark from one of the colliding hadrons can annihilate with an antiquark from the other hadron to give a single gluon, which produces a $c\bar{c}$ pair; (b) a gluon from each of the hadrons can couple to a c quark line, producing a $c\bar{c}$ pair; and (c) gluons from each of the hadrons can fuse to a single gluon, which produces a $c\bar{c}$ pair. In each case a color singlet is obtained via final-state interactions neglected in calculations. For diagram (a) the cross section is given by

$$\sigma = \int_{(2m_c)^2}^{(2m_D)^2} \frac{ds'}{s'}\, \sigma(q\bar{q} \to g \to c\bar{c})\tau \int_\tau^1 \frac{dx}{x}\, f(x) f\left(\frac{\tau}{x}\right), \qquad 6.4$$

where s' is the subenergy, $\tau = s'/s$, and $f(x)$ are quark distributions. One way to estimate σ in the integrand is to assume an analogy to Drell-Yan calculations (160) of $\mu^+\mu^-$ production. Then σ is given by

$$\sigma \approx \int_{(2m_c)^2}^{(2m_D)^2} ds' \frac{d\sigma'_{DY}}{ds'} \frac{\alpha_s^2}{\alpha^2} \frac{2}{3}, \qquad 6.5$$

where α_s is the strong coupling constant and σ'_{DY} is the Drell-Yan cross section calculated without the quark charges. Diagrams (b) and (c) can be calculated similarly. When all three types of diagrams are included, one obtains the correct ratio for ψ production in pp relative to p$\bar{p}$ scattering.

In Figure 20 the data for the production of ψ in proton-nucleon scattering are shown. The curve is a theoretical calculation of Carlson & Suaya (156a) using the indirect production ($\chi \to \psi$) approach. Other authors using this and other approaches have obtained similar results for ψ production. To test the basic hypotheses of the approaches discussed, one can examine whether they can account for the observed cross section for $\Upsilon(9.4)$ production. For the indirect production approach, one sees from Equation 6.3 that if the data are adjusted for e_Q^2, $m^{-\frac{14}{3}}$ and the branching ratio $B(\Upsilon \to \mu^+\mu^-)$, then the Υ data should lie on the same curve as the ψ data. In Figure 20 these adjusted data are shown with the symbol Υ and do, in fact, lie on the same curve. It was assumed that the quark associated with Υ had charge $-\frac{1}{3}$; otherwise the adjusted data points would lie a factor of about 16 lower (both e_Q^2 and B change by a factor of about 4). For the direct production approach, adjusted data points lie below the ψ data by a factor (according to one calculation) of about 5 for charge $-\frac{1}{3}$ quarks and 20 for $\frac{2}{3}$ charge quarks (only B is different for different charges). Since it is probably unreasonable to expect these models for Υ production to be accurate to better than an order of magnitude, these results do not distinguish the two approaches nor are they completely reliable determinations of the charge of the quark in Υ.

It is interesting to ask what are the highest mass vector mesons ($Q\bar{Q}$) that can be produced in hadronic collisions at existing and future accelerators and storage rings. In Table 13 [from Carlson (161)] it is assumed that at least 10 events per year must be observed. Of course, these results are only crude estimates, since extrapolation from 3-GeV particles (ψ) to particles of enormous mass is difficult. There could be unforeseen complications; one such complication suggested by Bjorken and by Nieh (162) is that very massive quarks could have very significant weak decay modes so that the branching ratio to $\mu^+\mu^-$ (or e^+e^-) would decrease. The weak decay width becomes a sizable fraction of the total when the quark mass becomes comparable to the W boson mass.

The production of charmed particles ($D^+ \equiv c\bar{d}$, $D^0 \equiv c\bar{u}$, etc.) in hadronic collisions has not yet been observed. However, various theoretical estimates suggest that the actual cross sections are not far below the present experimental limits. There are some simple methods to estimate cross sections at Fermilab energies. For example, one could guess

$$\frac{\sigma_D}{\sigma_\psi} \approx \frac{\sigma_K}{\sigma_\phi} \approx 10. \qquad 6.6$$

Table 13 The highest mass of $Q\bar{Q}$ mesons that can be produced in pp collisions at existing and proposed facilities (161)[a]

Facility	$s^{\frac{1}{2}}$	Luminosity	Mass (GeV)	
			$-\frac{1}{3}$ charge	$\frac{2}{3}$ charge
ISR (31+31)	62	10^{31}	21 (21)	26 (26)
TRISTAN (180+180)	360	10^{33}	76 (86)	92 (115)
ISABELLE (400+400)	800	10^{33}	87 (114)	108 (165)
FNAL (270+1000)	1040	10^{33}	90 (122)	112 (181)
POPAE (1000+1000)	2000	10^{33}	94 (140)	117 (218)
UNK (2000+2000)	4000	10^{33}	96 (151)	121 (243)
VBC (10^4+10^4)	2×10^4	10^{33}	98 (161)	124 (276)
VBA (fixed target)	140	10^{37}	89 (96)	96 (103)

[a] The mass calculations include corrections for the existence of weak decay modes (see text); the parenthetical numbers are without such corrections. The first seven facilities are storage rings with the energy of each colliding beam given.

This gives $\sigma_D \approx 1$ μb for pp scattering. Sivers (163) has suggested that one could assume that the appropriate transverse momentum scaling variable is $(p_\perp^2 + m^2)^{\frac{1}{2}}$ rather than $p_\perp$; then the cross section for D meson production might be related to that for pions with $p_\perp \approx 2$ GeV so that

$$\sigma_D \approx 0.1 \ln \left(\frac{s}{4m_D^2} \right) \exp (26 m_D s^{-\frac{1}{2}}) \text{ mb.} \qquad 6.7$$

This method requires an assumption about the charmed quark content of nucleons; for the above estimate $c/s = 0.2$ (where c and s are charmed and strange quark content, respectively) was assumed, which gives $\sigma_D \approx 10$ μb. But there is reason to believe c/s is perhaps an order of magnitude smaller, so that σ_D should be much smaller. The present experimental limit is $\sigma_D < 1.5$ μb at $s^{\frac{1}{2}} = 27$ GeV (164).

More sophisticated calculations have been carried out (151, 154, 156a,b, 159a,b, 165). These are usually extensions of the "direct production" approach to ψ production where the limits of integration over s' (such as in Equation 6.4) are changed to $4m_D^2$ and s. As for ψ production, the diagrams (a), (b), and (c) can all contribute. Babcock, Sivers & Wolfram (165) (among others) discuss the results of such QCD calculations and conclude that diagrams (b) and (c) are more important than (a). They also discuss higher order effects and argue that it is reasonable to neglect them for most purposes. With standard assumptions Babcock et al estimate for pp scattering $\sigma_D \approx 1$ μb at $s^{\frac{1}{2}} = 27$ GeV (also $\sigma_D \approx 10$ μb at $s^{\frac{1}{2}} = 54$ GeV and 100 μb for $s^{\frac{1}{2}} \geqq 200$ GeV). They argue that the present experimental limit on D production in hadronic collisions favors either smaller values of α_s (than expected from leptoproduction experiments) or gluon distributions that are more peaked toward small x.

For heavier mesons such as $Q\bar{u}$ and $Q\bar{d}$ where $Q\bar{Q} \equiv \Upsilon(9.4)$, most estimates (165) are that for a very large range of energies the cross sections for production of $Q\bar{d}$ or $Q\bar{u}$ mesons will be two orders of magnitude lower than those for D mesons. This clearly makes observation of such mesons very difficult in hadronic collisions.

In addition to the above calculations, which are based on behavior expected for $y \approx 0$, there have been calculations of peripheral production (for $x > 0.5$) of charmed particles. For example, the cross section (166) for $\pi^- p \rightarrow D^- C_0^+$ (C is a charmed baryon) has been estimated as 0.5 nb while triple Regge calculations (166) of $\pi p \rightarrow DX$ (for $x > 0.5$) give $\sigma \approx 60$ nb.

6.3   Photoproduction of ψ and Charm

One of the earliest papers on ψ (prior to its discovery) was written by Carlson & Freund (167), who discussed the photoproduction of a then hypothetical $c\bar{c}$ vector meson. The photoproduction of ψ is usually assumed to be a dominantly diffractive process that can be understood with a modified vector-dominance model (97). The modification allows for the $\gamma\psi$ coupling to be different at $q^2 = 0$ and $q^2 = m_\psi^2$. With this assumption the cross section (see 168) is

$$\frac{d\sigma}{dt}(\gamma N \to \psi N) = \frac{3\lambda^2}{\alpha M_\psi} \Gamma(\psi \to e^+ e^-) \frac{d\sigma}{dt}(\psi N \to \psi N), \qquad 6.8$$

where λ measures the variation of the $\gamma\psi$ coupling $g_{\gamma\psi}$ with q^2 and the off-mass-shell extrapolation of the invariant amplitude ($\lambda = 1$ for the "naive" vector-dominance model). Making use of the optical theorem, one finds:

$$\frac{d\sigma}{dt}\left(\gamma N \to \psi N\right)\bigg|_{t_{min}} = \frac{1}{e^{-bt_{min}}} \frac{3\Gamma(\psi \to e^+ e^-)}{16\pi\alpha M_\psi} \lambda^2(1+\rho^2)\sigma_{tot}^2(\psi N), \qquad 6.9$$

where $d\sigma/dt$ was assumed to have t dependence e^{-bt} [which is consistent with data (169) for $b = 2.9$ GeV^{-2}] and $\rho \equiv (\text{Re}\mathscr{A}/\text{Im}\mathscr{A}) \to 0$ as $s \to \infty$, with $\mathscr{A}$ the amplitude for $\psi N \to \psi N$. An independent determination of $\sigma_{tot}(\psi N)$ can be extracted from the observed A dependence (170) (where A is the effective number of nucleons per nucleus) of ψ photoproduction; experiments on Be and Ta give $\sigma_{tot}(\psi N) = 3.5 \pm 0.8$ mb at $E_\gamma = 20$ GeV. To avoid consideration of threshold factors and of ρ, we will assume that this value stays approximately constant up to higher energies ($E_\gamma \approx 80$ GeV). Next, an assumption about the value of λ is needed. If the value of the naive vector-dominance model ($\lambda = 1$) is taken, then for $\rho \approx 0$, $d\sigma/dt(t = 0) \approx 400$ nb/GeV^2, which is far above the experimental values (169, 171) of about 60 nb/GeV^2 at $E_\gamma \approx 80$ GeV. Some theoretical models give $\lambda \approx 0.5$, which gives $d\sigma/dt$ ($t = 0$) ≈ 100 nb/GeV^2; choosing $\lambda = 0.3$ or 0.4 gives 40 or 60 nb/GeV^2. Clearly, the naive vector-dominance model must be modified to account for ψ photoproduction.

From knowledge of ψ photoproduction, there are some immediate implications for the photoproduction of charmed particles. By use of unitarity, it can be shown (168) that for a given energy:

$$16\pi \frac{d\sigma}{dt}\left(\gamma N \to \psi N\right)\bigg|_{t_{min}} \leq (1+\varepsilon)^2(1+\rho^2)\left(\frac{q^{\psi N}}{q^{\gamma N}}\right)\sigma(\gamma N \to \text{charm})$$

$$\sigma(\psi N \to \text{charm}), \qquad 6.10$$

where ε measures violations of the OZI rule and

$$\frac{q^{\psi N}}{q^{\gamma N}} = \frac{[s-(m_p+m_\psi)^2]^{\frac{1}{2}}[s-(m_p-m_\psi)^2]^{\frac{1}{2}}}{s-m_p^2} \qquad 6.11$$

(which is 0.33, 0.72, 0.93 for $E_\gamma = 10, 20, 80$ GeV). Application of the OZI rule again implies that $\sigma_{tot}(\psi N) \approx \sigma(\psi N \to$ charm$)$. Using $E_\gamma = 20$ GeV data (169, 171) for $(d\sigma/dt)$ and $\sigma_{tot}(\psi N)$ in Equation 6.10, one finds

$$\sigma(\gamma N \to \text{charm}) \gtrsim 115 \text{ nb}/(1+\varepsilon)^2(1+\rho^2). \qquad 6.12$$

Since ε and ρ are presumably small, it is safe to say $\sigma(\gamma N \to$ charm$) \gtrsim 100$ nb.

It is possible to estimate the photoproduction of charm by QCD techniques. Gluons from the nucleon and the incoming photon can each couple to a c quark line producing a c$\bar{\text{c}}$ pair. This has been discussed by various authors (172, 173), who find $\sigma(\gamma N \to$ charm$)$ increasing from about 100 nb at $E_\gamma = 20$ GeV to about 400 nb at $E_\gamma = 80$ GeV. Some assumptions are required to apply perturbative QCD to this problem, but the approach is not implausible. Roughly speaking, the large mass of the charmed quark is expected to set the scale for the effective coupling constant. The use of QCD here is similar in some ways to its use in explaining the total width of the c$\bar{\text{c}}$ states, and it surely needs further theoretical analysis.

These results are in approximate agreement with a sum rule of Shifman, Vainshtein & Zakharov (173), and may be consistent with the reported observation (118) of a charmed antibaryon of mass 2.26 GeV (whose cross section has not been reported yet). This experiment of Knapp et al (118) saw evidence for what may have been $\Lambda_c^- \to \bar{\Lambda}\pi^-\pi^-\pi^+$.

7 NEUTRAL CURRENT INTERACTIONS

The weak interactions provide an important probe in the study of new quarks (and new leptons). The neutral current interactions are a crucial measure not only of the existence of new quarks but also of the structure of the gauge theories of weak and electromagnetic interactions. Were the neutral current predictions of the WS model to fail, one would be forced to consider other models. In fact, however, the WS model is in good agreement with most neutral current data, as will be discussed below. The importance of neutral current phenomenology can be seen in the successful prediction of the existence of the c quark from the absence of strangeness-changing neutral currents [via the GIM mechanism (4)]. For much of the study of neutral current interactions the neutrino is used as a probe; it is uniquely suited to this purpose, since it is the only particle that has only weak interactions.

It is possible, a priori, that c quarks (or b or t quarks) could be produced directly by neutral current processes in which u quarks were changed into c quarks. In the WS model (with all left-handed quarks in doublets, none in singlets) such charm (or other flavor) changing neutral currents are forbidden by the GIM mechanism. Two types of experiments indicate that the neutral weak boson Z^0 [of SU(2) × U(1) models] does not change charm: (a) e^+e^- annihilation experiments (174) find no $D^0\bar{D}^0$ mixing (such mixing should be found if charm-changing currents exist); (b) neutrino scattering experiments (17) also find no evidence of

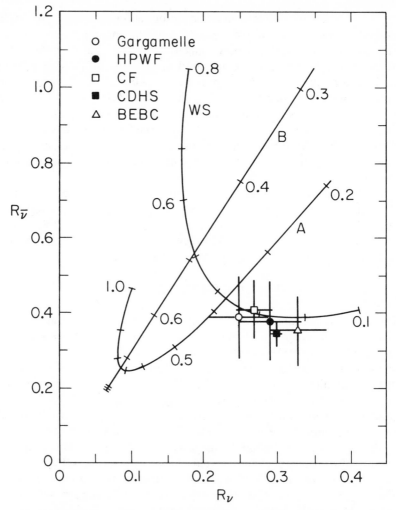

Figure 21 The ratio $\sigma(\nu N \to \nu + X)/\sigma(\nu N \to \mu + X)$ for antineutrinos vs that ratio for neutrinos. The tenth values of $\sin^2 \theta_W$ are shown with tick marks on the theoretical curves. The curve labeled A refers to the model with $(u\,b)_R$: the curve labeled B refers to the model with both $(u\,b)_R$ and $(t\,d)_R$. The data are from Reference 175.

$D^0\bar{D}^0$ mixing (which would lead to $\mu^-\mu^-$ events). In a model for a group other than SU(2) × U(1) it may be possible to have flavor-changing neutral currents if they occur via another boson that is quite heavy or does not have flavor-conserving couplings to light fermions. For SU(2) × U(1) (in which many models are possible) there is little likelihood that there would be t- or b-changing neutral currents, given that there are no strangeness- or charm-changing currents. In SU(2) × U(1) models with quarks of charges $\frac{2}{3}$ and $-\frac{1}{3}$ only, there are no flavor-changing neutral currents if all quarks of a given handedness are in doublets (or are all in singlets) of SU(2). If most quarks were in doublets but one were in a singlet, mixing among quarks would lead to flavor-changing neutral currents for all quarks of that charge (unless there were some reason why mixing was prevented) (115).

There are four types of neutrino experiments commonly used to test the diagonal (flavor-conserving) neutral current structure of gauge theories. These are inclusive scattering off heavy nuclei (175), elastic scattering off protons (176), semi-inclusive (single pion) scattering off heavy nuclei (177), and elastic scattering off electrons (178). The first three can be used to calculate the neutral current couplings of u and d quarks. In SU(2) × U(1) models these quark couplings are given by:

$$q_L = \tau_3^L - Q \sin^2 \theta_W$$
$$q_R = \tau_3^R - Q \sin^2 \theta_W$$

7.1

where L (R) refers to left-handed (right-handed), τ_3 is the weak isospin ($\pm\frac{1}{2}$ for quarks in doublets, 0 in singlets), Q is the charge of the quark, and θ_W is the Weinberg angle, which is a free parameter of this theory. The inclusive and elastic scattering results are usually reported as ratios of neutral current to charged-current cross sections for both neutrinos and antineutrinos; the semi-inclusive scattering experiments give ratios of π^+ to π^- (in the current fragmentation region). With these six numbers, the possible couplings (u_L, u_R, d_L, d_R) are severely limited. Some of the data are shown in Figures 21 and 22. Analyses have been done by many authors (179). An analysis of Hung & Sakurai (179) [who make use of conclusions of Sehgal (179)] finds that there are only two sets of couplings for u and d quarks allowed by the data. These are shown in Table 14 (where the uncertainties are always ± 0.15). Note that if all four signs are changed in set A or in set B, the resulting sets of couplings are, of course, equally allowed. If $\sin^2 \theta_W = 0.3$ is chosen, then the WS model predicts that u_L, d_L, u_R, d_R are 0.3, -0.4, -0.2, 0.1. This is very close to set A of allowed couplings. There may be other models such as the SU(2) × U(1) model with (u b)$_R$ (see Section 5) that have values similar to those of set B. Note, however, that the parameter θ_W is attributable only to a specific model, and other models may fit the data for different values of that parameter. If there is any shortcoming to the above analysis, it is that these results depend crucially on use of specific parton model assumptions in the analysis of the semi-inclusive data. Since that data was taken at very low energies where parton model assumptions could be questioned, it would be best to confirm the conclusions by independent means. A new analysis, which is near completion, by Abbott & Barnett (179) (based on very new data) will try to use independent methods to further isolate the allowed values of the neutral current couplings of u and d quarks.

There are three types of neutrino-electron elastic scattering experiments

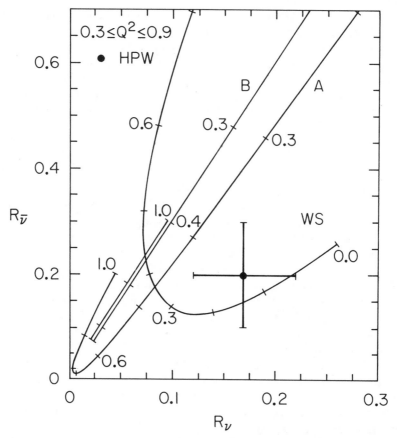

Figure 22 The ratio $\sigma(\bar{\nu}p \to \bar{\nu}p)/\sigma(\bar{\nu}p \to \mu^+n)$ vs the ratio $\sigma(\nu p \to \nu p)/\sigma(\nu n \to \mu^-p)$. The notation is the same as for Figure 21. The data shown are from D. Cline et al (176).

Table 14 Allowable couplings for u and d quarks (uncertainties are ± 0.15)

	u_L	d_L	u_R	d_R
A	$+0.29$	-0.40	-0.24	0
B	$+0.29$	-0.40	$+0.24$	0

(178) that have been reported: they are with ν_μ, $\bar{\nu}_\mu$, and $\bar{\nu}_e$ beams. With each cross section one can determine a locus of points in the $g_A - g_V$ plane consistent with the 90% confidence level upper and lower bounds for that cross section. Each of these is an annulus, and in Figure 23 the intersection of these three regions (which is shaded) is the allowed region. The WS model with $\sin^2\theta_W = 0.25$–0.3 lies within the lower part of the allowed region. Some other models lie in the upper part of the allowed region.

There are experiments testing weak neutral currents that do not involve neutrinos. These concern effects that arise from parity violation, which is

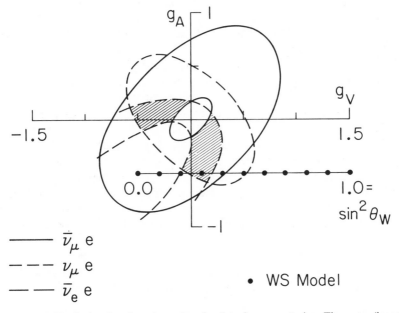

Figure 23 The limits placed on g_A and g_V by data for νe scattering. The outer (inner) lines indicate 90% confidence upper (lower) limits. The shaded regions are the overlap or allowed regions for g_A and g_V. The line with dots for tenth values of $\sin^2\theta_W$ is the prediction of the WS model. The data are from Reference 178 (the data of Reithler, which were not used, would give somewhat larger values of g_A and g_V).

possible in weak neutral currents in contrast to electromagnetic currents (these are purely vector and conserve parity). Among such experiments are those parity-violating transitions in heavy atoms (bismuth, thalium, cesium), in light atoms (hydrogen and deuterium), and in nuclei. There are also experiments that measure polarization asymmetries in electron-nucleon deep inelastic scattering and in $e^+e^- \to \mu^+\mu^-$.

The experiments involving bismuth are already reporting results that are consistent with zero parity violation. The Oxford group finds the measured optical rotation to be $(2.7 \pm 4.7) \times 10^{-8}$ rad, while the Washington group finds $(-0.7 \pm 3.2) \times 10^{-8}$ rad (180). The optical rotation measured by this type of experiment on heavy nuclei should be dominated by the interference term $A_{electron} V_{hadron}$ rather than by $V_{electron} A_{hadron}$. There is some controversy concerning the atomic and nuclear theory calculations (181, 182); however, the best estimates are that for the WS model the Oxford experiment should find -15×10^{-8} rad and the Washington experiment -12×10^{-8} rad. Within SU(2) $\times$ U(1) one can obtain a zero result for the bismuth experiment if the electron is given vector couplings ($A_e = 0$); the electron is vector if it has a coupling $(N_e e^-)_R$ in addition to $(\nu_e e^-)_L$ (where N_e is a heavy neutral lepton).

There are some models, notably $SU(2)^L \times SU(2)^R \times U(1)$ models (see Section 5), that expect both $A_e V_h$ and $V_e A_h$ to be zero (to first order). These models have two weak bosons (Z_V^0 and Z_A^0), one with purely axial-vector couplings to all fermions (except neutrinos) and one with purely vector couplings. There are experiments that are sensitive to both of the VA terms and have the added feature that they lack the theoretical difficulties of experiments on heavy nuclei (182). These experiments are performed on hydrogen and deuterium, and involve either atomic transitions or electron-nucleon deep inelastic scattering. While the theory is clear, it will be difficult for these experiments to obtain sufficient sensitivity to distinguish among various models. Were the absence of parity violation to be confirmed by these experiments (which expect results in the next year or two), it would be a serious problem for the WS-GIM model.

The neutral current interactions present a serious challenge to any gauge theory of quarks and leptons. A theory that could account for the wide range of data discussed here would be most impressive.

8 CONSERVATION LAWS

In the construction of gauge theories that incorporate more than four quarks (and leptons), the question of mixing among fermions must be considered again. It was already clear from Cabibbo mixing that the weak interaction eigenstates were not identical to the mass eigenstates. While a deep understanding of the cause of this mixing is still lacking, the phenomenological consequences of it should not be overlooked. One important consequence can be the breakdown of certain conservation laws. Two relevant conservation laws are those for CP (the product of charge conjugation and parity) and for muon number. The violation of these quantities is quite small: CP-violating decays of K^0 mesons are about 10^{-3} of CP-conserving decays ("milliweak"), and muon-number-violating decays of muons have never been observed and are less than 10^{-8} of muon-number-conserving decays. The understanding of such conservation laws and their breakdown is a crucial step in building a theory of quarks and leptons.

8.1   *CP Violation*

The theory of CP violation has been studied (183) for many years. A variety of approaches has been considered involving left-handed and sometimes right-handed currents. Here attention will be limited to the case of the WS model, although some results are applicable to other models. Consideration of CP violation in weak interactions involves not only the question of how it occurs, but also of why it is milliweak. In the WS model, the possibility of CP violation depends in an important way on the number of quarks. If there were only four quarks (u, c, d, s) and only left-handed currents, then CP would be completely conserved in the quark sector. Weinberg (184) and Sikivie (185) have proposed that CP violation could occur only in Higgs exchange in such models, which can automatically give a milliweak violation.

If there are six (or more) quarks, then one expects to find CP violation, which a priori need not be small. In contrast to the four-quark case where the weak coupling matrix has one parameter (the Cabibbo angle θ_C), the WS model with six quarks [discussed first by Kobayashi & Maskawa (137)] has four parameters. They can be taken to be four angles, θ_C, θ_1, θ_2, and δ, in terms of which the weak coupling matrix is:

$$\begin{pmatrix} C_C & -S_C C_2 & -S_C S_2 \\ S_C C_1 & C_C C_1 C_2 - S_1 S_2 e^{i\delta} & C_C C_1 S_2 + S_1 C_2 e^{i\delta} \\ S_C S_1 & C_C S_1 C_2 + C_1 S_2 e^{i\delta} & C_C S_1 S_2 - C_1 C_2 e^{i\delta} \end{pmatrix}, \qquad 8.1$$

where the rows correspond to the quarks u, c, and t, the columns to d, s, and b, and $C_C \equiv \cos\theta_C$, $C_1 \equiv \cos\theta_1$, etc. CP violation cannot be calculated since three angles are not known; however, there are experimental results that limit the possible values of θ_1 and θ_2 and allow some comment on the expected magnitude of CP violation.

In this generalized case, θ_C must still have the usual value ($\theta_C \approx 13°$). From the universality of quark and lepton couplings, Ellis et al (151) find that the $u\bar{b}$ coupling $\sin^2 \theta_C \sin^2 \theta_2 < 0.003$, so that $\sin^2 \theta_2 < 0.06$. Given this limit, the fact that charmed particles decay dominantly to strange particles leads to no useful limit on θ_1 (only $\sin^2 \theta_1 \lesssim 0.8$). Following the method of Gaillard & Lee (13), the $K_L - K_S$ mass difference can set some bounds on $\sin^2 \theta_1$ (151), depending on several factors: the c quark mass, the t quark mass, and the quantitative accuracy of the Gaillard-Lee estimate. If $\cos^2 \theta_2 \approx 1$, then

$$\sin^2 \theta_1 \approx \{a + [a^2 + (f-1)\eta b]^{\frac{1}{2}}\}b^{-1},$$

8.2

where $a \equiv \eta + \eta \ln \eta$, $b \equiv 1 + \eta + 2\eta \ln \eta$, $\eta \equiv m_c^2/m_t^2$, and f is a factor measuring the multiplicative deviation from the Gaillard-Lee estimate. Clearly if that estimate were exact ($f = 1$), then $\sin^2 \theta_1 = 0$ (note that a is negative). If $f = 2$, one finds $\sin^2 \theta_1 = 0.24$ for $\eta = 0.1$ ($m_t \approx 5$ GeV) and $\sin^2 \theta_1 = 0.07$ for $\eta = 0.01$ ($m_t \approx 15$ GeV). If $f = 5$, one finds $\sin^2 \theta_1 = 0.61$ ($\eta = 0.1$) and 0.25 ($\eta = 0.01$).

With this information plus a guess for δ, one can estimate (see 151, 186) the ratio of the CP-violating to the CP-conserving parts of the K^0 mass matrix:

$$|\varepsilon| \approx 2^{-\frac{1}{2}} \left| \frac{Im \, M_{12}^K}{\Delta m^K} \right| \approx 2^{\frac{1}{2}} \sin \delta \sin \theta_1 \sin \theta_2 \left(\frac{b \sin^2 \theta_1 - a}{b \sin^4 \theta_1 - 2a \sin^2 \theta_1 + \eta} \right),$$

8.3

where a, b and η are defined above and δ is the phase in the weak coupling matrix 8.1. If one chooses $\theta_C = \theta_1 = \theta_2 = \delta$ (which puts all angles below the experimental upper limits), then one finds the calculated $|\varepsilon|$ to be 10 times the observed value (187) (which is about 2×10^{-3}) for $\eta = 0.1$ and 40 times it for $\eta = 0.01$. Alternatively, given the observed CP violation and choosing intermediate values for θ_1 and θ_2, one can determine δ. For $\sin^2 \theta_2 = 0.03$ and $f = 1.5$ (which gives $\sin^2 \theta_1 = 0.15$ for $\eta = 0.1$ and 0.05 for $\eta = 0.01$), the values obtained are $\sin^2 \delta = 2 \times 10^{-4}$ for $\eta = 0.1$ and $\sin^2 \delta = 5 \times 10^{-5}$ for $\eta = 0.01$.

To summarize, the WS model with six quarks does give CP violation. By choosing the angles in the weak coupling matrix to be sufficiently small, one certainly can obtain the correct magnitude for CP violation. If, however, a random choice of angles is made (within the bounds described above), the predicted CP violation can be one or two orders of magnitude larger than the observed violation. While the magnitude of CP violation cannot be predicted accurately, the CP-violating terms are clearly much smaller than those for nonrare decays, so that the usual K^0 decay phenomenology is obtained qualitatively. It is possible, of course, that there are symmetry arguments or other reasons why θ_1, θ_2, and/or δ must be small.

This analysis also gives information concerning the coupling strengths for various charged-current terms useful in other sections of this review. Using the coupling matrix 8.1, the $u\bar{b}$ coupling squared is proportional to $\sin^2 \theta_C \sin^2 \theta_2$, which is less than 0.003 compared to the $u\bar{d}$ coupling. The $t\bar{d}$ coupling squared is proportional to $\sin^2 \theta_C \sin^2 \theta_1 \lesssim 0.03$. Furthermore, the ratios of couplings squared for $t\bar{s}/t\bar{d}$ and $c\bar{b}/u\bar{b}$ are both greater than 10 for most but not all angles θ_1 and θ_2. The small CP violation indicates that at least one of the angles θ_1, θ_2, and δ must be even smaller, but it does not indicate which one(s).

8.2   *Muon-Number Nonconservation*

Among the interesting tools for understanding the structure of the weak and electromagnetic interactions are experiments searching for processes such as $\mu \to e\gamma$, $\mu \to eee$, and $\mu^- N \to e^- N$. While the standard theories expect lepton number to be conserved, muon number may be violated in higher order diagrams. It is assumed that μ and ν_μ have muon-number one and all other particles have zero. Here, three means of finding muon-number violation in $SU(2) \times U(1)$ models are discussed.

In the context of the WS model (although it is applicable elsewhere) Bjorken & Weinberg (188) consider the interactions of leptons with Higgs scalars:

$$
H = -g_1 \overline{\begin{pmatrix} \nu_\mu \\ \mu^- \end{pmatrix}}_L \begin{pmatrix} \phi_1^+ \\ \phi_1^0 \end{pmatrix} \mu_R^- - g_2 \overline{\begin{pmatrix} \nu_e \\ e^- \end{pmatrix}}_L \begin{pmatrix} \phi_2^+ \\ \phi_2^0 \end{pmatrix} \mu_R^-
$$

$$
-g_3 \overline{\begin{pmatrix} \nu_\mu \\ \mu^- \end{pmatrix}}_L \begin{pmatrix} \phi_3^+ \\ \phi_3^0 \end{pmatrix} e_R^- - g_4 \overline{\begin{pmatrix} \nu_e \\ e^- \end{pmatrix}}_L \begin{pmatrix} \phi_4^+ \\ \phi_4^0 \end{pmatrix} e_R^- + \text{h.c.}
$$

8.4

where the ϕ_i are linear combinations (not necessarily independent) of several scalar fields of definite mass. Since the μ and e are defined as the physical states found in the diagonalization of the mass matrix, if there is only one Higgs doublet (as is sometimes assumed), then g_2 and g_3 must be zero. However, if there is more than one Higgs doublet, then in general it is possible that g_2 and/or g_3 are nonzero, and virtual Higgs scalars will give physical transitions between μ and e such as those shown in Figure 24. Because the Higgs coupling to the light leptons is so weak, the two-loop diagrams (Figure 24b), in general, dominate one-loop diagrams (Figure 24a):

$$
\frac{1 \text{ loop}}{2 \text{ loops}} \approx \frac{2\pi}{\alpha} \left(\frac{m_\mu}{m_H}\right)^2.
$$

8.5

Bjorken & Weinberg roughly estimate

$$\frac{\mu \to e\gamma}{\mu \to e\nu\bar{\nu}} \lesssim 10^{-8},$$

8.6

depending on the amount of mixing among the Higgs scalars.

No muon-number violation has been observed yet. The present experimental limit (90% confidence level) for $B(\mu \to e\gamma)$ is 3.6×10^{-9} (189).

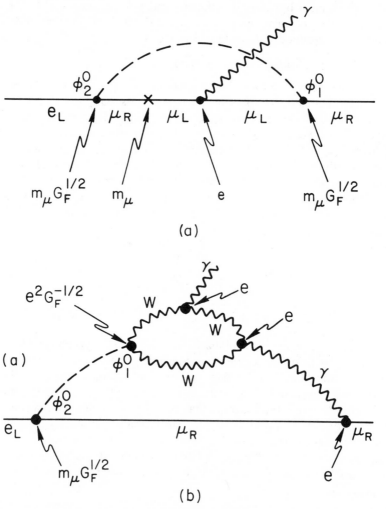

(a)

(b)

Figure 24 (a) One- and (b) two-loop diagrams in which virtual Higgs exchange leads to the decay $\mu \to e\gamma$. This figure was taken from Reference 188.

In the model, the decay $\mu \rightarrow 3e$ was expected to be very small. The decay $K_L \rightarrow \mu e$ is forbidden in lowest order (or one would get strangeness-changing neutral currents). They predict

$$\frac{\sigma(\mu^- N \rightarrow e^- N)}{\sigma(\mu^- N \rightarrow \nu N')} \sim 4 \times 10^{-9}, \qquad 8.7$$

where N is a nucleus, and the experimental limit is 1.6×10^{-8} (190).

In models that also have right-handed currents there is another source of muon-number violation. This source, discussed first by Cheng & Li and by Bilenkii et al (191), involves the mixing of massive neutral leptons that have right-handed couplings to the electron and muon, $(N'_e e^-)_R$ and $(N'_\mu \mu^-)_R$. In analogy with the Cabibbo mixing of the d and s quarks, they suggest:

$$N'_e = N_e \cos \phi + N_\mu \sin \phi$$
$$N'_\mu = -N_e \sin \phi + N_\mu \cos \phi. \qquad 8.8$$

Then clearly if one considers the simple one-loop diagram of Figure 25, there will be a GIM-like cancellation. The cancellation is not complete, to the extent that N_e and N_μ have unequal masses; the amplitude for this $\mu \rightarrow e\gamma$ process is proportional to

$$\cos \phi \sin \phi \, (m_{N_\mu}^2 - m_{N_e}^2). \qquad 8.9$$

Bjorken, Lane & Weinberg (192; this paper refers to much of the literature on muon-number violation) argue that the Higgs couplings that give masses and lead to the above mixing also cause small but finite mixing of the left-handed parts of N_e and N_μ with ν_e and ν_μ. This mixing is of order m_μ/m_N. There are, as a result, left-right diagrams in addition to the right-right diagram, Figure 25. These left-right terms have the same form as the right-right terms, but their amplitude is multiplied by -6. If the value of Expression 8.9 is 1 GeV^2, then (incorporating the Bjorken-Lane-Weinberg modification)

$$\frac{\mu \rightarrow e\gamma}{\mu \rightarrow e\nu\bar{\nu}} \approx 4 \times 10^{-10}. \qquad 8.10$$

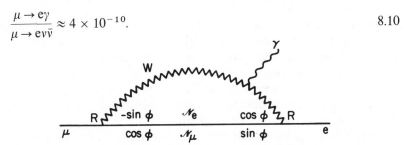

Figure 25   One of the diagrams in which N_e and N_μ exchange leads to the decay $\mu \rightarrow e\gamma$.

Cheng & Li estimate the branching ratio for $\mu \rightarrow e e \bar{e}$ to be about 10^{-11} [where the experimental limit (193) is 6×10^{-9}], and for $\mu^- N \rightarrow e^- N$ to be as large as 10^{-9}. For $m_{N_\mu}/m_{N_e} \approx 4$, they find the branching ratio for $K_L \rightarrow e\bar{\mu}$ to be about 10^{-10} [the experimental limit is 2×10^{-9} (194)].

Glashow (195) and Fritzsch (196) have shown that muon number can be violated in models without right-handed currents and with only one Higgs doublet. If the charged heavy lepton τ has a left-handed coupling to a massive neutral lepton N_τ, then N_τ can mix with v_e and v_μ. Decays such as $\mu \rightarrow e\gamma$ could occur in the same fashion as proposed by Cheng & Li and by Bilenkii et al, where Δm^2 is replaced with $m_{N_\tau}^2$.

The mixed states can be written as:

$$v'_e = v_e \cos \theta + N_\tau \sin \theta$$

$$v'_\mu = v_\mu \cos \phi + (-v_e \sin \theta + N_\tau \cos \theta) \sin \phi$$

$$N'_\tau = (N_\tau \cos \theta - v_e \sin \theta) \cos \phi - v_\mu \sin \phi. \qquad 8.11$$

Both angles can be shown to be small by the requirement of universality (seen through μ and β decays) and by the lack of v_e in v'_μ (v_μ do not produce electrons in scattering). If $\mu \rightarrow e\gamma$ were observed at the 10^{-10} to 10^{-9} level, then the smallness of the angles θ and ϕ requires that m_{N_τ} be quite large (much larger than m_τ in fact). Since $m_\tau < m_{N_\tau}$, the heavy lepton τ can only decay through the mixing of N_τ with v_e and v_μ.

It would be possible to rule out this mode of muon-number violation by measuring the lifetime of τ carefully, but it probably will be difficult to obtain a better experimental limit than the present value of about 10^{-11} sec. Since the angle ϕ is so small, it would be very rare for v_μ scattering to produce τ leptons.

A precise measurement of the conservation or nonconservation of muon number would be a valuable tool for studying the existence, mixing, and currents of new leptons.

9   ATTEMPTS AT A GRAND SYNTHESIS

Although charm and many predictions of gauge theories for neutral and charged currents seem well verified, the experimental evidence in favor of strong and weak gauge theories remains somewhat indirect. From a theoretical point of view, however, the importance of gauge theories cannot be overemphasized. They present the first possibility for a theoretically consistent description of both the weak and electromagnetic interactions, and the strong interactions.

In fact, if the gauge theory framework applies generally, then there is every reason to believe that within it, a grand unification of all three interactions can be attained. It is an important enterprise to begin working on this, even though it is some ways premature. New questions can be raised and a framework provided for studying unsolved problems such as mass generation and the existence of mixing angles like θ_c.

We do not enter into a detailed discussion of the many models proposed for grand unification. However, there are several features and problems necessarily common to any specific model and it is possible to discuss the subject in general terms. This will give the reader some appreciation of these general features and will also serve as a conclusion for the entire review.

First of all, it is worth repeating and underscoring the two assumptions that form the foundation for the approach to grand unification to be discussed here.

1. The weak and electromagnetic interactions are described by a spontaneously broken gauge theory based on some Lie group G_w, perhaps one of the several models we have discussed. The success of the WS model in dealing with neutrino neutral current interactions indicates that it will contain SU(2) × U(1) as a subgroup. If G_w is larger than SU(2) × U(1), then presumably some of the gauge bosons of G_w will be considerably heavier than M_W and M_Z. The quarks and leptons are assumed to fill out low dimensional representations of G_w, with the total number "reasonably" small. If G_w is simple (or semisimple in the form $G'_w \times G'_w$, with a reflection symmetry relating the two factors), then a single coupling constant is involved. If SU(2) × U(1) is a subgroup, θ_w will then be determined before proceeding on to grand unification.

With the fermion content arranged to eliminate triangle anomalies, this theory will be renormalizable, and therefore exhibit only logarithmic growth with energy. The coupling strength at laboratory energies will be of order $\alpha = \frac{1}{137}$. At higher energies E, the effective coupling constant $\alpha(E)$ (see Section 2) can change logarithmically but the effect is not significant until $\alpha \log (E/E_{lab})$ becomes of order unity.

2. The underlying theory of strong interactions is QCD (Equation 2.2). QCD is renormalizable, and providing that the total number of quark flavors f is less than 16.5 (Equation 2.15), it is asymptotically free. The running coupling constant $\alpha_s(-q^2)$ for $q^2 > 1$–2 GeV^2 takes the form

$$\alpha_s(-q^2)$$
$$= \frac{\alpha_\mu}{1 + \frac{11\alpha_\mu}{4\pi} \ln\left(\frac{-q^2}{\mu^2}\right) - \frac{\alpha_\mu}{\pi} \sum_{i=1}^{f} \int_0^1 dz\, z(1-z) \ln\left(\frac{m_i^2 - q^2 z(1-z)}{m_i^2 + \mu^2 z(1-z)}\right)},$$

9.1

which agrees with Equations 2.15 and 2.16 if $-q^2, \mu^2 \gg m_i^2$. If $-q^2, \mu^2 \ll m_i^2$ for some flavors, then those quarks can be seen to "decouple" (197) in $\alpha_s(-q^2)$. Even if $f > 16.5$, a temporary asymptotic freedom could sustain itself until a q large enough to vacuum polarize the seventeenth flavor is attained. Nevertheless, in the spirit of assumption 1, we shall take $f < 16.5$ so that the asymptotic freedom is truly asymptotic.

The essential notion in superunification is that both QCD and the weak and electromagnetic theory must be viewed as low energy theories. Since they are both renormalizable theories with only logarithmic variation in energy, the range of energies over which they can be viewed as independent theories is necessarily very large. Nevertheless, it is possible to imagine that at some extremely large energy, the strong, weak and electromagnetic interactions will be described by a single theory—a gauge theory, of course—based on some Lie group G, which contains $G_w \times SU(3)_c$.

If a spontaneous symmetry breakdown takes place at some extremely large mass scale M, then it is possible that some subset of the gauge bosons acquires a mass of order M leaving the subgroup $G_w \times SU(3)_c$ unbroken. The G_w and $SU(3)_c$ gauge bosons will remain massless at this level. At energy scales somewhat below M, the exchange of the superheavy bosons of mass M will be suppressed, and it will appear as if QCD and the weak and electromagnetic theory are two separate field theories. It is possible that the spontaneous symmetry breakdown is a multistep process. If G_w is larger than $SU(2) \times U(1)$, it could break down at some scale $M' \ll M$ (but $M' \gg M_W, M_Z$). The final step at mass scale M_W, M_Z would then leave only the $U(1)$ subgroup intact, with a massless photon.

That is the scenario in rough outline. Its actual implementation depends on finding the right group G and probably on understanding spontaneous symmetry breaking and the Higgs mechanism much more deeply than we do now. Nevertheless, several features and consequences of the program seem to be understood. We offer a list of those we consider to be most important and then conclude with a partial and subjective list of the many unsolved problems.

1. If the grand unification group G is simple (or semisimple in the form $G' \times G'$, with a reflection symmetry relating the two factors), the unified theory will involve a single coupling constant. It is then possible to estimate the order of magnitude of M, the unification mass scale (198, 199). The QCD effective coupling constant $\alpha_s(-q^2)$ is already small at $-q^2 = 10$ GeV^2. Estimates range from about 0.2 based on charmonium decay to about 0.5 based on analyses of electroproduction. This, however, is still much larger than the weak coupling constant (of order $\alpha = \frac{1}{137}$), and if these theories are to coalesce into a single-coupling-constant

theory, then something must bring the coupling strengths together. It is primarily the logarithmic decrease of $\alpha_s(-q^2)$ with $-q^2$ that does this. This decrease can bring $\alpha_s(-q^2)$ down to weak and electromagnetic strength at very high momentum scales, but where the effective weak and electromagnetic coupling strength still has not changed much from laboratory energies. The grand unification mass M can be estimated roughly by equating the effective coupling strengths. The result depends on several unknown factors, such as the starting value $\alpha_s(10 \text{ GeV}^2)$, the number of quark flavors, the contribution of Higgs bosons, and the mass scales characterizing the various possible levels in the symmetry-breaking chain. Most estimates (198, 199), however, have ranged between $M \approx 10^{16}$ GeV and $M \approx 10^{19}$ GeV, far beyond laboratory energies. One intriguing feature of these estimates is that they imply the existence of elementary particles with masses on the order of the Planck mass $G^{-\frac{1}{2}} = 1.22 \times 10^{19}$ GeV. This suggests that grand unification may necessarily involve the gravitational interaction, a possibility to which we return shortly.

2. The prototype for grand unification into a simple group is the $G = SU(5)$ model of Georgi & Glashow (200). It has a maximal subgroup structure of $SU(2) \times U(1) \times SU(3)_c$, so that the usual WS model and QCD are naturally incorporated. If experiments force us to go beyond $SU(2) \times U(1)$ for G_w, then some larger group will have to be used for G. Some possibilities are $SO(10)$ (201) and the exceptional groups E_6 and E_7 proposed by Gürsey and collaborators (125). The group, E_6 for example, contains $SU(3) \times SU(3) \times SU(3)_c$ and can accommodate six quarks and four charged leptons in two 27-plets. The possibility of $G_w = SU(3) \times SU(3)$ was already discussed in Section 5 and it might be an attractive possibility. However, the study of symmetry breakdown in the E_6 model indicates that some of the $SU(3) \times SU(3)$ bosons will become superheavy so that G_w is some proper subset of $SU(3) \times SU(3)$ (202).

3. Once $SU(2) \times U(1)$ is embedded in a larger simple group, the value of the Weinberg angle is determined. The value of $\sin \theta_w$ depends on the structure of G and on the energy scales characterizing the stages of symmetry breakdown to $SU(2) \times U(1)$. In the $SU(5)$ model, for example, the value of $\sin^2 \theta_w$ at the unification mass scale M is $\frac{3}{8}$. This will be reduced by renormalization effects at laboratory energies since the effective coupling constants of the $SU(2)$ and $U(1)$ subgroups scale differently with q. Numerical estimates lead to $\sin^2 \theta_w \approx 0.2$ (198, 199) and the range of values allowed by experiment is now $0.2 \leqq \sin^2 \theta_w \leqq 0.3$ (see Section 7). A similar prediction is obtained in the E_6 model (202). It is premature to take these estimates of $\sin^2 \theta_w$ too seriously, but they

might be useful in at least excluding some groups G. For example, the choice $G = E_7$ leads to $\sin^2 \theta_W = \frac{3}{4}$ at the grand unification mass M, and it would appear that this is too large to be brought into agreement with experiment by renormalization effects (203).

4. An important feature of grand unification is that quarks and leptons are placed together in single representations of G. Thus, there will be gauge bosons (sometimes called leptoquarks) that connect leptons to quarks and lead to the breakdown of separate lepton and quark conservation. This is a potential disaster since the experimental lower bound on the lifetime for such transitions is incredibly large. Quark nonconservation can lead to baryon nonconservation, and the lower limit on the proton lifetime is 2×10^{30} years (204)! However, if the leptoquark mass is on the order of the grand unification mass M predicted from renormalization group considerations, then lifetimes at least this long can be obtained. The decay of a proton into a lepton plus pions can proceed by single leptoquark exchange (205), so that the lifetime will be proportional to M^4/M_H^5 where M_H is some typical hadronic mass scale. Furthermore, the constant of proportionality might be expected to be of order $\alpha^{-2}(M^2) \geq 10^3$ (199). With $M_H \leq 1$ GeV and $M \geq 10^{16}$ GeV, one obtains proton lifetimes well in excess of 10^{30} years. Although it appears to be sufficiently suppressed, baryon nonconservation is a natural feature of many grand unification models[21] (205).

The grand unification scenario is attractive, but there are too many unanswered questions and loose ends to be sure that it is the wave of the future. A short list of some of these problems should point up the limitations of the present theoretical framework, both for dealing with grand unification and perhaps even for understanding the weak and electromagnetic interactions alone.

1. The nature of spontaneous symmetry breakdown is very poorly understood. The only known way of implementing the breakdown and the Higgs mechanism is by explicitly introducing multiplets of elementary Higgs fields into the Lagrangian. For the $SU(2) \times U(1)$ model, this required only one complex doublet and led to only one physical Higgs boson. However, for the large groups G required for grand unification, or even for larger weak and electromagnetic groups G_w, several very big multiplets of Higgs fields are needed to get a reasonable pattern of symmetry breaking (125, 201). The notion of fundamental Higgs fields becomes uneconomical, if not downright unpalatable.

A possibility with some attraction is that the spontaneous symmetry

---

[21] It is possible to push the decay to higher orders in the coupling constant. For an extensive discussion of grand unification and proton stability, see 205.

breakdown is "dynamical," that is, not induced by fundamental Higgs fields in the Lagrangian. The Goldstone bosons incorporated into the gauge fields would presumably be bound states of quarks. Whether and how this kind of dynamical symmetry breakdown takes place is unknown, and, in fact, it is hard to see what kind of force could produce the necessary binding. Even if dynamical breakdown is a possibility, in moving up from SU(2) × U(1) to G, it is perhaps reasonable to introduce Higgs fields at each stage. They may appear fundamental at one level if not at all energy scales.

2. There seems to be almost no understanding of quark and lepton masses and weak mixing angles. Hopefully, these fundamental questions will find their solution within the framework of spontaneous symmetry breakdown, but just how is far from clear. (For a recent discussion along with some speculation and references to other work, see 206.) Many ingredients will presumably enter the solution: the unifying groups G_w and G; spontaneous symmetry breakdown; and the effects of strong, weak, and electromagnetic renormalization. Empirical relations among these parameters, such as $\theta_c^2 \approx m_d/m_s$ and $m_e/m_\mu \approx m_u/m_c$, are tantalizing but they can be insidiously misleading.

3. Perhaps grand unification into a Lie group G, without also an incorporation of gravity, is impossible. The natural appearance of mass scales on the order of the Planck mass $G^{-\frac{1}{2}}$ suggests this possibility. At such extremely-small-distance scales, the gravitational interaction is probably comparable to the other fundamental forces, and it could play an important role, for example, in driving spontaneous symmetry breakdown. Just how this might work is very poorly understood, but at least one development offers promise. The concept of supersymmetry (207), which relates fermions and bosons, has recently been combined with the notion of local gauge invariance to produce a class of theories known as supergravity (208). These theories contain a graviton, one or more spin-$\frac{3}{2}$ particles known as gravitinos, and a host of lower spin particles. They have the possibility of being renormalizable (209), but whether or not realistic theories of grand (supergrand?) unification can be constructed along these lines is not yet clear. If supergravity is the road to grand unification, then the group G will not be a simple Lie group. Instead, it will be a graded (or super) Lie group, with an algebra containing both commutation and anticommutation relations.

It is important to think about these deep theoretical questions, but if the recent past is something of a guide, progress will come in more modest steps with experiment playing an important, and possibly leading, role. It is hard to think of another period in particle physics when so many important experiments were either under way or in the planning stages. Within the next few years we may well be able to offer plausible, if not completely accepted, answers to some questions of limited scope but of great import:

1. Is everything "in order" with charmonium and the charmed particles? The pseudoscalar $c\bar{c}$ states are especially puzzling.
2. To what extent is the upsilon a new, improved charmonium system? Will it give us important information about the quark-antiquark potential in the transition region from linearity to short distances ($\lesssim 1$ GeV^{-1})?
3. Will QCD remain a viable theory of strong interactions?
4. Will spontaneously broken gauge theories remain a viable theoretical framework for weak and electromagnetic interactions? The most urgent need here is for some evidence in favor of the existence of intermediate vector bosons—perhaps in the mass range 60–100 GeV. Do Higgs bosons, either elementary or composite, exist in an accessible mass range?
5. What is the correct gauge group of weak and electromagnetic interactions G_w? How many quarks and leptons are there and how do they fill out representations of G_w? The weak interaction properties of the τ lepton and the b quark and the question of parity violation in atomic physics are central here.
6. What about masses and mixing angles? Are the neutrinos ν_e, ν_μ, and ν_τ massless? Are muon and tau lepton number separately conserved at more stringent levels? Will the pattern of quark and lepton masses offer some clue to the understanding of mass generation?

If the near future provides us with answers to some of these questions, it may be an even more exciting era in particle physics than the recent past.

ACKNOWLEDGMENTS

We wish to thank many of our colleagues for conversations that were useful in preparation of this review, especially J. Bjorken, C. Carlson, E. Eichten, J. Ellis, G. Feldman, F. Gilman, S. Glashow, K. Gottfried, F. Gürsey, F. Martin, C. Quigg, J. Rosner, R. Suaya, and S. Weinberg. Two of us (T.A. and K.L.) wish to acknowledge the hospitality of the SLAC theory group and two of us (M.B. and K.L.) wish to acknowledge the hospitality of the Yale theory group. It has been a privilege for all three of us to have known and to have learned from Ben Lee.

Literature Cited

1. Gell-Mann, M. 1964. *Phys. Lett.* 8:214; Zweig, G. 1964. *CERN Rep. TH.401, TH.412*
2. Greenberg, O. W. 1978. *Ann. Rev. Nucl. Part. Sci.*: 28:327–86
3. Bjorken, J. D., Glashow, S. L. 1964. *Phys. Lett.* 11:255; Amati, D., et al. 1964. *Phys. Lett.* 11:190; Tarjanne, P., Teplitz, V. L. 1963. *Phys. Rev. Lett.* 11:447; Hara, Y. 1964. *Phys. Rev. B* 134:701; Maki, Z., Ohnuki, Y. 1964. *Prog. Theor. Phys.* 32:144
4. Glashow, S. L., Iliopoulos, J., Maiani, L. 1970. *Phys. Rev. D* 2:1285
5. Carithers, W. C. 1973. *Phys. Rev. Lett.* 30:1336, 31:1025; Fukushima, Y. 1976. *Phys. Rev. Lett.* 36:348
6. Cabibbo, N. 1963. *Phys. Rev. Lett.* 10:531
7. Weinberg, S. 1967. *Phys. Rev. Lett.* 19:1264; Salam, A. 1968. *Elementary Particle Physics: Relativistic Groups and Analyticity* (Nobel Symp. 8), ed. N. Svartholm, p. 367. Stockholm: Almquist & Wiksell.
8. Yang, C. N., Mills, R. 1954. *Phys. Rev.* 96:191
9. 't Hooft, G. 1971. *Nucl. Phys. B* 33:173; Lee, B. W. 1972. *Phys. Rev. D* 6:1188; Lee, B. W., Zinn-Justin, J. 1972. *Phys. Rev. D* 5:3132, 3137, 3155
10. Barish, B. C. et al. 1977. *Phys. Rev. Lett* 39:1595
11. Goldstone, J. 1961. *Nuovo Cimento* 19:154; Goldstone, J., Salam, A., Weinberg, S. 1962. *Phys. Rev.* 127:965
12. Higgs, P. W. 1964. *Phys. Rev. Lett.* 12:132, 13:508; 1966. *Phys. Rev.* 145:1156; Guralnik, G. S., Hagen, C. R., Kibble, T. W. B. 1964. *Phys. Rev. Lett.* 13:585; Englert, F., Brout, R. 1964. *Phys. Rev. Lett.* 13:321
13. Gaillard, M. K., Lee, B. W. 1974. *Phys. Rev. D* 10:897
14. Litke, A. et al. 1973. *Phys. Rev. Lett.* 30:1189; Tarnopolsky, G. et al. 1974. *Phys. Rev. Lett.* 32:432
15. Richter, B. 1974. *Proc. Int. Conf. High Energy Phys., 17th, London, 1–10 July 1974*, p. IV-37. Rutherford Laboratory: Sci. Res. Counc.
16. Burmester, J. et al. 1977. *Phys. Lett. B* 66:395
17. Benvenuti, A. et al. 1975. *Phys. Rev. Lett.* 35:1199, 1203, 1249; 1977. *Phys. Rev. Lett.* 38:1183; Barish, B. C. et al. 1976. *Phys. Rev. Lett.* 36:939; 1977. *Phys. Rev. Lett.* 39:981; von Krogh, J. et al. 1976. *Phys. Rev. Lett.* 36:710; Holder, M. et al. 1977. *Phys. Lett. B*

69:377, 70:396; Deden, H. et al. 1977. *Phys. Lett. B* 67:474; Baltay, C. et al. 1977. *Phys. Rev. Lett.* 39:62; Bosetti, P. et al. 1977. *Phys. Rev. Lett.* 38:1248; Ballagh, H. C. et al. 1977. *Phys. Rev. Lett.* 39:1650
18. Augustin, J.-E. et al. 1974. *Phys. Rev. Lett.* 33:1406
19. Aubert, J. J. et al. 1974. *Phys. Rev. Lett.* 33:1404
20. Abrams, G. S. et al. 1974. *Phys. Rev. Lett.* 33:1453
21. Braunschweig, W. et al. 1975. *Phys. Lett. B* 57:407; Feldman, G. J. et al. 1975. *Phys. Rev. Lett.* 35:821; Tanenbaum, W. M. et al. 1975. *Phys. Rev. Lett.* 35:1323
22. Goldhaber, G. et al. 1976. *Phys. Rev. Lett.* 37:255; Peruzzi, I. et al. 1976. *Phys. Rev. Lett.* 37:569
23. Herb, S. W. et al. *Phys. Rev. Lett.* 39:252; Innes, W. R. et al. 1977. *Phys. Rev. Lett.* 39:1240; Kephart, R. D. et al. 1977. *Phys. Rev. Lett.* 39:1440
24. Perl, M. L. et al. 1975. *Phys. Rev. Lett.* 35:1489
25. Perl, M. L. et al. 1977. *Phys. Lett. B* 70:487
26. Politzer, H. D. 1973. *Phys. Rev. Lett.* 26:1346; Gross, D., Wilczek, F. 1973. *Phys. Rev. Lett.* 26:1343
27. Greenberg, O. W. 1964. *Phys. Rev. Lett.* 13:598
28. Fritzsch, H., Gell-Mann, M. 1973. *Proc. Int. Conf. High Energy Phys., 16th, Chicago-Batavia, Ill., 1972*, ed. J. D. Jackson, A. Roberts, Vol. 2, p. 135. Natl. Accel. Lab.: Batavia, Ill.
29. Bellettini, G. et al. 1970. *Nuovo Cimento A* 66:243; Kryshkin, V. I. et al. 1970. *J. Exp. Theor. Phys.* 30:1037; Browman, A. et al. 1974. *Phys. Rev. Lett.* 33:1400
30. Cabibbo, N., Parisi, G., Testa, M. 1970. *Nuovo Cimento Lett.* 4:35
31. Rapidis, P. et al. 1977. *Phys. Rev. Lett.* 39:526
32. Kirkby, J. 1977. *1977 Proc. Int. Symp. Lepton Photon Interactions High Energies*, ed. F. Gutbrod, p. 3. Hamburg, Germany: DESY
33. Yamada, S. 1977. See Ref. 32, p. 69
34. Burmester, J. et al. 1977. *Phys. Lett. B* 66:395; Knies, G. 1977. See Ref. 32, p. 93
35. Abers, E. S., Lee, B. W. 1973. *Phys. Rep.* 9C:1; For a review of many features of QCD see Marciano, W. J., Pagels, H. R. 1978. *Phys. Rep. C* 36:137
36. Fadde'ev, L. D., Popov, V. N. 1967. *Phys. Lett. B* 25:29

37. Appelquist, T., Dine, M., Muzinich, I. J. 1977. *Phys. Lett. B* 69:231; 1978. *Phys. Rev. D* 17:2074
38. Feinberg, F. 1977. *Phys. Rev. Lett.* 39: 316; *MIT Rep. CTP-687*; Fischler, W. 1977. *Nucl. Phys. B* 129:157
39. Gell-Mann, M., Low, F. 1954. *Phys. Rev.* 95:1300; Callan, C. 1970. *Phys. Rev. D* 2:1541; Symanzik, K. 1970. *Commun. Math. Phys.* 18:227
40a. Bjorken, J. D. 1969. *Phys. Rev.* 179: 1547
40b. Nachtmann, O. 1977. See Ref. 32, p. 811
41. Wilson, K. 1964. *Cornell Rep. LNS-64-15.* Unpublished; 1969. *Phys. Rev.* 179: 1499
42. Adler, S. L. 1974. *Phys. Rev. D* 10:3714
43. Appelquist, T., Georgi, H. 1973. *Phys. Rev. D* 8:4000; Zee, A. 1973. *Phys. Rev. D* 8:4038
44. Poggio, E., Quinn, H., Weinberg, S. 1976. *Phys. Rev. D* 13:1958; Shankar, R. 1977. *Phys. Rev. D* 15:755
45. Field, R. D. 1978. *Phys. Rev. Lett.* 40: 997
46. Gaillard, M. K., Lee, B. W. 1974. *Phys. Rev. Lett.* 33:108; Altarelli, G., Maiani, L. 1974. *Phys. Lett. B* 52:351; Shifman, M. A., Vainshtein, A. I., Zakharov, V. I. 1977. *Nucl. Phys. B* 120:316
47. Appelquist, T., Politzer, H. D. 1975. *Phys. Rev. Lett.* 34:43
48. Yao, Y.-P. 1976. *Phys. Rev. Lett.* 36:653; Appelquist, T. et al. 1976. *Phys. Rev. Lett.* 36:768; Kinoshita, T., Ukawa, A. 1976. *Phys. Rev. D* 13:1573; 1977. *Phys. Rev. D* 15:1596; Poggio, E., Quinn, H. 1976. *Phys. Rev. D* 14:578
49. 't Hooft, G. 1977. *Deeper Pathways in High Energy Physics—Orbis Scientiae 1977,* ed. A. Perlmutter, L. F. Scott. New York: Plenum. 699 pp.
50. Polyakov, A. 1975. *Phys. Lett. B* 59:82; Belavin, A. et al. 1975. *Phys. Lett. B* 59:85; 't Hooft, G. 1976. *Phys. Rev. Lett.* 37:8
51. Wilson, K. 1975. *Phys. Rev. D* 10:2445
52. Kogut, J., Susskind, L. 1975. *Phys. Rev. D* 11:395
53. De Rújula, A., Glashow, S. L. 1975. *Phys. Rev. Lett.* 34:46
54. Chang, N.-P., Nelson, C. A. 1975. *Phys. Rev. Lett.* 35:1492
55. Chew, G. F., Rosenzweig, C. 1976. *Nucl. Phys. B* 104:290
56. Novikov, V. A. et al. 1977. *Phys. Rev. Lett.* 38:626; 1977. *Phys. Lett. B* 67:409
57. Gottfried, K. 1977. See Ref. 32, p. 667
58. Feldman, G. J., Perl, M. L. 1975. *Phys. Rep. C* 19:234; Schwitters, R. F., Strauch, K. 1976. *Ann. Rev. Nucl. Sci.*

26:89–149; Wiik, B. H., Wolf, G. 1977. *DESY Rep. 77/01*; Chinowsky, W. 1977. *Ann. Rev. Nucl. Sci.* 27:393–464
59. Feldman, G. J., Perl, M. L. 1977. *Phys. Rep. C* 33:285
60. Okubo, S. 1963. *Phys. Lett.* 5:165; Zweig, G. 1964. See Ref. 1, Zweig; Iizuka, J. 1966. *Prog. Theor. Phys.* 37–38: Suppl. 21
61. Braunschweig, W. et al. 1977. *Phys. Lett. B* 67:243
62. Chanowitz, M. S., Gilman, F. J. 1976. *Phys. Lett. B* 63:178; Lane, K. 1975. Presented at Meet. Div. Part. Fields, Am. Phys. Soc., Seattle, WA, Aug., 1975; Eichten, E. et al. 1976. *Phys. Rev. Lett.* 36:500
63. Eichten, E. et al. 1975. *Phys. Rev. Lett.* 34:369
64. Kang, J. S., Schnitzer, H. J. 1975. *Phys. Rev. D* 12:841, 2791; Harrington, B. J., Park, S. Y., Yildiz, A. 1975. *Phys. Rev. Lett.* 34:168, 706
65. Appelquist, T., Politzer, H. D. 1975. *Phys. Rev. D* 12:1404
66. Tryon, E. P. 1972. *Phys. Rev. Lett.* 28:1605; Kogut, J., Susskind, L. 1974. *Phys. Rev. D* 9:3501; Wilson, K. G. 1974. *Phys. Rev. D* 10:2445
67. Willemsen, J. F. 1974. *Proc. Summer Inst. Part. Phys., July 29–Aug. 10, 1974,* ed. M. C. Zipf, Vol. 1, p. 445. SLAC, Stanford Univ., CA
68. Eichten, E. et al. 1978. *Phys. Rev. D* 17:3090
69. Appelquist, T. et al. 1975. *Phys. Rev. Lett.* 34:365; Callan, C. G. et al. 1975. *Phys. Rev. Lett.* 34:52
70. Martin, A. 1977. *Phys. Lett. B* 67:330; Grosse, H. 1977. *Phys. Lett. B* 68:343
71. Gottfried, K. 1978. *Phys. Rev. Lett.* 40:598
72. Van Royen, R. P., Weisskopf, V. F. 1967. *Nuovo Cimento A* 50:617
73. Barbieri, R., Gatto, R., Kögerler, R. 1976. *Phys. Lett. B* 60:183; Barbieri, R., Gatto, R., Remiddi, E. 1976. *Phys. Lett. B* 61:465
74. Chanowitz, M. 1975. *Phys. Rev. D* 12: 918; Okun, L., Voloshin, M. 1976. *Moscow Rep. ITEP-95-1976*; Brodsky, S. J. et al. 1978. *Phys. Lett. B* 73:203
75. Eichten, E. et al. 1976. *Phys. Rev. Lett.* 36:500
76. Isgur, N. 1976. *Phys. Rev. Lett.* 36:1262
77. De Rújula, A., Jaffe, R. 1977. *MIT Rep. CTP-658*; Lane, K. 1976. Presented at Meet. Am. Phys. Soc., Stanford, CA, Dec. 22–24, 1976
78. Giles, R., Tye, S.-H. H. 1977. *Phys. Rev. D* 16:1079; Horn, D., Mandula, J. 1978. *Phys. Rev. D* 17:298

79. Wilczek, F. 1977. *Phys. Rev. Lett.* 39:1304
80. Ellis, J., Gaillard, M. K., Nanopoulos, D. V. 1976. *Nucl. Phys. B* 106:292
81. Harari, H. 1976. *Phys. Lett. B* 64:469
82. Jackson, J. D. 1977. *Rep. TH 2305-CERN*; 1978. *Proc. Rencontre Moriond, 12th*, Mar. 1977. In press
83. Duncan, A. 1976. *Phys. Rev. D* 13:2866; Appelquist, T. 1975. *Caltech Rep. CALT-68-499*
84. De Rújula, A., Georgi, H., Glashow, S. L. 1976. *Phys. Rev. D* 12:147
85. Johnson, K. 1978. Private communication. *MIT Rep.* In press
86. Celmaster, W., Georgi, H., Machacek, M. 1978. *Phys. Rev. D* 17:879, 886
87. Schnitzer, H. J. 1975. *Phys. Rev. Lett.* 35:1540; 1975. *Phys. Rev. D* 13:74; 1976. *Phys. Lett. B* 65:239
88. Pumplin, J., Repko, W., Sato, A. 1975. *Phys. Rev. Lett.* 35:1538
89. Henriques, A. B., Kellet, B. H., Moorhouse, R. G. 1976. *Phys. Lett. B* 64:85; Chan, L.-H. 1977. *Phys. Lett. B* 71:422; Carlson, C. E., Gross, F. 1978. *Phys. Lett. B* 74:404
90. Rapidis, P. A. et al. 1977. *Phys. Rev. Lett.* 39:526
91. Bacino, W. et al. 1977. *Phys. Rev. Lett.* 40:671
92. Chodos, A. et al. 1974. *Phys. Rev. D* 9:3471
93. Bars, I. 1976. *Phys. Rev. Lett.* 36:1521; Tye, S.-H. H. 1976. *Phys. Rev. D* 13:3416; Giles, R. C., Tye, S.-H. H. 1977. *Phys. Rev. D* 16:1079; Ng, T. J., Tye, S.-H. H. 1977. *Phys. Rev. D* 16:2468
94. Lane, K., Eichten, E. 1976. *Phys. Rev. Lett.* 37:477
95. Kogut, J., Susskind, L. 1975. *Phys. Rev. Lett.* 34:767; 1975. *Phys. Rev. D* 12:2821
96a. Weisskopf, V. F., Wigner, E. P. 1930. *Z. Phys.* 63:54, 65:18
96b. Gottfried, K. 1970. *Brandeis Univ. Summer Inst. Theor. Phys. 1967*, ed. M. Chrétien, S. Schweber, Vol. 2. New York: Gordon & Breach; Dashen, R. F., Healy, J. B., Muzinich, I. J. 1976. *Phys. Rev. D* 14:2773
97. Sakurai, J. J. 1969. *Phys. Rev. Lett.* 22:981
98. Le Yaouanc, A. et al. 1977. *Phys. Lett. B* 71:397
99. Goldhaber, G. et al. 1977. *Phys. Lett. B* 69:503
100. Okun, L. B., Voloshin, M. B. 1976. *Zh. Eksp. Theor. Fiz.* 23:369; 1976. *JETP Lett.* 23:333; De Rújula, A., Georgi, H., Glashow, S. L. 1977. *Phys. Rev. Lett.* 38:317; Bander, M.

et al. 1976. *Phys. Rev. Lett.* 36:695; Rosenzweig, C. 1976. *Phys. Rev. Lett.* 36:697
101. Machacek, M., Tomozawa, Y. 1978. *Ann. Phys.* 110:407
102. Quigg, C., Rosner, J. 1977. *Phys. Lett. B* 71:153
103. Cahn, R., Ellis, S. D. 1978. *Phys. Rev. D* 17:2338; Barnett, M. 1977. *Proc. Eur. Conf. Part. Phys., Budapest, Hungary, 4–9 July 1977*, ed. L. Jenik, I. Montvay. Budapest: CRIP. 995 pp. Carlson, C. E., Suaya, R. 1977. *Phys. Rev. Lett.* 39:908; Ellis, J. et al. 1977. *Nucl. Phys. B* 131:285; Hagiwara, T., Kazama, Y., Takasugi, E. 1978. *Phys. Rev. Lett.* 40:76
104. Eichten, E., Gottfried, K. 1977. *Phys. Lett. B* 66:286
105. Quigg, C., Rosner, J. L. 1978. *Phys. Lett. B* 72:462
106. Gaillard, M. K., Lee, B. W., Rosner, J. L. 1975. *Rev. Mod. Phys.* 47:277
107. Jackson, J. D. 1976. *Proc. Summer Inst. Part. Phys., Aug. 2–13, 1976*, ed. M. C. Zipf, p. 147. SLAC, Stanford Univ., CA
108. Feldman, G. J. 1977. *Rep. SLAC-PUB-2068*
109. Lane, K., Weinberg, S. 1976. *Phys. Rev. Lett.* 37:717; Fritzsch, H. 1976. *Phys. Lett. B* 63:419
110. Deshpande, N. G. et al. 1976. *Phys. Rev. Lett.* 37:1305
111. Coremans-Bertrand, G. et al. 1976. *Phys. Lett. B* 65:480; Hand, L. et al. 1975. *Rep. FNAL-Proposal-382*; Burhop, E. H. S. et al. 1976. *Phys. Lett. B* 65:299; Voyvodic, L. 1976. *Rep. Fermilab-FN-289*; Conversi, M. 1975. *Rep. CERN-NP-75-17*; Dine, M. et al. 1976. *Rep. Fermilab-Proposal-P490*
112. Einhorn, M. B., Quigg, C. 1975. *Phys. Rev. D* 12:2015
113. Ellis, J., Gaillard, M. K., Nanopoulos, D. V. 1975. *Nucl. Phys. B* 100:313
114. Altarelli, G., Cabibbo, N., Maiani, L. 1975. *Nucl. Phys. B* 88:285; 1975. *Phys. Lett. B* 57:277; Kingsley, R. L. et al. 1975. *Phys. Rev. D* 11:1919; 1975. *Phys. Rev. D* 12:106; Pais, A., Rittenberg, V. 1975. *Phys. Rev. Lett.* 34:707
115. Glashow, S. L., Weinberg, S. 1977. *Phys. Rev. D* 15:1958; Paschos, E. A. 1977. *Phys. Rev. D* 15:1966; Gaillard, M. K., Lee, B. W., Rosner, J. L. 1975. *Rev. Mod. Phys.* 47:277; Kingsley, R. L. et al. 1975. *Phys. Rev. D* 11:1919, 12:106; Okun, L. B., Zakharov, V. I., Pontecorvo, B. M. 1975. *Lett. Nuovo*

496 APPELQUIST, BARNETT & LANE

Cimento 13:218; De Rújula, A., Georgi, H., Glashow, S. L. 1975. *Phys. Rev. Lett.* 35:69
116. Brandelik, R. et al. 1977. *Phys. Lett. B* 70:132
117. Cazzoli, E. G. et al. 1975. *Phys. Rev. Lett.* 34:1125
118. Knapp, B. et al. 1976. *Phys. Rev. Lett.* 37:882
119. Wiss, J. et al. 1976. *Phys. Rev. Lett.* 37:1531
120. Brandelik, R. et al. 1977. *Phys. Lett. B* 70:387
121. Feller, J. M. et al. 1978. *Phys. Rev. Lett.* 40:274
122. Kirkby, J. 1977. See Ref. 32, p. 3
123. Barnett, M. 1975. *Phys. Rev. Lett.* 34:41; 1975. *Phys. Rev. D* 11:3246; 1976. *Phys. Rev. D* 13:671; Fayet, P. 1974. *Nucl. Phys. B* 78:14; Gürsey, F., Sikivie, P. 1976. *Phys. Rev. Lett.* 36:775; Ramond, P. 1976. *Nucl. Phys. B* 110:214
124. De Rújula, A., Georgi, H., Glashow, S. L. 1975. *Phys. Rev. D* 12:3589; Wilczek, F. A. et al. 1975. *Phys. Rev. D* 12:2768; Fritzsch, H., Gell-Mann, M., Minkowski, P. 1975. *Phys. Lett. B* 59:256; Pakvasa, S., Simmons, W. A., Tuan, S. F. 1975. *Phys. Rev. Lett.* 35:702
125. Gürsey, F., Sikivie, P. 1976. *Phys. Rev. Lett.* 36:775; 1977. *Phys. Rev. D* 16:816; Ramond, P. 1976. *Nucl. Phys. B* 110:214; 1977. *Nucl. Phys. B* 126:509; Gürsey, F., Ramond, P., Sikivie, P. 1976. *Phys. Rev. Lett. B* 60:177; 1975. *Phys. Rev. D* 12:2166; Gürsey, F., Serdaroglu, M. 1978. *Nuovo Cimento Lett.* 21:28
126. Horn, D., Ross, G. G. 1977. *Phys. Lett. B* 67:460
127. Adler, S. L. 1970. *Lectures on Elementary Particles and Quantum Field Theory*, ed. S. Deser, M. Grisaru, H. Pendleton. Cambridge, MA: MIT Press; Gross, D. J., Jackiw, R. 1972. *Phys. Rev. D* 6:477
128. Knies, G. 1977. See Ref. 32, p. 93
129. Benvenuti, A. et al. 1976. *Phys. Rev. Lett.* 36:1478; 37:189
130. Barish, B. C. et al. 1977. *Phys. Rev. Lett.* 39:741, 1595; 1977. See Ref. 32, p. 239; Holder, M. et al. 1977. *Phys. Rev. Lett.* 39:433; Steinberger, J. 1977. CERN Rep.; Schultze, K. 1977. See Ref. 32, p. 359; Bosetti, P. C. et al. 1977. *Phys. Lett. B* 70:273; Berge, J. P. et al. 1977. *Phys. Rev. Lett.* 39:382
131. Barish, B. C. et al. 1977. *Phys. Rev. Lett.* 38:577; Benvenuti, A. et al.

1977. *Phys. Rev. Lett.* 38:1110, 1183; 1978. *Phys. Rev. Lett.* 40:488; Holder, M. et al. 1977. *Phys. Lett. B* 70:393
132. Barger, V. et al. 1977. *Phys. Rev. Lett.* 38:1190; 1977. *Phys. Rev. D* 16:2141; Albright, C. H., Smith, J., Vermaseren, J. A. M. 1977. *Phys. Rev. Lett.* 38:1187; 1977. *Phys. Rev. D* 16:3182, 3204; 1978. *Phys. Rev. D* 18:108; Albright, C. H., Shrock, R. E., Smith, J. 1978. *Phys. Rev. D* 17:2383; Zee, A., Wilczek, F., Treiman, S. B. 1977. *Phys. Lett. B* 68:369; Barnett, M., Chang, L. N. 1977. *Phys. Lett. B* 72:233; Barnett, M., Chang, L. N., Weiss, N. 1978. *Phys. Rev. D* 17:2266; Langacker, P., Segre, G. 1977. *Phys. Rev. Lett.* 39:259; Barger, V. et al. 1977. *Phys. Lett. B* 70:329; 1977. *Phys. Rev. D* 16:3170; Cox, P. H., Yildiz, A. 1977. *Phys. Rev. D* 16:2897; Pakvasa, S., Sugawara, H., Suzuki, M. 1977. *Phys. Lett. B* 69:461
133. Barnett, M., Chang, L. N. 1977. *Phys. Lett. B* 72:233; Barnett, M., Chang, L. N., Weiss, N. 1978. *Phys. Rev. D* 17:2266; Langacker, P., Segre, G. 1977. *Phys. Rev. Lett.* 39:259; Albright, C. H., Smith, J., Vermaseren, J. A. M. 1978. *Phys. Rev. D* 18:108; Albright, C. H., Shrock, R. E., Smith, J. 1978. *Phys. Rev. D* 17:2383
134. Bletzacker, F., Nieh, H. T., Soni, A. 1977. *Phys, Rev. Lett.* 38:1241; Soni, A. 1977. *Phys. Lett. B* 71:435; Goldberg, H. 1977. *Phys. Rev. Lett.* 39:1598; Young, B.-L., Walsh, T. F., Yang, T. C. 1978. *Phys. Lett. B* 74:111; see Ref. 133, Barnett & Chang, Barnett, Chang & Weiss, Albright, Smith & Vermaseren
135. Smith, J., Vermaseren, J. A. M. 1978. *Phys. Rev. D* 17:2288; Barnett, M., Chang, L. N., Weiss, N. 1978. *Phys. Rev. D* 17:2266; Barger, V., Gottschalk, T., Phillips, R. J. N. 1978. *Phys. Rev. D* 17:2284
136. Barnett, M. 1977. *Deeper Pathways in High Energy Physics—Orbis Scientiae 1977*, ed. A. Perlmutter, L. F. Scott, p. 389. New York: Plenum
137. Kobayashi, M., Maskawa, K. 1973. *Prog. Theor. Phys.* 49:652; Harari, H. 1975. *Phys. Lett. B* 57:265; 1975. *Ann. Phys.* 94:391
138. Golowich, E., Holstein, B. R. 1975. *Phys. Rev. Lett.* 35:831; 1977. *Phys. Rev. D* 15:3472; Branco, G., Mohapatra, R. N. 1976. *Phys. Rev. Lett.* 36:926; Wilczek, F. A. et al. 1975. *Phys. Rev. D* 12:2768; Branco, G., Mohapatra, R. N., Hagiwara, T.

1976. *Phys. Rev. D* 13:680
139. Pati, J., Salam, A. 1974. *Phys. Rev. D* 10:275; Fritzsch, H., Minkowski, P. 1976. *Nucl. Phys. B* 103:61; Mohapatra, R. N., Sidhu, D. P. 1977. *Phys. Rev. Lett.* 38:667; De Rújula, A., Georgi, H., Glashow, S. L. 1977. *Ann. Phys.* 109:258; Beg, M. A. B., Zee, A. 1973. *Phys. Rev. Lett.* 30:675; 1973. *Phys. Rev. D* 8:1460; Beg, M. A. B. et al. 1977. *Phys. Rev. Lett.* 38: 1254; Beg, M. A. B. et al. 1977. *Phys. Rev. Lett.* 39:1054
140. Georgi, H., Weinberg, S. 1978. *Phys. Rev. D* 17:275
141. Segre, G., Weyers, J. 1976. *Phys. Lett. B* 65:243; Lee, B. W., Weinberg, S. 1977. *Phys. Rev. Lett.* 38:1237; Lee, B. W., Shrock, R. 1978. *Phys. Rev. D* 17:2410; Barnett, M., Chang, L. N. 1977. *Phys. Lett. B* 72:233; *SLAC Rep.* In preparation; Barnett, M., Chang, L. N., Weiss, N. 1978. *Phys. Rev. D* 17:2266; Langacker, P., Segre, G. 1977. *Phys. Rev. Lett.* 39:259; Langacker, P., Segre, G., Golshani, M. 1978. *Phys. Rev. D* 17:1402
142. Bjorken, J. D., Lane, K. 1978. *SLAC Rep.*
143. Pati, J. C., Salam, A. 1975. *Phys. Lett. B* 58:333; 1977. *Proc. Neutrino 76, Aachen, June 1976*, ed. H. Faissner, H. Reithler, P. Zerwas. Braunschweig: Vieweg
144. Han, M.-Y., Nambu, Y. 1965. *Phys. Rev.* 139:B1006
145. Barnett, M. 1977. *Proc. Part. Fields 1976*, ed. H. Gordon, R. F. Peierls, p. D77. Upton, NY: Brookhaven Nat. Lab.
146. Holder, M. et al. 1977. *Phys. Lett. B* 69:377
147. Altarelli, G., Parisi, G., Petronzio, R. 1976. *Phys. Lett. B* 63:183; Altarelli, G., Parisi, G. 1977. *Nucl. Phys. B* 126: 298; Barnett, M., Georgi, H., Politzer, H. D. 1976. *Phys. Rev. Lett.* 37:1313; Kaplan, J., Martin, F. 1976. *Nucl. Phys. B* 115:333; Barnett, M., Martin, F. 1977. *Phys. Rev. D* 16:2765; Zakharov, V. I. 1977. *Proc. Int. Conf. High Energy Phys., 18th, Tbilisi, July 1976*, ed. N. N. Bogoliubov et al, Vol. II, p. B69. Dubna, USSR: J. Inst. Nucl. Res.; Zee, A., Wilczek, F., Treiman, S. B. 1974. *Phys. Rev. D* 10:2881; Buras, A. J., Gaemers, K. J. F. 1977. *Phys. Lett. B* 71:106; Buras, A. J. 1977. *Nucl. Phys. B* 125:125; Buras, A. J. et al. 1977. *Nucl. Phys. B* 131:308; Fox, G. C. 1978. *Nucl. Phys. B* 134:269; Baluni, V., Eichten, E.

1976. *Phys. Rev. Lett.* 37:1181
148. Chang, L. N., Derman, E., Ng, J. N. 1975. *Phys. Rev. Lett.* 35:6, 1252; 1975. *Phys. Rev. D* 12:3539; Chang, L. N., Ng, J. N. 1977. *Phys. Rev. D* 16:3157; Pais, A., Treiman, S. B. 1975. *Phys. Rev. D* 12:3539; Chang, Derman, E. 1976. *Nucl. Phys. B* 110:40
149. Baltay, C. et al. 1977. *Phys. Rev. Lett.* 39:62
150. von Krogh, J. et al. 1976. *Phys. Rev. Lett.* 36:710; Schultze, K. 1977. See Ref. 32, p. 359
151. Ellis, J. et al. 1977. *Nucl. Phys. B* 131: 285; Ellis, J., Gaillard, M. K., Nanopoulos, D. V. 1976. *Nucl. Phys. B* 109:213
152. Cahn, R. N., Ellis, S. D. 1977. *Phys. Rev. D* 16:1484; Ali, A. 1977. *Rep. CERN-TH-2411*
153. Aubert, J. J. et al. 1975. *Nucl. Phys. B* 89:1; Blanar, G. J. et al. 1975. *Phys. Rev. Lett.* 35:346; Knapp, B. et al. 1975. *Phys. Rev. Lett.* 34:1044; Büsser, F. W. et al. 1975. *Phys. Lett. B* 56:482; Anderson, K. J. et l. 1976. *Phys. Rev. Lett.* 36:237; Antipov, Y. M. et al. 1976. *Phys. Lett. B* 60:309; Snyder, H. D. et al. 1976. *Phys. Rev. Lett.* 36:1415; Branson, J. G. et al. 1977. *Phys. Rev. Lett.* 38:1334; Cordon, M. J. et al. 1977. *Phys. Lett. B* 68:96; Bushnin, Y. B. et al. 1977. *Phys. Lett. B* 72:269; Cobb, J. H. et al. 1977. *Phys. Lett. B* 68:101; Amaldi, E., et al. 1977. *Lett. Nuovo Cimento* 19:152; Bamberger, A. et al. 1978. *Nucl. Phys. B* 134:1, 1978
154. Sivers, D. 1976. *Nucl. Phys. B* 106:95; Barnett, M., Silverman, D. 1975. *Phys. Rev. D* 12:2037; Gunion, J. 1975. *Phys. Rev. D* 12:1345; Green, M. B., Jacob, M., Landshoff, P. 1975. *Nuovo Cimento* 29:123; Donnachie, A., Landshoff, P. 1976. *Nucl. Phys. B* 112:233
155. Binkley, M. et al. 1976. *Phys. Rev. Lett.* 37:578
156a. Ellis, S., Einhorn, M., Quigg, C. 1976. *Phys. Rev. Lett.* 36:1263; Carlson, C. E., Suaya, R. 1976. *Phys. Rev. D* 14:3115; 1977. *Phys. Rev. D* 15:1416; 1978. *Phys. Rev. D* 18:760; Ellis, S., Einhorn, M. 1975. *Phys. Rev. D* 12:2007
156b. Cahn, R., Ellis, S. 1977. *Phys. Rev. D* 16:1484
157. Cordon, M. J. et al. 1977. *Phys. Lett. B* 68:96
158. Cobb, J. H. et al. 1978. *Phys. Lett. B* 72:497

159a. Fritzsch, H. 1977. *Phys. Lett. B* 67: 217; Halzen, F. 1977. *Phys. Lett. B* 69:105; Gaisser, T. K., Halzen, F., Paschos, E. 1977. *Phys. Rev. D* 15: 2577; Gluck, M., Owens, J. F., Reya, E. 1978. *Phys. Rev. D* 17:2324

159b. Ellis, J. et al. 1977. *Nucl. Phys. B* 131:285; Jones, L., Wyld, H. 1978. *Phys. Rev. D* 17:2332; Owens, J., Reya, E. 1978. *Phys. Rev. D* 17:3003

160. Drell, S., Yan, T.-M. 1970. *Phys. Rev. Lett.* 25:316

161. Carlson, C. E. Private communication

162. Bjorken, J. D. 1977. See Ref. 32, p. 960; Nieh, H. T. 1977. *Stony Brook Rep. ITP-SB-77-64*

163. Sivers, D. 1976. *Nucl. Phys. B* 106:96

164. Coremans-Bertrand, G. et al. 1976. *Phys. Lett. B* 65:480

165. Babcock, J., Sivers, D., Wolfram, S. 1978. *Phys. Rev. D* 18:162; Gaisser, T. K., Halzen, F., Kajantie, K. 1975. *Phys. Rev. D* 12:1968; Pilachowski, L., Tuan, S. F. 1975. *Phys. Rev. D* 11:3148; Barnett, M. 1975. *Phys. Rev. D* 12:3441; Gaisser, T. K., Halzen, F. 1975. *Phys. Rev. D* 11:3157; 1976. *Phys. Rev. D* 13:171; Hinchliffe, I., Llewellyn Smith, C. H. 1976. *Phys. Lett. B* 61:472; 1976. *Nucl. Phys. B* 114:45; Bourquin, M., Gaillard, J. M. 1976. *Nucl. Phys. B* 114:334; McKay, D. W., Young, B.-L. 1977. *Phys. Rev. D* 15:1282; Cox, P. H., Park, S. Y., Yildiz, A. 1977. *Phys. Lett. B* 70:317; Jones, L. M., Wyld, H. W. 1978. *Phys. Rev. D* 17:1782; Gustafson, G., Peterson, C. 1977. *Phys. Lett. B* 67:81; Halzen, F., Matsuda, S. 1978. *Phys. Rev. D* 17:1344

166. Barger, V., Phillips, R. 1975. *Phys. Rev. D* 12:2623; Field, R. D., Quigg, C. 1975. *Rep. Fermilab-75/15-THY*

167. Carlson, C. E., Freund, P. G. O. 1972. *Phys. Lett. B* 39:349

168. Sivers, D., Townsend, J., West, G. 1976. *Phys. Rev. D* 13:1234; Walsh, T. 1975. *Nuovo Cimento Lett.* 14:290; Boreskov, K. G., Ioffe, B. L. 1976. *Moscow Rep. ITEP-102-1976*; Ioffe, B. L., Okun, L. B., Zakharov, V. I. 1975. *Moscow-ITEP Rep.*; Aviv, R. et al. 1975. *Phys. Rev. D* 12:2862; Pumplin, J., Repko, W. 1975. *Phys. Rev. D* 12:1376; Horn, D. 1975. *Phys. Lett. B* 58:323; Humpert, B., Wright, A. C. D. 1977. *Phys. Rev. D* 15:2503; Humpert, B. 1977. *Phys. Lett. B* 68:66

169. Camerini, U. et al. 1975. *Phys. Rev. Lett.* 35:483

170. Anderson, R. L. et al. 1977. *Phys.* *Rev. Lett.* 38:263

171. Gittleman, B. et al. 1975. *Phys. Rev. Lett.* 35:1616; Knapp, B. et al. 1975. *Phys. Rev. Lett.* 34:1040; Nash, T. et al. 1976. *Phys. Rev. Lett.* 36:1233

172. Babcock, J., Sivers, D., Wolfram, S. 1978. *Phys. Rev. D* 18:162; Chen, M.-S., Kane, G. L., Yao, Y.-P. 1976. *Mich. Rep. UM HE 76-17*; Novikov, V. A. et al. 1978. *Nucl. Phys. B* 136:125; Jones, L. M., Wyld, H. W. 1978. *Phys. Rev. D* 17:759

173. Shifman, M. A., Vainshtein, A. I., Zakharov, V. I. 1976. *Phys. Lett. B* 65:255

174. Feldman, G. J. et al. 1977. *Phys. Rev. Lett* 38:1313

175. Blietschau, J. et al. 1977. *Nucl. Phys. B* 118:218; Benvenuti, A. et al. 1976. *Phys. Rev. Lett.* 37:1039; Merritt, F. S. et al. 1978. *Phys. Rev. D* 17:2199; Holder, M. et al. 1977. *Phys. Lett. B* 71:222, 72:254; Schultze, K. 1977. See Ref. 32, p. 359

176. Cline, D. et al. 1976. *Phys. Rev. Lett.* 37:252, 648; Lee, W. et al. 1976. *Phys. Rev. Lett.* 37:186; Sulak, L. R. et al. 1977. See Ref. 143, 1977; Pohl, M. et al. 1978. *Phys. Lett. B* 72:489

177. Kluttig, H., Morfin, J. G., Van Doninck, W. 1977. *Phys. Lett. B* 71: 446

178. Hasert, F. J. et al. 1973. *Phys. Lett. B* 46:121; Blietschau, J. et al. 1976. *Nucl. Phys. B* 114:189; Reines, F. et al. 1976. *Phys. Rev. Lett.* 37:315; Reithler, H. 1977. See Ref. 32, p. 343

179. Sehgal, L. M. 1977. *Phys. Lett. B* 71:99; Hung, P. Q., Sakurai, J. J. 1977. *Phys. Lett. B* 72:208; Abbott, L., Barnett, M. 1978. *Phys. Rev. Lett.* 40:1303; Barnett, M. 1976. *Phys. Rev. D* 14:2990; Albright, C. H. et al. 1976. *Phys. Rev. D* 14:1780; Barger, V., Nanopoulos, D. V. 1977. *Nucl. Phys. B* 124:426; Sidhu, D. P. 1976. *Phys. Rev. D* 14:2235; Bernabeu, J., Jarlskog, C. 1977. *Phys. Lett. B* 69:71; Langacker, P., Sidhu, D. P. 1978. *Phys. Lett. B* 74:233

180. Baird, P. E. G. et al. 1976. *Nature* 264:528; Sandars, P. G. H. See Ref. 32, p. 343

181. Feinberg, G. 1977. *Columbia Rep. CU-TP-111*; 1978. *Proc. Ben Lee Mem. Int. Conf., Batavia, Ill.*, Oct. 20–22, 1977. In press

182. Cahn, R. N., Kane, G. L. 1977. *Phys. Lett. B* 71:348; Marciano, W. J., Sanda, A. I. 1978. *Phys. Rev. D* 17: 3055

183. Sachs, R. G. 1963. *Ann. Phys.* 22:239;

Wolfenstein, L. 1964. *Phys. Rev. Lett.* 13:562; Mohapatra, R. N. 1972. *Phys. Rev. D* 6:2023; Pais, A. 1973. *Phys. Rev. D* 8:625; 1972. *Phys. Rev. Lett.* 29:1719; 1973. *Phys. Rev. Lett.* 30: 114; Lee, T. D. 1973. *Phys. Rev. D* 8:1226; 1974. *Phys. Rep. C* 9:143; Mohapatra, R. N., Pati, J. C., Wolfenstein, L. 1975. *Phys. Rev. D* 11:3319

184. Weinberg, S. 1976. *Phys. Rev. Lett.* 37:657

185. Sikivie, P. 1976. *Phys. Lett. B* 65:141

186. Maiani, L. 1976. *Phys. Lett. C* 62: 183; Pakvasa, S., Sugawara, H. 1976. *Phys. Rev. D* 14:305

187. Geweniger, C. et al. 1974. *Phys. Lett. B* 48:487; Messner, R. et al. 1973. *Phys. Rev. Lett.* 30:876

188. Bjorken, J. D., Weinberg, S. 1977. *Phys. Rev. Lett.* 38:622

189. Depommier, P. et al. 1977. *Phys. Rev. Lett.* 39:1113

190. Bryman, D. A. et al. 1972. *Phys. Rev. Lett.* 28:1469

191. Cheng, T.-P., Li, L.-F. 1977. *Phys. Rev. Lett.* 38:381; 1977. *Phys. Rev. D* 16: 1425; Bilenkii, S. M., Petkov, S. T., Pontecorvo, B. 1977. *Phys. Lett. B* 67:309

192. Bjorken, J. D., Lane, K., Weinberg, S. 1977. *Phys. Rev. D* 16:1474

193. Korenchenko, S. M. et al. 1976. *Sov. Phys. JETP* 43:1

194. Fitch, V. L. et al. 1967. *Phys. Rev.* 164:1711

195. Glashow, S. L. 1977. *Harvard Rep. HUTP-77/A008*

196. Fritzsch, H. 1977. *Phys. Lett. B* 67:451

197. Appelquist, T., Carazzone, J. 1975. *Phys. Rev. D* 11:2856

198. Georgi, H., Quinn, H., Weinberg, S. 1974. *Phys. Rev. Lett.* 33:451

199. Buras, A. J. et al. 1978. *Nucl. Phys. B* 135:66

200. Georgi, H., Glashow, S. L. 1974. *Phys. Rev. Lett.* 32:438

201. Fritzsch, H., Minkowski, P. 1975. *Ann. Phys.* 93:193; 1976. *Nucl. Phys. B* 103:61

202. Gürsey, F., Serdaroglu, M. 1978. *Nuovo Cimento Lett.* 21:28

203. Ramond, P. 1976. *Nucl. Phys. B* 110:214

204. Reines, F., Crouch, M. F. 1974. *Phys. Rev. Lett.* 32:493

205. Gell-Mann, M., Ramond, P., Slansky, R. 1977. *Los Alamos Rep. LA-UR-77-2059*

206. Weinberg, S. 1977. *Harvard Rep. HUTP-77/A057*

207. Gel'fand, Y., Likhtman, E. P. 1971. *Zh. Eskp. Teor. Fiz. Pis'ma Red.* 13:452 (in English 13:323); Wess, J., Zumino, B. 1974. *Phys. Lett. B* 49:54

208. Freedman, D., van Nieuwenhuizen, P., Ferrara, S. 1976. *Phys. Rev. D* 13:3214; Deser, S., Zumino, B. 1976. *Phys. Lett. B* 62:335

209. Grisaru, M., van Nieuwenhuizen, P., Vermaseren, J. 1976. *Phys. Rev. Lett.* 37:1662

Ann. Rev. Nucl. Part. Sci. 1978. 28:501–22
Copyright © 1978 by Annual Reviews Inc. *All rights reserved*

ISOTOPIC ANOMALIES IN   *5599
THE EARLY SOLAR SYSTEM

Robert N. Clayton

Enrico Fermi Institute, and Departments of Chemistry and of the
Geophysical Sciences, University of Chicago, Chicago, Illinois 60637

CONTENTS

1 INTRODUCTION

Measurements of isotopic abundances of primitive solar system materials
are relevant to two major astrophysical areas: (*a*) the origin and evolution
of stars and planetary systems, and (*b*) the synthesis of nuclei by stars and
the return of matter from stars to interstellar space. The age of the sun is
only a small fraction of the age of our galaxy; it was formed from more
dispersed matter some 4.5–5 billion years ago. Furthermore, the presence
of long-lived radioactivities in solar system matter proves the finite age
of these nuclei. In fact, much evidence exists that the time interval between

501

0066-4243/78/1201-0501$01.00

the last major nucleosynthetic event and the formation of the sun must have been short. There has, therefore, been interest for many years in seeking variations in isotopic abundances from one region of the solar system to another, which might carry the record of recent nucleosynthesis in the matter from which the sun and planets were formed. Until recently, however, searches for isotopic heterogeneities residual from nucleosynthesis proved fruitless, and the conclusion was generally drawn that solar system material had been thoroughly homogenized prior to formation of the sun and planets. This in turn led many authors to assume a high temperature stage in which all solar system matter was homogenized as a gas. Since 1972, several elements have indeed been found to have isotopic variations of nucleosynthetic origin. These include both radioactive and stable isotopes, so a time scale may be attached to the nuclear events that produced the variations. These "isotopic anomalies" and their astrophysical significance form the subject of this review.

The classic paper of Burbidge et al (1) first presented in detail the variety of types of nuclear processes necessary to explain the observed abundances of the nuclides. Since that time, many important advances have been made in the study of nucleosynthesis, both in determining the parameters of temperature, density, and initial composition for nucleosynthesis, and in identifying the astrophysical sites of these processes. For the very light elements, an initial "big bang" seems required for the production of deuterium and helium, and cosmic-ray spallation reactions in space are the dominant source of lithium, beryllium, and boron (2). Nuclear reactions within stars, burning either hydrostatically or explosively, account most reasonably for the observed abundances of the nuclides with $Z > 5$. However, it is commonly found that the various isotopes of an element require different nucleosynthetic processes, occurring under different conditions, on different time scales, and in different parts of a star. Hence, the relative isotopic abundances we observe are established by a mixing of materials derived from a variety of sites.

The principal means for return of the products of stellar nucleosynthesis to space is probably the explosion of supernovae. The ejecta from any given supernova are decelerated and mixed with the surrounding interstellar medium on a short time scale of 10^6 years (3). Thus it is unlikely that isotopic heterogeneities can be maintained in the interstellar gas for much longer times (e.g. the rotational period of the galaxy, which is on the order of 10^8 years). An important exception to this conclusion occurs if solid particles condense from the supernova ejecta on a time scale of a year or so (4, 5). These solids could have highly unusual isotopic abundances, which would not be appreciably modified by subsequent interactions with the interstellar gas, and which could, therefore contribute significantly to isotopic heterogeneity in a later star-forming event.

1.1 *Meteorites as Samples of the Solar System*

Many samples of solar system matter are available for study, either in the laboratory or by remote sensing methods. They include the sun and the solar wind, planets and their satellites, asteroids, comets, and meteorites. For the observation of isotopic variations of the magnitude discussed here, only laboratory study of samples has sufficient precision. The Earth and Moon have undergone extensive internal chemical processing that has obscured the evidence of any initial isotopic heterogeneity. Terrestrial and lunar materials provide baselines relative to which we may observe variability in other solar system objects.

The meteorites are our best source of information concerning the earliest history of the solar system. They are almost all ancient, having remained virtually unchanged from their time of formation about 4.6 billion years ago until the time of removal from their parent bodies and delivery to earth. Many meteorites show clear evidence of having gone through a high temperature stage within the parent body, which led to a metamorphic recrystallization, and in some cases to melting. However, some groups of meteorites, notably the carbonaceous chondrites, are heterogeneous agglomerations of particles that have not been extensively heated since their aggregation. These primitive meteorites are the ones most likely to contain unaltered presolar grains and primary solid condensates from the solar nebular gas. It is the carbonaceous chondrites in which the isotopic anomalies are most clearly seen.

Most of the interesting isotopic variations in meteorites have been found not by analyzing the meteorite as a whole, but by analyzing individual small fragments or minerals separated from it. In most cases, the large anomalies in submillimeter samples are diluted by material of normal isotopic composition in bulk rock samples. Thus the important advances in isotopic studies of the early solar system have come not primarily by development of new analytical techniques, but by recognition of which are the critical samples to be studied.

1.2 *Isotopic Variations in Meteorites*

Several processes contribute to variations in isotopic abundances in meteorites. Some of these fractionate isotopes on the basis of their mass differences; others create or destroy isotopes in nuclear reactions, including both radioactive decay and energetic particle reactions.

Gaseous diffusion is an example of a mass-dependent fractionation process that has probably played an important part in establishing differences in the isotopic compositions of the noble gases (6). Equilibrium isotopic fractionation between gas and solids is a factor in determining the isotopic abundance of oxygen in the various solar system bodies (7).

Radioactive decay is effective in modifying the isotopic abundances of the elements that are daughter products not only of long-lived parents still present (e.g. ^{238}U, ^{235}U, ^{232}Th, ^{147}Sm, ^{87}Rb, ^{40}K) but also of extinct radionuclides (e.g. ^{244}Pu, ^{129}I, ^{26}Al). All of these effects have been observed in meteorites. After liberation from their parent bodies, the meteorites are exposed to the galactic cosmic radiation for a period of 10^6–10^9 years, during which time substantial quantities of "cosmogenic" nuclides, both stable and radioactive, are produced by spallation reactions. If at any time in its history, an extraterrestrial object is exposed to the solar wind, its surface acquires an implanted solar component, which may have been isotopically modified by nuclear reactions in the sun. The most striking example of this is the hydrogen implanted in lunar soils; it is virtually devoid of deuterium, which is destroyed in the sun (8).

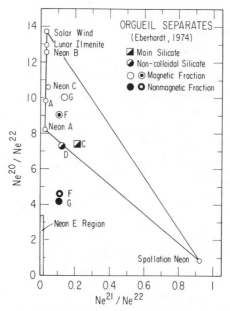

Figure 1 Three-isotope plot of meteoritic neon (from 14). Almost all samples fall within the triangle determined by the compositions of the solar wind, "planetary" neon (neon-A), and spallation-produced neon. The solar and planetary compositions may be related by a mass-dependent fractionation process. Some primitive meteorites contain an additional component, neon-E, which is enriched in ^{22}Ne by a factor of at least 10 with respect to the solar and planetary components, and which may be of presolar origin (11).

All of the types of isotopic variability described in the preceding two paragraphs have been extensively exploited in the study of extraterrestrial samples, and they have contributed greatly to our knowledge of the early history of the solar system (9). They are not, however, the subject of this review, which deals primarily with those additional isotopic variations not accounted for in terms of these known solar system processes and probably arisen from presolar nucleosynthesis.

Nucleosynthetic isotopic anomalies have been clearly recognized in oxygen, neon, magnesium, silicon, calcium, barium, neodymium, samarium, and possibly sulfur, mercury, and uranium. Suggestive but equivocal evidence exists for anomalies in carbon and nitrogen. Krypton and xenon have complicated isotopic variations, which may be due in part to nucleosynthetic processes. Although the largest part of the variations observed in magnesium isotopic abundances are due to a combination of mass-dependent fractionation and ^{26}Al decay, there is also a small nucleosynthetic effect. It is expected that many more anomalies will be found, since the search in the appropriate meteoritic samples has just begun. In the sections that follow, the details of the observations are discussed. The very existence of the anomalies implies a presolar super-nova source. Their association with live ^{26}Al implies a time interval of not more than 2×10^6 years between the supernova event and the condensa-tion of large solar system bodies. The effects in the heavy elements are primarily due to neutron-capture on an explosive time scale, consistent with a supernova event. The close temporal and spatial association of a supernova and new star formation suggests a causal relationship. Astronomical observations also support the likelihood of an active role of supernova ejecta in triggering star formation in an adjacent cloud (10).

2 LIGHT ELEMENTS IN CARBONACEOUS CHONDRITES

2.1 *Neon*

The nature and magnitude of the neon isotope anomaly [the so-called neon-E (11)] are shown in Figure 1. The three-isotope plot is especially useful for resolving isotope effects due to different causes (12). It has the properties that binary mixtures define straight lines between two end-member compositions and that mass-dependent fractionation processes follow a predictable trajectory. Neon in meteorites is a mixture of at least three components: (*a*) a "planetary" component incorporated from the nebular gas at the time of formation of the parent body; (*b*) a "solar" component implanted from the solar wind, presumably at the surface of the parent body; and (*c*) a "cosmogenic" component produced by spallation reactions of cosmic-ray nuclei on heavier target nuclei after separation of the meteoroid from the parent. Any mixtures of these three components have isotopic compositions that lie within the triangle of Figure 1. A commonly used technique to resolve these mixtures into their components is stepwise heating of the meteorite in vacuum, which releases differently sited neon atoms at different temperatures (13). Stepwise heating of some of the primitive low temperature meteorites revealed the existence of another neon component, enriched in ^{22}Ne, which has been labeled neon-E. This component is physically separable from the others, in that its abundance can be manyfold enhanced by concentration of its host mineral phase (14). Neon-E has also been found in another meteorite class, the "unequilibrated ordinary chondrites" in which it appears to be contained within graphite (15).

The exact composition of neon-E is not yet known. It will be particularly interesting to learn whether it is an isotopic mixture, or consists of ^{22}Ne alone. If the latter is the case, it is possible that 2.6-year ^{22}Na may be implicated in its origin. Sodium-22 might have been produced within the solar system by (p,n) reaction on ^{22}Ne (16). Of course, if the ^{22}Na remains in the gas phase, it decays to ^{22}Ne with no net isotopic effect. If, however, it is condensed into a solid phase before decay, the solid will subsequently carry a ^{22}Ne excess. It is not known whether any such rapid condensation process occurred within the solar system.

An alternative hypothesis for the origin of Ne-E, also involving a ^{22}Na precursor, has been proposed by D. D. Clayton (17). He suggested that solid grains bearing newly synthesized ^{22}Na were formed in the expanding envelope of a supernova. In this case, a time scale for condensation on the order of a year is expected, but it has not been demonstrated that an element as volatile as sodium will condense at all under these conditions. In the scenario proposed by Clayton (17), the ^{22}Na nucleosynthesis may have occurred long before formation of the solar system, with the ^{22}Ne subsequently transported into the protosolar cloud within solid dust particles.

A third possibility for the origin of Ne-E is synthesis in a supernova event immediately prior to the collapse of the protosolar cloud, followed by injection of isotopically anomalous neon into the cloud. This model is suggested by the observations of the isotopes of oxygen and magnesium, discussed below. In this case, no special role for ^{22}Na is required, as the composition of Ne-E may be produced directly in the supernova (18).

Srinivasan & Anders (19) have observed neon-E in the same microscopic fraction of a meteorite sample as isotopically anomalous krypton and xenon. The isotopic patterns of the two heavy rare gases are unique, showing enhanced abundances of ^{128}Xe, ^{130}Xe, ^{132}Xe, and ^{80}Kr and

^{82}Kr. This is the first instance of a correlation between the neon-E anomaly and any other nuclear effect, and may lead to a better understanding of its origin. The suggestion that it is derived from dust ejected from red giant stars (19), is discussed below in Section 4.4.

2.2 Oxygen

The isotopic patterns of oxygen are less complex than those of neon, and appear to require only two components to account for all of the observations. This is primarily because oxygen is an abundant element in meteorites, whereas neon is very rare. Thus, small nuclear effects, such as galactic cosmic-ray reactions, have a negligible effect on the isotopic abundances of oxygen, but are readily seen in neon, even though the numbers of atoms involved are similar.

The effects of two major processes leading to variability in the isotopic abundances of oxygen are illustrated in Figure 2. The "Terrestrial" line is the locus of compositions of all natural terrestrial materials: rocks, waters, atmosphere, etc. Variations in the isotopic compositions of terrestrial samples arise from mass-dependent differences in the rates and equilibria of chemical and physical processes operating on the Earth. For example, limestones precipitated from ocean water have an $^{18}O/^{16}O$ ratio about 3% greater than that of the water because of the equilibrium isotopic fractionation between calcium carbonate and water. The corresponding enhancement in the $^{17}O/^{16}O$ ratio is 1.5%: approximately

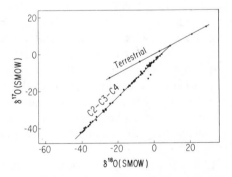

Figure 2 Three-isotope plot of oxygen from terrestrial samples and from the high temperature minerals of primitive meteorites (types C2, C3, and C4 carbonaceous chondrites). The variation in terrestrial materials is due to mass-dependent fractionation in geological processes. The meteorite samples define a mixing line that extrapolates to nearly pure ^{16}O. The ^{16}O-rich component is of presolar origin, and constitutes up to 5% of the minerals analyzed (24). The ordinate is the $^{17}O/^{16}O$ ratio, measured in parts-per-thousand deviation from an arbitrary terrestrial standard (ocean water). The abscissa is the $^{18}O/^{16}O$ ratio relative to the same standard.

half as great for a one-mass-unit change as for a two-mass-unit change. A detailed treatment of the mass-dependence of isotopic fractionation in a three-isotope system is given by Matsuhisa et al (20). The theoretical slope of the "Terrestrial" fractionation line is about 0.52 for the typical range of natural abundance variations found on Earth.

If the solar system had condensed from a homogeneous gas cloud, as was widely believed a few years ago, all samples from planets and meteorites would also have oxygen isotopic abundances that fell on the same fractionation line. In fact almost all meteorites are displaced from this line, either above it or below it, indicating the involvement of nuclear processes as well as the fractionation process (21). The nature of the nuclear processes is best understood from the observations of the high temperature condensate minerals of the primitive carbonaceous chondrites of classes C2, C3, and C4. The minerals are believed to have condensed from the solar nebular gas (with or without an intervening liquid state) at temperatures above $1450°K$ (22) and may represent the oldest available macroscopic samples of the solar system (23). Their oxygen isotopic compositions are plotted as the line C2-C3-C4 in Figure 2. A straight line on the three-isotope graph results from binary mixtures of two end-member components. In this instance, one component appears to have a composition similar to that found in the earth and many meteorites; the other is enriched in ^{16}O, which thus produces the low observed values of $^{17}O/^{16}O$ and $^{18}O/^{16}O$. The slope of the mixing line is very nearly unity (0.94) (24). A line of unit slope corresponds to a constant $^{17}O/^{18}O$ ratio, with variable amounts of ^{16}O, i.e. the mixing line extrapolates very nearly to pure ^{16}O—an observation similar to that discussed above for neon in which a component close to pure ^{22}Ne is found. The largest ^{16}O excess observed to date is 5%, which is typical of spinel and pyroxene grains in the Allende meteorite. The observation of oxygen isotopic variations in all of the major meteorite classes indicates that the ^{16}O excesses seen so prominently in the primitive meteorites are also present, at lower concentrations, in all meteorites, i.e. isotopic heterogeneity was not an isolated local phenomenon, but pervaded the solar system (21).

Clayton et al (25) considered the possibility that the ^{16}O-rich component might have been produced within the early solar system from destruction of ^{17}O and ^{18}O by high energy particle irradiation. They concluded that the proton fluence necessary to produce the effects illustrated in Figure 2 was two to three orders of magnitude larger than the upper limits set by the isotopic homogeneity observed in several other elements. They also suggested that some minor differences in oxygen isotopic compositions, such as between different classes of the ordinary chondrites (21) might have been caused by particle irradiation within the solar nebula.

Since production of the ^{16}O-rich component within the solar system is apparently ruled out, other sources must be found. Oxygen-16 is produced both by thermonuclear reactions from ^{4}He under hydrostatic conditions within stable stars and by explosive carbon-burning in supernova explosions (26, and references therein). In either case, the effective method for return of the products of nucleosynthesis to space for use in a later generation of stars is by means of ejection from an exploding supernova (27). Presumably most of the ^{16}O of which solar system matter is made was produced in this manner and was mixed in interstellar space or in the protosolar cloud with ^{17}O and ^{18}O produced in other sources. If the additional ^{16}O-rich component found in the meteorites was derived from a supernova source, its inhomogeneous distribution in the solar nebula implies either that it was transported in solid ^{16}O-rich presolar grains that escaped vaporization (28) or that the supernova was so close to the solar nebula that condensation could take place prior to isotopic homogenization (29). The available evidence for oxygen alone does not provide a clear distinction between these two possibilities, but the association of the ^{16}O excess with minerals containing radiogenic ^{26}Mg (see next section) supports the second alternative.

2.3 Magnesium

The observation of large ^{16}O excesses in primitive meteorites, apparently derived from a supernova, led naturally to a search for correlated isotopic anomalies in other light elements, such as magnesium and silicon, that are chemically combined with the oxygen in the meteorites. Such nucleosynthetic anomalies have not yet been observed for these elements (except in samples C1 and EK 1-4-1, discussed below). However, magnesium in the Allende inclusions does have substantial excesses of ^{26}Mg, apparently from *in situ* decay of ^{26}Al (half-life 720,000 years). This effect is, therefore, an extinct radioactivity, rather than an isotopic anomaly in the sense used in this review. However, it merits discussion here, since its interpretation bears critically on the understanding of the origins of the anomalies in other elements. This comes about because the short lifetime of ^{26}Al severely limits the time interval between nucleosynthesis and solar system condensation.

Aluminum and magnesium are major elements in most of the minerals found in high temperature condensates in primitive meteorites (30). If ^{26}Al was present in the solar nebula at the time of condensation, it must have crystallized along with stable ^{27}Al in these minerals, with an initial ratio $(^{26}\text{Al}/^{27}\text{Al})_0$ at the time of crystallization. Subsequently, ^{26}Al decayed to ^{26}Mg, which is measurable today as an enhancement of the ^{26}Mg/^{24}Mg ratio. The magnitude of this enhancement is proportional to the ratio of aluminum to magnesium in the mineral. Thus, for a set of minerals of different Al/Mg ratios, formed at the same time from a reservoir of uniform ^{26}Al/^{27}Al ratio:

$$^{26}\text{Mg}/^{24}\text{Mg} = (^{26}\text{Mg}/^{24}\text{Mg})_0 + (^{27}\text{Al}/^{24}\text{Mg})(^{26}\text{Al}/^{27}\text{Al})_0,$$

where the subscript 0 indicates the initial ratios. The ratios ^{26}Mg/^{24}Mg and ^{27}Al/^{24}Mg are measurable; hence the equation defines a straight line with intercept $(^{26}$Mg/^{24}Mg$)_0$ and slope $(^{26}$Al/^{27}Al$)_0$. An example of such a plot is shown in Figure 3 from Steele et al (31). Five inclusions (four from Allende, one from a similar meteorite, Leoville), measured in three laboratories, have yielded essentially identical results: $(^{26}$Mg/^{24}Mg$)_0$ the same as normal terrestrial magnesium, and $(^{26}$Al/^{27}Al$)_0 = 5 \times 10^{-5}$ (32–35). If the relative production rates of the aluminum isotopes $(^{26}$Al/^{27}Al) in explosive carbon burning is taken as 10^{-3} (36), and if no allowance is made for ^{26}Al decay, then the fraction of the aluminum atoms in the meteorite samples derived from the nucleosynthetic event is about 5%, which happens to be the same as the ^{16}O excess found in the same samples. There is no a priori reason to expect these numbers to be the same, since chemical stoichiometry must determine to some extent the degree of incorporation of the exotic components. It can be concluded,

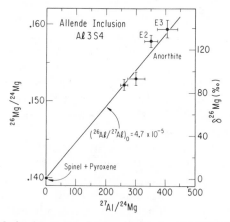

Figure 3 ^{26}Al-^{26}Mg isochron plot of minerals from a single inclusion in the Allende meteorite (from 31). The linear relationship between excess ^{26}Mg and the Al/Mg elemental ratio results from the *in situ* decay of ^{26}Al. The slope of the line is a measure of the ratio of ^{26}Al/^{27}Al at the time of solidification of the minerals. The existence of substantial amounts of live ^{26}Al within the early solar system implies a time interval between nucleosynthesis and solar system condensation of not more than one or two million years (33).

however, that the ^{26}Al was condensed into the minerals within a few half-lives of its production, i.e. within one or two million years. This is the critical conclusion from the ^{26}Mg observations: The time interval between the last major nucleosynthetic event and the condensation of solid bodies in the solar system was on the order of 10^6 years, rather than 10^8 years as had previously been believed on the basis of the extinct radioactivities of ^{129}I and ^{244}Pu (37). Two principal inferences follow from this conclusion: (a) that a supernova explosion took place in the neighborhood of the solar nebula just prior to its collapse and condensation, and may have triggered the collapse (29); and (b) that ^{26}Al was present in the early solar system at sufficiently high concentrations to serve as a major heat source for the melting and differentiation of planetesimals (33, 38).

3 ALLENDE INCLUSIONS C1 AND EK 1-4-1

Two small samples from the Allende meteorite are so exceptional and remarkable that they merit special consideration. They have anomalous isotopic compositions for almost every element that has been analyzed. In particular, they contain nuclear anomalies in the heavy elements, as well as the light element effects seen in other samples. These two inclusions, C1 and EK 1-4-1, are rounded, about one centimeter in diameter, and are composed primarily of relatively coarse crystals of the minerals melilite, pyroxene, and spinel. In visual appearance, mineralogy, and major and minor element chemical composition, these inclusions are not recognizably different from hundreds of others seen in the Allende meteorite. Only their isotopic compositions are strikingly different. They were first recognized by their oxygen isotope abundances (24, 39), which have been modified by a mass-fractionation process superimposed on the commonly observed ^{16}O addition. The magnesium isotopes were found to have a large mass fractionation accompanied by a small nuclear effect (40). The heavy isotopes of silicon are also enriched by a mass-fractionation process (41, 42). These fractionations are all on the order of 1.2–3% per mass unit, and probably result from a kinetic effect of a transport process in the gas phase at high temperature prior to condensation of the solid minerals. Apparently these materials were favorably located in the solar nebula to collect a whole host of nucleosynthetic debris. At the time of this writing, these two samples have been found to contain nuclear anomalies in oxygen, magnesium, silicon, calcium, strontium, barium, neodymium, and samarium. It seems certain that anomalies in other elements will be found as they are sought. None of the effects seen is large: isotope ratio variations are mostly

$< 1\%$. However, with the precision attainable by modern mass-spectrometric methods, isotopic variations on the order of one part in 10^4 are readily measurable. In principle, one would expect to be able to derive a detailed scenario of nucleosynthesis, since the number of anomalous abundances far exceeds the number of known nucleosynthetic processes. However, this has not yet been accomplished, and it may be that even in these small samples, we have derived material from different zones of a supernova, or perhaps from more than one supernova. The details of the observations are described, element by element, in the following sections.

There exists inherent ambiguity in the interpretation of isotopic abundance patterns if more than one isotope has a modified abundance. The problem is even worse when the nuclear effects being observed are similar in size to the mass-dependent fractionations that occur in the mass-spectrometric measurement itself. Several approaches to this problem have been attempted; each involves basic assumptions that might well be incorrect. One possibility is to minimize the number of isotopes with anomalous relative abundances, and thus to attribute some special processes to those isotopes that remain anomalous (43). A consequence of this procedure is that it sometimes leads to negative anomalies, i.e. to cases in which the meteoritic abundance of a particular nuclide is less than the terrestrial abundance. Unless some special circumstance provides for a depletion mechanism for the meteoritic material (such as holdup in the vapor phase of a long-lived radioactive precursor), a negative anomaly in the meteorite implies that the rest of the solar system has been uniformly contaminated by an excess of the nuclide in question. Thus the number of atoms involved and the magnitude of the implied astrophysical processes are immensely greater for negative anomalies than for positive ones. This observation leads to an alternative assumption in the normalizing procedure, namely that there are no negative anomalies. This assumption, accompanied by additional assumptions of the relative magnitudes of isotopic excesses, derived from theories of nucleosynthesis, has led to plausible interpretations of the isotopic anomalies in samarium and neodymium (44, 45), which have complex patterns.

3.1 Oxygen, Magnesium, and Silicon

The oxygen isotopic abundances in C1 and EK 1-4-1 can be interpreted as the result of a mass fractionation superimposed on an ^{16}O-rich material of the type found in other Allende high temperature inclusions (39). The magnesium and silicon patterns are similar to one another: large mass-fractionation effects enriching the heavier isotopes, combined with much smaller nuclear effects that enrich the middle isotope (^{25}Mg and ^{29}Si) or deplete the light or heavy isotopes. The mass-fractionation effects are larger in Cl than in EK 1-4-1; the nuclear effects are larger in EK 1-4-1 (as are the nuclear effects in all the elements described in the following sections). The isotopes ^{24}Mg, ^{25}Mg, ^{26}Mg, ^{29}Si, and ^{30}Si are all produced in approximately solar relative abundances in explosive carbon burning (36), but the proportions are somewhat sensitive to the initial density, temperature, and neutron excess, so that any single supernova (or small number of them) need not yield the solar abundances. It is surprising that larger silicon isotope anomalies have not been observed, since ^{28}Si is expected to be derived primarily from the oxygen-burning zone of a supernova (46), and therefore mixing of different zones is necessary to produce solar system abundances.

3.2   Calcium

Calcium is one of the elements most likely to display isotopic variability if material is derived from different regions of a single supernova. Both ^{40}Ca and ^{42}Ca can be formed in their solar system abundances (relative to ^{28}Si) in explosive oxygen burning (46). The rare isotopes ^{43}Ca, ^{46}Ca, and ^{48}Ca may be produced by neutron and proton capture reactions on seed nuclei (primarily ^{36}Ar and ^{40}Ca) during explosive carbon burning (47). The solar system abundance of ^{44}Ca has not been accounted for by any nucleosynthetic scheme. Explosive silicon burning produces only about one tenth of the observed abundance (46). Clayton (4), noting that ^{44}Ca is probably produced mainly via ^{44}Ti (half-life 47 years), suggested that ^{44}Ca/^{40}Ca variations could result if a chemical separation of calcium from titanium took place immediately after a supernova explosion. The calcium isotope anomalies in EK 1-4-1 and Cl bear no obvious relationship to these nucleosynthetic schemes. Lee et al (48), after normalizing to the terrestrial ^{44}Ca/^{40}Ca ratio to remove fractionation effects, found a 0.17 ± 0.02% excess of ^{42}Ca and a 1.35 ± 0.06% excess of ^{48}Ca. Sample Cl is deficient in ^{48}Ca by 0.27±0.03%, and showed no measurable effects on the other isotopes.

As was discussed above, the choice of isotopes for normalization is arbitrary, and qualitatively different conclusions may result from alternative choices. Lee et al (48), using a double-spike technique by addition of a tracer enriched in ^{42}Ca and ^{48}Ca, found the ^{44}Ca/^{40}Ca ratio in EK 1-4-1 to be 0.72% greater than the terrestrial value. They assumed this difference to be the consequence of a mass-dependent isotopic fractionation, and corrected the other ratios proportionally. However, in view of the complexity of the nucleosynthesis of calcium, it is quite likely that the ^{44}Ca/^{40}Ca ratio has been altered by nuclear processes as well, so that the normalization process used would distort the abundance pattern of all of the other isotopes. Although a complete understanding of the calcium isotope anomalies is not yet at hand, it can be said at the very least that two or more isotopes have nonterrestrial abundances, and that the departures from terrestrial values are on the order of 1–2%.

3.3   Strontium

Strontium is an important element for investigation of early solar system processes because of the presence of variable abundances of radiogenic ^{87}Sr (from 10^{10}-year ^{87}Rb). Samples from the Allende meteorite are more primitive (lower $^{87}Sr/^{86}Sr$ ratio) than any other known meteoritic materials (23). However, the $^{87}Sr/^{86}Sr$ ratios vary from sample to sample, and are not linearly correlated with the Rb/Sr ratio, as they would be if they had remained as isolated closed systems, so that it has not been possible to determine an age for the Allende high temperature inclusions by this method. It was suggested (49) that there might be significant variations in abundances of the nonradiogenic strontium isotopes, but a search for such variations proved negative (23). However, analyses of Allende samples EK 1-4-1 and C1 have indeed yielded anomalous isotopic abundances, with apparent ^{84}Sr deficiencies of $0.38 \pm 0.03\%$ and $0.1 \pm 0.02\%$, respectively (50). The observations were interpreted as excesses of ^{86}Sr, ^{87}Sr, and ^{88}Sr, which are produced primarily in the s-process (1): neutron capture on a time scale slow compared to β decay.

3.4   Barium

Isotopic abundances in meteoritic barium have been found to be the same as terrestrial except in EK 1-4-1 and C1 (43). If the ratio of a pair of dominantly s-process nuclei ($^{134}Ba/^{138}Ba$) is chosen for normalization, EK 1-4-1 has excesses of ^{135}Ba and ^{137}Ba of $0.13 \pm 0.01\%$ and $0.12 \pm 0.01\%$ respectively, with ^{130}Ba and ^{132}Ba having terrestrial abundances. The pattern is consistent with an excess of r-process nuclei, as is seen in neodymium and samarium, described below. However, sample C1, when similarly normalized, has an anomaly only at ^{135}Ba, and in this case ^{135}Ba is deficient by $0.02 \pm 0.01\%$. The possibility exists that this deficiency of ^{135}Ba may be related to the fact that its nucleosynthetic precursor is long-lived ^{135}Cs (half-life 2×10^6 years). If barium and cesium were chemically separated prior to decay of ^{135}Cs, a ^{135}Ba deficiency would result.

3.5   Neodymium and Samarium

The rare earth elements have long been recognized for their importance in testing theories of nucleosynthesis. Because of the very close similarities of their chemical properties, they move together in nature as a group, with only minor variations in their relative abundances. Thus it is possible to observe, with good precision, the solar system abundances of the nuclides over a range of some 40 mass units. The dominant nucleosynthetic processes in this mass region are neutron capture (r- and s-process) and proton capture (p-process) (1).

The s-process occurs within stars during hydrostatic burning (51, 52), using neutrons derived from reactions such as $^{22}Ne(\alpha, n)^{25}Mg$. The abundances of s-process nuclides in the region of barium and the light rare earths are enhanced relative to those of neighboring elements because of their low neutron-capture cross sections associated with magic neutron number 82.

The r-process requires neutron capture on iron seed nuclei on a time-scale of seconds (1) and is therefore associated with a catastrophic event, presumably a supernova explosion. A less extreme version of neutron capture, in which capture times are comparable to β-decay lifetimes (the n-process) (53), may be important in some supernova explosions. Stable r-process nuclei result from β decay of extremely neutron-rich nuclei produced in multiple neutron capture by iron seed nuclei. Some nuclides, such as ^{142}Nd, ^{144}Sm, ^{148}Sm, ^{150}Sm, cannot be produced by the r-process because of shielding by a stable lower-Z isobar. Thus mixtures of s-process material with r-process material in variable proportions should lead to isotopic abundance variations.

Substantial isotope anomalies have been observed in neodymium and samarium in sample EK 1-4-1 (43, 44). As has been described above, the magnitudes of the effects on individual isotopes depends on the choice of normalizing procedure. An attractive procedure is to choose a pair of isotopes that are either pure s-process or pure r-process, and assume that they are enriched or depleted by the same factor in any mixture of r- and s-material. This pair can then be used to determine the mass-fractionation correction for all of the isotopes. The reference isotope may then be chosen to make all anomalies positive. This process was applied to the observations for neodymium, by choosing the two pure r-process nuclei, ^{148}Nd and ^{150}Nd (45), and to observations for samarium, by choosing the two pure s-process nuclei ^{148}Sm and ^{150}Sm (44). In both cases, the observed isotopic variations are very well matched by excesses of r-process nuclei in the amount of 0.34–0.38%. The abundance pattern of the excess r-process matter closely matches the observed solar system abundances of r-process nuclides from $A = 143$ to $A = 154$. The generality of these interpretations can be explored by a search for r-process excesses in the neighboring elements, cerium and gadolinium, and in heavier elements. The anomalies in barium (in EK 1-4-1) are also apparently due to an excess of r-process nuclides (43, 45).

4 POSSIBLE ISOTOPIC ANOMALIES IN OTHER ELEMENTS

4.1 *Carbon*

Since carbon has only two stable isotopes, there is no unequivocal way to distinguish between the effects of mass fractionation and the effects of nuclear reactions in modifying its isotope ratio. Variations in $^{13}C/^{12}C$ ratio of the order of 9% have been observed among the different carbon compounds within individual carbonaceous meteorites, with ^{13}C enriched in carbonates and ^{12}C enriched in reduced carbon compounds (54–56). This range is large compared to terrestrial samples of similar chemistry, but is not beyond that which might be obtained by equilibrium or kinetic processes at low temperatures.

4.2 *Nitrogen*

Nitrogen also has only two stable isotopes, and thus suffers the same ambiguities of interpretation as carbon. In the case of nitrogen, however, the range of observed variations of the $^{15}N/^{14}N$ ratio in meteorites is much greater, i.e. about 24% (57). Evidence for large variations in nitrogen isotopic abundances in solar system materials also comes from measurements of lunar soils, which themselves show a range of 24% (not including the much larger ^{15}N excesses due to cosmic-ray spallation reactions) (58, 59). The combination of lunar and meteoritic data extends the range of variation of the $^{15}N/^{14}N$ ratio to about 30% Although it is tempting to attribute a part of this large variability to nuclear effects, it must be recognized that extensive mass fractionation may take place in the gas phase because of, for example, escape from a gravitational field. Such a process has been invoked to account for 75% excess of ^{15}N in the atmosphere of Mars (60).

4.3 *Sulfur*

A search for isotopic anomalies in meteoritic sulfur was carried out several years ago (61), and showed only small mass-fractionation effects, and no evidence for nucleosynthetic heterogeneities. More recently, Rees & Thode (62) reported a small enhancement of ^{33}S (0.1% above the terrestrial value) in an acid residue of the Allende meteorite. The authors could not rule out the possibility that the observed effect could have resulted from some unknown contamination in the mass spectrum.

4.4 *Krypton and Xenon*

Isotopic abundances of Kr and Xe, particularly Xe, have provided a wealth of information about processes in the early solar system. The abundances are affected by a number of physical processes: (a) mass-dependent fractionation, (b) production by spallation reactions involving cosmic rays, (c) decay of ^{129}I to ^{129}Xe, and (d) production by fission of uranium, ^{244}Pu, and possibly a superheavy element (63). Whether isotopic variations of nucleosynthetic origin have been observed is a matter of controversy.

Because of the chemical inertness of the noble gases, and their very low background levels, it is possible to extract and analyze very small quantities of these elements. Furthermore, the weakness of their chemical interaction with other elements results in their having little "preference" of one site over another, so that their specific location within a heterogeneous solid matrix may be determined by the process by which they were implanted into that matrix. For example, a xenon atom formed in the fission of ^{244}Pu within a mineral grain remains at the end of its recoil track, either within the same crystal as the parent plutonium atom or in an adjacent one if the track crosses a grain boundary. Thus atoms of different origin, e.g. from β decay, from fission, or from spallation, are usually situated in chemically and structurally different locations. Then, by chemically or physically sampling these different locations, one can measure the different components individually. As a consequence of these special attributes of the noble gases, it is possible to measure effects involving quantities of material so minute that they would be unobservable in other elements.

A recent review by Podosek (9) discusses in some detail the isotopic variations of krypton and xenon in the solar system. The present discussion is limited to the meteoritic component variously labeled as carbonaceous chondrite fission xenon or xenon-X (64, 65). It occurs in low temperature phases in the primitive carbonaceous chondrites (63) rather than in the high temperature phases in which other heavy element isotopic anomalies are found. Since these meteorites are mechanical mixtures of materials of low temperature and high temperature origin, formed in chemically different systems, there need be no close genetic association of the nuclear effects in the different fractions. This anomalous xenon is enriched both in the lightest isotopes, ^{124}Xe, ^{126}Xe, and ^{128}Xe, and in the heaviest isotopes, ^{134}Xe and ^{136}Xe, by as much as a factor of two (63). A striking observation is that the light isotope enrichment (L-Xe; 66) and the heavy isotope enrichment (H-Xe; 66) are in a constant (or very nearly constant) ratio for samples from many meteorites, even though different processes seem required for their production. In one proposed scenario, the light isotope enrichment is attributed to a general mass-dependent fractionation process affecting all isotopes; the heavy isotope enrichment is then attributed to an addition of fission products (63). However, the relative abundances of the excess heavy isotopes do not correspond to the products of fission of any known nuclide. This fact, in addition to the chemical properties of the fissioning element inferred from its associations in the meteorites, led to the conclusion that the element in question was superheavy, with $Z \approx 115$ (67). This interpretation has not yet been confirmed by observations of other fission products, and a search for radiation-damage fission tracks has yielded negative results (68).

An alternative hypothesis for the origin of the xenon anomalies is the presence of a mixture of excess p-process and r-process isotopes, giving rise to the L-Xe and H-Xe enrichments, respectively (69). The observed isotopic abundances of the H-Xe component are greatest for ^{136}Xe, and decrease with decreasing N, whereas the opposite slope is expected for a pure r-process. Black (69) suggested a modified r-process with a somewhat longer time scale, allowing a significant amount of β decay, and shifting the peak in the abundance curve (due to magic neutron number 82) from $A = 130$ to $A = 136$. He then proposed production of both L-Xe and H-Xe (or P-Xe and R-Xe in his notation) in a supernova explosion, and transport to the cool regions of the solar nebula within dust grains. Blake & Schramm (70) have criticized this model on the grounds that the time scale is too long for an exploding object, and have proposed an alternative neutron-capture scheme (their n-process) based on a supernova time scale.

Cameron & Truran (29) call upon a combination of s-process in a helium-burning shell and neutron capture in explosive carbon burning to produce the H-Xe, and p-process in the region of explosive oxygen burning to produce the L-Xe.

A model with similar nuclear physics but different astrophysics has been proposed by Manuel & Sabu (71), who suggest not only that the xenon anomalies are derived from a combination of r-process and p-process in a supernova, but that the solar system in its entirety was assembled from the debris of the supernova.

The isotopic variations in krypton in carbonaceous chondrites are smaller and less diagnostic than those in xenon, with an enrichment in the heavier isotopes correlated with the enrichment in X-Xe (63). The effects may be attributable either to fission or to neutron-capture processes, but to date, the krypton data have proved less restrictive on hypotheses of origin than have the xenon data.

4.5 Mercury

Some very large isotopic variations have been reported for mercury in several meteorites, including both primitive and metamorphosed types (72–74). The analyses were done by neutron activation, followed by stepwise heating in vacuum to release mercury from various sites within the meteorites. Of the seven stable isotopes of mercury, this technique allowed measurement only of the $^{202}Hg/^{196}Hg$ ratio. This ratio was found to be the same as the terrestrial value for 37 stony meteorites in an earlier survey (75), which did not employ the stepwise thermal release procedure. The stepwise release allows measurement of individual small fractions of mercury that may be located in different parts of the meteorite, and is thus able to reveal small fractions of anomalous composition that would be diluted beyond recognition in the bulk meteorite analysis. Jovanovic & Reed (73) found enhancement in the $^{202}Hg/^{196}Hg$ ratio ranging from about 15% up to a factor of 22. Although these enrichments are well above the statistical errors of the measurements, and no obvious systematic errors or contamination have been recognized, the experiments suffer the weakness that the effects are not reproducible in replicate analyses of the same meteorite.

A mass-spectrometric search for isotopic anomalies in mercury (76) in two of the meteorites studied by Jovanovic & Reed (73) revealed no anomalies for the isotopes with $A > 198$. Unfortunately ^{196}Hg could not be measured accurately in this study because of hydrocarbon interference.

Although the lack of reproducibility and of independent confirmation of the isotopic anomalies in mercury has left much of the meteoritical community skeptical about their real existence, nucleosynthetic theories to explain the effects are available should they be confirmed. Jovanovic & Reed (74) discuss both the possibility of excesses in ^{202}Hg, derived from a ^{202}Pb precursor (half-life 3×10^5 years), and of deficiencies in ^{196}Hg and other p-process nuclei. Cameron & Truran (29) also discussed production of ^{202}Pb by a p-process in the oxygen zone of a supernova.

4.6 Uranium

Isotopic abundances of uranium are especially important for determining the time of operation of the r-process for solar system matter (77). The terrestrial $^{238}U/^{235}U$ ratio of 137.8 has been repeatedly observed in terrestrial and lunar samples (e.g. 78). However, a recent study of uranium isotopes in stony meteorites showed a surprising variability, with $^{238}U/^{235}U$ ratio ranging from 106.8 to 137.5 (79). Furthermore, acid-leaching experiments showed that the low ratios were due to even greater ^{235}U enrichments in the residues. The lowest value ($^{238}U/^{235}U = 40.2$) was found in a 0.07% residue from an ordinary chondrite, Barwell. Any addition of freshly synthesized r-process nuclei would give an effect in the observed direction, since the production rates of ^{235}U and ^{238}U are almost equal (80). It is of considerable importance that these measurements be confirmed and extended to more specific meteorite samples.

5 CONCLUSIONS

Isotopic measurements of meteorites have shown that the solar nebula was not completely homogenized prior to formation of the planetary system. Variability of isotopic abundances for many elements results from addition to the protosolar nebula of matter probably produced in a single nucleosynthetic event just prior to the gravitational collapse and condensation of the solar nebula. This event may have been a supernova explosion that initiated the collapse.

The amount of freshly synthesized matter added to the solar system cannot be determined unambiguously from the available data, since the magnitudes of the observed isotopic anomalies provide only lower limits as a result of the unknown degree of homogenization of the exotic component with local matter. In some meteorite samples, this lower limit is as great as 5%, as in the case of oxygen and probably aluminum. Additions of exotic matter at these concentrations do not have a significant effect on the elemental abundances in the solar system. Their importance in solar system history is twofold: (a) the dynamical effects on the proto-solar cloud associated with a supernova explosion in its immediate neighborhood, and (b) the effect of ^{26}Al as a heat-source for the melting and igneous differentiation of small planetismals, thus significantly altering the chemical and physical properties of these building blocks of the planets.

When more data are available, the patterns of isotopic anomalies within individual small samples will be reducible to specific contributions from specific nucleosynthetic processes, probably from a single event. This will add immeasurably to our understanding of the various nucleosynthetic processes and their astrophysical sites, since the observational evidence previously available necessarily came from an average over a variety of physical parameters. The first steps toward this goal have already been made with the isotopic data on the heavy elements, which show excesses of neutron-rich nuclides characteristic of some kind of rapid neutron-capture process. Much more will be accomplished when meteorite analyses are done on more elements and on samples with different exposure to the exotic component.

ACKNOWLEDGMENTS

This work was supported by grants EAR 74-19038 (NSF) and NGL 14-001-169 (NASA). I have benefitted from discussions with Professors E. Anders, L. Grossman, and D. N. Schramm of the University of Chicago.

Literature Cited

1. Burbidge, E. M., Burbidge, G. R., Fowler, W. A., Hoyle, F. 1957. *Rev. Mod. Phys.* 29:547–650
2. Reeves, H. 1974. *Ann. Rev. Astron. Astrophys.* 12:437–69
3. Reeves, H. 1972. *Astron. Astrophys.* 19:215–23
4. Clayton, D. D. 1975. *Nature* 257:36–37
5. Lattimer, J. M., Schramm, D. N., Grossman, L. 1978. *Astrophys.* 219:230–49
6. Krummenacher, D., Merrihue, C. M., Pepin, R. O., Reynolds, J. H. 1962. *Geochim. Cosmochim. Acta* 26:231–49
7. Onuma, N., Clayton, R. N., Mayeda, T. K. 1972. *Geochim. Cosmochim. Acta* 36:169–88
8. Epstein, S., Taylor, H. P. 1970. *Proc. Apollo 11 Lunar Sci. Conf., Geochim. Cosmochim Acta,* Suppl. 1, pp. 1085–96
9. Podosek, F. 1978. *Ann. Rev. Astron. Astrophys.* 16:293–334
10. Herbst, W., Assousa, G. 1978. *Astrophys. J.* 217:473–87
11. Black, D. C. 1972. *Geochim. Cosmochim. Acta* 36:377–94
12. Reynolds, J. H., Turner, G. 1964. *J. Geophys. Res.* 69:3263–81
13. Jeffrey, P. M., Reynolds, J. H. 1961. *Z. Naturforsch.* 16a:431–32
14. Eberhardt, P. 1974. *Earth Planet. Sci. Lett.* 24:182–87
15. Niederer, F. Eberhardt, P. 1977. *Meteoritics* 12:327–31
16. Heymann, D., Dziczkaniec, M. 1976. *Science* 191:79–81
17. See Ref. 4
18. Arnould, M., Nørgaard, H. 1978. *Astron. Astrophys.* 64:195–213
19. Srinivasan, B., Anders, E. 1978. *Science* 51–56
20. Matsuhisa, Y., Goldsmith, J. R., Clayton, R. N. 1978. *Geochim. Cosmochim. Acta* 42:173–82
21. Clayton, R. N., Onuma, N., Mayeda, T. K. 1976. *Earth Planet. Sci. Lett.* 30:10–18
22. Grossman, L., Larimer, J. 1974. *Rev. Geophys. Space Phys.* 12:71–101
23. Gray, C. M., Papanastassiou, D. A., Wasserburg, G. J. 1973. *Icarus* 20:213–39
24. Clayton, R. N., Onuma, N., Grossman, L., Mayeda, T. K. 1977. *Earth Planet. Sci. Lett.* 34:209–224
25. Clayton, D. D., Dwek, E., Woosley, S. E. 1977. *Astrophys. J.* 214:300–15
26. Clayton, D. D., Woosley, S. E. 1974. *Rev. Mod. Phys.* 46:755–71
27. Arnett, W. D. 1978. *Astrophys. J.* 219:1008–16
28. Clayton, R. N., Grossman, L., Mayeda, T. K. 1973. *Science* 182:485–88
29. Cameron, A. G. W., Truran, J. W. 1977. *Icarus* 30:447–61
30. Grossman, L. 1972. *Geochim. Cosmochim. Acta* 36:597–619
31. Steele, I. M., Smith, J. V., Hutcheon, I. D., Clayton, R. N. 1978. *Lunar and Planetary Science IX*, pp. 1104–6. Houston: Lunar Planet. Inst.
32. Lee, T., Papanastassiou, D. A., Wasserburg, G. J. 1976. *Geophys. Res. Lett.* 3:109–12
33. Lee, T., Papanastassiou, D. A., Wasserburg, G. J. 1977. *Astrophys. J.* 211:L107–10
34. Lorin, J. C., Shimizu, N., Christophe-Michel Lévy, M., Allègre, C. J. 1977. *Meteoritics* 299–300
35. Bradley, J. G., Huneke, J. C., Wasserburg, G. J. 1978. *J. Geophys. Res.* 83:244–54
36. Arnett, W. D. 1969. *Astrophys. J.* 157:1369–80
37. Hohenberg, C. M., Munk, M. N., Reynolds, J. H. 1967. *J. Geophys. Res.* 72:3139–77
38. Urey, H. C. 1955. *Proc. Natl. Acad. Sci. USA* 41:127–44
39. Clayton, R. N., Mayeda, T. K. 1977. *Geophys. Res. Lett.* 4:295–98
40. Wasserburg, G. J., Lee, T., Papanastassiou, D. A. 1977. *Geophys. Res. Lett.* 4:299–302
41. Clayton, R. N., Mayeda, T. K., Epstein, S. 1978. See Ref 31, pp. 186–88
42. Yeh, H., Epstein, S. 1978. See Ref. 31, pp. 1289–91.
43. McCulloch, M. T., Wasserburg, G. J. 1978. *Astrophys. J.* 220:L15–19
44. Lugmair, G., Marti, K., Scheinin, N. B. 1978. See Ref. 31, pp. 672–74
45. Clayton, D. D. 1978. Submitted to *Astrophys. J. Lett.*
46. Woosley, S. E., Arnett, W. D., Clayton, D. D. 1973. *Astrophys. J. Suppl. Ser.* 26:231–312
47. Howard, W. M., Arnett, W. D., Clayton, D. D., Woosley, S. E. 1972. *Astrophys. J.* 175:201–16
48. Lee, T., Papanastassiou, D. A., Wasserburg, G. J. 1978. *Astrophys. J.* 220:L21–25
49. Cameron, A. G. W. 1973. *Nature* 246:30–32
50. Papanastassiou, D. A., Huneke, J. C., Esat, T. M., Wasserburg, G. J. 1978. See Ref. 31, pp. 859–61
51. Cameron, A. G. W. 1954. *Phys. Rev.* 93:932

52. Fowler, W. A. 1961. In *Modern Physics for the Engineer*, Vol. 2, ed. L. W. Ridenour pp. 177–239. New York: McGraw-Hill
53. Blake, J. B., Schramm, D. N. 1976. *Nature* 263:707–8
54. Clayton, R. N. 1963. *Science* 140:192–93
55. Belsky, T., Kaplan, I. R. 1970. *Geochim. Cosmochim. Acta* 34:257–78
56. Krouse, H. R., Modzeleski, V. E. 1970. *Geochim. Cosmochim. Acta* 34:459–74
57. Kung, C. C., Clayton, R. N. 1978. *Earth Planet. Sci. Lett.* 38:421–35
58. Becker, R. H., Clayton, R. N., Mayeda, T. K. 1976. *Proc. Lunar Sci. Conf. 7th*, pp. 441–58. Oxford: Pergamon
59. Becker, R. H., Clayton, R. N. 1977. *Proc. Lunar Sci. Conf. 8th*, pp. 3685–3704. Oxford: Pergamon
60. McElroy, M. B., Yung, Y. L., Nier, A. O. 1976. *Science* 194:70–72
61. Hulston, J. R., Thode, H. G. 1965. *J. Geophys. Res.* 70:3475–84
62. Rees, C. E., Thode, H. G. 1977. *Geochim. Cosmochim. Acta* 41:1679–82
63. Lewis, R. S., Srinivasan, B., Anders, E. 1975. *Science* 190:1251–62
64. See Ref. 12
65. Manuel, O. K., Hennecke, E. W., Sabu, D. D. 1972. *Nature Phys. Sci.* 240:99–101
66. Pepin, R. O., Phinney, D. 1978. See Ref. 31, pp. 882–84
67. Anders, E., Higuchi, H., Gros, J., Takahashi, T., Morgan, J. W. 1975. *Science* 190:1262–71
68. Fraundorf, P., Flynn, G. J., Shirck, J. R., Walker, R. M. 1977. *Earth Planet. Sci. Lett.* 37:285–95
69. Black, D. C. 1975, *Nature* 253:417–19
70. See Ref. 53
71. Manuel, O. K., Sabu, D. D. 1975. *Trans. Mo. Acad. Sci.* 9:104–22
72. Reed, G. W. Jr., Jovanovic, S. 1969. *J. Inorg. Nucl. Chem.* 31:3783–88
73. Jovanovic, S., Reed, G. W. Jr. 1976. *Earth Planet. Sci. Lett.* 31:95–100
74. Jovanovic, S. Reed, G. W. Jr. 1976. *Science* 193:888–91
75. Reed, G. W. Jr., Jovanovic, S. 1964. *J. Geophys. Res.* 72:2219–28
76. Von Helden, J., Begemann, F. 1976. *Meteoritics* 11:297
77. Schramm, D. N., Wasserburg, G. J. 1970. *Astrophys. J.* 162:57–69
78. Rosholt, J. N., Tatsumoto, M. 1970. *Geochim. Cosmochim. Acta*, Suppl. 1, pp. 1499–1502
79. Arden, J. W. 1977. *Nature* 269:788–89
80. Seeger, P. A., Fowler, W. A., Clayton, D. D. 1965. *Astrophys. J. Suppl. Ser.* 11:121–66

Ann. Rev. Nucl. Part. Sci. 1978. 28:523–96

SELF-CONSISTENT CALCULATIONS OF NUCLEAR PROPERTIES WITH PHENOMENOLOGICAL EFFECTIVE FORCES[1]

×5600

P. Quentin[2]

Theoretical Division, Los Alamos Scientific Laboratory, University of California, Los Alamos, New Mexico 87545

H. Flocard

Division de Physique Théorique,[3] IPN, BP No. 1, 91406 Orsay, France

CONTENTS

[1] The US Government has the right to retain a nonexclusive, royalty-free license in and to any copyright covering this article.

[2] Permanent address: Division de Physique Théorique, IPN, BP No. 1, 91406 Orsay, France.

[3] Laboratoire associé au CNRS.

523

INTRODUCTION AND SUMMARY

A basic assumption of low energy nuclear physics consists of viewing the nucleus as a collection of nucleons interacting through a nucleon-nucleon potential. The natural framework for such a study is therefore the ordinary (nonrelativistic) quantal many-body problem. In view of its intermediate number of degrees of freedom, the atomic nucleus constitutes a very exciting avatar of this general problem. Intermediate is to be understood in the following way. On the one hand, the number of nucleons in an average nucleus is sufficiently large to allow a sensible description of most of the global nuclear properties in terms of a semiclassical approximation. On the other hand, this number is small enough that a limited yet significant set of overall nuclear properties (like deformation properties) can only be explained from a purely quantal description. In an alternative way of expressing this essential feature of the nuclear structure, one may emphasize the coexistence and the interplay in the nucleus of both the individual particle and the collective degrees of freedom.

An exact microscopic treatment of the many-body nuclear problem as defined above is not yet possible. Some approximations are thus necessary. Yet to pertain to the studied system, these approximations must compulsorily allow a correct coupling of the single-particle and the collective motions.

By analogy with atomic physics, it was quite natural to apply to nuclei the concepts of the independent-particle (or Hartree-Fock) approximation. This has indeed been the assumption behind the shell model description of nuclear structure. The successes of the latter, however, have not been easily understood at first. A mean field description seemed contradictory with an attractive short-range free nucleon-nucleon interaction, which becomes highly repulsive at very close distances. Later, the validity of the independent-particle approximation was found to be a consequence of the Pauli exclusion principle: By preventing particle diffusion into already occupied states, it leads to a nucleonic mean free path much larger than a characteristic internucleonic distance. A quantitative check of these ideas has been provided by infinite nuclear matter calculations: Nucleonic wave functions are indeed little affected by the interaction except in a small region.

The independent-particle approximation consists in finding the solution of the Schrödinger equation governing the dynamics of the system in a restriction of the Hilbert space: the ensemble of Slater determinants. The price to pay for such an approximation is the definition of an effective Hamiltonian allowing a reproduction (as good as possible) of the properties of the physical system when described in such an unphysically limited subspace. There are essentially two groups of effective nucleon-nucleon interactions. In the first, such an interaction in finite nuclei is built up (after Brueckner) from approximate nuclear matter calculations. These interactions lead to a fair reproduction of static properties of magic nuclei. Their main appeal lies in the fact that they are almost completely derived microscopically from the interaction between free nucleons; they are, however, not completely without ad hoc renormalization. The second approach is purely phenomenological and is the subject of this review. It consists in assuming, a priori, an analytical form for the effective interaction. Such a choice will first be oriented by numerical practicability, but it will also be compelled by the gross features of the microscopically determined effective interactions. One of the most important constraints of the latter type is the density dependence of the interaction. Its inclusion has been the key for the successful reproduction within the Hartree-Fock approximation of both total and single-particle binding energies.

Having defined the analytical structure of the force, one adjusts its parameters by fitting a limited set of experimental properties. The latter are generally static properties of magic nuclei; sometimes nuclear matter properties viewed as an indirect source of experimental information are also considered in the fit procedure.

Solving the stationary Schrödinger equation in the Hartree-Fock approximation is equivalent to searching for a Slater determinant that yields a minimal expectation value for the effective Hamiltonian. Solving the time-dependent Schrödinger equation in the same approximation also amounts to a minimization problem but, in that case, for the action integral. In both cases one obtains a Schrödinger-like set of equations for the single-particle orbitals building up the Slater determinant. The particularity of this so-called Hartree-Fock set of equations is their highly nonlinear character, which is understood easily if one contemplates the mean field governing the motion of the individual nucleons, as determined by the wave functions of these nucleons themselves. This self-consistency property is one of the main sources of computational difficulty in practical Hartree-Fock calculations.

It is rather well known that there is at least one type of correlation that must be included in order to obtain a realistic description of many nuclei in their ground states and of almost all nuclei when they deform. These correlations are of the pairing type. The Hartree-Fock-Bogolyubov approximation takes into account these correlations. In comparison with the Hartree-Fock ansatz for uncorrelated systems, it corresponds to the same degree of approximation to the many-body problem, but for correlated systems. In other words, both are self-consistent models, the former constituting an independent particle approach, the latter an independent quasiparticle approach. Even though static Hartree-Fock-Bogolyubov calculations have been achieved recently, they are by far more time consuming than corresponding Hartree-Fock calculations. Consequently one has generally performed instead Hartree-Fock plus Bardeen-Cooper-Schrieffer (BCS) calculations, which constitute a much simpler (though satisfactory in most cases) approach. We refer to them in what follows as self-consistent calculations.

As seen in extensive studies for nuclei all over the periodic table, self-consistent calculations give binding energies almost as close to experimental values as those obtained from liquid drop fits. At the same time, they provide a fair reproduction of the single-particle level ordering near the Fermi level. The extraction of odd (or odd-odd) nuclei spectroscopic information from Hartree-Fock single-particle states is possible through rather crude model assumptions. They compare, however, in a satisfactory qualitative way with experimental data for both spherical and deformed nuclei. The nucleonic density constitutes another output of these calculations. Its monopole part has been found to well reproduce measured quantities such as rms charge radii or elastic electron scattering cross sections. Quadrupole and hexadecapole components of this nucleonic density are also in impressive agreement with relevant experimental data [static quadrupole moment of first 2^+ levels, $B(E2)$, $B(E4)$, parametrized optical potentials, etc].

Upon constraining the static solutions of our variational problem to given values of one (or a few) collective variable, one is able to map a potential energy curve (or surface) for the corresponding collective motion. The most familiar examples of such deformation energy curves are undoubtedly the fission barriers of actinide nuclei. Self-consistent calculations do indeed reproduce the experimentally well-known double-humped barrier, but cannot yet accurately reproduce experimental fission barrier heights. The latter, however, are strongly contingent upon specific dynamical assumptions and a definite conclusion about the validity of the approach for this process cannot be drawn in the absence of fully consistent static and dynamical calculations.

In some nuclei, two classes of low energy levels coexist, each one corresponding to a definite intrinsic configuration. At both edges of a major shell, one often finds such a situation caused by an approximate degeneracy of prolate- and oblate-shaped equilibrium solutions. Upon varying the number of nucleons, one may encounter a ground-state transition from one intrinsic shape to the other. Self-consistent calculations have reproduced (or predicted) such systematical trends in a qualitative way. More quantitative assessments (such as pinpointing the critical number of nucleons that characterize a ground-state shape transition, for instance) presently exceed the possibilities of the method.

It may be appropriate at this point to recall the limitation of the self-consistent approach, which are the obvious counterparts of its practicability. Some are simply drawbacks of any phenomenological parameterization within a restrictive formal framework. The versatility of the approach is contingent upon a wide choice of basic data to which the model parameters are to be fitted. However, the issue is often nothing but a compromise. Specific difficulties are direct consequences of the description of the stationary many-body state by an independent-particle wave function. Among them one may single out the incorrect treatment of center-of-mass motion and the fact that the model wave functions are generally not eigenvalues of the angular momentum.

There have been numerous substitutes for self-consistent calculations. They are characterized as phenomenological parametrizations either of the total binding energies or of the mean field. To the former group belongs the liquid drop model, based on a semiclassical description of the nuclei. It has been supplemented by the proper inclusion of the so-called shell-effect corrective energy. Within the Hartree-Fock framework, the validity of the Strutinsky method to evaluate this fluctuating part of the binding energy has been demonstrated. On the other hand it has been shown that the single-particle states deduced from phenomenological one-body potential were indeed rather close to those obtained in self-consistent calculations. One of the main advantages of the latter, however (justifying their comparatively larger computing time), is that they simultaneously provide both the single-particle (fluctuating) and the semiclassical (smoothly varying) properties of nuclei, thus avoiding possible inconsistencies that result from a dual approach of the nuclear structure.

We have referred, so far, to self-consistent descriptions of nuclei in their ground states, or at least in low excited states. The Hartree-Fock approximation at finite temperature is a natural extension of this formalism to high excitation energies. This approximation is nothing but the independent-particle limit of the grand canonical thermodynamical equilibrium problem. Even though its validity range is rather limited, it provides a nice insight into average properties of excited nuclei.

The independent-particle approximation to the nonstationary Schrödinger equation is referred to as the time-dependent Hartree-Fock (TDHF) approximation. Recently nontrivial solutions of the TDHF equations have been numerically calculated. These calculations have essentially been performed for the study of two colliding nuclei. They are not yet very extensive but in a few cases they have yielded a qualitative agreement between calculated and experimental fusion cross sections.

Long before most general TDHF solutions were available, some particular solutions were studied. One group of such solutions has been obtained within the random phase approximation (RPA), which is the small amplitude motion limit of the TDHF formalism. Upon calculating consistently both the single-particle wave functions and the coherent particle-hole excitations, low energy spectroscopic properties of magic nuclei have been reproduced in a rather satisfactory way. Another limiting case of the TDHF formalism is met when the characteristic velocities associated with nuclear collective motions are small with respect to those corresponding to individual-particle degrees of freedom. The study of such adiabatic solutions is of particular interest for large amplitude collective motions (as in the fission process); indeed, in the small amplitude case the latter formalism has been shown to be equivalent to the RPA.

From the preceding, one sees that an extensive study of TDHF solutions for various initial conditions will, in the near future, allow a critical discussion of current assumptions concerning the dynamics of nuclear collective motion and its interplay with single-particle motion.

1 PHENOMENOLOGICAL EFFECTIVE FORCES

1.1 *Effective Forces*

It is the aim of this paper to review the methods and the results of calculations in finite nuclei that make an explicit use of phenomenological effective interactions within the Hartree-Fock approximation (1–3). It is therefore out of the scope of the present work to cover topics generally referred to as nuclear matter calculations. Neither do we attempt to write at length on the application of the latter to finite nuclei through a set of relevant assumptions. On these matters the reader can find comprehensive review articles in (4, 5). In this section we put into proper perspective the purely phenomenological approach that constitutes our subject.

Upon assuming a priori the nuclear dynamics to be governed by two-body nucleon-nucleon interactions, the first task is to evaluate this static interaction potential from experimental data pertaining to two-nucleon systems. As a result, it appears that this does not yield an unambiguous potential. For instance, it has been impossible to determine uniquely its short-range repulsive part. Therefore this does not allow any assessment of the validity of a perturbative treatment of the many body problem: (a) On the one hand, it is indeed possible to build up nucleon-nucleon interactions that simultaneously yield reasonable fits of two-nucleon data and rather small, second order corrections in a standard perturbation approach (it is fair to add, however, that such calculations do not match the level of practicability allowing them to cover the whole chart of nuclides); (b) On the other hand, when dealing with harder free nucleon-nucleon potentials, the treatment of the many-body problem can be kept within a suitable, first order (Hartree-Fock) approximation.

In the latter case, one should allow after Brueckner (6–10) that whenever two nucleons interact, they do it to infinite order. Consequently the interaction to be used results from a partial resummation of the perturbative series. This interaction (also referred to as the G or K matrix) is called effective because it is acting in a restrictive subspace and dressed in such a way as to mock up the effects of the neglected part of the Hilbert space. In the present case, one intends to describe the many-body nuclear states by an independent-particle wave function. The corresponding effective interaction is thus designed to reproduce nuclear properties in the best way possible within the Hartree-Fock approximation.

Calculations of such effective interactions have been performed in nuclear matter. To deal with finite nuclei, one needs to make further approximations. Among other possible approaches the method we discuss here is the one relying upon the local density approximation (11). As a net result, it has been shown (see 12–18) that one could indeed extract out of the free nucleon-nucleon interaction an effective interaction giving a fair account of static properties of spherical nuclei (like binding energies and spatial nuclear densities). This approach is not completely free from phenomenology, though: some phenomenological piece must be added to the derived effective interaction in order to improve the binding energy and the nuclear size, otherwise badly reproduced. In doing so, one is supposed to correct for the oversimplifying assumptions introduced in the many-body treatment of the nuclei considered.

As a fully phenomenological alternative to this method, one may adjust the parameters of a given analytical form for the effective interaction in such a way as to reproduce a given set of experimental data (see Section 1.2.4). Compared with the more microscopically founded calculations mentioned above, such a phenomenological approach is credited with practical simplicity. After all, this must be considered its primary justification. Nevertheless the density-dependence character of the effective force apparent in (12–18) is an important feature to be compulsorily included in any phenomenological approach. It is well known that without a density dependence of the two-body interaction, Hartree-Fock calculations are unable to simultaneously reproduce total and individual-particle binding energies (see 19–23). We therefore restrict our review to forces that include this density dependence.

1.2 Current Parametrizations

1.2.1 CENTRAL PART   The various parametrizations of the central part of the effective interaction may be classified according to their range. One class of forces are of zero range (delta type) plus corrective (second order) velocity-dependent terms acting in the relative S- and P-wave subspaces. They all include a density dependence in the delta force; this dependence, however, is either linear[4] or involves a power of the density smaller than one. It has been explicitly shown (24) that there exists a strong correlation between the incompressibility modulus in nuclear matter and the exponent of the density dependence. In view of some recent experimental results (see Section 7.3), it seems that this exponent should be significantly smaller than one ($\frac{1}{6}$–$\frac{1}{3}$). Spin-isospin symmetry effects are taken into account through the use of convenient exchange admixtures.

A force of this type was first proposed by Skyrme (26, 27) and interest in it was later revived by Vautherin & Brink (28–30) and Moszkowski (31). Table 1 sketches references that describe the main properties of the various parameterizations currently in use (30–41).

The second class of interactions refers to finite range forces. They are generally of a Gaussian type (42–49). In one case (44, 45) the tail of the

---

[4] Under the condition that the studied Slater determinant has no spin-vector part of its one-body reduced density, a linear density-dependent two-body delta interaction is equivalent to a three-body delta interaction (25). However, when the previous condition is no longer valid this equivalence fails (see, for instance, the discussion of Section 7.3).

Table 1 References that describe the main properties of various parametrizations of the effective interactions of the Skyrme type

Density dependence ($\rho^{\alpha}\delta$)				
$\alpha = 1$	$\alpha < 1$	Also in corrective terms	With more exchange operators[a]	Fourth order corrective terms
30, 32, 34–36	31, 33, 37–41	34, 36	33, 35, 40, 41	39

[a] Here are listed the forces that introduce exchange operators for terms other than the density-independent delta part.

mesonic one-boson exchange potential (OBEP) potential of Reference 50 has been included. Some interactions also introduce velocity-dependent terms (43–45). Moreover one interaction (48, 49) has been explicitly fitted to yield realistic pairing matrix elements (see Section 3). Finally, all these forces include a zero range density-dependent force.

Such a classification according to the range reflects the different methods of solving the Hartree-Fock equations. As discussed in Section 2, forces of the first class are much more easily handled in practical calculations than forces of the second class. But the counterpart of this simplicity is the implied bold mocking up of finite range effects. Actually the practical difficulties with finite range forces come from their exchange contributions. Consequently some authors either neglect the latter (51) or treat them within a local approximation of the density matrix expansion (DME) type (52–54) (see Section 1.3.1).

1.2.2 SPIN-ORBIT AND TENSOR PARTS When a two-body spin-orbit interaction is used, it is generally done within the zero range limit approximation (55–58). On the other hand, upon microscopically deriving the effective interaction from a realistic interaction (between two free nucleons) one finds a rather important tensor component (e.g. see 15). Some phenomenological effective interactions have indeed included it in their parametrization (43, 44, 59). However, its phenomenological interest seems rather dubious (45, 59) since in practice its effect can apparently be yielded by other parts of the force.

1.2.3 APPROXIMATIONS OF THE COULOMB INTERACTION Because of its range, the Coulomb interaction is not very easy to handle in practical Hartree-Fock calculations, at least as far as its exchange contribution is concerned. Actually one may either use an integral representation (60) or perform a finite expansion (61) of the Coulomb interaction, both in terms of Gaussian potentials. But more frequently, one makes some simplifying assumptions. One of them consists in assuming that the ratio of the direct to the exchange Coulomb energy is identical in finite nuclei to what it is in plane-wave nuclear matter. This is called the Slater approximation (62). One may refine this approximation by combining it with a kind of local density approximation, which leads (52, 63) to the following Coulomb exchange energy:

$$E_{\text{coul exch}} = -\frac{3e^2}{4}\left(\frac{3}{\pi}\right)^{\frac{1}{3}} \int [\rho_p(\mathbf{r})]^{\frac{4}{3}} \, d^3\mathbf{r}, \qquad 1.1$$

where ρ_p is the proton density. This yields a local contribution to the Hartree-Fock Hamiltonian. An alternative rough approximation to the Coulomb exchange Hartree-Fock Hamiltonian has also been proposed (64). All these approximations have been checked (65) and found to be in qualitative agreement with exact calculations. It turns out, however, that the exchange energy of Equation 1.1 is almost constant upon elongating a Slater determinant (see, for instance, 66). This was recently found to be incorrect from a schematic model calculation using harmonic oscillator wave functions: the Coulomb exchange energy of closed shell nuclei possesses a deformation dependence very similar to that of the direct part (P. Quentin, unpublished).

1.2.4 DETERMINATION OF PARAMETERS   Once the analytical form of the effective interaction is chosen, its parameters must be fitted. The central part of the force may usually be adjusted by reproducing saturation properties (total binding energies and charge radii) of some finite nuclei. In practice the best choices are magic nuclei whose spherical symmetry makes simpler the numerous calculations involved in such a fit. Nuclear matter properties, which are supposed to be deduced from bulk properties of actual nuclei, are also used sometimes. The problem of such a procedure is that these rather indirect data are not known with a sufficient accuracy (or in a sufficiently model-independent fashion) to pinpoint precise values of the force parameters. They provide, however, useful indications of the physical range to which the search for optimal parameter values must be restricted.

The strength of the spin-orbit force is usually determined either by fitting spin-orbit splitting energies or by trying to reproduce as well as possible single-particle level orderings in heavy nuclei. This is currently done for magic nuclei. This procedure, however, is not completely free from ambiguities. There are strong reasons not to equate blindly experimental with Hartree-Fock single-particle energies. These reasons can be formulated in terms of Brueckner rearrangement effects (e.g. 67 and references therein). It seems that the coupling of individual degrees of freedom with the vibrational collective motions contributes largely to this rearrangement (68). For deformed nuclei, where the long-range correlations are mostly effective in building up a deformed mean field, it is believed that the comparison between experimental and Hartree-Fock single-particle energies should be more meaningful (see the discussion of Figures 3, 4 in Section 4).

1.3   Other Related Approaches

1.3.1   THE DENSITY MATRIX EXPANSION   Negele & Vautherin (52, 53) have proposed a truncated expansion of the one-body reduced density matrix represented in center-of-mass and relative coordinates. The leading term is just the plane-wave nuclear matter density. When used to evaluate the exchange contribution of finite range forces, this approach leads to an energy density (see Equations 2.9), which presents all the computational simplicity of Skyrme-like forces. (As a matter of fact, a Skyrme force with density-dependent parameters is then produced.) This provides some microscopic foundation to the phenomenological Skyrme force insofar as one may find its origin in the effective force that Negele (15) and others derived from nuclear matter G (or K) matrix.

Actual calculations using this expansion have been performed for both spherical (37, 52–54, 69) and deformed (145 and M. Brack, P. Quentin, unpublished) nuclei. Recently a slightly different truncated expansion was proposed (70) and the link between this expansion and a semiclassical expansion clarified (70, 71, 293).

1.3.2   THE ENERGY DENSITY FORMALISM   In the energy density formalism approach (72–74, 295) one starts from an energy density (see Equations 2.9) that depends on the nuclear density $\rho(\mathbf{r})$ and the kinetic energy density $\tau(\mathbf{r})$. The latter is evaluated in terms of $\rho(\mathbf{r})$ in the Thomas-Fermi approximation. The nuclear ground state corresponds to the density minimizing the energy. It is quite clear that the rough treatment of the kinetic energy and of the velocity-dependent parts of the underlying effective Hamiltonian does not correctly take into account the nuclear surface. Moreover, this method is evidently unable to reproduce the rapid fluctuations (upon varying the nucleon numbers or the deformation) known as shell effects.

Further approaches referred to also (and unfortunately) as energy density formalism calculations (75–79), have worked out a method very close indeed to the Hartree-Fock approximation: the densities $\rho(\mathbf{r})$ and $\tau(\mathbf{r})$ have been actually computed from single-particle states eigensolutions of one-body Schrödinger equations, which are the conditions for the energy to be stationary. Instead of defining an effective force (as in standard Hartree-Fock calculations) this approach starts from an analytical expression of the expectation value of the effective Hamiltonian. The fact that the latter is expressed in terms of an integral over an energy density involving only a few spatial density functions makes the present formalism very close indeed to the Skyrme force approach.

1.3.3   THE SELF-CONSISTENT K-MATRIX MODEL   The self-consistent K-matrix model was initiated by Meldner (80). It consists in the parametrization of a phenomenological interaction entering the definition of the mean field potential (81, 82). It is a self-consistent approach since the potential part of the so constructed Hartree-Fock Hamiltonian depends through the nuclear density upon the solutions of the one-body Schrödinger equation. As far as both numerical difficulty and physical content are concerned, this method is also very close to the standard Hartree-Fock approach with Skyrme-like forces.

2   THE TECHNIQUES OF HARTREE-FOCK CALCULATIONS

2.1   General Remarks

In this section we present some of the most common methods of solving the (static) Hartree-Fock equations. As discussed in Sections 3 and 6, the same methods, with minor changes, are relevant for the Hartree-Fock + BCS and the temperature-dependent Hartree-Fock problems.

The general properties of the Hartree-Fock (HF) equations are described in numerous papers and textbooks; we outline them insofar as it allows us to introduce our notations. The fundamental quantity here is the expectation value E of the effective Hamiltonian $\mathscr{H}$:

$$E = \langle \Phi | \mathscr{H} | \Phi \rangle. \tag{2.1}$$

Here Φ is a Slater determinant built from the single-particle wave functions $\{\psi_i, i = 1, \ldots, A\}$, where A is the total number of nucleons. Considering E as a function of the ψ_i's and their complex conjugates, we obtain the HF equations as the conditions for E to be stationary under independent variations of the ψ_i^*'s. These variations are subject to the additional constraints that each wave function ψ_i should remain normalized. The Lagrange parameters associated with these constraints are nothing but the single-particle energies $\{e_i\}$. (In the Hartree-Fock + BCS and the temperature-dependent Hartree-Fock cases, though, the Lagrange parameters are the ratios of the single-particle energies to the corresponding occupation probabilities.) The Hartree-Fock equations form a coupled set of one-body static Schrödinger equations

$$h\psi_j = e_j\psi_j. \tag{2.2}$$

The coupling between these equations for different j values comes from the dependence of the Hartree-Fock Hamiltonian h on the set $\{\psi_i\}$. This dependence is also responsible for the nonlinearity of these equations.

For the most general type of effective Hamiltonians $\mathcal{H}$ the Hartree-Fock equations in projection on the **r**-basis constitute a set of integro-differential equations. The kinetic energy part of $\mathcal{H}$ gives the differential operator $(h^2/2m)\mathbf{V}^2$ contained in h. The two-body interaction provides both a local potential through its direct contribution and an integral operator through its exchange contribution. For some simple interactions, however, the latter can be arranged in such a way that the Hartree-Fock equations are purely differential at the cost of introducing an **r**-dependent effective nucleonic mass in the differential operator, namely: $-\mathbf{V}\cdot[h^2/2m^*(\mathbf{r})]\mathbf{V}$. We now refer to these interactions as being of the Skyrme type.[5]

The mathematical theory of Hartree-Fock equations is poorly developed. Nevertheless, physicists have produced a rather large store of numerical techniques, and, despite this lack of mathematical foundations, the quality of the current numerical results may be tested. Besides, the confidence in the latter is reinforced by the identity of answers yielded by very different numerical methods. This explains why, in what follows, we present methods without attempting to justify them from a mathematical point of view but discussing tests of convergence when available.

The Hartree-Fock equations are solved iteratively. In a first class of methods each iteration consists of two steps. One first constructs the Hamiltonian h from the wave functions provided by the preceding iteration; one then searches the eigenfunctions ψ_i of h and selects those corresponding to the lowest eigenvalues. This method makes apparently no use of the minimal property of E achieved by the Hartree-Fock solution, which is what is actually sought. The underlying hope is that the filling of the Slater determinant with the wave functions corresponding to the lowest eigenvalues will lead to an absolute minimum of E. This is not always the case, in particular when the single-particle level density near the Fermi surface is large. (However, in such a case the Hartree-Fock-Bogolyubov or the Hartree-Fock + BCS approximations are more suitable than the simple Hartree-Fock approximation.) As a convergence check, one may compare the expectation value of $\mathcal{H}$ with the following expression, which is equivalent when self-consistency is reached:

$$E' = \sum_{i=1}^{A} \tfrac{1}{2}(e_i + T_i) + E_R, \qquad\qquad 2.3$$

[5] From a practical point of view, it is convenient to introduce in this category interactions for which the exchange term vanishes or is neglected.

where T_i is the expectation value of the kinetic energy operator and the rearrangement energy E_R is a well-known correction arising from the use of density-dependent forces.

A second class of methods uses more explicitly the variational aspect of the Hartree-Fock theory. They are an application of the steepest descent method to the subspace of Slater determinants (83). We describe below the imaginary time method, which bears many similarities with this variational method, but is also related to the methods mentioned above, in that it attempts to compute the lowest energy levels of the Hartree-Fock Hamiltonian and to use the corresponding eigenfunctions as the occupied states of the Slater determinant.

An alternative classification of the numerical methods is used in the brief review that follows. We distinguish between the solutions of the Hartree-Fock equations either in the **r**-representation or through their projection onto another (truncated) basis.

2.2   Solution of the Hartree-Fock Equations in the Configuration Space

It is convenient to single out the methods used in cases where spherical symmetry is assumed. In these cases, the Hartree-Fock equations reduce readily to a set of one-dimensional equations for which a large number of techniques are available. This is not the case for two-dimensional (2D) or three-dimensional (3D) cases where the calculations have so far been restricted to Skyrme-like forces only. In general, the solution in **r**-space of the Hartree-Fock equations lies within the framework of a finite difference method, resulting in a coupling of the values of the wave functions ψ_i at points of a discrete mesh. For a spherical Slater determinant, one may afford to work with a mesh size small enough to make errors from this source negligible. This is no longer true in 2D or 3D calculations where the necessity of reducing the number of mesh points leads to mesh intervals as large as ~ 1 fm. A thorough study of the consequences of such a large mesh size is thus unavoidable.

For the more general two-body interaction the spherical Hartree-Fock equations reduce to[6]

$$\frac{\hbar^2}{2m}\left(-\frac{d^2 u_i}{dr^2} + \frac{l_i(l_i+1)}{r^2}u_i\right) + \int_0^\infty \Gamma(r,r')\, u_i(r')\, dr' = e_i u_i(r), \qquad 2.4$$

where $u_i(r)/r$ is the radial part of the wave function with angular momentum l_i and single-particle energy e_i. The difficulties in solving Equation

---

[6] To simplify the notation we have included the local potential generated by the direct part of the two-body interaction into the integral operator Γ.

2.4 are due to the integral operator with kernel Γ, introduced by the two-body force. Brueckner, Gammel & Weitzner (84) replace this equation by a differential one including a first order derivative, solved by means of the Runge-Kutta algorithm (e.g. 85). In another method (86), Equation 2.4 is transformed into an ordinary Schrödinger equation for which more efficient algorithms like the one proposed by Numerov (e.g. 87) may be used. To do so, the integral operator is replaced by the local potential

$$V_i(r) = \frac{1}{u_i(r)} \int_0^\infty \Gamma(r,r') \, u_i(r') \, dr'. \qquad 2.5$$

Some care should be exerted in the vicinity of the zeros of u_i; in practice one linearly interpolates V_i in this critical region to eliminate the singularity. As in the method of Reference 84 the Hartree-Fock equations are solved iteratively.

For a Skyrme-like interaction the radial Hartree-Fock equation becomes (30, 88)

$$\left[-\frac{d}{dr} \frac{\hbar^2}{2m^*(r)} \frac{d}{dr} + \frac{l_i(l_i+1)}{r^2} + V(r) \right] u_i = e_i u_i. \qquad 2.6$$

The two-body interaction is responsible for the occurrence of the local potential $V(r)$ and, as already pointed out, for the effective mass $m^*(r)$. Since Equation 2.6 involves first and second order r-derivatives of u_i, the Runge-Kutta method can be used. It is, however, convenient to slightly modify the Hartree-Fock equations (89) to make them suitable for the Numerov algorithm. Upon introducing the new function

$$v_i = (m/m^*)^{\frac{1}{2}} u_i, \qquad 2.7$$

Equation 2.6 becomes

$$-\frac{\hbar^2}{2m} \frac{d^2 v_i^2}{dr^2} + \mathscr{V}_i(r) \, v_i + e_i \left(1 - \frac{m^*}{m} \right) v_i = e_i v_i$$

$$\mathscr{V}_i(r) = \frac{m^*}{m} \left[\frac{l_i(l_i+1)}{r^2} + V \right] - \frac{\hbar^2}{2m} \left[(m^*)^{\frac{1}{2}} \frac{d}{dr} \frac{1}{m^*} \frac{d}{dr} (m^*)^{\frac{1}{2}} \right]. \qquad 2.8$$

An energy-dependent piece $e_i[1 - (m^*/m)]$ appears now in the differential equation. Given a potential V and an effective mass m^*, one solves in a few iterations Equations 2.8 for v_i, from an initial guess of e_i.

Let us turn now to the problems connected with the solutions of 2D or 3D Hartree-Fock equations for Skyrme-like interactions (which is the only case where results are available so far). With such interactions the energy (Equation 2.1) can be written as a function of the density and kinetic energy density (30, 88):

$$E = \int H(\rho, \tau) \, d^3 \mathbf{r}$$

$$\rho = \sum_{i=1}^{A} |\psi_i|^2 \qquad\qquad 2.9$$

$$\tau = \sum_{i=1}^{A} |\nabla \psi_i|^2,$$

where H is the Hamiltonian density. (This expression is only valid for a spin-saturated independent particle state with no spin-vector one-body density[7] and for a purely central effective interaction.) In the case of a central Skyrme interaction supplemented by the direct part of the Coulomb interaction one obtains the following functional form for H and an $N = Z$ nucleus (with identical proton and neutron wave functions):

$$H(r) = \frac{\hbar^2}{2m} \tau + \alpha_1 \rho^2 + \alpha_2 \rho^3 + \alpha_3 \rho\tau + \alpha_4 \rho \nabla^2 \rho + \frac{e^2}{8} \rho \int \frac{\rho(\mathbf{r}')}{|\mathbf{r} - \mathbf{r}'|} \, d^3 \mathbf{r}',$$

$$2.10$$

where the functions ρ and τ are calculated at the position $\mathbf{r}$ unless otherwise specified. We give in Table 2 an example of the accuracy achieved by a 3D calculation for a Slater determinant built from spherical harmonic oscillator states. It is apparent that the kinetic energy is the major source of inaccuracy. In self-consistent calculations, the effect of an error on a part of the Hartree-Fock Hamiltonian may result in errors on the other parts. This is illustrated in Table 3. The too weakly repulsive kinetic energy term leads to unduly small nuclei. As a consequence the attractive term proportional to α_1 becomes much too large. When calculating heavy nuclei, the spatial structure of the wave functions gets more and more complicated and the inaccuracy of the kinetic energy term is increased. A detailed study in this case of the accuracy of the finite difference method remains to be done.

In such 2D or 3D calculations it is also necessary to ensure the consistency between the discrete mesh approximations of the energy and of the Hartree-Fock Hamiltonian; in other words, the latter should be derived by a direct variation of the former, where the variational quantities are the values of the ψ_i at each mesh point. The last major problem of such calculations consists in finding the N lowest eigen-

---

[7] When this is not the case, see Reference 90.

Table 2 Comparison of analytical and calculated values of different contributions (in MeV) to the total energy of a ^{16}O nucleus[a]

Mesh size (Δx) (fm)	K	A_1	A_2	A_3	A_4	C
Exact	205.29	−480.60	93.17	27.82	36.15	15.36
0.7	205.03	−480.62	93.18	27.66	36.06	15.42
1.0	204.31	−480.60	93.16	27.40	35.80	15.49

[a] The entries A_i ($i = 1, \ldots, 4$) correspond to the contribution to $\int H(\mathbf{r}) \, d^3\mathbf{r}$ of terms proportional to α_i in Equation 2.10, K to the kinetic energy, and C to the direct part of the Coulomb energy. The SIII Skyrme force (32) has been used with spherical harmonic oscillator wave functions ($m\omega/\hbar = 0.275$ fm^{-2}). The quantity Δx represents the mesh size of the 3D calculations. The integrals have been performed according to the trapeze method and the kinetic energy operator has been exactly calculated up to the order Δx^4.

Table 3 Comparison of analytical and calculated values of different contributions (in MeV) to the total energy of a ^{40}Ca nucleus[a]

Mesh size (Δx) (fm)	K	A_1	A_2	A_3	A_4	C	R
Exact	650.49	−1710.43	434.00	135.28	94.03	79.27	3.41
0.7	652.42	−1720.34	438.61	135.66	94.69	79.50	3.39
1	654.39	−1741.75	449.09	137.03	86.66	80.12	3.38

[a] In this case, the Slater determinants are obtained from Hartree-Fock calculations using the SIII Skyrme force (32) either with a spherical differential equation code (exact) or with a finite difference method code. The rms radii R of the solutions are also given (in fm).

functions of the discrete Hartree-Fock Hamiltonian matrix h, which has a rather large dimension (up to several thousands). We mention two algorithms that both rely on the sparsity of the involved matrices (a large number of their matrix elements vanish, which allows fast matrix multiplication) to reach a numerical solution at a reasonable computing cost.

The first method (91) is based on the Lanczos algorithm and amounts to a diagonalization of a small matrix by building up a tridiagonal representation of h.

Another method is referred to as the imaginary time technique[8] (92 and M. Weiss, H. Doubre, H. Flocard, unpublished). It originates in the fact that if λ is a sufficiently small number, multiplication of a trial Slater determinant by $(1 - \lambda h)$ leads (after a suitable renormalization)

[8] This is not a completely new method. Its rediscovery in the Hartree-Fock context was suggested by the numerical solution of time-dependent Hartree-Fock equations of motion. This is the reason for its name. An alternative presentation of the same method may be found in Reference 83.

to an increase in the weight of the lowest eigenstates of h in the resulting Slater determinant. An iterative procedure consisting of successive applications of $(1 - \lambda h)$ leads to the exact eigenvalues of h and at the same time ensures a decrease of the total energy from one iteration to the next.

2.3 Solution of the Hartree-Fock Equations by Projection onto a Truncated Basis

The projection of the Hartree-Fock equations on an orthonormal basis of dimension M leads to a standard diagonalization problem: for all i, α

$$\sum_{j=1}^{M} h_{ij} C_j^\alpha = e_\alpha C_i^\alpha, \qquad 2.11$$

where i, j, and α designate basis and Hartree-Fock single-particle states, respectively, and

$$\begin{aligned} h_{ij} &= \langle i | h | j \rangle \\ C_j^\alpha &= \langle j | \alpha \rangle. \end{aligned} \qquad 2.12$$

The choice of the basis is determined by two requirements, which are sometimes conflicting. First, the basis states should be close enough to the actual solutions to minimize their number. Second, the analytical properties of these basis states should allow a rapid computation of the matrix elements h_{ij}. The latter can be written for a two-body interaction v as:

$$h_{ij} = \langle i | T | j \rangle + \sum_{i=1}^{A} \langle ik | \tilde{v} | jl \rangle \langle l | \rho | k \rangle, \qquad 2.13$$

where T is the kinetic energy operator, ρ the one-body reduced density matrix, and $\tilde{v}$ the antisymmetrized interaction v. The practical difficulty of solving the diagonalization problem (Expression 2.11) lies in the computation and the storage of the matrix elements of $\tilde{v}$. For harmonic oscillator basis wave functions ϕ_i there are two rather simple ways of handling the calculation of a quantity such as $\int\int\phi_1(\mathbf{r}) \, \phi_2(\mathbf{r}') \, v(\mathbf{r}-\mathbf{r}') \, \phi_3(\mathbf{r}) \, \phi_4(\mathbf{r}') \, d^3\mathbf{r} \, d^3\mathbf{r}'$, where v is a central interaction. First, one may use Moshinsky coefficients (60, 93, 94) to transform the product $\phi_1\phi_2$ (or $\phi_3\phi_4$) into a product of oscillator wave functions for relative and center-of-mass coordinates, which thus allows a trivial integration with respect to the latter. Second, one may transform the product $\phi_1\phi_3$ (or $\phi_2\phi_4$) into a limited weighted sum over oscillator wave functions of the variable $\mathbf{r}$ (or $\mathbf{r}'$); when v is of a Gaussian type[9] integral equations for the oscillator

---

[9] This can be easily generalized to other forces like Coulomb or Yukawa forces (60).

wave functions will eliminate the integral involving the interaction (95, 96). In both cases one ends up with a finite sum over pretabulated coefficients. It seems, however, that the second method is much faster to use.

In the case of a Skyrme-like force one avoids the time-consuming building up of matrix elements of h through Equation 2.13; instead (once the density functions are known) they can be directly evaluated by numerical integration methods of the Gaussian type (see 97, 98) as

$$h_{ij} = \int \phi_i^*(\mathbf{r}) \left[-\nabla \frac{\hbar^2}{2m^*(\mathbf{r})} \nabla + V(\mathbf{r}) \right] \phi_j(\mathbf{r}) \, d^3\mathbf{r}. \qquad 2.14$$

Up to moderate nuclear deformation, a harmonic oscillator well is a sufficiently good approximation of the Hartree-Fock mean field to allow the use of a rather limited basis size in the expansion of the single-particle solutions, as seen in Table 4. It is clear that truncating the basis of expansion yields upper bounds for the Hartree-Fock total energy. This is in contrast with the calculations performed in the configuration space, which have been always found to produce lower bounds to it.

Such bases are also determined by some parameters. To restrict their influence on the Hartree-Fock results, one retains the values of the parameters that minimize the total energy. This proves in general to be a difficult thing to do, particularly when the basis depends upon two or more parameters.

When necking appears in the nuclear density or in scattering configurations, it is clearly more appropriate to use a two-center basis. If axial symmetry is imposed, the resultant wave functions are products of

Table 4 Hartree-Fock total energies E (in MeV) of two magic nuclei calculated with the SIII Skyrme force (32) as functions of the number N of major oscillator shells included in the basis[a]

N	^{40}Ca		^{208}Pb	
	b	E	b	E
5	0.58	-333.8	—	—
7	0.58	-335.0	0.425	-1500.3
9	0.58	-335.2	0.47	-1593.7
11	0.58	-336.2	0.49	-1607.7
Exact	—	-337.0	—	-1618.7

[a] For each basis size the oscillator parameter $b = (m\omega/\hbar)^{\frac{1}{2}}$ is given (in fm^{-1}). The "exact" calculation corresponds to a calculation in the $\mathbf{r}$-space for spherically symmetric solutions as described in Section 2.2.

a longitudinal wave function along the symmetry axis (say the z axis) and a two-dimensional transverse wave function. The latter is usually expanded on cylindrical oscillator wave functions. Three methods have been proposed to define the basis for longitudinal wave functions. The first uses the eigenfunctions of a two-center harmonic oscillator (99, 100). The second takes the functions associated with the orthonormal polynomials defined in the whole z axis by the two-center weight function $\exp\left[-(z+z_0)^2\right] + \exp\left[-(z-z_0)^2\right]$, where $2z_0$ represents somehow the distance between the two centers (101). The third method consists in orthonormalizing the biorthogonal basis made of two displaced oscillator bases (D. Gogny, H. Berger, unpublished).

The diagonalization procedure is considerably facilitated by all possible symmetries of the solutions commuting with the effective Hamiltonian (e.g. 102). In such cases the matrix h_{ij} splits up in a block diagonal form. That is why the axial symmetry is rather costly to release and therefore has been, with few exceptions, generally imposed.

3   PAIRING CORRELATIONS

3.1   *The Hartree-Fock-Bogolyubov Approximation*

Apart from a limited number of magic nuclei in the immediate vicinity of their spherical equilibrium configuration, pairing correlations are known to play an important role in the definition of nuclear static properties. In this respect, the approximation that is equivalent to the Hartree-Fock approximation is referred to as the Hartree-Fock-Bogolyubov (HFB) approximation (103–107, 294). It is not our intention to give a comprehensive review of this approach, which may be found elsewhere (e.g. 108). We simply sketch here its most important features, aiming in particular at providing a formal framework for further simplifying assumptions.

The Bogolyubov transformation (103) defines a set of new fermion (quasiparticle) operators from the original particle creation operators $\{c_i^\dagger\}$. The HFB trial wave function Φ is chosen to be an independent quasiparticle state. The approximate variational solution of the static Schrödinger equation satisfies a set of equations (HFB equations) that are coupled nonlinear equations defining the coefficients of the Bogolyubov transformation. Apart from the ordinary one-body reduced density matrix (whose matrix elements are defined—as usual—as expectation values for the HFB state of the product $c_j^\dagger c_i$, see Equation 6.3), another tensor plays an important role here, the so-called pairing tensor κ defined as

$$\kappa_{ij} = \langle \Phi | c_j c_i | \Phi \rangle. \qquad\qquad 3.1$$

One further defines (with obvious notation) a Hartree-Fock-like potential $\mathscr{V}$ and a pairing potential Δ by:

$$\mathscr{V} = \mathrm{tr}\,(\rho\tilde{v}),$$
$$\Delta = \tfrac{1}{2}\,\mathrm{tr}\,(\kappa\tilde{v}),$$

3.2

where $\tilde{v}$ is the antisymmetrized effective two-body interaction. The energy E of the HFB approximate solution of the static Schrödinger equation is:

$$E = \mathrm{tr}\,T\rho + \tfrac{1}{2}[\mathrm{tr}\,(\mathscr{V}\rho) - \mathrm{tr}\,(\Delta\kappa)].$$

3.3

The solution of the HFB equations is far more difficult than the solution of the HF equations. This is mostly because of a doubling of the dimensionality of the coupled equation system to be solved. Another practical difficulty arises from the fact that the HFB wave function is not an eigenstate of the particle operator. To describe a given nucleus the solution is generally contrained to integer values of the average nucleonic numbers N and Z. This imposes therefore at each step of the iterative process an adjustment of the relevant Lagrange parameters (chemical potentials). To go further and avoid spurious components of undesired N or Z values one must make a projection (at least) after variation onto states of good nucleonic numbers. Satisfactory and rather fast approximations for the latter have been developed (109–111).

Because of their complexity, numerical solutions of the full HFB equations for all types of nuclei (spherical and deformed, light and heavy) were obtained only recently by Gogny (48, 49). Earlier calculations (111, 112) restricted themselves to more schematic forces, or to light nuclei, or to the approximation of an inert core, or finally to Hartree-Bogolyubov calculations (neglecting the exchange contribution of the interaction).

3.2   The Hartree-Fock + BCS Approximation

A particular case of the general Bogolyubov transformation has been introduced by Bogolyubov (113) and Valatin (114). The corresponding independent quasiparticle state is the well-known BCS wave function. As a substitute to HFB calculations, one may first perform Hartree-Fock calculations and, with the particle basis so defined, solve the BCS equations, which thus yields the optimal quasiparticle states. This is a well-defined approximation to HFB calculations [particularly in the context of the Bloch-Messiah theorem (115)].

It may be noted, however, that in the fit of most of the effective interactions v [with the exception of the Gogny force (48, 49), which has been explicitly designed to be used in HFB calculations] one has not paid any attention to the realistic (or unrealistic) character of pairing matrix elements

$$G_{\lambda\mu} = \langle\lambda\bar{\lambda}|\tilde{v}|\mu\bar{\mu}\rangle, \qquad 3.4$$

where λ and μ are Hartree-Fock states. In such cases, one has sometimes computed these matrix elements in a nonconsistent way, i.e. from an interaction [as the one proposed in (116, 117) for instance] which is not the one used to determine the Hartree-Fock single-particle states. The total energy (Equation 3.3) is thus evaluated from the former interaction as far as the term $\mathrm{tr}\,(\Delta\kappa)$ is concerned and from the latter for the remainder. In an alternative (and more frequent) approach, one has made further approximations discussed below.

3.3   Two Simple Substitutes

A first simplifying assumption to the approach described in Section 3.2 consists in neglecting the state dependence of the matrix elements $G_{\lambda\mu}$ of Equation 3.4, namely for all λ, μ

$$G_{\lambda\mu} = -g. \qquad 3.5$$

The total energy E to be minimized is then (up to appropriate constraints)

$$E = \mathrm{tr}\left[\left(T+\frac{1}{2}\mathcal{V}\right)\rho\right] - \frac{g}{4}\left\{\sum_{\lambda}\left[v_{\lambda}(1-v_{\lambda}^2)^{\frac{1}{2}}\right]\right\}^2 \qquad 3.6$$

with the usual notation for the Bogolyubov-Valatin amplitudes v_{λ}. Varying the single-particle wave functions ψ_{λ} and the amplitudes v_{λ} (under the usual constraints that the ψ_{λ}'s remain normalized and that the nucleon numbers are conserved) yields the standard Hartree-Fock and BCS equations (98). The former equations, however, are defined in terms of a density matrix ρ corresponding to a correlated system (and no longer to a Slater determinant). It is well known that the BCS equations diverge in this case. This implies a truncation of the particle states active in such a BCS treatment, together with a suitable normalization of g.

In another related approach, instead of defining g, one starts from a given value of the pairing gap Δ (i.e. a state-independent gap). The variation (under appropriate constraints) of

$$E' = \mathrm{tr}\left[\left(T+\frac{1}{2}\mathcal{V}\right)\rho\right] - \Delta\sum_{\lambda}v_{\lambda}(1-v_{\lambda}^2)^{\frac{1}{2}} \qquad 3.7$$

leads to the same equations as before. Note, however (98), that E' includes twice as much pair-condensation energy (the term $-\frac{1}{2}\,\mathrm{tr}\,(\Delta\kappa)$ in Equation 3.3) as E. Therefore, in this case, the total binding energy will be given by the quantity

$$E'' = \text{tr}\left[\left(T + \frac{1}{2}\mathscr{V}\right)\rho\right] - \frac{\Delta}{2}\sum_{\lambda} v_{\lambda}(1 - v_{\lambda}^2)^{\frac{1}{2}} \qquad 3.8$$

These bold assumptions concerning the state independence of g (or Δ) facilitate considerably the numerical solutions of variational equations in the case of a correlated system. In turn, this method encounters a rather serious difficulty: When varying the intrinsic configuration (e.g. the overall deformation), there are no guidelines to determine the variation of the pairing parameter. If one may adjust it to reproduce the experimental odd-even local mass differences, one is obliged to guess what it becomes when the nuclear system is no longer in its equilibrium state. This should be borne in mind when discussing (as we do in Section 4.4) relative energies deduced from two such self-consistent solutions, like the so-called prolate-oblate energy differences or the fission barrier heights.

3.4 Pairing in Dynamical Calculations

A derivation of the time-dependent HFB equations may be found for instance in (118). As in the TDHF case (see Section 7.1), one assumes that the many-body wave function remains an independent quasiparticle state at all times. The time-dependent HFB equations are formally similar to the TDHF Equation 7.2 upon replacing h and ρ by the two 2×2 matrices

$$\begin{pmatrix} \rho & -\kappa \\ \kappa^* & 1-\rho^* \end{pmatrix}, \quad \begin{pmatrix} T-\lambda+\mathscr{V} & \Delta \\ -\Delta^* & -T^*+\lambda-\mathscr{V}^* \end{pmatrix},$$

respectively, where λ is the chemical potential. The adiabatic limit of this formalism has been formally studied (M. Baranger, M. Vénéroni, unpublished, and 118, 119). So far the solution of the full time-dependent HFB equation has not been attempted. Substitutes identical to those used in static calculations (see Section 3.3) have been proposed (120) and recently used in practical calculations (121).

4 STATIC NUCLEAR PROPERTIES

4.1 Preliminary Remarks

This section is devoted to a brief presentation of some results of static self-consistent calculations. To be specific, let us define the type of calculations we wish to report: (a) They are performed with phenomenological effective density-dependent interactions valid for the whole chart

of nuclides. (*b*) They are performed within the Hartree-Fock approximation, with an appropriate treatment of pairing correlations if needed.
(*c*) They are performed with a reasonable numerical accuracy (in
particular, if the single-particle solutions are projected onto a truncated
basis, the parameters and the size of the latter have been suitably
chosen). Even with these restrictions, the relevant literature is rather
abundant. That is why we do not attempt to be exhaustive, but rather
we select the most typical contributions.

Before doing that, however, we would first like to put the subject into
a proper perspective by emphasizing some limitations of the description
of the many-body wave function by a Slater determinant (or a BCS wave
function). The nuclear state so defined is in general an eigenstate neither
of the angular momentum **J** nor of the center-of-mass linear momentum
P. As a consequence the expectation value of the effective Hamiltonian
includes spurious rotational and translational kinetic energies. This is
sometimes corrected for by subtracting from the Hamiltonian (or from
its expectation value) the following approximated kinetic energy operators
(or expectation values)

$$E_{\text{rot}} = (\hbar^2/2I)\,\mathbf{J}^2 \qquad\qquad 4.1a$$

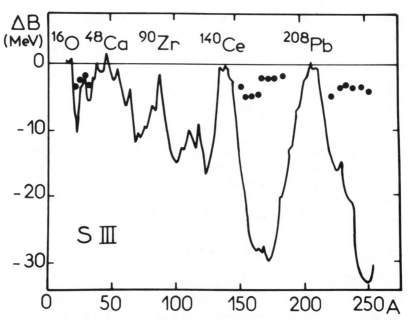

Figure 1   Differences ΔB between spherical Hartree-Fock + BCS and experimental total
binding energies (from 32). The SIII Skyrme interaction has been used. Black dots correspond to deformed equilibrium solutions.

$$E_{transl} = \mathbf{P}^2/2mA, \hspace{4cm} 4.1b$$

where I is the moment of inertia, m the nucleonic mass, and A the total number of nucleons.

As a consequence of the approximation made concerning $\mathbf{J}$, it is clear, in particular, that such self-consistent calculations do not describe deformed nuclear ground states but instead the so-called intrinsic states (defined, for instance, in Reference 122). As a matter of fact, the formal framework to build out of a Slater determinant eigenstates of $\mathbf{P}$ and $\mathbf{J}$ is well known: it is the double projection method of Peierls & Thouless (123). Systematic calculations according to it, however, are far beyond the numerical possibilities of the moment. Calculations using the simple projection method of Peierls & Yoccoz (124) actually do exist for light nuclei. They are not guaranteed, however, to yield correct inertial parameters (123).

4.2 *Energies*

4.2.1 TOTAL BINDING ENERGIES
Figures 1 and 2 show the difference between experimental (125) and calculated binding energies (positive quantity) along some path in the (N,Z) plane. In these figures, two effective forces have been used: the Skyrme force with the set SIII of

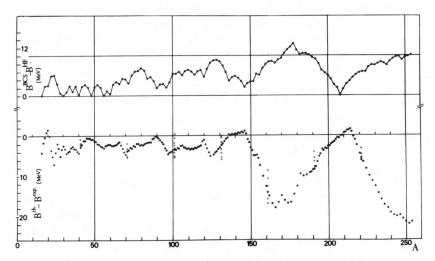

Figure 2 Differences between theoretical and experimental binding energies for nuclei distributed throughout the periodic table [Figure 3 of (75)]. The upper part of the figure represents the gain in energy due to the pairing correlations. Note that the full circles and the stars characterize increases of A obtained by the addition of one or two neutrons and one or two protons, respectively. The empty circles (squares) correspond to proton (neutron) rich isobars. The set F1 of parameters has been used.

parameters (32) and the force underlying the so-called energy density formalism of Beiner & Lombard (75) with the set F1 of parameters. For spherical nuclei they lead to an agreement of about 2 MeV between experimental and theoretical total energies. For deformed nuclei such differences are much larger: This is due to the fact that spherical symmetry is also imposed in this case. Upon calculating the deformed equilibrium state one must add the spherical barrier height to the spherical binding energy. When this is done (see Figure 1) the disagreement is reduced to about 5 MeV. Part of this discrepancy can be attributed to the spurious rotational energy [see Section 4.1 and the estimations of (66, 126–128)]. As a result, binding energies are reproduced within 2–3 MeV on the whole chart of nuclides with the SIII Skyrme force. Recently Tondeur (78, 79) proposed a parametrization of the energy density formalism (see Section 1.3.2) leading to binding energies for light and medium spherical nuclei as close as approximately 1 MeV to experimental ones, which thus matches the accuracy level of liquid drop (or droplet) model fits.

Such impressive successes, which—let us insist on it—are not obtained at the price of a poor reproduction of other static properties, provide some confidence in predictions of atomic masses far from the beta-stability valley, or in the superheavy element region (45, 129–135). As an illustration of such extrapolations we single out the calculations of Reference 136 made for some neutron-rich sodium isotopes recently studied experimentally (137). These self-consistent calculations using the SIII Skyrme effective force yielded an amazingly good reproduction of binding energies of nuclei up to very exotic species (such as the $N = 22$ sodium isotope). But they also provided a convincing explanation for the anomalous behavior of the two-neutron separation energy around the neutron number $N = 20$ in terms of a shape transition (from a quasispherical to a prolate shape) arising from the crossings of the 1d $\frac{3}{2}$ and 1f $\frac{7}{2}$ neutron subshells.

To conclude this section, let us mention calculations of static Hartree-Fock solutions that are not symmetric under time reversal. In particular one has studied the influence on the total energy of the time-reversal odd part of the Hartree-Fock Hamiltonian. One of these calculations (138) concerned odd nuclei or odd-odd nuclei, where the last particle was exactly included in the Slater determinant, i.e. not as a time-reversal doublet with occupancy 0.5 (the so-called filling approximation used in particular in Reference 137). The other calculations (139) were made under a constraint on the component j_x of the angular momentum. Upon varying the corresponding Lagrange multiplier one obtains an approximate nuclear spectrum (corresponding to a quasirotational band).

4.2.2 SINGLE-PARTICLE SPECTRA   Spherical single-particle spectra arising from self-consistent calculations are not directly comparable with
spectroscopic information deduced from odd nuclei in the vicinity of a
magic nucleus (see the discussion of Section 1.2.4). However, upon
assuming that the rearrangement effects might result in nothing but a

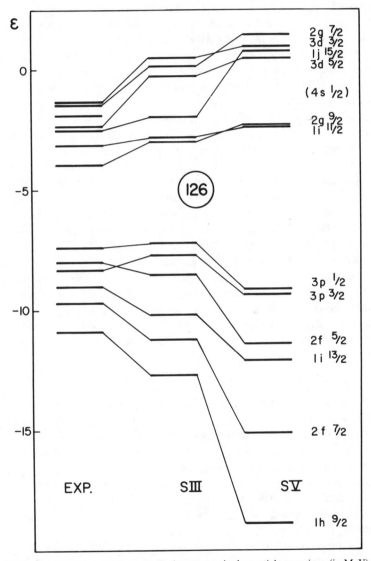

Figure 3 Experimental and Hartree-Fock neutron single-particle energies ε (in MeV) near
the Fermi surface of ^{208}Pb (from 32). The SIII and SV Skyrme interactions have been used.

compression of the spectrum one has generally tried to reproduce the experimental level orderings (found for instance in 140). Figure 3 gives an example of such a comparison. One definite difference between the two Skyrme forces in use there, lies in the value of the effective mass m^* in nuclear matter (usually expressed in units of the nucleonic mass). Indeed, of the five parameters of the standard central Skyrme force, four were more or less determined by the saturation properties of finite nuclei (32). The remaining one may be related to the saturation mechanism itself (velocity dependence vs density dependence). Because of the compression effect recalled above, its value cannot be determined by a single comparison between experimental and theoretical spherical spectra.

As already mentioned in Section 1.2.4, the situation is completely different in deformed nuclei: Comparison of calculated single-particle energies with odd-nuclei spectroscopic data, through rough model assumptions [like the rotor plus quasiparticle ansatz, see (122)], is more justified. Within the Nilsson model approach, careful fits of the mean field parameters have been performed in particular in the rare earth region (141). In Figure 4, the neutron spectrum so obtained is compared with the spectra resulting from self-consistent calculations using the SIII and SIV Skyrme forces, for the deformed ground state of the ^{168}Yb nucleus. As a result it appears that the SIII force ($m^* \sim 0.75$) has probably a more realistic velocity dependence than the SIV force ($m^* \sim 0.5$). More direct comparisons with experimental data have been recently achieved (M. Meyer, J. Letessier, J. Libert, P. Quentin, unpublished). Within the rotor plus quasiparticle approach (where the quasiparticle states were determined self-consistently) nuclear spectra of a variety of odd nuclei have been produced in good agreement with experimental data.

4.3   Densities

4.3.1   MONOPOLE PART   Even though recent progress has been made in the knowledge of the neutron density, the bulk of experimental information concerning the monopole part of the nuclear density comes from the charge distribution. Data pertaining to this problem concern either specific moments of this distribution (as in muonic atom experiments) and the variation of them (as isotopic shift measurements) or more detailed information (electron scattering results). Below we consider only two such pieces of data: rms charge radii and elastic electron scattering cross sections.

The charge distribution must be obtained by convoluting the point proton density arising from self-consistent calculations with some reason-

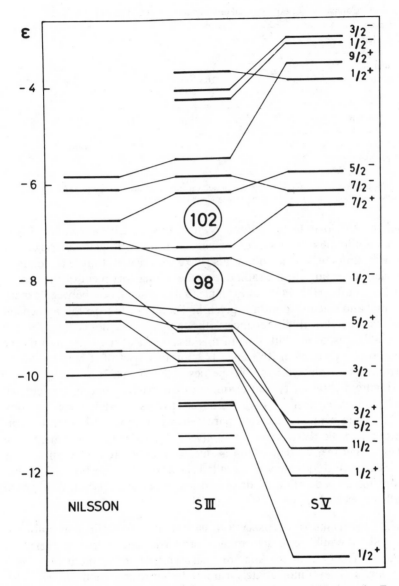

Figure 4 Comparison of neutron single-particle energies ε (in MeV) near the Fermi surface of ^{168}Yb, obtained in self-consistent and modified harmonic oscillator calculations [Figure 6 of (146)]. The Nilsson levels of (141) correspond to $\hbar\dot{\omega}_0 = 41A^{-\frac{1}{3}}$ MeV and their origin has been chosen in order to equate the $\Omega^\pi = \frac{7}{2}^+$ level obtained in this calculation, with the one coming from the SIII Skyrme force.

Table 5 Comparison of experimental and calculated charge rms radii (in fm)[a]

Element	^{16}O	^{20}Ne	^{24}Mg	^{28}Si	^{32}S	^{40}Ca	^{48}Ca
Calculated	2.69	2.97	3.10	3.17	3.28	3.48	3.53
Experimental	2.73	2.91	3.03	3.14	3.24	3.49	3.48

Element	^{56}Ni	^{90}Zr	^{152}Sm	^{162}Dy	^{184}W	^{208}Pb
Calculated	3.79	4.32	5.16	5.29	5.46	5.57
Experimental	3.75	4.27	5.09	5.21	5.37	5.50

[a] The SIII Skyrme force has been used. Relevant experimental references are to be found in (32, 66). Center-of-mass and neutron corrections have not been included.

able proton form factor. Moreover, neutrons can give a significant contribution to these cross sections in particular through their electric charge distribution (142). One should also take into account (at least for light nuclei) corrections to the radii due to the wrong treatment of the center-of-mass motion (15, 143). An example (32) of comparison between experimental and calculated radii is given in Table 5. A fair overall agreement is observed. Local discrepancies, though, may be noticed as for the ^{40}Ca-^{48}Ca isotopic shift. Part of this disagreement may be cured by the neutron corrections of Reference 142, another part of it may be due to erroneous isospin-symmetry properties of the force (33).

Figure 5 shows a typical example of comparison between calculated (32) and experimental (144) elastic electron scattering cross sections. Within the standard Skyrme parametrization, one finds that low m^* forces (like the SV Skyrme force) give a slightly better agreement with these experimental data. This is due to a better reproduction of the nuclear surface. [It may be recalled, however, (see Section 4.2.2) that from the consideration of deformed single-particle spectra, one would rather prefer a high (~ 0.75) m^* value.]

4.3.2 EQUILIBRIUM DEFORMATION PROPERTIES Experimental information about equilibrium deformation properties comes from a variety of sources. The most direct data are measurements of Q_{20} moments of first 2^+ levels. They unfortunately cannot be compared immediately with the results of self-consistent calculations (without projection of the angular momentum). As a matter of fact within the unified model (122) one may relate these data to the intrinsic state nuclear distributions attainable by such calculations. Within the same model framework one may also extract similar information from $B(E2)$ and $B(E4)$ measurements. More indirect data come from the deformed parameters either of optical

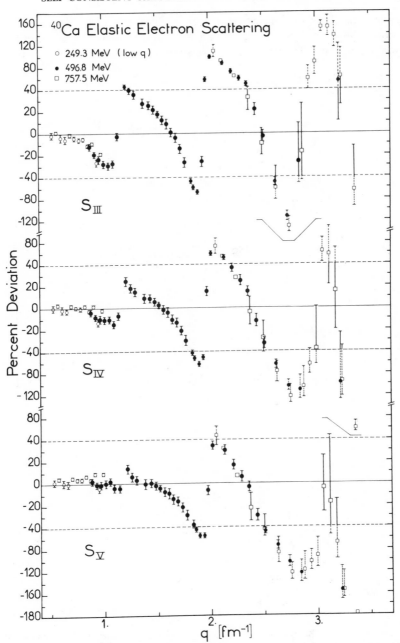

Figure 5 Percent deviation defined by $200\,(\sigma_{\text{theor}} - \sigma_{\text{exp}})/(\sigma_{\text{theor}} + \sigma_{\text{exp}})$ as a function of momentum transfer q in the case of ^{40}Ca [Figure 2a of (32)]. The experimental data are taken from (144). The SIII, SIV, and SV Skyrme forces have been used.

potentials (for hadron scattering) or of model nuclear density (from muonic atom x rays). In the latter cases calculational results (multipole moments of the density) must be translated in terms of β_2, β_4, ... para-

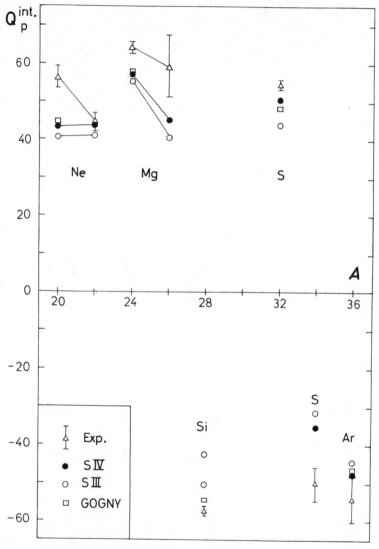

Figure 6 Intrinsic charge quadrupole moments of some s-d shell nuclei [Figure 1 of (146)]. Moments Q_p^{int} are expressed in fm^2 vs the nucleon number A. For the experimental results, absolute values are extracted from $B(E2)$ data, whereas the signs are taken from static measurements. The SIII, SIV Skyrme forces and the D1 Gogny force have been used.

Table 6 Comparison between experimental and calculated intrinsic quadrupole moments (Q_{20}) (in barn) of the charge distribution for some rare earth nuclei[a]

Element	^{152}Sm	^{158}Gd	^{162}Dy	^{166}Er	^{168}Yb	^{174}Yb	^{178}Hf	^{184}W
Calculated	5.84	7.11	7.49	7.79	7.61	7.81	7.12	6.19
Experimental	5.90	7.36	7.20	7.70	7.39	7.58	6.82	6.15

[a] The SIII Skyrme force has been used. Experimental data come from $B(E2)$ measurements (167).

meters à la Bohr-Mottelson for an equivalent sharp-edged[10] liquid drop.

Deformed nuclei in the s-d shell have been studied in great detail within the self-consistent approach considered here (42, 47, 98, 148–153, 160). To restrict the discussion, let us limit ourselves mostly to $4n$ nuclei ($N = Z =$ even number). Whereas ^{20}Ne, ^{24}Mg, and ^{32}S have been found experimentally to be prolate in their ground states, ^{28}Si and ^{36}Ar are actually oblate. As seen in Figure 6, this trend is reproduced by calculations using Skyrme forces (SIII and SIV) and the Gogny force D1, when one assumes that the ground state is represented by the local equilibrium solution with the lowest energy. These axially symmetrical solutions are also stable against γ deformation (153). The systematic behavior of β_4 parameters is also very well reproduced (149): a change of sign between ^{20}Ne and ^{24}Mg as well as between ^{24}Mg and ^{32}S.

Self-consistent calculations in the rare earth region are by far more scarce (98, 145, 154, 155). Again the calculated systematics of both the β_2 and β_4 parameters is in excellent agreement with experimental data (see for instance Table 6). In the actinide region (146), whereas the quadrupole data are as well reproduced as in lighter nuclei, the calculated hexadecapole properties show some systematical disagreement with the Coulomb excitation data of Reference 156 for curium isotopes, as seen in Figure 7.

The very fact that self-consistent calculations using phenomenological effective forces (whose parameters have been for most of them fitted on saturation properties of spherical nuclei) yield such excellent results for equilibrium deformation properties is a proof a posteriori of the physical validity of the whole approach.

4.4   Deformation Energy Curves

4.4.1 TRANSITIONAL NUCLEI   In the low energy spectrum of some nuclei, two classes of levels corresponding to two different intrinsic states (namely to two different deformations) are frequently observed. This is

[10] Introducing a diffuse density by convolution with a monopole form factor does not affect the multipole moments (66, 147).

mostly the case at the end of a major shell (e.g. in the neutron deficient mercury isotopes) or at its beginning (e.g. the $N = 88$–90 shape transition). A microscopic description of this phenomenon (consistent with the assumptions of the unified model) would involve the following steps: (*a*) calculation of the adiabatic curve (or surface) of potential energy by

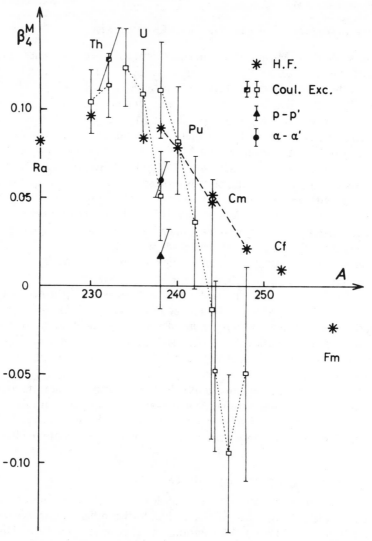

Figure 7 Hexadecapole β_4 deformation parameters of the mass distribution of some actinide nuclei [Figure 5 of (146)]. Experimental values come from Coulomb excitation, p-p′, and α-α′ data. The SIII Skyrme interaction has been used for the calculated values (HF).

self-consistent calculations under constraints on one (or a few) suitable collective variable; (b) calculation of mass parameters in the adiabatic limit of the TDHF approximation (see Section 7.4); and (c) solution of the Schrödinger equation associated with the Bohr Hamiltonian (157) defined in such a microscopic way, after its ad hoc requantification.

Clearly many rough assumptions are involved in all the previous steps. Besides, this program has not yet been fully completed in particular as far as step (b) is involved. Calculations of potential energy surfaces in such cases (155, 158, 168) are simply indicative: They assess the possibility of a shape coexistence (in particular of the existence of a ground-state transition). Nevertheless they can hardly reach a quantitative level. As an example, Figure 8 shows deformation energy curves for samarium isotopes calculated (155) within the approximate Hartree-Fock + BCS method discussed in Section 3.3. When going from 86 to 92 neutrons the prolate minimum deepens (thus substantiating the concept of a transition from a quasispherical to a prolate shape), but this transition is quite progressive and does not show any dramatic change between $N = 88$ and $N = 90$. The neutron-deficient odd mercury isotopes present a ground-state transition (from an oblate or a quasispherical shape to a prolate one between $N = 107$ and $N = 105$) whose signature is an

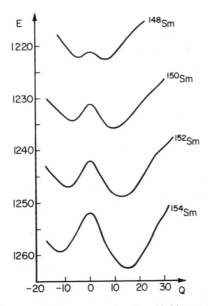

Figure 8 Deformation energy curves near the $N = 88$–90 transition region for some samarium isotopes (from 155). Energies E are expressed in MeV and mass quadrupole moments Q in barn. The D1 Gogny force has been used.

anomalous isotopic shift as in the $N = 88$–90 region. Self-consistent calculations (158 and J. Letessier, H. Flocard, P. Quentin, unpublished) actually yield two almost degenerate equilibrium solutions corresponding to the desired charge radii difference. They are, however, too imprecise (see in particular the discussion about the pairing parameter in Section 3.3) to pinpoint a precise value for the location of any ground-state transition.

The previous calculations were restricted to axial symmetry. It is, however, likely that whenever an oblate and a prolate solution are found almost degenerate, either they are unstable against γ deformation or (at least) the potential energy surface is very shallow in this direction. As a result of non-self-consistent (159) and self-consistent (42, 47, 153, 160, 161) calculations, the latter case is the most frequently encountered, which overemphasizes the importance of a correct dynamical treatment of the problem (as sketched above).

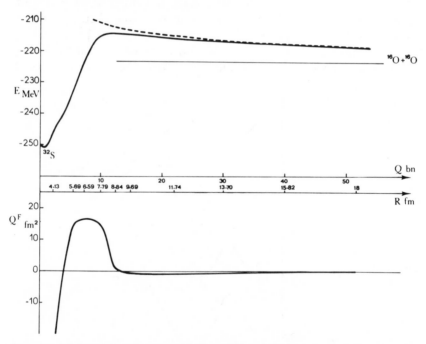

Figure 9 Adiabatic interaction potential of two ^{16}O nuclei [Figure 1 of (101)]. The upper part of the figure displays as a function of the total mass quadrupole moment Q the self-consistent energy curve (*solid line*). The Coulomb energy curve of two rigid ^{16}O corresponds to the dashed line. The horizontal line represents the energy of two infinitely separated ^{16}O. A scale in the distance R of centers of mass is given below the Q axis. The lower part of the figure represents the variation of the quadrupole moment of the fragments Q^F along the deformation path. The SIII Skyrme force has been used.

4.4.2   ADIABATIC SCATTERING POTENTIALS   If a self-consistent solution is constrained to be composed of two disconnected pieces and their inter-distance varied, adiabatic scattering potentials that yield valuable information about the Coulomb barrier may be built up. Such calculations (99–101, 162) have been performed for identical ions, which thus implies parity symmetry. (Axial symmetry is also imposed.) Figure 9 shows a potential so obtained for the collisions of two ^{16}O nuclei (101). This figure also provides some useful information on the Coulomb polarization of each ion. One serious problem arises from the fact that the motion of the center of mass is incorrectly taken into account. The kinetic energy (Equation 4.1b) does not correct in a suitable way the center-of-mass spurious motion of this kind of nuclear system. Its blind inclusion throughout the scattering potential curve results in an incorrect asymptotic value of the total energy (see Figure 9).

4.4.3   FISSION BARRIERS   The fission barrier of an actinide nucleus constitutes the best-known example of an adiabatic potential energy curve. Its double-humped character has been recognized to be a consequence of quantal shell effects (see the discussion of Section 5.2). "Experimental" barrier heights and widths are deduced from data pertaining to the penetration of these potential barriers. They are thus dependent upon specific dynamical assumptions concerning this collective motion. A microscopic account of such phenomena would start with the steps (*a*) and (*b*) described in Section 4.4.1. As in the case of transitional nuclei, however, this program is not yet completed: At the moment only a few self-consistent calculations of fission barriers have been performed for actinide (163, 164) and superheavy elements (130, 131).

Figure 10 shows the fission barrier of ^{240}Pu calculated with the SIII Skyrme force (163). Figure 11 displays the deformation energy curve above the scission point for the asymmetric fission of ^{236}U computed within the self-consistent K-matrix model of Reference 82 as obtained by R. Cusson and D. Kolb (unpublished). On Figure 12 the fission barrier of the hypothetical ^{298}114 superheavy nucleus is given from self-consistent calculations performed with the SIII Skyrme force (130).

In Figure 10 it appears clear that the corresponding calculation does not yield the experimental height of (for instance) the second fission barrier. There are reasons for the lowering of this barrier as well as for its raising. In the former category one can mention: the imposed left-right reflection symmetry (responsible in the calculations of Reference 165 for a 5-MeV decrease of the barrier); the slightly too small basis size (by comparing calculations made with 13 and 15 major oscillator shells, one may estimate the underbinding to ~2 MeV); the nonuniformity of the spurious rotational energy correction as the nucleus elongates (an estimate according to Equation 4.1a leads to a lowering of the isomeric state energy of ∼1 MeV). Among the reasons in favor of a higher barrier one may list: the probably unrealistic variation of the pairing parameter *g* (defined in Equation 3.5 and assumed to be proportional to the nuclear surface); the use of an approximate exchange Coulomb energy according to Equation 1.1 that may yield an overbinding of elongated nuclei as discussed in Section 1.2.3.

The disagreement (if there is any left) may also be a result of a wrong surface tension of the implicit liquid drop properties of the SIII Skyrme

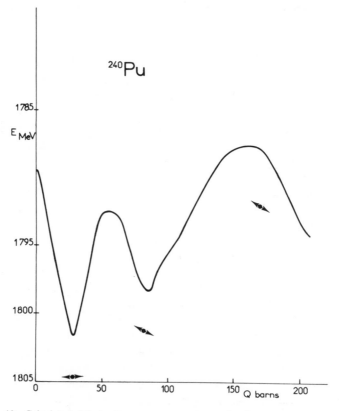

Figure 10   Calculated deformation energy curve over the fission barrier [Figure 1 of (163)]. The basis includes 13 oscillator major shells. Energies are expressed in MeV. Mass quadrupole moments are given in barn. Three calculated points with a larger basis (15 oscillator major shells) are also shown. Double arrows represent the calculated slopes at these points. They are found to be close to those of the curve. The shift in energy is uniform within 1 MeV. It corresponds to the extra binding energy due to a better representation of the tails of the wave functions. The SIII Skyrme force has been used.

force. However, upon extracting the liquid drop part of the energy—
as discussed in Section 5.4—one finds (see Figure 14) a fission barrier
in perfect agreement with the one corresponding to the same collective
path and deduced from the Myers-Swiatecki liquid drop parametrization (166). Since the latter has been fitted to reproduce fission barrier
heights in the actinide region, this seems to rule out any explanation
relying on wrong liquid drop properties of the force in use.

In the present state of the art for such calculations, it is difficult to
judge self-consistent calculations pertaining to low energy fission data
as successful (or unsuccessful). In order to do so, it would be necessary
to achieve (among other things) some dynamical calculations in a self-
consistent way.

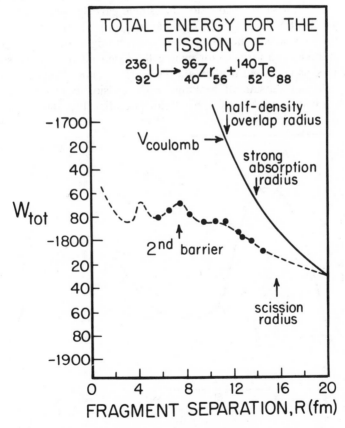

Figure 11 Total energy (in MeV) for an asymmetric fission of ^{236}U (from R. Cusson,
D. Kolb, unpublished). The self-consistent K-matrix model of (82) has been used.

5 RELATED APPROXIMATIONS

5.1 *Phenomenological Single-Particle Potentials*

Since the recognition of the existence of the nuclear shell structure (169, 170), the difficulty of a microscopic derivation of the mean field has often been avoided by directly parametrizing the single-particle well in a phenomenological way. The further discovery of the importance of deformation effects in the mean field (171, 172) gave rise to a very successful and yet simple parametrization known as the Nilsson model (173, 174). It was later improved by small corrective terms and is now referred to as the modified harmonic oscillator model (175). Several other successful parametrizations are also of common use. The assignment of a Woods-Saxon (176) shape to the single-particle well was widely developed by the Kiev-Basel-Copenhagen group (177). Another parametrization constructs the potential by convoluting a square well potential with a Yukawa form factor, which thus yields a reasonable surface diffuseness, as proposed by Nix and co-workers (178). Finally, in situations involving substantial necking or in collision configurations one may use a two-center harmonic oscillator potential (179, 180), as done by

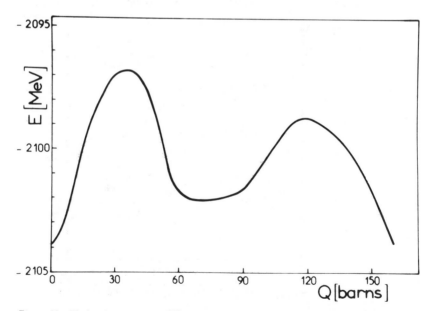

Figure 12   Fission barrier of the $^{298}114$ nucleus [Figure 5 of (130)]. The total HF energy E is plotted vs the mass quadrupole moment Q. The SIII Skyrme force has been used.

Andersen, Dickmann & Dietrich (181), and Greiner and co-workers (182).

In this section we present a comparison between single-particle properties obtained within a phenomenological approach (modified harmonic oscillator) and the Hartree-Fock method (using the SIII Skyrme force). As shown previously (see Figures 3 and 4), apart from the consequences of a slightly different ordering of spherical levels, the deformation dependence of the single-particle levels is quite similar in both cases. In particular, states (generally with a high value of Ω) that are almost purely of harmonic oscillator character are consistently found in both types of calculations. Another example of quantitative good agreement is shown on Figure 13. This is singularly noticeable for such an exotic

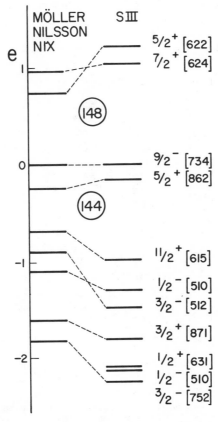

Figure 13 Comparison of the self-consistent single-neutron level scheme at the second minimum of ^{240}Pu with the one of Reference 183 (from 163).

deformation [isomeric fission state of ^{240}Pu (163, 183)], which is far outside the region where the modified harmonic oscillator parameters have been adjusted.

Let us turn now to the comparison of single-particle wave functions (P. Möller, P. Quentin, unpublished). Both single-particle states have been expanded onto the axially deformed harmonic oscillator basis. First the assignations of so-called asymptotic quantum numbers (184) are coherent in both cases (upon labeling each single-particle state by its higher component). It appears, however, that the Hartree-Fock field introduces a substantially larger major shell mixing with respect to the modified harmonic oscillator model (even at high $|\varepsilon_4|$ values). As an

Table 7 Some single-particle properties calculated in the modified harmonic oscillator model (lower line) and in Hartree-Fock calculations using the SIII Skyrme interaction (upper line)[a]

Level	$Q(\text{fm}^2)$	$r(\text{fm})$	$|C^\uparrow|^2$ (%)
$\frac{1}{2}^+$ [640]	33.2	6.355	54
	37.7	6.497	54
$\frac{7}{2}^-$ [514]	−3.3	5.825	10
	−2.9	5.914	16
$\frac{1}{2}^-$ [521]	23.4	5.963	36
	23.1	6.050	49
$\frac{7}{2}^+$ [633]	15.9	6.363	85
	17.5	6.449	83
$\frac{5}{2}^-$ [512]	−1.0	5.893	93
	−0.8	5.919	73
$\frac{9}{2}^+$ [624]	4.1	6.327	93
	4.9	6.394	90
$\frac{1}{2}^-$ [510]	0.6	5.925	89
	−0.2	5.924	78
$\frac{3}{2}^-$ [512]	−0.2	5.962	11
	−0.7	5.926	20
$\frac{11}{2}^+$ [615]	−9.5	6.249	98
	−9.3	6.316	96
$\frac{7}{2}^-$ [503]	−19.9	5.713	98
	−21.0	5.760	95
$\frac{9}{2}^-$ [505]	−20.8	5.700	02
	−21.2	5.760	04

[a] For each level labeled with the asymptotic quantum numbers of Mottelson & Nilsson, quadrupole moments Q, radii r, and the sum of the square of spin-up components $|C^\uparrow|^2$ are shown for neutron states near the Fermi level in the ^{184}W nucleus.

example, Table 7 presents a comparison of some spectroscopic properties. They concern the ground-state neutron single-particle states around the Fermi level of a deformed nucleus at the end of the rare earth region. The agreement between the two sets of results leads to the conclusion that at the same equilibrium deformation the phenomenological single-particle wave functions are very good approximations of the self-consistent ones.

5.2   The Strutinsky Energy Theorem

It is a rather surprising fact that most of the binding energy of the nucleus comes from a semiclassical description of the nucleonic motion. A phenomenological parametrization of the latter is known as the liquid drop model (185, 186). The greatest discrepancy in the liquid drop binding energies have been found to be approximately 15 MeV, with fluctuations in the error clearly correlated with neutron and proton magic numbers (166). This energy defect representing the irreducible quantal component of the binding energy has been called shell effect energy. Strutinsky was the first to propose (187, 188) a microscopic derivation of it within the Hartree-Fock approximation. Here, we sketch the Strutinsky arguments referred to as the "Strutinsky energy theorem" (189; see also 210).

Let us consider the Hartree-Fock reduced density matrix ρ and total energy $E(\rho)$. The latter may be written with obvious notations

$$E(\rho) = \operatorname{tr} T\rho + \tfrac{1}{2} \operatorname{tr} \operatorname{tr} \rho v\rho \qquad\qquad 5.1$$

Let us assume now that we know $\bar{\rho}$ as a semiclassical approximation of ρ. We may define (through $\bar{\rho}$) a semiclassical approximation $[\bar{\mathscr{V}} = \operatorname{tr} \bar{\rho}v]$ of the Hartree-Fock potential. This potential is indeed what the phenomenological mean fields discussed in Section 5.1 are likely to reproduce. The semiclassical character of $\bar{\rho}$ makes $\bar{\mathscr{V}}$ vary smoothly with the number of particles, the deformation of the solution, etc. This is assumed in these models but is not achieved for the original Hartree-Fock potential. The solution of the Schrödinger equation for the one-body Hamiltonian $T + \bar{\mathscr{V}}$ is characterized by a reduced density matrix $\hat{\rho}$ and a set of single-particle energies $\hat{e}$. Let us assume that we know also $\bar{\hat{\rho}}$ as the semiclassical approximation of $\hat{\rho}$. This implies in particular that the sum of occupied single-particle energies $\hat{e}$

$$\sum \hat{e} = \operatorname{tr} \left[\hat{\rho}(T + \bar{\rho}v) \right] \qquad\qquad 5.2$$

has a semiclassical approximation $\overline{\sum \hat{e}}$ defined as

$$\overline{\sum \hat{e}} = \operatorname{tr} \left[\bar{\hat{\rho}}(T + \bar{\rho}v) \right]. \qquad\qquad 5.3$$

Now the Strutinsky energy theorem states that $E(\rho)$ is given by

$$E(\rho) = E(\bar{\rho}) + \left(\sum \hat{e} - \overline{\sum \hat{e}}\right) + \mathcal{O}_2, \qquad 5.4$$

where $\mathcal{O}_2$ is a quantity of second order in the operator $(\rho - \bar{\rho})$ or some related operators as specified later.

The proof of Equation 5.4 follows from the fact that E is stationary around the self-consistent solution ρ:

$$E(\rho) = E(\hat{\rho}) - A(\rho - \hat{\rho}), \qquad 5.5$$

with

$$A(\rho) = \tfrac{1}{2} \operatorname{tr} \operatorname{tr} \rho v \rho. \qquad 5.6$$

Upon adding and subtracting $\operatorname{tr} \operatorname{tr} (\hat{\rho} v \bar{\rho}) + \overline{\sum \hat{e}} + E(\bar{\rho})$ in Equation 5.5 one finds after a tedious but straightforward calculation that the total energy $E(\rho)$ takes the form of Equation 5.4 with $\mathcal{O}_2$ given by

$$\mathcal{O}_2 = \operatorname{tr} T(\overline{\hat{\rho}} - \bar{\rho}) + \operatorname{tr} \operatorname{tr} \bar{\rho} v(\overline{\hat{\rho}} - \bar{\rho}) - A(\rho - \hat{\rho}) - A(\hat{\rho} - \bar{\rho}). \qquad 5.7$$

Now Strutinsky assumes that $\rho - \bar{\rho}$ and $\rho - \hat{\rho}$ are small quantities of the same order, and that $\bar{\rho} - \overline{\hat{\rho}}$ is of second order in $\rho - \bar{\rho}$. These assumptions are quite reasonable if one recalls that the semiclassical density $\bar{\rho}$ is expected to be a very good approximation of ρ as demonstrated by its contribution to $E(\rho)$. On the other hand, it is likely that the difference between the semiclassical approximations to ρ and $\hat{\rho}$ are at least of second order as compared to the "shell effect" difference $\rho - \bar{\rho}$. However, it remains that all those assumptions deserve a careful check in a variety of nuclear structure calculations, and Section 5.4 is indeed devoted to it.

If these smallness assumptions are valid, one finds in Equation 5.4 a very convenient shortcut to the lengthy Hartree-Fock calculations. For that purpose, one has used a liquid drop model to evaluate $E(\bar{\rho})$ and a phenomenological mean field to determine $(\sum \hat{e} - \overline{\sum e})$. Such a duality in the parametrization is not at all free from inconsistency as discussed in Section 5.4. This approach, however, has been credited with a considerable number of successes. Among them, it has allowed Strutinsky to propose an explanation of some intriguing experimental data on nuclear fission (190–192) in terms of a double-humped fission barrier. The impressive quantitative agreement obtained from such calculations with experimental data as fission barrier heights (e.g. 175, 177, 178, 181, 193) may, of course, be partially due to the phenomenological flexibility allowed by the various parametrizations inherent to the method.

5.3   *The Techniques of the Semiclassical Approximation*

It has been assumed so far that given any Hamiltonian $\mathscr{H}$, a sufficiently good semiclassical approximation to its Hartree-Fock solution can be found. However in the most general situation such an approximation is not yet available. One has thus been obliged to find reasonable substitutes for it. A comprehensive review of this subject would exceed by far the limits of this paper and for more details and further references the reader is invited to consult, for instance, Reference 194. We simply give here some guidelines.

The first and yet more widely used approach is the single-particle energy averaging method proposed by Strutinsky (195). The accuracy of this method is limited by the approximate inclusion of unbound states to generate a reasonable average level density at the Fermi level (196, 197). An alternative method would consist in averaging the single-particle level density over the nucleon number as recently discussed (198). Another interesting proposal (199) originates in the disappearance of shell effects for sufficiently high temperatures, where the nuclear system behaves like a Fermi degenerate gas (see our discussion in Section 6.5). When extrapolating back to zero temperature the parabolic relation between excitation energy and entropy, one finds that the energy corresponding to a zero entropy is nothing up to a minus sign but the shell effect energy calculated according to the energy averaging method. An analytical proof of this numerical result has been given in References 200 and 201. One may also use an asymptotic expansion of the smooth part of the sum $\sum \hat{e}$ in powers of $A^{\frac{1}{3}}$ to determine, by difference with the exact sum, the shell effect energy (e.g. 202).

In all the preceding approaches, the semiclassical character of the quantities under consideration is not explicitly apparent, even though qualitatively obvious. This is not the case for the method initiated by Bhaduri & Ross (203) in which the exact sum $\sum \hat{e}$ is expanded for simple model cases in powers of the Planck's constant and its smooth part obtained simply by truncating the expansion. This method is identical to the Strutinsky energy averaging (204) in the case of infinite potentials. For finite potentials including spin-orbit forces, the numerical agreement between the two methods is gratifying (205) and proves that the first Strutinsky ansatz (195) to generate semiclassical quantities was indeed quantitatively correct.

5.4   *Numerical Checks of the Strutinsky Energy Theorem*

The intensive use of the Strutinsky energy theorem, and the impressive successes that many calculations according to it have met, necessitate a thorough study of its numerical validity. Such calculations have been achieved by three groups (206–208). In Reference 208 the smooth density matrix $\bar{\rho}$ has been calculated through a large pairing gap approximation, which is not a semiclassical approximation of ρ but rather a mocked-up solution of the Hartree-Fock problem at finite temperature (see 209 for discussion). Moreover it is also not clear how relevant the underlying self-consistent calculations (210) were for actual nuclei, since they were first order Hartree-Fock calculations using the realistic Tabakin nucleon-nucleon interaction (211). An explicit analytical expression of the second order shell correction energy has been proposed in Reference 206 and numerical calculations have been performed in the lead region with standard parameters for both the mean field and the Migdal scattering amplitudes entering the formalism. As a result, the second order shell correction has been found to lie between 0 and 2 MeV.

In Reference 207 and subsequent papers (209, 212, 213) another point of view has been adopted. From exact Hartree-Fock calculations, the Strutinsky smoothed density matrix $\bar{\rho}$ is computed:

$$\bar{\rho} = \sum_v n_v \, |v\rangle \langle v|,$$

where $|v\rangle$ are the Hartree-Fock single-particle states and n_v the Strutinsky smoothed occupation numbers defined in (177, 214). In view of the discussion of Section 5.3, such a density $\bar{\rho}$ may be considered a reasonably good semiclassical approximation of the Hartree-Fock solution ρ.

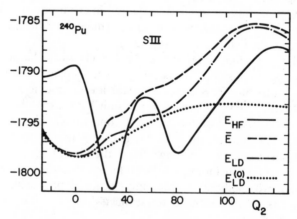

Figure 14 Deformation energy curves for the ^{240}Pu nucleus (from 213). Hartree-Fock (E_{HF}), Strutinsky smoothed ($\bar{E}$), and liquid drop model (E_{LD}) energies are shown. The curve E_{LD} corresponds to a liquid drop having the same moments $\bar{Q}_2$ and $\bar{Q}_4$ as in $\bar{E}$, whereas the curve $E_{LD}^{(0)}$ is obtained along the liquid drop fission valley. Energies are expressed in MeV and mass quadrupole moments in barn. The SIII Skyrme interaction has been used.

From ρ and $\bar{\rho}$, one computes $E(\rho)$, $E(\bar{\rho})$, and the first order shell correction $\sum\hat{e}-\sum\tilde{e}$ (see Equation 5.4). The sum of second and higher order terms $\mathcal{O}_2$ is then obtained by difference. Pairing correlations generally must be included and their contributions to the first order shell energy are in fact calculated according to the prescription of Reference 177.

The first question one would like to answer is whether or not the energy $E(\bar{\rho})$ behaves like a liquid drop energy. As a matter of fact, as seen in Figure 14 in the case of the ^{240}Pu nucleus this energy is minimal at sphericity and presents a fission saddle point at the expected deformation. One may be worried by the existence of wiggles around the ground state and the first fission barrier. They are, however, very easily explained by the influence of shell effects on the actual Hartree-Fock path along which we are evaluating the liquid-drop-like energy $E(\bar{\rho})$. The latter statement is substantiated by the appearance of similar wiggles on the liquid drop fission barrier, parametrized as in Reference 166, and following the Hartree-Fock path in the (Q_2, Q_4) plane.

The next question concerns the magnitude of the $\mathcal{O}_2$ term in Equation 5.4. As seen in Figure 15, they range from 1 to 3 MeV. This result is consistent with the findings for the second order terms alone (206).

Brack & Quentin (212) have pointed out that if $\bar{\rho}$ is determined self-consistently, the maximum amplitude of the sum of second and higher correction energies can be reduced to 0.6 MeV. This self-consistent determination of the semiclassical density matrix has been independently suggested by Tyapin (215) and leads to the numerical solution of equations very similar to the Hartree-Fock equations at finite temperature with the important difference, though, that in the present case no excitation energy is brought into the nucleus through the averaging process. The quality of the convergence of Equation 5.4 in this case is exemplified in Figure 16 and in Table 8 upon varying both the nuclear deformation and the nucleon number.

As a conclusion, it appears that the Strutinsky energy theorem provides us with a very rapidly convergent truncation scheme for the Hartree-Fock energy. On the other hand, it should be borne in mind that the impressive smallness of dropped terms in the truncated Equation 5.4 has

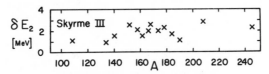

Figure 15 Higher order shell corrections $\delta E_2 = \mathcal{O}_2$ for 14 nuclei in their ground state [Figure 4 of (207)]. The Skyrme SIII interaction has been used. Pairing effects have been included in all deformed nuclei.

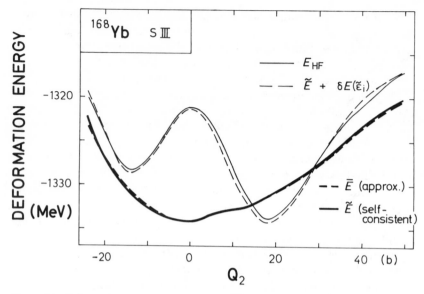

Figure 16 Deformation energy curves for the ^{168}Yb nucleus [Figure 2 of (212)]. The thin solid line is the Hartree-Fock energy E_{HF} and the heavy dashed line is the average energy $\bar{E} = E(\bar{\rho})$. The heavy solid line is the self-consistent average energy $\tilde{E}$ introduced in (212). Finally, the thin dashed line is the approximation $\tilde{E} + \delta E(\tilde{\varepsilon}_i)$, where $\delta E(\tilde{\varepsilon}_i)$ is the first order shell correction deduced from the self-consistent average density matrix $\tilde{\rho}$ of (212). The Skyrme SIII interaction has been used.

Table 8 Sum of the second and higher order terms neglected in the Strutinsky method[a]

Nucleus	$\mathcal{O}_2$	$\mathcal{O}_2^{sc}$
^{16}O	1.6	0.0
^{40}Ca	3.3	0.5
^{56}Ni	5.1	0.6
^{90}Zr	2.3	0.6
^{114}Sn	0.8	0.2
^{168}Yb	2.0	0.3
^{208}Pb	2.3	0.5

[a] The column $\mathcal{O}_2$ corresponds to the standard version of the Strutinsky energy theorem, whereas the column $\mathcal{O}_2^{sc}$ corresponds to the same quantity when the semiclassical density $\bar{\rho}$ is determined self-consistently.

only been obtained through a consistent treatment of both the liquid drop part of the energy and the averaged mean field. Such a consistency is not at all guaranteed when each of the pieces of the ensemble is separately determined. Therefore one should be aware of the possibilities of fluctuations in the energy terms not included in the Strutinsky approximation to the total energy. Such a fluctuating behavior as a function of deformation, for instance, could be of interest when determining small relative energies as in transitional nuclei (213).

An interesting prospect of practical significance can be deduced from the results of Reference 212. If one would know the semiclassical approximation $\bar{\rho}$ of ρ without going through the lengthy calculations of Reference 212 (actually as lengthy as Hartree-Fock calculations themselves), one would have only to fold $\bar{\rho}$ with the effective force to get the averaged mean field, which would provide a rather inexpensive and yet accurate shortcut to self-consistent calculations. The problem so far is to derive practical methods leading to reasonable estimates of $\bar{\rho}$ in particular in the surface region. Some recent works within the so-called extended Thomas-Fermi approximation (216, 217) constitute encouraging progress (218–220) in this direction.

5.5 The Expectation Value Method

In the present state of the numerical art, it is almost impossible to handle extensive Hartree-Fock calculations of heavy nuclei giving up all possible spatial symmetries (as axial and left-right reflection symmetries). This has provided a motivation for a non-self-consistent but microscopic approximation proposed in (221, 222), and recently further improved (223). It consists essentially in computing the expectation value of the effective Hamiltonian for a Slater determinant stemming from a phenomenological mean field. The key to the success of such an approximation lies obviously in the choice of the mean field. In the case of the Skyrme effective force, Brack (223) has proposed the following procedure: (a) First, make a spherical Hartree-Fock calculation for the given number of nucleons and with the chosen effective interaction. Such calculations are indeed rather easy (see Section 2.2). (b) Then fit the parameters of a Woods-Saxon (176) potential to reproduce the central part of the Hartree-Fock field. Make a similar adjustment for the spin-orbit form factor, the Coulomb field, and the effective mass. (c) Let all these quantities deform according to a reasonable scaling law (97). (d) Finally, diagonalize the one-body Hamiltonian so obtained, and retain the solution corresponding (for a given basis size) to the basis parameters that yield the minimal expectation value of the energy operator.

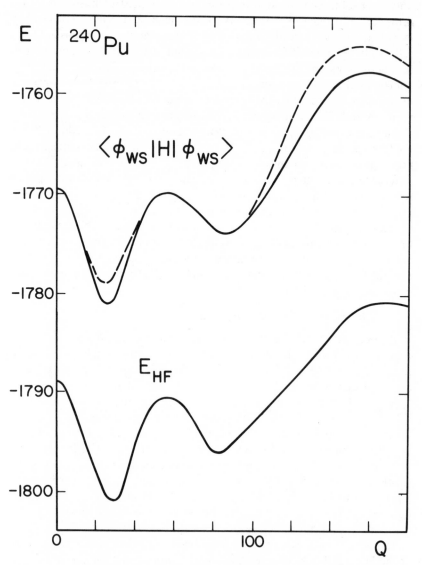

Figure 17   Comparison of the fission barriers of ^{240}Pu calculated self-consistently (E_{HF}) and within the expectation value method ($\langle \phi_{ws} | \mathscr{H} | \phi_{ws} \rangle$) (from 223). Dashed portions of upper curves are obtained with $h = 0$, with the notation of (177). The solid upper curve corresponds to a minimization of the energy in the (c,h) plane for each fixed value of Q_2. Energies are expressed in MeV and mass quadrupole moments in barn. The SIII Skyrme interaction has been used.

Upon doing that, actinide fission barrier heights, for instance, are reproduced within 1–2 MeV as illustrated in Figure 17. Such an accuracy may be considered as rather poor for the purpose of comparison with experimental data. It is, however, quite sufficient in completely un-explored regions of the nuclear chart to sketch interesting gross pro-perties yielded by a given effective interaction (e.g. 131).

6 HARTREE-FOCK CALCULATIONS AT FINITE TEMPERATURE

6.1 Introductory Remarks

Thermodynamical concepts have long been used by Frenkel (224) and many others (225–228) to describe nuclear excitations. Extensive experi-mental studies of nuclear level densities have further substantiated such an approach, exhibiting in particular a nice global agreement between experimental data and very rough model estimates (e.g. 229, 230). Recently, theoretical descriptions of excited nuclei have been achieved in the independent-particle limit of the grand canonical problem. Such calculations, referred to as Hartree-Fock calculations at finite tempera-ture, are briefly reviewed in this chapter.

The use of grand canonical equilibrium thermodynamics for the des-cription of excited nuclei is only meaningful if at least two conditions are fulfilled. First of all, given a value for the chemical potential, the nuclear temperature T should be small enough to avoid occupation probabilities for unbound states that are too large and lead to the non-equilibrium process of particle emission; under normal circumstances this would fix a limit for T of about 5 MeV for nuclear systems. Second the excitation energy should be reasonably shared among all individual degrees of freedom. As a consequence, for energies where collective excitations are known to play an important role, such Hartree-Fock calculations cannot yield more than average properties of excited nuclei.

6.2 The Hartree-Fock Approximation at Finite Temperature

The thermodynamical equilibrium of a grand canonical ensemble cor-responds to a situation of maximal entropy S under constraints on both the number of particles N and the total energy $\mathscr{H}$ operators. In such a case the thermodynamical potential Ω defined (231) as

$$\Omega = \langle \mathscr{H} \rangle - TS - \mu \langle N \rangle \qquad\qquad 6.1$$

(where T is the temperature and μ the chemical potential) is stationary. The formal solution of the former variational problem is given in terms of the density matrix D by

$$D = Z^{-1} \exp\left[-(\mathscr{H} - \mu N)/kT\right], \qquad 6.2$$

where Z is the grand partition function.

The Hartree-Fock approximation at finite temperature is obtained by restricting the variational space to the subspace of independent-particle wave functions. In this case the equilibrium solution is given by replacing in Equation 6.1 the effective Hamiltonian $\mathscr{H}$ by its one-body reduction (232). The latter is also called the Hartree-Fock Hamiltonian $h(\rho)$, where ρ is the corresponding one-body reduced density matrix defined in a given basis $\{i,j,\ldots\}$ by:

$$\langle i|\rho|j\rangle = \mathrm{tr}\,(Dc_j^\dagger c_i). \qquad 6.3$$

As in the zero temperature case, the nonlinearity of the solution D stems from its definition (Equation 6.2) in terms of a Hamiltonian $h(\rho)$ depending through ρ on D itself.

The equilibrium solution is then obtained by solving Hartree-Fock equations exactly similar to those considered in Section 2.1 but where the density ρ defined in Equation 6.3 is

$$\rho = \sum f_v |v\rangle\langle v| \qquad 6.4$$

[where $|v\rangle$ is the eigenstate of $h(\rho)$ corresponding to the single-particle energy e_v, and f_v is its occupation probabilities]. For our fermion states, the latter are given by

$$f_v = \{1 + \exp\left[(e_v - \mu)/kT\right]\}^{-1}. \qquad 6.5$$

6.3 *Practical Solution of the Hartree-Fock Equations at Finite Temperature*

In practice the problem is very similar indeed to the Hartree-Fock + BCS calculations discussed in Section 3.3. One has just to replace the pairing occupation probabilities by the Fermi ones defined in Equation 6.5. The Lagrange multiplier μ is fixed as in the pairing case, by imposing the value of the average number of nucleons

$$\sum_v f_v = N. \qquad 6.6$$

Upon varying T, considered here as an external parameter, one can study the variation of the energy $\langle\mathscr{H}\rangle$, the single-particle energies e_v and the entropy S as functions of the temperature and therefore extract the thermodynamical properties of the system.

6.4   *The Temperature Dependence of the Effective Interaction*

A generalization of the Brueckner approximation to the case of finite temperature was derived long ago by Bloch & De Dominicis (233, 234), and Kohn & Luttinger (235) (see also 236). But practical calculations for infinite nuclear (actually neutronic) matter have been achieved only very recently (237).

Turning now to our subject, which deals with finite nuclei and phenomenological effective forces, the actual parametrization of the temperature dependence of such forces remains to be done. This is why Hartree-Fock calculations at finite temperature have so far ignored such a dependence, i.e. have included at all excitation energies the zero temperature phenomenological interaction.

6.5   *Results*

Practical calculations have been performed by two groups (238–240), both using the Skyrme effective force parametrization. The main concern of such calculations was the actual importance of a self-consistent variation of both the wave functions and the occupation probabilities, to assess the validity of the non-self-consistent one-body approach in which a zero temperature Hartree-Fock spectrum is heated (e.g. 230, 241).

As a result, no sizable self-consistent effect (238) has been found. The single-particle Hartree-Fock spectrum is significantly constant up to the maximal temperature under study (i.e. ~ 5–6 MeV). Consequently the occupation probabilities and thus the entropy,

$$S = -k \sum_v [f_v \log f_v + (1-f_v) \log (1-f_v)], \qquad 6.7$$

are likely to be comparable in both the self-consistent and the non-self-consistent approaches described above. Moreover the correct Hartree-Fock excitation energy at temperature T_0,

$$E_{ex}^{(1)}(T_0) = \text{tr} \{[D_{HF}(T = T_0) - D_{HF}(T = 0)]\mathscr{H}\}, \qquad 6.8$$

is found very close indeed to the approximated one-body excitation energy using the zero temperature spectrum

$$E_{ex}^{(2)}(T_0) = \sum_v \{[f_v(T = T_0) - f_v(T = 0)]e_v(T = 0)\}. \qquad 6.9$$

As pointed out by Brack & Quentin (242) the validity of the approximation (Equation 6.9) is closely related to the validity of the Strutinsky approximation discussed earlier. Roughly speaking, both approximations are found similar upon replacing the "cold" Strutinsky smoothing by a "hot" one. The smallness of higher order terms in the former approxima-tion gives a hint for the validity of the latter. This is exemplified in Figure 18 where one sees that for a given entropy the two excitation energies $E_{\text{ex}}^{(1)}$ and $E_{\text{ex}}^{(2)}$ are identical.

In Figure 18 one also notices the parabolic dependence of the excita-tion energy as a function of the entropy at least for temperatures higher than 2–3 MeV, reflecting the degenerate Fermi gas character of our finite system as soon as shell effects are washed out by the thermal smearing of the single-particle level density.

As a further consequence the shape of a nucleus in its thermodynamical equilibrium state changes with the temperature. The existence of deformed nuclei is due to shell effects. When they disappear with increasing values of T, the equilibrium state becomes less and less deformed as shown in Figure 19 for the nucleus ^{168}Yb.

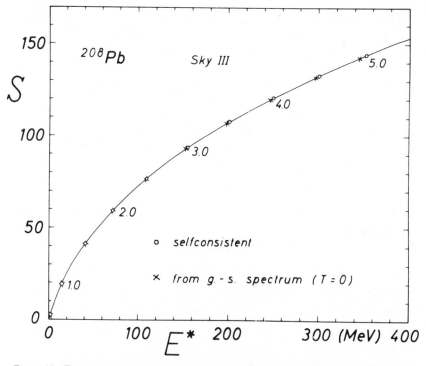

Figure 18   Entropy S vs excitation energy E^* ($E_{\text{ex}}^{(1)}$ in the text) of ^{208}Pb, calculated at the temperatures indicated (in MeV) along the curve [Figure 2 of (238)]. Circles show the self-consistent values, crosses the approximations derived from the fixed ground-state spectrum (see Equation 6.9). The SIII Skyrme force has been used.

In analysis of fission experiments that use the liquid drop model, it is important to know the temperature dependence of the model parameters. Some of them can be deduced from studies of nuclear matter at finite temperature. In this case and for a Skyrme effective Hamiltonian one obtains rather simple expressions for all thermodynamic quantities of interest as functions of the temperature and the chemical potential (243). One may then easily determine the thermodynamic properties of hot nuclear matter, for instance through Van der Waal's isotherms (pressure vs density) as illustrated in Figure 20. For those coefficients of the liquid drop model not attainable by infinite medium calculations, Hartree-Fock calculations were performed (243) at various temperatures and for several nuclei. Fitting the binding energies per particle by polynomials in $A^{-\frac{1}{3}}$ (A is the nucleon number) yields the temperature dependence of parameters such as the surface tension. This fitting procedure is meaningful only if the range of temperatures under consideration is beyond the critical temperature of complete disappearance of shell effects.

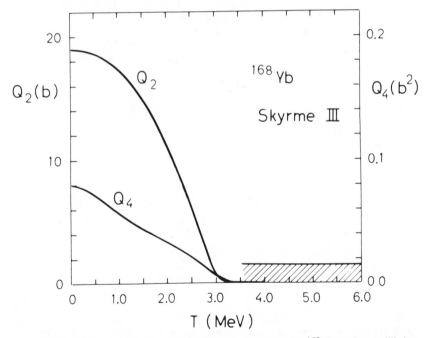

Figure 19 Mass quadrupole and hexadecapole moments of ^{168}Yb at its equilibrium deformation vs temperature [Figure 5 of (239)]. The shaded area represents the region where these moments are found at high temperature (continuum truncation effects give rise to this numerical inaccuracy). The SIII Skyrme force has been used.

In conclusion, Hartree-Fock calculations (at a finite temperature) provide a useful tool to investigate gross properties of excited nuclei. It should be borne in mind, however, that they have not yet included the temperature dependence of the effective interaction. If this dependence is negligible, one of the most important practical outputs of these calculations will be to justify the current non-self-consistent approaches.

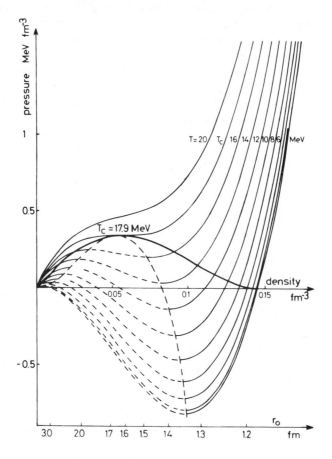

Figure 20 Equation of state for nuclear matter with the SIII Skyrme force [Figure 2 of (243)]. The solid lines show the pressure of nuclear matter as function of the density for various fixed temperatures (isotherms). The unstable states of the homogeneous system are indicated by the dashed lines. The heavy line gives the pressure and density for the liquid-vapor phase transition. The scale on the axis below shows the nuclear-radius constant that corresponds to the density.

7 NUCLEAR DYNAMICS WITHIN THE TIME-DEPENDENT HARTREE-FOCK APPROXIMATION

7.1 Generalities

The time-dependent Hartree-Fock (TDHF) approximation was introduced by Dirac (3) in the context of atomic physics. It was explicitly used in nuclear physics for the first time by Ferrell (244). Its small amplitude motion limit was later recognized (245, 246) as nothing but the random phase approximation (RPA) introduced in Reference 247. On the other hand, its adiabatic limit has been widely studied in the Inglis cranking method (248, 249) and in more elaborated approaches (118, 250–255). However, it is only recently that Bonche, Koonin & Negele (51) have provided for the first time numerical solutions of the full TDHF equations.

The TDHF approximation assumes that during its time evolution, the nuclear wave function Φ remains a Slater determinant. From the minimization of the "action" integral (for a given effective Hamiltonian $\mathscr{H}$),

$$I = \int_{t_1}^{t_2} \left\langle \Phi \left| \mathscr{H} - i\hbar \frac{\partial}{\partial t} \right| \Phi \right\rangle \, dt, \qquad 7.1$$

one obtains the TDHF equations:

$$[h(\rho),\rho] = i\hbar \frac{\partial}{\partial t} \rho, \qquad 7.2$$

where ρ is the one-body reduced density matrix and $h(\rho)$ the corresponding one-body reduction of $\mathscr{H}$. In terms of the single-particle wave functions ψ_j composing the Slater determinant Φ, Equation 7.2 reads

$$h(\rho)|\psi_j\rangle = i\hbar \frac{\partial}{\partial t}|\psi_j\rangle. \qquad 7.3$$

The nonlinear system of Equations 7.3 is the one to be solved in practical calculations.

One of the theoretical problems related to the use of the TDHF approximation in nuclear dynamics is the possible dependence of the effective force with respect to the total energy of the nuclear system. So far, one has generally used interactions reproducing gross properties of cold nuclear matter, thus flatly neglecting such an energy dependence. However, given an effective interaction, it has been recently demonstrated, upon performing RPA calculations in a consistent way (see

Section 7.3), that one could reproduce at the same time static properties and low energy spectroscopic data up to giant multipole resonance energies (34, 36, 256, 257, 258 and references therein). The attitude has thus been to try the same interactions (except for some technical simplifications discussed below) higher and higher in energy, until some manifest breakdown in such a blind extrapolation appears.

7.2   Recent Progress

The practical solution of Equations 7.3 represents a considerable numerical task, which has only been afforded so far at the price of some simplifying restrictions on the effective force, the imposed symmetries, and the numerical accuracy. The effective interaction currently in use was proposed in Reference 51. It consists of a zero range, two-body force (including a linear density-dependent term) plus the direct part of a Yukawa interaction to simulate surface effects. In general a spin-isospin fourfold degeneracy is assumed for each single-particle state, which together with the absence of spin-orbit or tensor components contributes to the numerical simplicity by reducing the number of independent solutions of Equations 7.3. The Coulomb interaction is included for its direct part only, with an effective charge of $e/2$ allotted to both proton and neutron single-particle states.

In practical TDHF calculations, it has appeared very important to guarantee the unitarity of the numerical approximation of the time-evolution operator U. In the first TDHF calculations (51) this was done by using the Crank-Nicholson (259) approximation of U. Recently a truncation of the Taylor expansion of U to the first few terms has proven to be satisfactorily accurate (92).

Apart from a very preliminary and highly schematic work on nuclear fission (121), most of the TDHF calculations presently performed have been devoted to the study of nuclear collisions. In the first of them (51), the authors studied the collision of two slabs (a medium infinite in two directions and finite in the third). This calculation must be credited first with clearly proving the numerical feasibility of TDHF calculations. Moreover, upon varying the incident energy, a broad variety of phenomena emerged within the same model framework, such as elastic and quasielastic scattering, resonance, fusion, and deep inelastic scattering.

A second generation of calculations handled the time evolution of 2D wave functions. In one case (P. Bonche, B. Grammaticos, A. Jaffrin, unpublished; see, however, a brief account in 260) the collision of two pucks and of two infinite rods has been studied. Even though such calculations were capable of a wider variety of initial conditions (allowing any finite impact parameter) than the previous ones, the geometrical peculiarity they involved restricted their physical output. Other calculations assumed an axial symmetry of the nuclear system (261, 262). To account for the impact parameter influence on the dynamics, the nuclear system was plunged into an external rotating field corresponding to a given l value. Such two-dimensional TDHF calculations in a rotating frame have been performed for systems involving in the entrance channels two identical nuclei: ^{16}O or ^{40}Ca (263), as well as two different nuclei: ^{12}C+^{14}N (264). Recently (indeed after the implementation of three-dimensional TDHF codes) another 2D approximation has been proposed in the case of collision events (296). It consists of neglecting the motion normal to the scattering plane by considering separable TDHF single-particle wave functions. This approximation has been shown (296, 297) to be highly successful when compared with some exact 3D calculations discussed below.

A third generation of calculations gives up the restriction to 2D systems and deals with the time evolution of 3D solutions. A calculation of this type was published as early as 1976 (265). It was, however, performed with an oversimplified nuclear interaction and omitted Coulomb interaction. As a matter of fact, three-dimensional TDHF calculations using interactions of the same quality as the one used in Reference 51 have only recently been achieved (92, 266).

One of the most appealing features of TDHF calculations is that they provide us with a series of snapshots of the evolving nuclear systems. In Figure 21, for instance, one can visualize the succession of various stages of the reaction of a symmetrical nuclear system (^{40}Ca+^{40}Ca): polarization before contact, contact, separation, excitation of the final nuclei. But the output of TDHF calculations goes far beyond that. Given an incident laboratory energy and an impact parameter one may give, in the case of a scattering event, a quantitative account of the energy loss during nuclear contact. Upon summing geometrical cross sections for all impact parameters leading to fusion, one may also estimate the total fusion cross section at a given energy. This has been done in Reference 266 and yields an amazingly good agreement with experimental data (267, 268) in the reaction ^{16}O+^{16}O for bombarding energies ranging from 20 to 120 MeV, as can be seen in Figure 22.

For the reaction ^{40}Ca+^{40}Ca near 300-MeV laboratory energy, two-dimensional TDHF calculations within the rotating frame approximation (263) have not yielded fusion events for intermediate values of the impact parameter ($25 \lesssim l \lesssim 70$) in contradistinction with recently measured total fusion cross sections (269). Such a qualitative disagreement could originate either in the restrictive 2D assumptions made in Reference 263 or in the TDHF approximation itself. As a matter of fact, the 3D calcula-

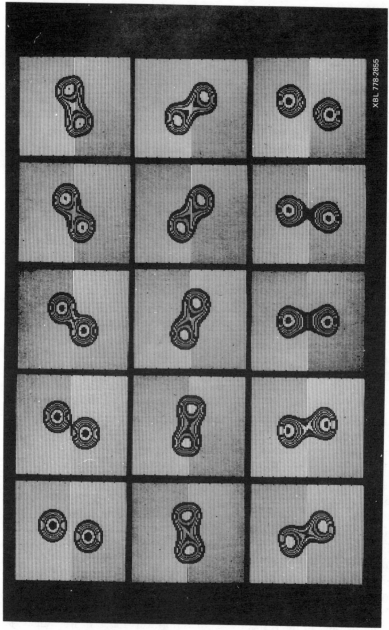

XBL 778-2855

Figure 21 Typical example of a TDHF collision of two ^{40}Ca nuclei ($E_{lab} = 278$ MeV and $l = 75\ \hbar$) giving rise to a scattering event. The time interval between two snapshots is 1.6×10^{-22} sec.

tions of Reference 266 support the former explanation: the existence of calculated fusion events has clearly demonstrated the importance in that case of the tangential degrees of freedom, which were neutralized in the calculations of Reference 263.

Another interesting feature of the results of References 92, 265, and 266 is the existence of a low-l cutoff. Even though experimental data on this issue are still missing, it is noticeable that the TDHF calculations in this case lead to conclusions in contradistinction with recent liquid drop calculations (R. Nix and A. Sierk, unpublished).

7.3   The Limiting Case of Small Amplitude Motion

In this section we do not intend to give a full account of RPA calculations in nuclear physics but rather to illustrate the possibility of describing with the same effective interaction both static properties and low energy excitation properties in spherical nuclei. We therefore limit ourselves here to those RPA calculations (34, 36, 256–258, 270) that are fully self-consistent in the sense that both single-particle states and coherent particle-hole excitations are determined from the same effective inter-action. As a matter of fact, this self-consistency has been explicitly demonstrated (258) to be of great importance for the description of coherent phenomena such as giant multipole resonances in nuclei.

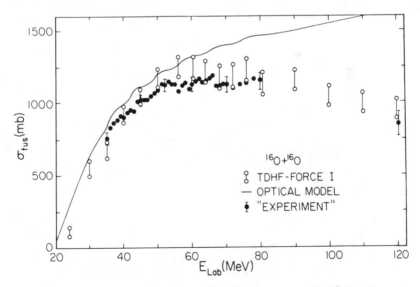

Figure 22   Fusion excitation function for $^{16}O + ^{16}O$ [Figure 3 of (266)]. The data come from (267, 268). The optical model results come from (292). Calculations have been performed with the force (*I*) introduced in (51).

All the various RPA calculations quoted above used various parametrizations of the Skyrme force or they made slight modifications in it
by adding a linear density dependence of the velocity-dependent parts of
the force [for the S-wave part (34) or for both the S- and the P-wave
parts (36)]. Two of them, however (258, 270), have also used other forces,
one proposed by Pearson, Rouben & Saunier (271), the other by Gogny
(48, 49). In (34, 256, 257) the Coulomb interaction and the spin-spin
and spin-orbit parts of the nuclear interaction have not been taken into
account for the determination of the particle-hole interaction; this was
not the case in (36, 258, 270).

We have already mentioned (Section 1.2.1) the equivalence of a
linearly density-dependent delta force (acting in the spin-triplet subspace)
with a three-body delta force for Slater determinants having no spin-
vector component of the one-body reduced density matrix (25). It has
been shown either by RPA calculations (270) or by evaluating the
relevant Landau-Migdal parameter (35) that a three-body force leads to
spin instability, whereas this is not the case if one instead considers this
part of the interaction as a density-dependent force (272, 273).

One of the practical problems in RPA calculations is the occurrence
of particle-hole excitations with particle states in the continuum. Some
calculations approximate these latter states either by diagonalizing a
Hartree-Fock Hamiltonian supplemented by an infinite wall located at a
sufficiently large distance from the nuclear surface (34, 256) or by diagonal-

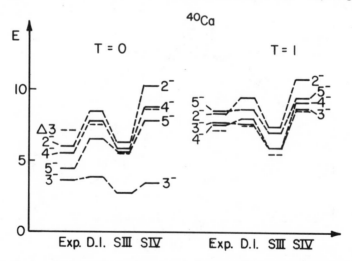

Figure 23 Low-lying negative parity states (in MeV) of ^{40}Ca (from 258). The quantity
$\Delta 3$ is the energy difference $\varepsilon_{f\frac{7}{2}} - \varepsilon_{d\frac{3}{2}}$. The SIII and SIV Skyrme forces and the D1 Gogny
force have been used.

izing the exact Hartree-Fock Hamiltonian in a harmonic oscillator representation (258, 270). In two calculations (36, 257), however, continuum states have been correctly taken into account. It is obvious that only the latter calculations can give access to a width of giant multipole resonances.

We now present some selected results of such fully self-consistent RPA calculations. Blaizot, Gogny & Grammaticos (274) performed a systematic study of the giant monopole resonance with various interactions. They determined that the value of the incompressibility modulus K, consistent with the experimental results of Marty et al (275, 276) and interpreted as E0 resonances, would be around 210–220 MeV. This would definitely discard, for that mode, the standard Skyrme force where for reasonable values of the effective mass K ranges well above 300 MeV. However, such a clear-cut deficiency does not seem to extend to other multipolarities [see, for example, the calculated E1 strength functions in ^{16}O (36, 257) as compared with the experimental one (277)]. Figure 23 reproduces the low-lying negative parity states of ^{40}Ca from the calculations of Reference 258. Both self-consistent calculations performed with the SIII Skyrme interaction and the D1 Gogny interaction yield approximately the same reasonable agreement with experimental data (278).

As a conclusion it thus seems possible that effective forces can successfully describe ground-state properties as well as low energy spectroscopic data. This is, of course, only true insofar as the latter enter in some way or another in the fitting procedure of the force parameters.

7.4 *The Adiabatic Limit*

The adiabatic limit of the TDHF approximation (ATDHF) plays an important role in nuclear physics since it is one of the basic implicit assumptions of the unified model description of low energy collective motion (171, 172, 279) that has met with a considerable success (e.g. 122). By definition, in an adiabatic collective motion the characteristic velocities associated with a few collective variables are much smaller than the ones corresponding to independent-particle motion in the mean field. The TDHF approximation is indeed capable of providing a consistent coupling between these two types of modes and constitutes therefore an appropriate framework to study such phenomena.

There have been a number of works dealing with the ATDHF approximation (118, 248–255, 298). Baranger & Vénéroni (255) recently made a comprehensive study and we adopt their notations. Given a TDHF solution characterized by its one-body reduced density matrix ρ, they proved (280), under a restrictive condition we discuss later, that a unique set of hermitian time-reversal even operators ρ_0 and χ can always be defined (see also the somewhat lengthy proof of 281):

$$\rho = e^{i\chi}\rho_0 e^{-i\chi}. \tag{7.4}$$

Since ρ satisfies the projector identity, ρ_0 does also and thus represents a Slater determinant.

One may formally expand ρ in Equation 7.4 in powers of χ:

$$\rho = \sum_q \rho_q \tag{7.5}$$

with

$$\rho_q = \frac{i^q}{q!}\underbrace{\left[\chi, \left[\chi, \ldots [\chi, \rho_0] \ldots \right]\right]}_{q \text{ commutators}} \tag{7.6}$$

Upon inserting a truncated expansion (Equation 7.6) in the TDHF Equation 7.2, one obtains for the time-reversal odd and time-reversal even part of it:

$$[h_0, \rho_1] + [h_1, \rho_0] = i\hbar\dot\rho_0 \tag{7.7a}$$

$$[h_0, \rho_0] + [h_2, \rho_0] + [h_1, \rho_1] + [h_0, \rho_2] = i\hbar\dot\rho_1, \tag{7.7b}$$

where the dot stands for the time derivative and h_q is the qth order part of $h(\rho)$. These equations are referred to as the ATDHF equations.

So far, one may wonder what such a formal expansion in χ has to do with adiabaticity. As a matter of fact, one may view the ATDHF equations as a set of canonical Hamilton's equations where the matrix elements of ρ_0 and χ play the role of coordinates and momenta, respectively (254). Truncating an expansion of the Hartree-Fock energy $E(\rho)$ to second order in the momentum χ is indeed consistent with the general meaning of adiabaticity and yields to:

$$E(\rho) \simeq E(\rho_0) + K, \tag{7.8}$$

where $E(\rho_0)$ is viewed as a potential energy and K (which is quadratic in χ) is given (254) by:

$$K = (\hbar/2)\,\mathrm{tr}\,(\chi\,\dot\rho_0). \tag{7.9}$$

The latter is actually the adiabatic kinetic energy.

To recover the actual meaning of adiabaticity in the context of nuclear collective motion, we must further assume that the whole dynamics is restricted to the time evolution of a limited set of collective variables. For the sake of simplicity, we consider that the dynamics is only governed by a single collective variable Q. Solving the Equations

7.7 amounts then to finding the adiabatic path $\rho_0(Q)$, the Q dependence of the operator χ and then deduce the time-evolution law for Q. However, the solution of ATDHF coupled equations has not yet been attempted (it seems at least as complicated as solving the exact TDHF equations). Instead an adiabatic path $\rho_0(Q)$ is postulated and, from Equation 7.7a, one may deduce $\chi(Q)$ or equivalently the adiabatic mass $M(Q)$ defined by (see Equation 7.9)

$$M(Q) = \frac{2K}{\dot{Q}^2} = \frac{\hbar}{\dot{Q}}\,\mathrm{tr}\,\chi\,\frac{\partial\rho_0}{\partial Q}. \qquad\qquad 7.10$$

One ansatz for the adiabatic path can be provided by a series of Hartree-Fock calculations under constraint on the expectation value Q of the operator $\hat{Q}$ (298, 299), where it was shown (282) that a convenient way to solve Equation 7.7a was to redo the previous constrained Hartree-Fock calculations with a supplementary time-reversal odd constraining field

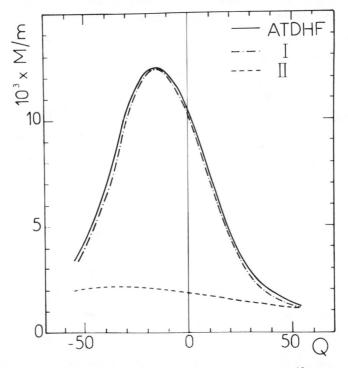

Figure 24   Ratio of the isoscalar quadrupole adiabatic inertial mass of ^{12}C to the nucleon mass m (in fm^{-2}) as a function of the mass quadrupole moment Q (fm^2) [Figure 2 of (282)]. Labels I and II correspond respectively to the Inglis cranking formula and to the scaling value (see text). The SIII Skyrme force has been used.

$-i\hbar\dot{Q}\left[\partial\rho_0/\partial Q,\rho_0\right]$. This may be viewed as a generalization to modes other than collective rotational motion of the result of Thouless & Valatin (283). A resulting adiabatic mass $M(Q)$ is shown in Figure 24 for the isoscalar quadrupole mode in ^{12}C. The bell-shape variation of $M(Q)$ is easily understood as the structure of the dominant particle-hole excitation energy (284) and might be related to the existence of a deformed intrinsic ground state for this nucleus, whereas its potential energy curve presents a (shallow) minimum at sphericity.

The Inglis cranking method (248, 249) is actually the non-self-consistent version of the previous approach, i.e. neglecting the piece h_1 of the Hartree-Fock Hamiltonian stemming from the time-reversal odd part of ρ. Given an adiabatic path ρ_0, ρ_1 may be computed by (see Equation 7.7a):

$$[h_0,\rho_1] = i\hbar\dot{\rho}_0, \qquad\qquad 7.11$$

which together with Equation 7.10 readily yields the well-known cranking formula for $M(Q)$. It can be seen in Figure 25 that the Inglis cranking

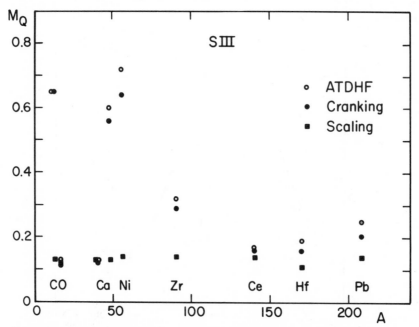

Figure 25   Systematics of isoscalar quadrupole adiabatic inertial masses as a function of the nucleon number A. Masses are divided by $mA^{\frac{5}{3}}$, (where m is the nucleonic mass) and are expressed in fm^2. They all correspond to a spherical shape and have been calculated with the SIII Skyrme force. Values obtained within the Inglis cranking and the scaling (see text) approximations are also reported.

formula leads to masses that are not more than $\sim 20\%$ smaller than self-consistent masses. However it may be noticed that the previous results have been obtained for one specific parametrization of the Skyrme force (SIII). A change in the force (and thus its velocity-dependence content) results in almost constant ATDHF masses but largely varying cranking masses (M. J. Giannoni, F. Moreau, P. Quentin, unpublished).

The uniqueness of Expression 7.4 for ρ is contingent upon a specific assumption about the spectrum of χ, which must lie within the $-\pi/4 < \chi < \pi/4$ open interval (280). As a consequence, such an operator χ has no particle-hole matrix elements, where particle (hole) refers to unoccupied (occupied) states of the Slater determinant ρ_0. This restrictive condition is not fulfilled in a variety of physical situations, as for instance the uniform translation of a Slater determinant (283) or collective motions involving odd nuclei (285).

If we assume that the TDHF density matrix ρ can be written as in Equation 7.4 where χ is a local operator, the TDHF Equations 7.2 can be transformed into an equation of state for the nuclear medium, supplemented by a continuity equation and an Euler's equation corresponding to an irrotational flow (286). It may be noticed that this result does not imply any hypothesis on the smallness of χ nor on the particle-hole character of χ. It is indeed possible to demonstrate (285) that a local χ operator with only particle-hole matrix elements would vanish identically. Assuming χ to be small, a special simple class of local χ operators has been deduced from Equation 7.7a in the case where the collective variable is a scaling variable (282), yielding simple expressions (90, 282, 287–290) for the adiabatic mass—which is actually in this case an irrotational flow inertia parameter.

Moreover, Baranger & Vénéroni showed (291) that the RPA approximation, which was already known as a limiting case of the TDHF approximation, corresponded also to a particular solution of ATDHF Equations 7.7. The latter provide a unified framework for a variety of apparently disconnected approximations (Inglis cranking, Thouless-Valatin formalism, irrotational model, RPA). It is therefore of primary importance to check to what extent they constitute a satisfactory approximation of the TDHF equations for given initial conditions. Such a check implies essentially that one is able to extract χ (and thus ρ_0) from ρ. A tractable method to do so, starting from the Baranger-Vénéroni identity (280)

$$e^{4i\chi} = (2\rho - 1)(2\bar{\rho} - 1) \qquad\qquad 7.12$$

(where $\bar{\rho}$ is the time reverse of ρ), has been recently worked out (285).

CONCLUSIONS

The self-consistent calculations presented here produce nuclear static properties in amazingly good overall agreement with experimental data. This is all the more impressive because the number of model parameters is actually very small. In most cases the quantitative agreement with experiment is comparable with that yielded by approaches enjoying a much larger phenomenological flexibility through ad hoc parameter adjustments. It is therefore reasonable to believe that the assumptions made here include the major part of the physics involved.

These calculations have also given a firmer basis to a number of more phenomenological approaches. The latter are generally reasonably good approximations to Hartree-Fock calculations when they meet specific consistency conditions (for the model parameters) that are hard to assess a priori. It is our opinion that such drawbacks of current phenomenological approaches constitute a sufficient motivation for the more lengthy self-consistent calculations. This is particularly true for static nuclear properties far beyond experimental knowledge, where the predictive power of self-consistent calculations is enhanced by the inclusion *ab initio* of a proper extrapolation mechanism.

Turning now to dynamical properties of nuclei, the results of calculations assuming an independent-particle motion (TDHF or its approximations) are much scarcer. Nevertheless one may already conclude that they qualitatively reproduce experimental data, at least for processes involving excitation energies less or equal to those encountered in giant multipole excitations. At higher energies, it is probably premature to assess the validity of the approach in a definite way. It seems established, however, that the formal framework is capable of describing the bulk of phenomena like those observed during heavy ion collision with incident energy per particle in the 2–5-MeV range, for instance.

In the last few years, an increase in the knowledge of the nucleon-nucleon effective interaction correlative with a development of fast and reliable numerical techniques has resulted in considerable progress in the quality of the self-consistent calculations of nuclear properties. In spite of the drastic limitation of the Hilbert space that they imply, such calculations will apparently constitute an increasingly useful tool in forthcoming investigations of both the statics and the dynamics of atomic nuclei.

ACKNOWLEDGMENTS

Our first steps in the nuclear self-consistent calculations wonderland were guided by D. Vautherin and M. Vénéroni. Without the long and profitable years of close collaboration with them, we would probably have not been able to achieve this work. We also acknowledge innumerable discussions with many colleagues that have deeply enriched our understanding of the field under study here. Among them we would like to express our appreciation to M. Beiner, M. Brack, D. M. Brink, P. Bonche, R. Y. Cusson, M. J. Giannoni, D. Gogny, A. K. Kerman, J. Letessier, U. Mosel, P. Möller, Nguyen Van Giai, G. Ripka, D. W. L. Sprung, and M. S. Weiss. P.Q. is grateful to J. R. Nix for the hospitality enjoyed at the theoretical division of the Los Alamos Scientific Laboratory. Partial funding from the US Department of Energy (under contract W-7405-Eng-36) is also gratefully acknowledged.

Literature Cited

1. Hartree, D. 1928. *Proc. Cambridge Philos. Soc.* 24: 89–132
2. Fock, V. 1930. *Z. Phys.* 61: 126–48
3. Dirac, P. A. M. 1930. *Proc. Cambridge Philos. Soc.* 26: 376–85
4. Bethe, H. A. 1971. *Ann. Rev. Nucl. Sci.* 11: 93–244
5. Sprung, D. W. L. 1972. *Adv. Nucl. Phys.* 5: 225–343
6. Brueckner, K. A., Levinson, C. A., Mahmoud, H. M. 1954. *Phys. Rev.* 95: 217–28
7. Brueckner, K. A., Levinson, C. A. 1955. *Phys. Rev.* 97: 1344–52
8. Brueckner, K. A. 1955. *Phys. Rev.* 97: 1353–66
9. Goldstone, J. 1957. *Proc. R. Soc. London, Ser. A* 239: 267–79
10. Hugenholtz, N. M. 1957. *Physica (Utrecht)* 23: 481–532
11. Brueckner, K. A., Gammel, J. L., Weitzner, H. 1958. *Phys. Rev.* 110: 431–45
12. Brueckner, K. A., Lockett, A. M., Rotenberg, M. 1961. *Phys. Rev.* 121: 255–69
13. Köhler, H. S. 1965. *Phys. Rev. B* 137: 1145–57
14. Köhler, H. S. 1965. *Phys. Rev. B* 138: 831–46
15. Negele, J. W. 1970. *Phys. Rev. C* 1: 1260–1321
16. Campi, X., Sprung, D. W. L. 1972. *Nucl. Phys. A* 194: 401–42
17. Nemeth, J., Vautherin, D. 1970. *Phys. Lett. B* 32: 561–64
18. Nemeth, J., Ripka, G. 1972. *Nucl. Phys. A* 194: 329–52
19. Volkov, A. B. 1965. *Nucl. Phys.* 74: 33–58
20. Brink, D. M., Boeker, E. 1967. *Nucl. Phys. A* 91: 1–26
21. Nestor, C. W., Davies, K. T. R., Krieger, S. J., Baranger, M. 1968. *Nucl. Phys. A* 113: 14–26
22. Pearson, J. M., Saunier, G. 1968. *Phys. Rev.* 173: 991–94
23. Saunier, G., Pearson, J. M. 1970. *Phys. Rev. C* 1: 1353–64
24. Campi, X., Sprung, D. W. L. 1972. See Ref. 16, Sec. 2.2.1, pp. 408–9 and 429
25. Vautherin, D., Brink, D. M. 1972. *Phys. Rev. C* 5: 626–47, Pt. II, p. 628
26. Skyrme, T. H. R. 1956. *Philos. Mag.* 1: 1043–54
27. Skyrme, T. H. R. 1959. *Nucl. Phys.* 9: 615–34
28. Vautherin, D. 1969. *Thèse d'état*. Orsay, France. Unpublished
29. Vautherin, D., Brink, D. M. 1970. *Phys. Lett. B* 32: 149–53
30. Vautherin, D., Brink, D. M. 1972. See Ref. 25, pp. 626–47
31. Moszkowski, S. A. 1970. *Phys. Rev. C* 2: 402–14
32. Beiner, M., Flocard, H., Nguyen Van Giai, Quentin, P. 1975. *Nucl. Phys. A* 238: 29–69
33. Köhler, H. S. 1976. *Nucl. Phys. A* 258: 301–16
34. Liu, K. F., Brown, G. E. 1976. *Nucl. Phys. A* 265: 385–415
35. Bäckman, S.-O., Jackson, A. D., Speth, J. 1975. *Phys. Lett. B* 56: 209–11
36. Krewald, S., Klemt, V., Speth, J., Faessler, A. 1977. *Nucl. Phys. A* 281: 166–206
37. Treiner, J., Krivine, H. 1976. *J. Phys. G* 2: 285–307
38. Ehlers, J. W., Moszkowski, S. A. 1972. *Phys. Rev. C* 6: 217–27
39. Sinha, B., Moszkowski, S. A. 1978. *Nucl. Phys. A* 302: 237–56
40. Lassey, R. K., Volkov, A. B. 1971. *Phys. Lett. B* 36: 4–8
41. Lassey, R. K. 1972. *Nucl. Phys. A* 192: 177–99

42. Žofka, J., Ripka, G. 1971. *Nucl. Phys. A* 168:65–96
43. Krieger, S. J. 1970. *Phys. Rev. C* 1:76–84
44. Rouben, B., Pearson, J. M., Saunier, G. 1972. *Phys. Lett. B* 42:385–88
45. Rouben, B., Brut, F., Pearson, J. M., Saunier, G. 1977. *Phys. Lett. B* 70:6–8
46. Manning, M. R. P., Volkov, A. B. 1967. *Phys. Lett. B* 26:60–62
47. Lassey, R. K., Manning, M. R. P., Volkov, A. B. 1973. *Can. J. Phys.* 51: 2522–49
48. Gogny, D. 1973. In *Proc. Intern. Conf. Nucl. Phys.*, ed. J. De Boer, H. J. Mang, Vol. 1, p. 48. Amsterdam: North-Holland. 739 pp.
49. Gogny, D. 1975. In *Nuclear Self-Consistent Fields*, ed. G. Ripka, M. Porneuf, pp. 333–52. Amsterdam: North-Holland. 450 pp.
50. Bryan, R. A., Scott, B. L. 1964. *Phys. Rev. B* 135:434–50
51. Bonche, P., Koonin, S., Negele, J. W. 1976. *Phys. Rev. C* 13:1226–58
52. Negele, J. W., Vautherin, D. 1972. *Phys. Rev. C* 5:1472–93
53. Negele, J. W., Vautherin, D. 1975. *Phys. Rev. C* 11:1031–41
54. Sprung, D. W. L., Vallières, M., Campi, X., Che-Ming Ko. 1975. *Nucl. Phys. A* 253:1–19
55. Blin-Stoyle, R. J. 1955. *Philos. Mag.* 46:973–81
56. Bell, J. S., Skyrme, T. H. R. 1956. *Philos. Mag.* 1:1055–68
57. Skyrme, T. H. R. 1959. *Nucl. Phys.* 9:635–40
58. Scheerbaum, R. R. 1969. PhD thesis. Cornell Univ., Ithaca, NY. Unpublished
59. Stancu, F., Brink, D. M., Flocard, H. 1977. *Phys. Lett. B* 68:108–12
60. Quentin, P. 1972. *J. Phys. (Paris)* 33: 457–63
61. Griffin, R. E., Volkov, A. B., Lassey, R. K. 1973. *Can. J. Phys.* 51:2054–62; Sect. 4, pp. 2059–62
62. Slater, J. C. 1951. *Phys. Rev.* 81:385–90
63. Gombás, P. 1952. *Ann. Phys. (Leipzig)* 10:253–64
64. Kolb, D., Cusson, R. Y. 1972. *Z. Phys.* 253:282–88
65. Titin-Schnaider, C., Quentin, P. 1974. *Phys. Lett. B* 39:397–400
66. Quentin, P. 1975. *Description self-consistante des propiétés statiques de déformation nucléaire à partir d' interactions effectives nucléon-nucléon.* Thèse d'état. Orsay, France. Unpublished
67. Köhler, H. S. 1974. *Phys. Scr. A* 10:81–83
68. Hamamoto, I., Siemens, P. 1976. *Nucl. Phys. A* 269:199–209
69. Negele, J. W. 1975. See Ref. 49, pp. 113–43
70. Campi, X., Bouissy, A. 1978. *Phys. Lett. B* 73:263–66
71. Jennings, B. K. 1978. *Phys. Lett. B* 74:13–14
72. Brueckner, K. A., Buchler, J. R., Jorna, S., Lombard, R. J. 1968. *Phys. Rev.* 171:1188–95
73. Lombard, R. J. 1973. *Ann. Phys. (NY)* 77:380–413
74. Stock, H. 1975. *Nucl. Phys. A* 237:365–81
75. Beiner, M., Lombard, R. J. 1974. *Ann. Phys. (NY)* 86:262–305
76. Beiner, M., Lombard, R. J., Mas, D. 1975. *Nucl. Phys. A* 249:1–28
77. Beiner, M., Lombard, R. J., Mas, D. 1976. *At. Data Nucl. Data Tables* 17:450–54
78. Tondeur, F. 1978. *Nucl. Phys. A* 303: 185–98
79. Tondeur, F. 1977. *Thèse. Univ. Libre, Bruxelles, Belgique.* Unpublished
80. Meldner, H. 1969. *Phys. Rev.* 178: 1815–26
81. Kolb, D., Cusson, R. Y., Harvey, M. 1973. *Nucl. Phys. A* 215:1–23
82. Cusson, R. Y., Trivedi, H. P., Meldner, H. W., Weiss, M. S., Wright, R. E. 1976. *Phys. Rev. C* 14:1615–29
83. Bonaccorso, A., Di Toro, M., Russo, G. 1977. *Phys. Lett. B* 72:27–29
84. Brueckner, K. A., Gammel, J. L., Weitzner, H. 1958. See Ref. 11, Sect. 6, p. 445
85. Romanelli, M. J. 1960. In *Mathematical Methods for Digital Computers*, ed. A. Ralston, H. Wilf, pp. 110–20. New York: Wiley. 293 pp.
86. Vautherin, D., Vénéroni, M. 1967. *Phys. Lett. B* 25:175–78
87. Noble, B. 1964. *Numerical Methods*, Vol. 2, Sect. 10.9, pp. 316–21. Edinburgh: Oliver & Boyd. 372 pp.
88. Vautherin, D., Brink, D. M. 1970. *Phys. Lett. B* 32:149–53
89. Nguyen Van Giai. 1972. *Thèse d'état.* Orsay, France. Unpublished
90. Engel, Y. M., Brink, D. M., Goeke, K., Krieger, S. J., Vautherin, D. 1975. *Nucl. Phys. A* 249:215–38
91. Hoodboy, P., Negele, J. W. 1977. *Nucl. Phys. A* 288:23–44
92. Flocard, H., Koonin, S. E., Weiss, M. S. 1978. *Phys. Rev. C* 17:1682–99
93. Mutukrishnan, R. 1967. *Nucl. Phys. A* 93:417–35
94. Tuerpe, D. R., Bassichis, W. H., Kerman, A. K. 1970. *Nucl. Phys. A* 142:49–62
95. Copley, L. A., Volkov, A. B. 1966. *Nucl. Phys.* 84:417–23
96. Gogny, D. 1975. *Nucl. Phys. A* 237: 399–418
97. Damgaard, J., Pauli, H. C., Pashkevich, V. V., Strutinsky, V. M. 1969. *Nucl. Phys. A* 135:432–44

98. Vautherin, D. 1973. *Phys. Rev. C* 7: 296–316
99. Passler, K. H., Zint, P. G., Mosel, U. 1973. *Phys. Lett. B* 47:419–21
100. Zint, P. G., Mosel, U. 1975. *Phys. Lett. B* 58:269–72
101. Flocard, H. 1974. *Phys. Lett. B* 49: 129–32
102. Ripka, G. 1968. *Adv. Nucl. Phys.* 1(3): 189–92
103. Bogolyubov, N. N. 1958. *Dokl. Akad. Nauk SSSR* 119:244–46 (Engl. transl.: *Sov. Phys. Dokl.* 3:292–94)
104. Bogolyubov, N. N., Soloviev, B. G. 1959. *Dokl. Akad. Nauk SSSR* 124: 1011–14 (Engl. transl.: *Sov. Phys. Dokl.* 4:143–46)
105. Bogolyubov, N. N. 1959. *Usp. Fiz. Nauk* 67:549–80 (Engl. transl.: *Sov. Phys. Usp.* 2:236–54)
106. Baranger, M. 1961. *Phys. Rev.* 122: 992–96
107. Baranger, M. 1961. *Phys. Rev.* 130: 1244–52
108. Mang, H. J. 1975. *Phys. Rep. C* 18: 325–68
109. Dietrich, K., Mang, H. J., Pradal, J. H. 1964. *Phys. Rev. B* 135:22–34
110. Fellah, M., Hammann, T. F., Medjadi, D. E. 1973. *Phys. Rev. C* 8:1585–92
111. Goeke, K., Garcia, J., Faessler, A. 1973. *Nucl. Phys. A* 208:477–502
112. Banerjee, B., Mang, H. J., Ring, P. 1973. *Nucl. Phys. A* 215:366–82
113. Bogolyubov, N. N. 1958. *Zh. Eksp. Theor. Fiz.* 34:58–65 (Engl. transl.: *Sov. Phys. JETP* 34:41–46)
114. Valatin, J. G. 1958. *Nuovo Cimento* 7:843–57
115. Bloch, C., Messiah, A. 1962. *Nucl. Phys.* 39:95–106
116. Bleuler, K., Beiner, M., De Tourreil, R. 1967. *Nuovo Cimento B* 52:45–62
117. Bleuler, K., Beiner, M., De Tourreil, R. 1967. *Nuovo Cimento B* 52:149–86
118. Baranger, M., Kumar, K. 1968. *Nucl. Phys. A* 122:241–72
119. Krieger, S. J., Goeke, K. 1974. *Nucl. Phys. A* 234:269–84
120. Błocki, J., Flocard, H. 1976. *Nucl. Phys. A* 273:45–60
121. Negele, J. W., Koonin, S. E., Möller, P., Nix, J. R., Sierk, A. J. 1978. *Phys. Rev. C* 17:1098–1115
122. Bohr, Å., Mottelson, B. R. 1975. *Nuclear Structure*, Vol. 2. New York: Benjamin. 748 pp.
123. Peierls, R. E., Thouless, D. J. 1962. *Nucl. Phys.* 38:154–76
124. Peierls, R. E., Yoccoz, J. 1957. *Proc. Phys. Soc. London* 70:381–87
125. Wapstra, A. H., Grove, N. B. 1971. *Nucl. Data Tables* 9:265–355
126. Flocard, H., Quentin, P., Vautherin, D. 1973. *Phys. Lett. B* 46:304–7, Table 4, p. 305
127. Zaralingam, A., Negele, J. W. 1977. *Nucl. Phys. A* 288:417–28
128. Vallières, M., Lie, S. G., Sprung, D. W. L. 1978. *Can. J. Phys.* In press
129. Vautherin, D., Vénéroni, M., Brink, D. M. 1970. *Phys. Lett. B* 33:381–84
130. Beiner, M., Flocard, H., Vénéroni, M., Quentin, P. 1974. *Phys. Scr. A* 10:84–89
131. Brack, M., Quentin, P., Vautherin, D. 1978. In *Int. Symp. Super-Heavy Elements.* New York: Pergamon. In press
132. Vallières, M., Sprung, D. W. L. 1977. *Phys. Lett. B* 67:253–56
133. Köhler, H. S. 1978. See Ref. 131
134. Meldner, H. 1967. *Ark. Fys.* 36:583–98
135. Lombard, R. J. 1976. *Phys. Lett. B* 65:193–95
136. Campi, X., Flocard, H., Kerman, A. K., Koonin, S. E. 1975. *Nucl. Phys. A* 251:193–205
137. Thibault, C., Klapisch, R., Rigaud, C., Poskanzer, A. M., Prieels, R., Lessard, L., Reisdorf, W. 1975. *Phys. Rev. C* 12:644–57
138. Passler, K. H. 1976. *Nucl. Phys. A* 257:253–63
139. Passler, K. H., Mosel, U. 1976. *Nucl. Phys. A* 257:242–52
140. Bohr, Å., Mottelson, B. R. 1969. *Nuclear Structure*, Vol. 1, Sect. 3.2, pp. 317–32. New York: Benjamin. 471 pp.
141. Gustafson, C., Lamm, I.-L., Nilsson, B., Nilsson, S. G. 1967. *Ark. Fys.* 36:613–27
142. Bertozzi, W., Friar, J., Heisenberg, J., Negele, J. W. 1972. *Phys. Lett. B* 41:408–14
143. Quentin, P. 1975. See Ref. 49, pp. 297–329; Sect. 2.3, pp. 304–6
144. Sinha, B. B. P., Peterson, G. A., Whitney, R. R., Sick, I., McCarthy, J. S. 1973. *Phys. Rev. C* 7:1930–38
145. Negele, J. W., Rinker, G. 1977. *Phys. Rev. C* 15:1499–1514
146. Quentin, P. 1975. See Ref. 49, pp. 297–329
147. Davies, K. T. R., Nix, J. R. 1976. *Phys. Rev. C* 14:1977–94
148. Titin-Schnaider, C., Quentin, P. 1974. *Phys. Lett. B* 49:213–16
149. Flocard, H., Quentin, P. 1975. Preprint *IPNO/TH 75-11*, Orsay, France
150. Lassey, R. K., Volkov, A. B. 1972. *Phys. Lett. B* 39:169–72
151. Krieger, S. J., Moszkowski, S. A. 1972. *Phys. Rev. C* 5:1990–94
152. Lee, H. C., Cusson, R. Y. 1972. *Ann. Phys. (NY)* 72:353–427
153. Grammaticos, B. 1977. *Applications de*

la théorie du champ self-consistent en physique nucléaire. Thèse d'état. Orsay, France. Unpublished

154. Flocard, H., Quentin, P., Vautherin, D. 1973. See Ref. 126, pp. 304–7
155. Dechargé, J., Girod, M., Gogny, D. 1975. *Phys. Lett. B* 55:361–64
156. Bemis, C. E. Jr., McGowan, F. K., Ford, J. L. C. Jr., Milner, W. T., Stelson, P. H., Robinson, R. L. 1973. *Phys. Rev. C* 8:1466–80
157. Bohr, Å. 1952. *K. Dan. Vidensk. Selsk., Mat. Fys. Medd.* 26, No. 14. 40 pp.
158. Cailliau, M., Letessier, J., Flocard, H., Quentin, P. 1973. *Phys. Lett. B* 46:11–14
159. Götz, U., Pauli, H. C., Adler, K., Junker, K. 1972. *Nucl. Phys. A* 192:1–38
160. Curry, P. D., Sprung, D. W. L. 1973. *Nucl. Phys. A* 216:125–44
161. Girod, M., Grammaticos, B. 1978. *Phys. Rev. Lett.* 40:361–64
162. Cusson, R. Y., Hilko, R., Kolb, D. 1976. *Nucl. Phys. A* 270:437–70
163. Flocard, H., Quentin, P., Vautherin, D., Vénéroni, M., Kerman, A. K. 1974. *Nucl. Phys. A* 231:176–88
164. Kolb, D., Cusson, R. Y., Schmitt, H. W. 1974. *Phys. Rev. C* 10:1529–47
165. Ledergerber, T., Pauli, H. C. 1973. *Nucl. Phys. A* 207:1–32
166. Myers, W. D., Swiatecki, W. J. 1966. *Nucl. Phys.* 81:1–60
167. Stelson, P. H., Grodzins, L. 1965. *Nucl. Data Tables A* 1:21–102
168. Flocard, H., Quentin, P., Kerman, A. K., Vautherin, D. 1973. *Nucl. Phys. A* 203:433–72
169. Haxel, O., Jensen, J. H. D., Suess, H. E. 1949. *Phys. Rev.* 75:1766
170. Mayer, M. G. 1949. *Phys. Rev.* 75:1969–70
171. Rainwater, J. 1950. *Phys. Rev.* 79:432–34
172. Bohr, Å. 1951. *Phys. Rev.* 81:134–38
173. Mottelson, B. R., Nilsson, S. G. 1955. *Phys. Rev.* 99:1615–17
174. Nilsson, S. G. 1955. *Mat. Fys. Medd. Dan. Vid. Selsk.* 29, No. 16. 68 pp.
175. Nilsson, S. G., Tsang, C. F., Sobiczewski, A., Szymanski, Z., Wycech, S., Gustafson, C., Lamm, I.-L., Möller, P., Nilsson, B. 1969. *Nucl. Phys. A* 131:1–66
176. Woods, R. D., Saxon, D. S. 1954. *Phys. Rev.* 95:577–78
177. Brack, M., Damgaard, J., Jensen, A. S., Pauli, H. C., Strutinsky, V. M., Wong, C. Y. 1972. *Rev. Mod. Phys.* 44:320–405
178. Bolsterli, M., Fiset, E. O., Nix, J. R., Norton, J. L. 1972. *Phys. Rev. C* 5:1050–77
179. Cherdantsev, P. A., Marshalkin, V. E. 1966. *Izv. Vyssh. Uchebn. Zaved. Fiz.*

3:155–61 (Engl. transl.: *Sov. Phys. J.* 9: No. 3, 101–5
180. Demeur, M., Reidemeister, G. 1966. *Ann. Phys. (Paris)* 1:181–87
181. Andersen, B. L., Dickmann, F., Dietrich, K. 1970. *Nucl. Phys. A* 159:337–66
182. Scharnweber, D., Greiner, W., Mosel, U. 1971. *Nucl. Phys. A* 164:257–78
183. Möller, P., Nilsson, S. G., Nix, J. R. 1973. From Fig. 12 of Vandenbosch, R. 1974. In *Physics and Chemistry of Fission 1973.* Vol. 1, pp. 251–70. Vienna: IAEA. 578 pp.
184. Mottelson, B. R., Nilsson, S. G. 1959. *K. Dan. Vidensk. Selsk. Mat. Fys. Skr.* 1, No. 8. 105 pp.
185. Von Weizsäcker, C. F. 1935. *Z. Phys.* 96:431–58
186. Bethe, H. A., Bacher, R. F. 1936. *Rev. Mod. Phys.* 8:82–229
187. Strutinsky, V. M. 1967. *Nucl. Phys. A* 95:420–42
188. Strutinsky, V. M. 1968. *Nucl. Phys. A* 122:1–33
189. Bethe, H. A. 1971. See Ref. 4, Sect. 15f, pp. 225–29
190. Polikanov, I. S. M., Druin, V. A., Karnaukhov, V. A., Mikheev, V. L., Pleve, A. A., Skobelev, N. K., Subbotin, V. G., Ter-Akop'yan, G. M., Fomichev, A. A. V. A. 1962. *Zh. Eksp. Teor. Fiz.* 42:1464–71 (Engl. transl.: *Sov. Phys. JETP* 15:1016–21)
191. Paya, D., Derrien, H., Fubini, A., Michaudon, A., Ribon, P. 1967. In *Nuclear Data for Reactors*, Vol. 1. p. 128. Vienna: IAEA. 437 pp.
192. Migneco, E., Theobald, J. P. 1968. *Nucl. Phys. A* 112:603–8
193. Mosel, U., Schmitt, H. W. 1971. *Nucl. Phys. A* 165:73–96
194. Brack, M., Quentin, P. 1975. See Ref. 49, pp. 353–94; Sect. 3, pp. 363–68
195. Strutinsky, V. M. 1966. *Yad. Fiz.* 3:614–25 (Engl. transl.: *Sov. J. Nucl. Phys.* 3:449–57)
196. Ross, C. K., Bhaduri, R. K. 1972. *Nucl. Phys. A* 188:566–76
197. Tondeur, F. 1972. *J. Phys. (Paris)* 33:825–28
198. Strutinsky, V. M., Ivanjuk, F. A. 1975. *Nucl. Phys. A* 255:405–18
199. Ramamurthy, V. S., Kapoor, S. S., Kataria, S. K. 1970. *Phys. Rev. Lett.* 25:386–90
200. Kolomiets, V. M. 1973. *Yad. Fiz.* 18:288–94 (Engl. transl.: 1974. *Sov. J. Nucl. Phys.* 18:147–50)
201. Bhaduri, R. K., Das Gupta, S. 1973. *Phys. Lett. B* 47:129–32
202. Sobiczewski, A., Gyurkovich, A., Brack, M. 1977. *Nucl. Phys. A* 289:346–64

203. Bhaduri, R. K., Ross, C. K. 1971. *Phys. Rev. Lett.* 27:606–9

204. Jennings, B. K. 1973. *Nucl. Phys. A* 207:538–44

205. Jennings, B. K., Bhaduri, R. K., Brack, M. 1975. *Nucl. Phys. A* 253:29–44

206. Bunatian, G. G., Kolomiets, V. M., Strutinsky, V. M. 1972. *Nucl. Phys. A* 188:225–58

207. Brack, M., Quentin, P. 1973. See Ref. 183, Vol. 1, pp. 231–48

208. Bassichis, W. H., Kerman, A. K., Tuerpe, D. R. 1973. *Phys. Rev. C* 8:2140–43

209. Brack, M., Quentin, P. 1975. See Ref. 49, pp. 353–94; Sect. 4.4, pp. 387–88

210. Bassichis, W. H., Kerman, A. K., Tsang, C. F., Tuerpe, D. R., Wilets, L. 1972. In *Magic without Magic: John Archibald Wheeler, a Collection of Essays in Honor of His 60th Birthday*, ed. J. R. Klauder, pp. 15–46. San Francisco: Freeman. 491 pp.

211. Tabakin, F. 1964. *Ann. Phys. (NY)* 30:51–94

212. Brack, M., Quentin, P. 1975. *Phys. Lett. B* 56:421–23

213. Brack, M., Quentin, P. 1976. In *Atomic Masses and Fundamental Constants*, ed. J. H. Sanders, A. H. Wapstra, Vol. 5, pp. 257–63. New York: Plenum. 681 pp.

214. Brack, M., Pauli, H. C. 1973. *Nucl. Phys. A* 207:401–24

215. Tyapin, A. S. 1974. *Yad. Fiz.* 19:263–74. (Engl. transl.: *Sov. J. Nucl. Phys.* 19:129–34)

216. Brack, M., Jennings, B. K., Chu, Y. H. 1976. *Phys. Lett. B* 65:1–4

217. Jennings, B. K., Bhaduri, R. K. 1975. *Nucl. Phys. A* 237:149–56

218. Chu, Y. H., Jennings, B. K., Brack, M. 1977. *Phys. Lett. B* 68:407–11

219. Bhaduri, R. K. 1977. *Phys. Rev. Lett.* 39:329–32

220. Durand, M., Brack, M., Schuck, P. 1977. Preprint *ILL-77 DU 248*, Inst. Laue Langevin, Grenoble, France

221. Ko, C. H., Pauli, H. C., Brack, M., Brown, G. E. 1973. *Phys. Lett. B* 45:433–36

222. Ko, C. H., Pauli, H. C., Brack, M., Brown, G. E. 1974. *Nucl. Phys. A* 236:269–301

223. Brack, M. 1977. *Phys. Lett. B* 71:239–42

224. Frenkel, J. 1936. *Phys. Z. Sowjetunion* 9:533–36

225. Bohr, N., Kalchar, F. 1937. *Mat. Fys. Medd. Dan. Vid. Selsk.* 14, No. 10. 40 pp.

226. Bethe, H. 1937. *Rev. Mod. Phys.* 9:69–244

227. Landau, L. 1937. *Phys. Z. Sowjetunion* 11:556–65

228. Weisskopf, V. 1937. *Phys. Rev.* 52:295–303

229. Bohr, Å., Mottelson, B. R. 1969. See Ref. 140, Sect. 2.1, pp. 152–56, and Appendix 2B, pp. 281–93

230. Huizenga, J. R., Moretto, L. G. 1972. *Ann. Rev. Nucl. Sci.* 22:427–64

231. Landau, L. D., Lifshitz, E. M. 1969. *Statistical Physics*, Sect. 35, pp. 101–3. Reading: Addison Wesley. 484 pp. 2nd ed.

232. Des Cloizeaux, J. 1968. In *Many Body Physics*, ed. C. De Witt, R. Balian, pp. 1–36. New York: Gordon & Breach. 426 pp.

233. Bloch, C., De Dominicis, C. 1958. *Nucl. Phys.* 7:459–79

234. Bloch, C., De Dominicis, C. 1959. *Nucl. Phys.* 10:181–96, 509–26

235. Kohn, W., Luttinger, J. M. 1960. *Phys. Rev.* 118:41–45

236. Bloch, C. 1965. In *Studies in Statistical Mechanics*, ed. J. De Boer, G. E. Uhlenbeck, Vol. 3, pp. 1–211. Amsterdam: North-Holland. 388 pp.

237. Buchler, J.-R., Coon, S. A. 1977. *Astrophys. J.* 212:807–15

238. Brack, M., Quentin, P. 1974. *Phys. Lett. B* 52:159–62

239. Brack, M., Quentin, P. 1974. *Phys. Scr. A* 10:163–69

240. Mosel, U., Zint, P.-G., Passler, K. H. 1974. *Nucl. Phys. A* 236:252–68

241. Jensen, A. S., Damgaard, J. 1973. *Nucl. Phys. A* 210:282–96

242. Brack, M., Quentin, P. 1975. See Ref. 49, pp. 399–401

243. Sauer, G., Chandra, H., Mosel, U. 1976. *Nucl. Phys. A* 264:221–43

244. Ferrell, R. A. 1957. *Phys. Rev.* 107:1631–34

245. Ehrenreich, H., Cohen, M. H. 1959. *Phys. Rev.* 115:786

246. Goldstone, J., Gottfried, K. 1959. *Nuovo Cimento* 13:849–52

247. Bohm, D., Pines, D. 1953. *Phys. Rev.* 92:609–25

248. Inglis, D. R. 1954. *Phys. Rev.* 96:1059–65

249. Inglis, D. R. 1956. *Phys. Rev.* 103:1786–95

250. Belyaev, S. T. 1965. *Nucl. Phys.* 64:17–54

251. Villars, F. 1972. In *Dynamical Structure of Nuclear States, Proc. 1971 Mont Tremblant Int. Summer Sch.*, ed. D. J. Rowe, L. E. H. Trainor, S. S. M. Wong, T. W. Donelly, pp. 3–37. Univ. Toronto Press. 585 pp.

252. Villars, F. 1975. See Ref. 49, pp. 3–17

253. Villars, F. 1977. *Nucl. Phys. A* 285: 269–96
254. Brink, D. M., Giannoni, M. J., Vénéroni, M. 1976. *Nucl. Phys. A* 258:237–56
255. Baranger, M., Vénéroni, M. 1978. *Ann. Phys. (NY).* In press
256. Bertsch, G., Tsai, S. F. 1975. *Phys. Rep. C* 18:125–58
257. Liu, K. F., Nguyen van Giai 1976. *Phys. Lett. B* 65:23–26
258. Blaizot, J. P., Gogny, D. 1977. *Nucl. Phys. A* 284:429–60
259. Crank, J., Nicholson, P. 1947. *Proc. Cambridge Philos. Soc.* 43:50–67
260. Bonche, P. 1976. *J. Phys. (Paris)* 37, Colloque C5:213–21
261. Koonin, S. 1976. *Phys. Lett. B* 61: 227–30
262. Maruhn, J., Cusson, R. Y. 1976. *Nucl. Phys. A* 270:471–88
263. Koonin, S. E., Davies, K. T. R., Maruhn-Rezwani, V., Feldmeier, H., Krieger, S. J., Negele, J. W. 1977. *Phys. Rev. C* 15:1359–74
264. Maruhn-Rezwani, V., Davies, K. T. R., Koonin, S. E. 1977. *Phys. Lett. B* 67:134–38
265. Cusson, R. Y., Smith, R. K., Maruhn, J. 1976. *Phys. Rev. Lett.* 36:1166–69
266. Bonche, P., Grammaticos, B., Koonin, S. E. 1978. *Phys. Rev. C* 17:1700–5
267. Fernandez, B., Gaarde, C., Larsen, J. S., Pontopiddan, S., Videbæck, F. 1978. Preprint. Niels Bohr Inst., Copenhagen, Denmark
268. Conjeaud, M., Harar, S., Saint Laurent, F., Menet, J., Loiseaux, J. M., Viano, J. B. 1978. Preprint. Saclay, France
269. Doubre, H., Gamp, A., Jacmart, J. C., Poffé, N., Roynette, J. C., Wilczyński, J. 1978. *Phys. Lett. B* 73:135–38
270. Saunier, G., Le Tourneux, J., Rouben, B., Padjen, R. 1973. *Can. J. Phys.* 51:629–33
271. Pearson, J. M., Rouben, B., Saunier, G. 1973. See Ref. 270, Appendix, p. 633
272. Chang, B. D. 1975. *Phys. Lett. B* 56:205–8
273. Blaizot, J. P. 1976. *Phys. Lett. B* 60:435–38
274. Blaizot, J. P., Gogny, D., Grammaticos, B. 1976. *Nucl. Phys. A* 265:315–36
275. Marty, N., Morlet, M., Willis, A., Comparat, V., Frascaria, R. 1975. *Proc. Int. Symp. Highly Excited States Nuclei.* Kernforsch. Jülich Rep., Jül.-Conf.-16, Vol. 1, p. 17
276. Marty, N., Morlet, M., Willis, A.,

Comparat, V., Frascaria, R. 1976. Preprint *IPNO/PhN-76-03.* Inst. Phys. Nucl. Orsay
277. Berman, B. L., Fultz, S. C. 1975. *Rev. Mod. Phys.* 47:713–61
278. Lederer, C. M., Hollander, J. M., Perlman, I. 1967. *Table of Isotopes.* p. 174. New York: Wiley. 594 pp. 6th ed.
279. Bohr, Å., Mottelson, B. R. 1953. *K. Dan. Vidensk. Selsk. Mat. Fys. Medd.* 27, No. 16. 174 pp.
280. Baranger, M., Vénéroni, M. 1978. See Ref. 255, Sect. 8.6
281. Ring, P., Schuck, P. 1977. *Nucl. Phys. A* 292:20–28
282. Giannoni, M. J., Moreau, F., Quentin, P., Vautherin, D., Vénéroni, M., Brink, D. M. 1976. *Phys. Lett. B* 65:305–8
283. Thouless, D. J., Valatin, J. G. 1962. *Nucl. Phys.* 31:211–30
284. Quentin, P. 1977. *Quatrième Session d'Etudes Biennale de Physique Nucléaire, La Toussuire, Rep. LYCEN 7702,* Vol. 2, pp. C11.1–12
285. Bonche, P., Quentin, P. 1978. *Phys. Rev. C.* In press
286. Giannoni, M. J., Vautherin, D., Vénéroni, M., Brink, D. M. 1976. *Phys. Lett. B* 63:8–10
287. Vautherin, D. 1975. *Phys. Lett. B* 57:425–28
288. Flocard, H., Vautherin, D. 1975. See Ref. 49, pp. 35–36
289. Grammaticos, B. 1975. *Phys. Lett. B* 57:306–8
290. Caurier, E., Grammaticos, B. 1976. *Contrib. Int. Winter Meet. Nucl. Phys., Bormio, Italy, 14th, Rep. DPh-T/76-2.* Saclay, France
291. Baranger, M., Vénéroni, M. 1978. See Ref. 255, Sect. 6.5
292. Siemssen, R. H. 1971. *Proc. Argonne Symp. Heavy-Ion Scattering, Rep. ANL-7837,* p. 145. Argonne Natl. Lab.
293. Bhaduri, R., Sprung, D. W. L. 1978. *Nucl. Phys. A* 297:365–72
294. Valatin, J. G. 1961. *Phys. Rev.* 122: 1012–20
295. Wilets, L. 1958. *Rev. Mod. Phys.* 30:542–49 and references therein
296. Sandhya Devi, K. R., Strayer, M. R. 1978. *Phys. Lett. B* 77:135–40
297. Koonin, S., Flanders, B., Flocard, H., Weiss, M. 1978. *Phys. Lett. B* 77:13–17
298. Goeke, K., Reinhard, P.-G. 1978. *Ann. Phys. (NY)* 112:328–55
299. Goeke, K. 1977. *Phys. Rev. Lett.* 38: 212–15

AUTHOR INDEX

597

CUMULATIVE INDEXES

CONTRIBUTING AUTHORS VOLUMES 19-28

615

CHAPTER TITLES, VOLUMES 19-28

618   CHAPTER TITLES

Neutron Stars   G. Baym, C. Pethick   25:27-77
Status and Future Directions of the World Pro-
  gram in Fusion Research and Development   R. L. Hirsch   25:79-121
Nuclear Safeguards   T. B. Taylor   25:407-21
Global Consequences of Nuclear Weaponry   J. C. Mark   26:51-87
NUCLEAR PROPERTIES, RADIOACTIVITY, AND NUCLEAR REACTIONS
Mass Separation for Nuclear Reaction Studies   R. Klapisch   19:33-60
The Three-Nucleon Problem   R. D. Amado   19:61-98
Coupled-Channel Approach to Nuclear Reactions   T. Tamura   19:99-138
Parity and Time-Reversal Invariance in Nuclear
  Physics   E. M. Henley   19:367-432
Nuclear Mass Relations   G. T. Garvey   19:433-70
Coulomb Energies   J. A. Nolen Jr., J. P. Schiffer   19:471-526
Muonic Atoms and Nuclear Structure   C. S. Wu, L. Wilets   19:527-606
Low-Energy Photonuclear Reactions   F. W. K. Firk   20:39-78
Alpha-Particle Transfer Reactions   K. Bethege   20:255-88
Weak Coupling Structure in Nuclei   A. Arima, I. Hamamoto   21:55-92
Theory of Nuclear Matter   H. A. Bethe   21:93-244
Three Fragment Fission   I. Halpern   21:245-94
Isospin Impurities in Nuclei   G. F. Bertsch, A. Mekjian   22:25-64
Calculation of Fission Barriers for Heavy and
  Superheavy Nuclei   J. R. Nix   22:65-120
Electromagnetic Transitions and Moments in
  Nuclei   S. Yoshida, L. Zamick   22:121-64
Perturbation of Nuclear Decay Rates   G. T. Emery   22:165-202
Nuclear Level Densities   J. R. Huizenga, L. G. Moretto   22:427-64
The Nucleon-Nucleon Effective **Range** Expansion
  Parameters   H. P. Noyes   22:465-84
Symmetries in Nuclei   K. T. Hecht   23:123-61
Linear Relations Among Nuclear Energy Levels   D. S. Koltun   23:163-92
Intermediate Structure in Nuclear Reactions   C. Mahaux   23:193-218
Production Mechanisms of Two-To-Two Scatter-
  ing Processes at Intermediate Energies   G. C. Fox, C. Quigg   23:219-313
The Many Facets of Nuclear Structure   A. Bohr, B. R. Mottelson   23:363-93
Time Description of Nuclear Reactions   S. Yoshida   24:1-33
Resonance Fluorescence of Excited Nuclear Levels
  in the Energy Range 5-11 MeV   B. Arad, G. Ben-David   24:35-67
The Theory of Three-Nucleon Systems   Y. E. Kim, A. Tubis   24:69-100
Shell-Model Effective Interactions and the Free
  Nucleon-Nucleon Interaction   T. T. S. Kuo   24:101-50
Post-Fission Phenomena   D. C. Hoffman, M. M. Hoffman   24:151-207
Meson-Nucleus Scattering at Medium Energies   M. M. Sternheim, R. R. Silbar   24:249-77
Reactions Between Medium and Heavy Nuclei and
  Heavy Ions of Less than 15 MeV/amu   A. Fleury, J. M. Alexander   24:279-339
Proton-Nucleus Scattering at Medium Energies   J. Saudinos, C. Wilkin   24:341-77
Knock-Out Processes and Removal Energies   A. E. L. Dieperink, T. de
                                                       Forest Jr.   25:1-26
Preequilibrium Decay   M. Blann   25:123-66
Electron Scattering and Nuclear Structure   T. W. Donnelly, J. D. Walecka   25:329-405
Blocking Measurements of Nuclear Decay Times   W. M. Gibson   25:465-508
The Variable Moment of Inertia (VMI) Model and
  Theories of Nuclear Collective Motion   G. Scharff-Goldhaber, C. B.
                                                       Dover, A. L. Goodman   26:239-317
Excitation of Giant Multipole Resonances through
  Inelastic Scattering   F. E. Bertrand   26:457-509
Spontaneously Fissioning Isomers   R. Vandenbosch   27:1-35
The Weak Neutral Current and Its Effects in
  Stellar Collapse   D. Z. Freedman, D. N.
                                                       Schramm, D. L. Tubbs   27:167-207
Delayed Proton Radioactivities   J. Cerny, J. C. Hardy   27:333-51